Raising the Bar
The Competencies of Specialists in Gifted Education

Proefschrift

Ter verkrijging van de graad van doctor aan de Universiteit van Hasselt
door Eleonora Wilhelmina Josepha Maria van Gerven.
Geboren op 31 maart 1965 te Wormer, Nederland

Promotor
Prof. dr. T. Kieboom
Copromotor
Prof. dr. ir. K. Venderickx
Internationaal lid doctoraatscommissie
Dr. M. Sutherland
Doctoraatsjury
Prof. dr. K. Struyven
Prof. dr. A. Bakx
Dr. C. Deitz
Mrs. W. Behrens, MA(Ed)

SES
School voor
Educatieve Studies
UHASSELT

Raising the Bar
The Competencies of Specialists in Gifted Education

Dr. Eleonoor van Gerven

Colofon

NUR 190, 841

ISBN: 978-94-0361-786-2

Copy editing: Geraldine Glover
Final editing: Marita Weener
Cover and inside design and lay-out: Sabrina Wakker
Production support: Sabine Kokee

Voor Eugenio

Die mij aanmoedigde om onredelijk te durven zijn en dat vol te durven houden.

"The reasonable man adapts himself to the world around him;
the unreasonable one persists in trying to adapt the world to himself.
Therefore all progress depends on the unreasonable man."

George Bernard Shaw

Content

1 General Introduction

1.1 Professional and Social Relevance

In my daily practice as a teacher educator and as a developer of continuous professional development for teachers in the domain of gifted education, I need a theoretical underpinning to the curriculum that I teach. I approach education from an inclusive perspective.

According to Dutch and Flemish law, teachers are expected to deliver education that stimulates all the students in their classroom to reach for their zone of proximal development. Just like any other student, gifted students are entitled to be educated at that level of challenge; not only academically, but also in other domains of human development. Therefore, gifted students are no exception to what should be considered a teacher's core task. As a teacher educator, it is my job to prepare teachers for that task. Hence, for me, it is important that teachers, graduating from the courses that I teach, feel competent in organising education for gifted students within the context of a regular classroom and in integrating their education into the education of all the students at school.

The courses I teach, match the so-called Dublin Descriptors for higher education (Bologna Working Group on Qualifications Frameworks, 2005; European Commission, 2013). Indicators for competent professional behaviour can be deduced from these descriptors. Combined, these indicators underpin the framework for curriculum construction that I use to develop programmes for educating teachers into becoming specialists in gifted education.

The exit level for graduates of our programme "Specialist in Gifted Education" can be positioned at master's level. During the programme, we place a strong emphasis on a professional approach to evidence-informed teaching, underpinned by current educational theories. As I prepare these teachers for teaching in an inclusive context, interplay between theory and practice outside of the field of gifted education and theory and practice inside the field of gifted education is necessary in the programme itself.

Choices over the didactical approach to the programme are underpinned with different theories about learning and teacher education, that is, theories by Ambrose and Sternberg (2016), Biesta (2007, 2010, 2012, 2014, 2018a, 2018b), Hattie (2013), Korthagen (2004, 2017), Lunenberg et al. (2014), Marzano and Kendall (2007), Maslow (1954), and Vygotsky (1978). These theories offer a framework for both a social-creative constructivist didactical approach and the instructional strategies that encourage teachers to reflect on their professional behaviour. The educational objectives in this programme are based on Marzano and Kendall's New Taxonomy (2007). The strategies used, match the development of skills fit for education in 21st century teaching (European Commission, 2013). Encouragement of the reflective attitude and professional ethical behaviour is supported by working with Korthagen's concept of core reflection (King & Lau-Smith, 2013; Korthagen, 2017) and Biesta's concept of pedagogical virtuosity (Biesta, 2012).

As soon as teacher education programmes focus on the development of a specialism, content-bound educational objectives – aimed at developing that specialism – become mandatory. Research into the theoretical context of gifted education, reveals a broad spectrum of educational needs that are considered to be specific to gifted learners (Bakx et al., 2016; Dai & Chen, 2014; Haenen & Mol-Lous, 2014; Johnsen et al., 2017; Kreger-Silverman, 2013; Plucker et al., 2017; Subotnik et al., 2011; Tomlinson et al., 2009). Based on these educational needs, it is assumed that specific knowledge and skills are required to be a specialist in gifted education (Johnsen et al., 2016). Research into the understandings of stakeholders in the USA, undertaken by Johnsen et al. (2016) to underpin the standards of the NAGC (National Association for Gifted Children), confirms this assumption.

The supposition that teachers need to develop different competencies if they want to educate gifted students successfully is supported by professional communities already working with gifted students (Johnsen et al., 2016).[1] In The Netherlands and Flanders, teachers often express how they would be happy to work with gifted students, if only they knew how to work with them. They state that, although they are fully qualified teachers (graduating at bachelor's degree level), they do not feel qualified to work with this specific group of students (Haenen & Mol-Lous, 2014; Houkema et al., 2018). This raises questions: in what specific aspects of gifted education do these teachers experience a lack of competency, and what would they need to feel enabled to respond in a way that matches the educational needs of gifted students in the context of current educational paradigms?

Between 1997 and 2007, I trained approximately 1000 teachers in short teacher education programmes on gifted education. Before they enrolled in these programmes, they completed a submission form and stipulated the knowledge and skills they wanted to acquire during these courses. In 2008 this resulted in a "Knowledge and skills list", published as an appendix in *Slim beleid*, a practitioner's book for teachers and school management in how to design and implement school policy on gifted education (van Gerven, 2008). In this book, tasks that emerged logically from developing and implementing school policy on gifted education were matched with the competencies mentioned in the "Knowledge and skills list". Stepping stones for school policy were analysed based on a literature study of current literature at that time (Eyre, 2001, 2007; Heller et al., 2000; Renzulli, 1985; Sternberg, 2002). In doing so, those competences that were distinguished were presented thematically.

In 2009, I started developing the teacher education programme "Specialist in Gifted Education" and reanalysed the existing data. The data were organised into seven domains and combined with the seven domains of competencies underpinning the regular professional standard for teachers. This combination resulted in a matrix with 49 competencies specific to specialists in gifted

---

[1] In January 2020, the World Council for Gifted and Talented Children installed a subcommittee to prepare a position paper on teaching standards for gifted education. Expert delegates from 17 countries hold positions on this committee.

education. For every competency, knowledge and skills indicators, taken from the original "Knowledge and skills list", were described, making it possible to assess whether or not a competence is developed. In 2011, the latest version of this matrix was published in *Begaafd begeleiden* (van Gerven & Hoogenberg, 2011).
Characteristic of these competencies is that they are contextually bound and seen as changeable over time (Ceulemans et al., 2016; Lunenberg et al., 2014; Merriënboer et al., 2002). In 2018, I concluded that it was time to stop and explore how the idea of the existing matrix dating back to the timeframe 2008-2011 should be adjusted or maybe even reinvented. My urge to do so was ignited by the impact of changes that I had observed in Dutch and Flemish education in general and developments in gifted education in particular. A new matrix would contribute to a well-balanced curriculum for different teacher education programmes about gifted education in general and for courses to become a Specialist in Gifted Education in particular.

1.2 A Brief Context of the Position of Giftedness in Dutch and Flemish Education

According to studies on giftedness and gifted education, gifted students have specific educational needs, resulting not only from their academic abilities but also as the consequences of particular characteristics (Freeman, 2010; Gagné, 2010a; Kieboom & Venderickx, 2017a, 2017b; Kreger-Silverman, 2013; Leavitt, 2017; Peters et al., 2014; Pfeiffer, 2008; Plucker et al., 2017; Subotnik et al., 2011). Perspectives on what these characteristics are and how they influence educational needs vary enormously (Dai & Chen, 2014). It is up to the classroom teacher to meet the needs of every student as effectively as possible within the context of an inclusive classroom (Florian & Black-Hawkins, 2011). This includes the needs of gifted students (Dekker, 2014a), but how this is done depends on how schools and scholars define these needs and define gifted education (Dai & Chen, 2014; Morris Miller, 2008).
In the Dutch and Flemish framework for inclusive education, the structure of a pyramid represents different levels of student support (Ainscow et al., 2006; Audenaert et al., 2015; Clijsen et al., 2007). This pyramid is regularly represented in three layers. However, due to the way schools have organised their student support system, it can best be presented in five layers (van Gerven, 2015). The base of the pyramid represents the largest group of students: based on the similarity of their needs, these students can be clustered in such a way that the teacher can meet their needs through the standard curriculum and standard instructional and pedagogical strategies, without additional interventions or support. The top of the pyramid represents the smallest group of students: compared to the needs of other students, students at this level of support have additional educational needs. In current Dutch and Flemish educational practice, these are often students exhibiting (characteristics of) learning disabilities or developmental problems. In this pyramid of support, the education of gifted students is positioned between those top and bottom layers at levels two, three or four (van Gerven, 2015) (see Chapter 2, Figure 4).

According to the way educational needs are clustered and instructional strategies are used, meeting the educational needs of gifted students is seen as different from the regular strategies used for students at level one (Bakx et al., 2016; Houkema et al., 2018; Kuipers, 2017; Schrover, 2015). However, complex interventions are not necessary to meet the needs of gifted students in all cases (Bakx et al., 2016; Janson, 2017; Roberts & Inman, 2015; Tomlinson et al., 2009; Winstanley, 2010). If possible, classroom teachers respond to these needs themselves (level two) without the additional support of the (special) educational needs coordinator ((S)ENCo) (Peters & Oveross, 2020). Identically to other students, however, there are gifted students whose educational needs exceed the meaningful responses that can be provided by their teacher at level two. The needs of these students become more complex, making it reasonable to assume that the regular classroom teacher requires support to meet these students' needs. For those students and their teachers, at levels three or four, support is provided. For intervention at these levels, that is, special projects or part-time peer-group education or on-the-job coaching for teachers, the specialist in gifted education becomes involved.

The organisation of the support structure in Dutch and Flemish education is such that it creates a situation where there are hardly any educational facilities for gifted students with additional needs that exceed level four, and who do not belong to the group of so-called twice-exceptional learners. Giftedness in itself is not a classification as mentioned in the DSM-5 and, therefore, is rightfully placed outside the area of special educational needs or learning and developmental disorders. Full-time segregated education for gifted students in The Netherlands and in Flanders is not included at the top level of the pyramid of support. That level is reserved for students with specific learning or developmental disabilities. This puts full-time segregated education for gifted learners outside the formal organisation of this support structure. However, in The Netherlands, as a response to the heavy task load of regular classroom teachers, schools are increasingly looking for opportunities to create full-time segregated settings for gifted students to meet their educational needs. This development is seen by some as contradictory to the policy of inclusion as presented by the Dutch government and as described in Dutch educational law (Wet passend onderwijs, 2012). As a result, gifted students and the question of how to address their educational needs in an inclusive environment now has higher priority on the educational agenda. Remarkably, coinciding with an increasing number of opportunities for full-time segregated education for the gifted (Daeter, 2012), there has been a noticeable development, where school boards who have provided these services for some years have concluded that education for gifted students should become more integrated into regular education again. These boards have concluded that the number of students enrolling on full-time programmes is increasing rapidly and that consequently the available resources are not sufficient. As the available resources will not increase, the number of students has to decrease, and a two-step strategy has been chosen. The first step is to set higher admission standards; the second step is to increase the professional skills of regular classroom teachers. Sometimes there is even a third step, which excludes twice-exceptional students from being admitted into these segregated groups. To meet the needs of these students, there is sometimes the additional solution of separate full-time segregated programmes for twice-exceptional students, which match the structure of the pyramid of support at level five. Currently, this option is being increasingly explored by local school councils.

In Flanders, gifted education has only recently been explicitly mentioned on the educational agenda (Vlaamse Regering, 2019). Segregated schools for the gifted are not formally accepted, although there are initiatives for segregated schools. The legal construction is almost identical to that of The Netherlands, that is, regular schools set up special classes for full-time gifted education. In Flanders most interventions can be placed at levels two and three in the previously described pyramid of support: either implemented by classroom teachers or implemented in the context of special projects or part-time peer-group education.

The pyramid of support can only exist if the bottom layer is widespread and carefully anchored in the educational system in general (Ainscow et al., 2006). This means that, for effective student support, the centre of gravity must be placed at the base layer of the pyramid, directly in the hands of the classroom teacher. Consequently, when the intensity of student support that is needed to meet a student's educational needs cannot be provided at the originally designated level of support, it has to be provided by a professional at the next level (Peters & Oveross, 2020). Alternatively, the professional at the next level provides specific teacher support, in such a way that the gifted learner is still served by the regular classroom teacher in the regular group (Amsing et al., 2009; Kuipers, 2017; LBBO, 2019; Pameijer & van Beukering, 2007; van Gerven & Hoogenberg, 2011; van Meersbergen & de Vries, 2017). In situations where there are too many cases calling on the top levels of the pyramid and basic support is insufficient for a larger group of students, the shape of the pyramid changes into the shape of a fir tree, and the support system becomes unstable and feeble (van Gerven & Weterings-Helmons, 2014).

In daily practice in education, it is easy to recognise this disproportionate pressure at the top levels of the pyramid. Both teachers from in-school special programmes and peer groups of teachers working in programmes organised at council level, report that a large percentage of the students who enrol in their programmes could be relatively easily catered for in the regular classroom (Houkema et al., 2018). Regular classroom teachers nominate these students because they feel that the educational needs of these students differ so much from what they are capable of offering in the regular classroom, and that this justifies the student's participation in these programmes. There appears to be a gap between the needs of gifted students and the services that regular classroom teachers can offer them (Houkema et al., 2018).

In The Netherlands, this has resulted in increasing governmental pressure to bring in gifted education at a higher level of services. Schools feel the pressure of the Dutch school inspectorate to get high results for as many students as possible. Schools are experiencing an increased focus on results aimed mostly at the domains of language arts and spelling, (comprehensive) reading and maths (Blok et al., 2013). Excellence has become a keyword. Education is supposed to aim for excellence, and the government has made the concepts of excellence and giftedness mutually exchangeable (Dekker, 2014b; Slob, 2018a, 2018b).

1.3 Objectives

This research project aimed to develop a set of coherent recommendations offering a framework for the professional standards of specialists in gifted education. This framework is presented as a competency matrix that can be used within the context of the inclusive approach of primary education in The Netherlands and Flanders. In The Netherlands and Flanders, a specialist in gifted education is likely to have core tasks based in levels two, three and four of the pyramid of support. These tasks can best be compared with the tasks of special educational needs coordinators (SENCo). Therefore, the matrix should match the competencies needed at these levels. This implies that the matrix should not only aim for competencies in educating gifted learners, but also for competencies that support teachers of gifted learners. This is a similar approach to that used for other educational specialists such as SENCo's, specialists in reading, specialists in maths, and specialists in complex behaviour and learning disabilities (LBBO, 2019, n.d.-a). The matrix can be used by institutes for teacher education to design courses for gifted education both at undergraduate and postgraduate level. The matrix can also be used by a specialist in gifted education as a framework or outline for professional behaviour, and, as such, the matrix can be used as an instrument to assess the professional competences of current and future specialists in gifted education.
To translate the competency matrix into daily practice in education, I consider it necessary to have a set of concrete examples of knowledge and skills. These examples are indicators of competent behaviour, and a specialist in gifted education can use these indicators to demonstrate how their professional behaviour and attitudes can be recognised in their actions. Indicators of competences in knowledge and skills can be deduced by analysing existing international literature within and outside of the field of gifted education. In this systematic literature review, empirical research is considered to be equally relevant as the more practically based literature the international community of practitioners uses to design their strategies for gifted education. Teachers of the gifted, specialists in gifted education, and parents of gifted children use the more practically based literature to form their opinion on what they consider as good for gifted education and hence, what are relevant knowledge and skills for specialists in gifted education. To obtain the necessary and widespread support, this research is completed with an empirical section where stakeholders of gifted education are asked to select what, in their eyes, are the most relevant and important examples of competent professional behaviour.
Both the competency matrix and the matching indicators for knowledge and skills are not intended to be used unidirectionally or as a simple checklist. Hence, the empirical part of the research is not focused on the development of a list usable with a Likert-like scale. The research aimed for the development of a knowledge and skills profile that, according to the stakeholders, provides the best match for professional behaviour and approaches of the specialist in gifted

education within the current overall educational paradigms.
Competencies and indicators are stipulated, that offer a specialist in gifted education the professional and discretional freedom to mould these indicators to match both the context within which they are applied and the individual specialist's perspective of giftedness. As stated before, the Dublin Descriptors form the underlying framework for professional behaviour in education. Professionals should be taught to form their own critical and professional opinion on what is a good perspective of gifted education. Therefore, it is out of the question to direct and prescribe what is to be seen as *the* right perspective of gifted education or the *best* interventions.

1.4 Research Question

Based on the previously described context, the main question of this research is:

> *Which competencies are necessary to provide gifted education that meets the needs of gifted students in an educational environment in The Netherlands and Flanders, which knowledge and skills are representative of these competencies, and what is needed to develop these competencies?*

This main question can be divided into five partial questions.

1. Which developments and shifts in educational theories in The Netherlands and Flanders influence the professional context of a specialist in gifted education?
   a. Which developments and shifts in educational theories in The Netherlands and Flanders influence the professional context of primary school teachers in general?
   b. How can we currently position gifted education in The Netherlands and Flanders?
   c. Which specific topics are, therefore, recognisably of influence on the professional actions and approach of the specialist in gifted education?
2. Which developments in The Netherlands and Flanders regarding teacher education influence the professional actions and approach of the specialist in gifted education?
   a. What should be reflected in the educational level of the specialist in gifted education and how does this match the international demands for teacher education to which both The Netherlands and Flanders are committed?
   b. Which factors in current perspectives on teacher education influence the professional standards that can be set for specialists in gifted education?
   c. Which professional standards of teachers in general, interact with the professional standards set for specialists in gifted education?
   d. What are the criteria which should be met when stipulating a competency, and how can these criteria form a framework for assessing the quality of the competencies?

3. Which knowledge and skills for the specialist in gifted education are pointed out by the international professional community and international research on gifted education as essential for the optimal development of gifted students?
   a. What advice is available in international literature for the shaping of gifted education?
   b. How does this advice fit into the Dutch and Flemish perspective on education in general?
4. How can advice in international research be translated into the Dutch and Flemish educational context?
   a. What knowledge and skills can be deduced from international research on gifted education, considering the international educational standards for specialist teachers?
   b. How can these indicators be stipulated to meet the Dublin Descriptors for professional behaviour?
5. What knowledge and skills indicators are considered by Dutch and Flemish stakeholders as the most relevant and important examples of competent professional behaviour of the specialist in gifted education?
   a. Do different groups of stakeholders differ in their emphasis on specific indicators, and if so, what differences can be found?
   b. Are there noticeable differences between Dutch and Flemish stakeholders regarding their emphasis on specific knowledge and skills indicators, and if so, what differences can be found?

1.5 Outline of the Dissertation

This dissertation consists of three parts: (1) a wide theoretical and practical context for the matrix; (2) the actual construction of the matrix; and (3) overarching conclusions and recommendations.

In part 1 (Chapters 2 and 3), the theoretical context of gifted education in The Netherlands and Flanders and the theoretical context of teacher education in the Netherlands and Flanders are described. In doing so, I address research questions 1 and 2. Based on this theoretical context I build the outline for the new competency matrix.

In Chapter 2, the position of the education of gifted students in primary schools in the Dutch and Flemish educational system is explained using the context of five significant paradigm shifts in education. These paradigm shifts are not limited to The Netherlands and Flanders. Since the late eighties, these paradigms have determined the international discourse of perspectives on education. By clarifying the interdependency of these five paradigm shifts, the complexity of teaching as an occupation becomes visible. The stratification in the depth of interventions, arising from an internationally accepted continuum of support, makes it apparent that at each level of intervention other competences are required. This stratification does not change when it comes to educating the gifted. It merely creates a context for a specifically aimed development of competences in educating the gifted. Although the context of competence

development is described, it is not the purpose of this chapter to describe the specific competences at each level. It sketches the outline for the framework to develop the description of these competencies.
In Chapter 3, the need for a professional standard for specialists in gifted education is demonstrated. The context for this standard of professionalism is rooted in the European striving for high-quality education. In The Netherlands and Flanders, general teachers are educated at bachelor's degree level (level 6). Teacher-specialists and/or specialists in education are educated at master's degree level (level 7). As specialists in gifted education operate as teacher-specialists, then the prerequisite for this should be an initial teaching qualification, with additional studying at master's degree level to become a specialist in their field of expertise. We explore the demands that are conditional for teacher education at this level: the Dublin Descriptors, the notion of competence and a model for intrapersonal adaptation as the generic approach for teacher-specialists. The word "skilledge" is introduced to express the result of continuous contextual professional development for teacher-specialists. The necessity for a professional standard for specialists in gifted education influences the conditions of education of teachers in this domain of expertise. As a consequence, a professional standard for gifted education should reflect an integrated approach, combining both competence in giftedness and general teaching competencies.
Part 2 is based on an additional literature review study and an empirical study using a questionnaire to better understand teacher-specialists' understanding of the competencies of specialists in gifted education. The literature review study provides the content for the competency matrix and for the matching knowledge and skills indicator list. This provides the answers for research questions 3 and 4. The empirical study provides the data for answering research question 5.
Chapter 4 describes the methodology used in this second part of the dissertation. Firstly, the outline for the theoretical review is described. The chapter starts with a description of the intended use of literature to construct the content for the actual matrix and the knowledge and skills list. It describes how literature selection took place and how a narrative review of literature completed the systematic review of literature. Secondly, the outline for the empirical research is described. In this part of the chapter, statistical approaches are described and the thresholds for relevant differences, statistical significance and practical significance are defined. Chapter 4 concludes with an analysis of the population of respondents participating in the conducted empirical research. It was decided to include the final competency matrix and the knowledge and skills indicator list into several appendices for Chapter 4.
Chapters 5-9 each focus on a different domain of the competency matrix. In each chapter the results of the literature review are first presented, then the data analyses of the empirical part of the research. This is the point where the most detailed level of data analysis, per competency of each domain and per group of stakeholders, is presented; displaying data within the context in which they were observed. At this level, conclusions could not be far-reaching. Trends could be observed, but conclusions can only be considered as transferable within a specific group of stakeholders and cannot be generalised about the wider

group of stakeholders or even to both countries. Each chapter comes with a separate appendix, including all the tables being used in that specific chapter. Chapter 5 focuses on the first domain of the matrix: theories of giftedness. Chapter 6 addresses domain 2: seeing the educational needs of gifted learners. Chapter 7 describes domain 3: understanding the educational needs of gifted learners. Chapter 8 addresses domain 4: meaningful responses to the educational needs of gifted learners. Chapter 9 describes the results for domain 5: assessing the responses to interventions in gifted education.

In Chapter 10, the overarching results of the empirical research are presented. General trends observed in the empirical data and reported in their respective chapters are integrated and analysed at a more superficial level. This offers the opportunity to draw conclusions based on an overview of the entire matrix and the complete list of knowledge and skills indicators.

In part three of this thesis, Chapter 11, theory and practice are integrated. In this chapter the results of the research are summarised and discussed, perspectives are offered on the limitations of these results, directions outlined for further research and implications for teacher education programmes are considered.

Part 1

Positioning Gifted Education in Primary Schools in the Dutch and Flemish Educational System

2 Educational Paradigm Shifts and Effects on Educating Gifted Students

2.1 Introduction: The Interdependency of Educational Paradigms

The complexity of being a competent teacher increases rapidly. High quality in education is necessary for every nation to stay or become a significant player in the world economy. Only the best educational concepts are the major keys to social progress and the improvement of prosperity. To match the quality of education with social and economic developments, worldwide new educational paradigms have been, are being and will be introduced. This process is part of our history of education, and also part of the future of education.
According to Kuhn (1962), a paradigm provides a model for the analysis and interpretation of what occurs in reality by a community of practitioners. Hence, the use of a paradigm determines both one's assumptions and one's critical thinking and sets contextual boundaries for problem solving. Teachers can be considered as a community of practitioners (Enthoven & de Bruijn, 2010). As they reflect on their practice (Korthagen, 2017; Loughran, 2002), their practice is influenced by their critical reflections of both theory and daily experiences (Biesta, 2007, 2010, 2012; Carrier, 2015; Cooper et al., 2017; Nelson & Campbell, 2017).
A paradigm can be distinguished from a trend. A trend reflects a general movement in a situation or in the way people behave and does not provide a model for analyses and interpretation. Trends can be explained by looking through the lens of a paradigm to critically review the change in a situation.
A paradigm shift arises when a dominant paradigm is considered to be incompatible with new phenomena and facilitates the adoption of a theory or new paradigm (Kuhn, 1962).
At this time in our western society in education, five significant paradigm shifts can be distinguished that affect Dutch and Flemish gifted education. Its process started with the shift to a society based on the concept of inclusion: a society no longer based on the participation of *most* people but on the participation of *all* people. Changes in the educational system are therefore conditional. From striving for an inclusive society, emerges the strive for an inclusive approach to education (Borland, 2005; Mol-Lous, 2011; Rodriguez, 2019). To make this work, we need to shift the way we teach and support (Borland, 2005; Clijsen et al., 2007; Mol-Lous, 2011; van Meersbergen & de Vries, 2017). The concept of responding to educational needs is the second paradigm shift. It replaces the achievement-only oriented pedagogy in education, with a change and solution-focused pedagogy (Haenen & Mol-Lous, 2014), in which *educational needs* replace the concept of *special educational needs* (Oliver, 1996). *Learning* instead of achieving becomes the core business of education. If learning for all is accepted as the central concept in education, then the needs of gifted students should be addressed as well.
A consequence of the paradigm shift in the perspectives of giftedness that has gradually taken place is its influence on the current education of the gifted.

D. Matthews & Folsom (2009) describe this third paradigm shift as the development from a mystery to a mastery perspective. In modern society, giftedness is no longer considered as static but as the result of a developmental process (Freeman et al., 2010; Gagné, 2010a; Lo & Porath, 2017; Lo et al., 2019; Sternberg, 2009; Subotnik et al., 2011).

The fourth paradigm shift is a development towards curricular and instructional strategies based on social-constructivism, often referred to as 21st century skills. From the high speed changes in our modern society, has emerged the need to equip people with skills that make it possible for them to adjust themselves to the demands of future society (Marzano & Heflebower, 2012; Schleiger, 2012; Voogt & Pareja Roblin, 2010). Technological, economic and social developments require that we put a strong emphasis on equipping our children with the skills to gather and manipulate knowledge in a way that will be effective in a future society, whatever that future might be (Vermeulen & Vrieling, n.d.).

The fifth, and final, paradigm shift affecting Dutch and Flemish education that is introduced in this chapter is the shift towards evidence-informed teaching. In the last decade, the increasing value of the direct translation of research results into educational practice can be observed (Mol-Lous, 2011; Van den Berg & Suasso, n.d.; Vandenberghe, 2005). This is caused by a growing awareness that in education, many changes were implemented consecutively without the support of evidence that they would provide an enhancement to previous approaches (Onderwijsraad, 2006).

The dynamic of each of the five independent paradigm shifts and their influence on teachers shows the complexity of teaching as a profession in general and gifted education in particular. However, when one realises that these paradigm shifts cannot be seen purely on their own merit but always have to be seen through the interplay of one with another, an even more complex dimension becomes visible. The complexity of these changes is illustrated by the interdependency of the 'Big Five' as shown in Figure 1.

2.2 The First Paradigm Shift: The Inclusive Approach

In education, inclusion is about accepting that every student has an individual educational profile with personal educational needs (Oostdam, 2009; Pameijer & van Beukering, 2007; Peters & Oveross, 2020; van Meersbergen & de Vries, 2017). Teachers learn to see, understand and respond to those needs in a way that enables their students to reach towards their next zone of proximate development (Mol-Lous, 2011). Discussion of the concept of the inclusive approach is relevant within the context of developments in gifted education in The Netherlands and Flanders. In both countries, (part-time) segregated education for the gifted has become increasingly popular (Keeman, 2010; Vlaams Parlement, 2019). This seems contradictory to the inclusive approach and shows a growing polarisation between stakeholders in gifted education.

Figure 1

The Interdependency of "the Big Five"

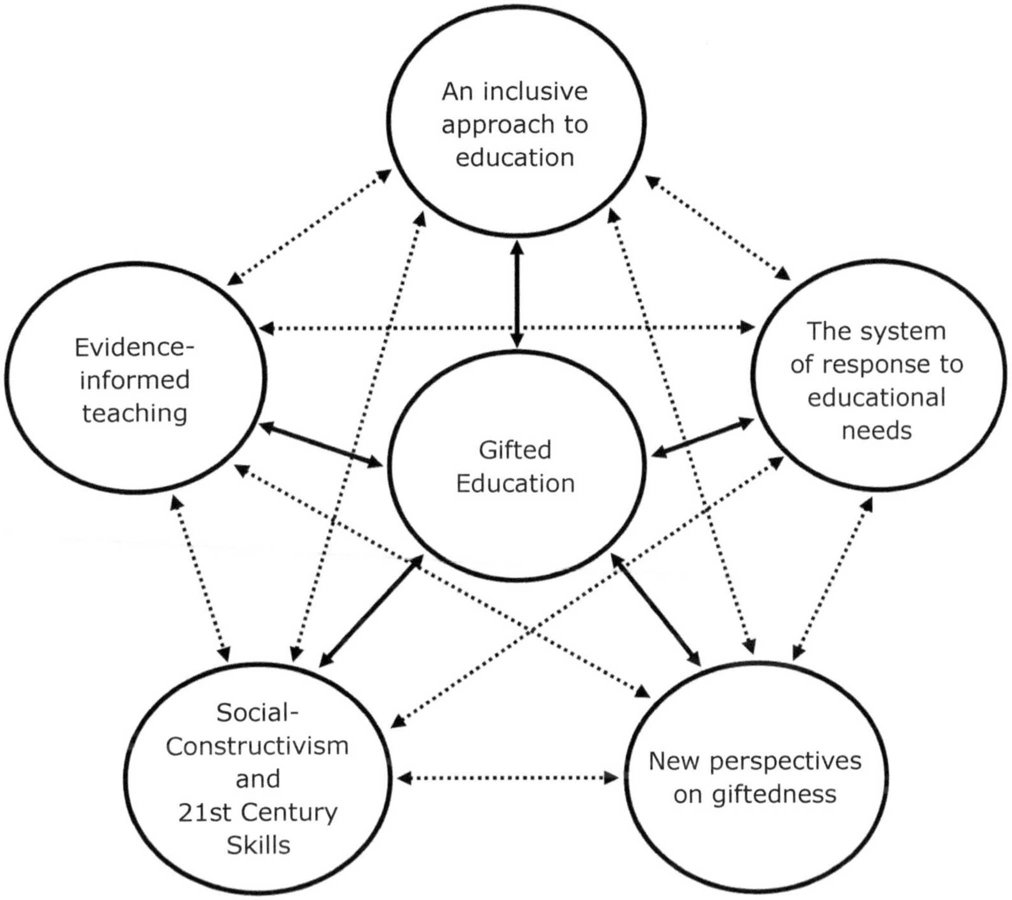

Note. Adapted from "The Student's Ecological System" by E. van Gerven, 2015, *Les 2.1. De Cirkel van Zorg en de Ecologie van de Leerling. Post-HBO Registeropleiding tot Specialist Begaafdheid.* Copyright 2015 by E. van Gerven.

2.2.1 The Inclusive Approach

When it comes to education, the difficulty with the concept of inclusion is that an overall definition that is accepted worldwide is lacking. "Inclusion means different things to different people" (Armstrong et al., 2010, p. 29). Inclusion is often defined by what one is doing (Florian & Black-Hawkins, 2011), so instead of one definition, there are as many descriptions of inclusion as there are describers. Ainscow et al. (2006) differentiate between two kinds of descriptions: descriptive and prescriptive definitions. A descriptive definition reports on how inclusion is used in practice, while a "prescriptive definition indicates the way we use the concept and would like it to be used by others" (Ainscow et al., 2006, p. 14).

The underpinning concept of the inclusive approach is a social model of disabilities (Oliver, 1996). In this model, disabilities are seen as a result of the way that society is organised. Logically, reasoning from this perspective means that if thresholds and barriers for individuals with impairments are taken away, only educational needs rather than special educational needs are left. Therefore special educational needs are not defined by the inability of the individual, but by our concept of normality and the boundaries that are set for what is normal and what is not (Borland, 2005; Oliver, 1996). Embracing an inclusive approach means that differences between students' didactical and pedagogical needs have to be accepted (Dijksma, 2007; Mol-Lous, 2011). Florian and Black-Hawkins (2011) state that the perspective of educational needs still refers to the argument that some children "necessarily require something 'different from' or 'additional to' that which is ordinarily available" (p. 815). They state that an inclusive practice refers to a situation where teachers extend and enrich their classroom activities in such a way that partici-pation in these activities becomes available for all students in their classroom.

It is fundamental to appreciate that a change and solution-focused approach based on an inclusive approach is not something one takes on as a project. Projects are defined by reaching a specific goal or target. When it comes to an inclusive approach, one will discover that it is an ongoing process. Its success depends highly on the translation of theory into practice. Ainscow et al. (2006) describe inclusion as:

> it [inclusion] is concerned with all children and young people in schools; it is focused on the presence, participation, and achievement; inclusion and exclusion are linked together such that inclusion involves the active combating of exclusion; and inclusion is seen as a never-ending process. Thus, an inclusive school is one that is on the move, rather than one that has reached a perfect state. (p. 25)

In 1994, The Netherlands and Flanders joined the Salamanca agreement. This agreement specifically mentions gifted children: "Schools should accommodate all children regardless of their physical, intellectual, social, emotional, linguistic or other conditions. This should include disabled and gifted children, street and working children" (Unesco, 1994, p. 6).

Both The Netherlands and Flanders have passed laws on inclusion. In The Netherlands, it is common to refer to the act on "Appropriate Education" (Wet

passend onderwijs, 2012) as a synonym for inclusion. Since 2014, in Flanders, reference has been to the so-called "M-Decreet" (Decreet betreffende maatregelen voor leerlingen met specifieke onderwijsbehoeften, 2014); in the forseeable future, this decree will be replaced with the so-called "Begeleidingsdecreet" (Vlaamse Regering, 2019). With these laws, both governments want to guarantee sufficient education for every child in its environment, or as they call it "in a home near environment" (Amsing et al., 2009; Wet passend onderwijs, 2012; Mol-Lous, 2011).
In The Netherlands, standards for the school inspectorate explicitly include gifted students (P. Wolters, 2009). In the current Flemish M-Decreet there is no explicit mention of gifted students (Vlaams Parlement, 2019). The new Begeleidingsdecreet (support decree) that is currently being prepared is expected to be introduced in 2021-2022. In the new decree explicit attention will be given to gifted and highly able learners (Vlaamse Regering, 2019). The notion of educational needs as the basis for a differentiated and inclusive pedagogy is currently dominant in The Netherlands and Flanders (Mol-Lous, 2011; Pameijer & van Beukering, 2007; van Meersbergen & de Vries, 2017). As a result, the concept of gifted education is seen, in both countries, as a meaningful response to the educational needs of gifted students.

2.2.1.1 Segregated Education for Gifted Students

While the influence of this paradigm shift is towards inclusion, it is important to note that there has been an increasing, fast-growing focus on full-time segregated education for the gifted in both countries. This reflects the seemingly contradictory perspective on the position of education for the gifted in that inclusive educational context. Having said that, it should also be noted that, on their governmental website, the Flemish Department of Education states that full-time segregated education for the gifted is not available (Vlaams Ministerie van Onderwijs en Vorming, n.d.). Only part-time segregated education for gifted students using pull-out programmes is recognised. The following therefore only applies to The Netherlands.
Special primary schools for the gifted were a rarity in The Netherlands in the 20th century (de Boer et al., 2013). At the turn of the 21st century, a few special schools for full-time segregated education for the gifted were introduced (Hoogeveen et al., 2004). They were all funded privately. Most of these schools closed almost as fast as they began. In 2007, a private initiative reintroduced full-time segregated education for the gifted at primary school level (Daeter, 2012; Keeman, 2010). Ignited by their personal experiences, the founders of these Leonardo schools stated that gifted students are better served in a full-time segregated situation of a homogeneous group with students of equal intellectual capacities (Daeter, 2012; Leonardo-onderwijs voor hoogbegaafde kinderen, n.d.). The Leonardo school founders did not consult international research. However, several studies of special programmes for gifted students in the United States, published between 1988 and 2004, showed that there are indeed positive effects to be observed. These studies were recently referred to by Kitsantas et al. (2017).

In The Netherlands, studies by Hoogeveen et al. (2004), Doolaard and Oudbier (2010), and also by de Boer et al. (2013), stated that, although these positive effects were observed in the United States, there was insufficient evidence that the same would apply in The Netherlands. They recommended further research on this topic. A more recent study by Hornstra et al. (2017) on "The effects of full-time and part-time high-ability programs on developments in students' achievement emotions" shows no significant positive effect on students' achievement emotions amongst participants of full-time programmes. The jury is still out concerning the effects of full-time segregated education for the gifted in The Netherlands.
Since 2007 the number of Dutch special schools for the gifted has increased rapidly (Daeter, 2012; Houkema et al., 2018; Keeman, 2010). Although the government prefers an inclusive approach to a segregated approach (de Boer et al., 2013), in The Netherlands full-time segregated education for the gifted is formally recognised. As a way to implement the Act "Appropriate Education" for their gifted students, schools have the right to install full-time segregated classes for the gifted (Dekker, 2014a, 2014b; Rijksoverheid, n.d.). The founders of the Leonardo schools argued adroitly that gifted students could and should be seen as special needs students (Keeman, 2010). They compared the intensity of the educational needs of gifted students with the intensity of the educational needs of traditionally perceived students with special needs. They argued, quite successfully to the general public, that if a teacher cannot respond as optimally as desired for traditional special needs students, it could hardly be expected that this same teacher would be able to do so for gifted students (Keeman, 2010; Mijland, 2010).
In the Dutch discussion on *appropriate education*, these arguments make sense to many people. Parents, teachers, school management teams, and members of school boards argue that aiming for inclusion in day-to-day practice focuses worryingly on how to integrate traditional special needs students in the ordinary classroom (Mol-Lous, 2011; Oostdam, 2009). Advocates for full-time segregated education for the gifted, argue that the effort that is required to make the basic concept of inclusion work, hardly leaves any room to invest equally in yet another group of students with special needs (Smits, 2019). They also argue that, during their initial teacher training courses, teachers do not acquire the necessary skills to cater effectively for the needs of gifted students in regular schools. Research into the quality of the competency of Dutch teachers seems to confirm this claim (de Boer et al., 2013; Mol-Lous, 2011).
The competency matrix for gifted education published in 2011 by van Gerven and Hoogenberg provides an overview of relevant competencies. This matrix is underpinned by the general teaching competencies required of Dutch and Flemish teachers in primary education. The matrix shows a large overlap between general teaching competencies and competencies for gifted education as they are suggested in international literature on gifted education (van Gerven & Hoogenberg, 2011). This overlap shows that the argument that Dutch and Flemish teachers are not equipped to cater for gifted students in their classroom can at least be disputed.

Since complex adjustments are required to respond to the differing educational needs of every student, one cannot argue that being gifted on its own merit is a valid argument for education in a full-time segregated setting (Borland, 2005; Lo & Porath, 2017). Putting the inclusive approach into practice, requires teachers to stretch their current professional abilities; reorganising their actions in a way that meets the student's needs based on a perspective of diversity (Adema et al., 2009; de Boer et al., 2013; Florian & Black-Hawkins, 2011; Haenen & Mol-Lous, 2014; Mol-Lous, 2011; Oostdam et al., 2007; Waslander, 2011).

2.3 The Second Paradigm Shift: Responding to Educational Needs

Educational interventions should be a direct response to the educational needs of individual students. Response to Intervention (RtI) is a translation of this needs-based approach (Coleman & Johnsen, 2011; Mol-Lous, 2011). In The Netherlands and Flanders, this concept is transformed into a change and solution-focused pedagogy. This approach is less dependable in its standardised test results, compared to the general interpretation of RtI. It creates ample opportunities for a broad range of diverse pedagogical and didactical strategies. Seven underpinning assumptions are needed to make this approach successful (van Meersbergen & de Vries, 2017):

- A student's individual educational needs are placed in a central position.
- One can only tune towards a student's needs if one keeps in mind how the interaction between persons and situations affects those needs.
- The teacher makes the difference.
- Aiming for opportunities and chances dominates the intervention strategies.
- Educating is a collaborative process between everyone who is involved.
- All interventions are specifically designed to reach the set goals.
- Interventions are systematically applied and transparent for everyone involved.

In The Netherlands and Flanders, the concept of a differentiated response to educational needs is not limited to the academic curriculum; it also includes the social-emotional curriculum. Pameijer and van Beukering (2007) pointed out that a differentiated curriculum can only be successful if a differentiated pedagogic strategy supports it. The ability to differentiate requires, therefore, not only managerial and didactic skills but also an extended repertoire of pedagogic and social skills (Bakx, 2015, Meijer, 2004, Pameijer, 2017; Pameijer & Denys, 2018, Schuman & de Vries, 2020; van der Meer, 2016). To make an approach like this work, teachers need to be watching for the student's potential, for factors stimulating the actualisation of this potential and for challenges or limitations in the actualisation of potential (Weterings, 2017). Within the bounds of reality and the bounds of what is reasonable, teachers should be able to respond to those needs based on their professionalism (Pameijer & van Beukering, 2007; Pameijer, 2017; Schuman & de Vries, 2020). However, if a teacher feels that they have hardly any knowledge or have not

developed the right level of skills for educating gifted students, it is not likely that such a teacher will be successful in finding the meaningful educational responses needed by their gifted students (de Boer et al., 2013; Segers & Hoogeveen, 2012; Troxclair, 2013; Vreijs et al., 2018).
A teacher's response to a student's educational needs and a student's response to the interventions is an ongoing process (Pameijer & van Beukering, 2007; Schuman & de Vries, van Meersbergen & de Vries, 2017). As educational interventions are now led by the individual needs of students, the long-used Procrustean approach as a concept of education should be dismissed. The idea of curricular confection with one-size-fits-all strategies cannot underpin teaching approaches any longer (Borland, 2005; Doornbos & Bergman, 1991; Oostdam, 2009; Stevens, 1997). Although basic standards can be stipulated, responding to educational needs implies that within a regular group there will be differences in the way students can meet those standards (Sutherland & Stack, 2014; Tomlinson et al., 2009). There may be students to whom the standards are set far too high, as there may also be students who can exceed these standards easily (Winstanley, 2010). A change in policy should offer enough room to include the education of the gifted in the regular classroom and the regular school. However, this is all based on the assumption that an inclusive approach means that we accept that every student is different (Armstrong et al., 2010; Clijsen et al., 2013; Florian & Black-Hawkins, 2011; Oostdam, 2009). It is also based on the assumption that teachers have enough means, opportunities, and the competency to respond appropriately for each individual student.

2.3.1 The Need for Differentiation

Dutch educational law (article 8) explicitly states that schools should facilitate ongoing continuous progress for all students (Wet primair onderwijs BES, 2016). In article 1.3 of Flemish educational law (Decreet betreffende gelijke onderwijskansen-1, 2012) the government refers to striving for optimal learning and developmental chances for all students. In the current Flemish coalition agreement (Vlaamse Regering, 2019) this is not only confirmed, but is also applied to gifted learners. The educational process must match the best possible way to develop a student's potential. Differentiation is a basic condition to create options for an inclusive approach in education (Mol-Lous, 2011; Oostdam, 2009) as it creates opportunities to match the curriculum and learning experiences to the learners (Roberts & Inman, 2013). Therefore, the ability to differentiate is one of the most important teachers' competencies, described in the Dutch Teacher Qualification Standard as well as in the Flemish Standard (Onderwijscoöperatie, 2014a, 2014b; Vlaamse Onderwijsraad, 2017).
Effective differentiation is intentional and requires effective assessment. Good assessment directs teachers' decisions on how to differentiate. It makes it possible to set educational goals for individual students, which then makes it possible for the teacher to plan matching actions (Hattie, 2013, 2014; Marzano, 2007; Wiliam, 2011). Consequently, the ability to differentiate creates ample opportunities for educating the gifted in the regular classroom (Janson, 2017). "Tuning" is the magic word required when striving to achieve an optimal match

Figure 2

The Equilateral "Classroom" Triangle

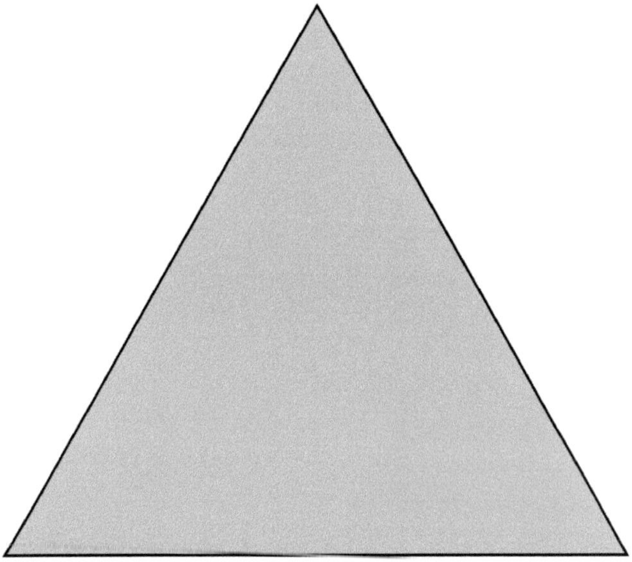

Note. From *Knapzak Praktijkgidsen: Uitdagend Onderwijs* (p. 62) by E. van Gerven, 2014, Leuker.nu. Copyright 2014 by E. van Gerven.

(Kramer & Wildeboer, 2017; Pameijer & van Beukering, 2007; van Meersbergen & de Vries, 2017). Without it, an intervention cannot be successful, because a match between the student's needs and the curriculum would be coincidental. It is important to realise that a differentiated curriculum is meaningless if the curriculum offered is just a selection of different assignments, without the process of tuning the curriculum to the student's needs (Chandra Handa, 2009). In that case, a match between the student's needs and the curriculum *is* coincidental. If a student's needs are not met, the student is still excluded from an education based on the principles of inclusion. Therefore, differentiation is conditional for inclusion, but that does not apply the other way around. Although responding to individual educational needs sounds logical – it might provide an individual student with an optimal learning environment and with optimal learning opportunities – and differentiation might be the key to all of this, it might also be a good suggestion to listen to the voice of reality. The change and solution-focused approach does not come without limitations to what is possible in day-to-day practice at schools. It only works by the grace of a realistic balance between that which is optimally workable, and that which is optimally desirable (Mol-Lous, 2011; Oostdam, 2009). It is the permanent trade-off between the reality of day-to-day teaching and wishful thinking about the opportunities every student should have. This balance applies to the "classroom triangle": the equilateral triangle between the teacher, the entire group of students, and the individual student (van Gerven, 2014). In the equilateral triangle, an equilibrium exists between all three actors. Each actor has the same operational space. Honouring individual differences in the developmental potential of all children in the group secures equal opportunities. Within the same context, the teacher's "individual professional space" is as important as the student's opportunities to develop. This concept of the educational equilateral triangle, as shown in Figure 2, is an analogy to the "golden triangle" developed by Broersen and Spreij (2009).

Figure 3

When the Focus on a Single Student has Become Too Intense

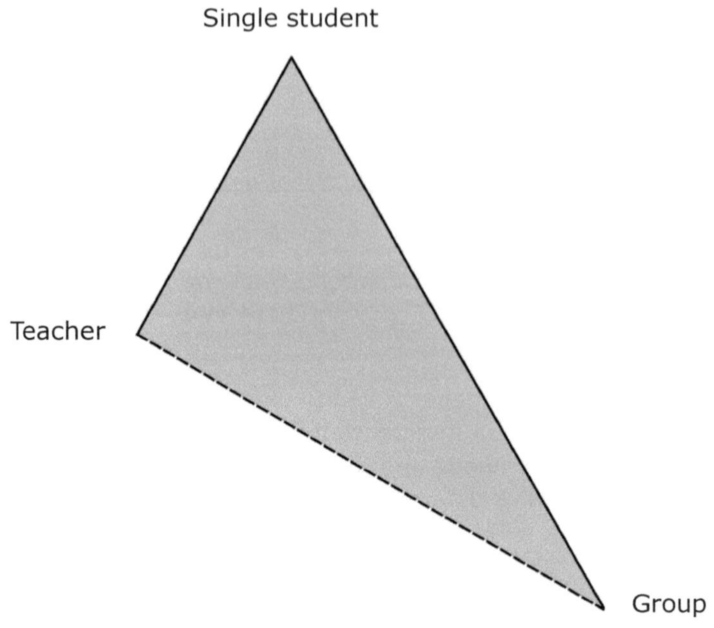

Note. From *Knapzak Praktijkgidsen: Uitdagend Onderwijs* (p. 63) by E. van Gerven, 2014, Leuker.nu. Copyright 2014 by E. van Gerven.

If, as a result of differentiation (either at the time of the intervention, in the depth of the intervention, or in the diversity of the intervention), a teacher has to focus intensely on one student, the equilibrium is disturbed. This is symbolised, in Figure 3, by the reduced distance on the triangle between the teacher and the individual student. As a result, all three actors have to give up some of their "personal space". In practice, this means that the teacher has less time to divide amongst the other students. It also means that the teacher has fewer opportunities to use as many different didactical approaches as might be fruitful for the other students. It might also be less fulfilling for the teacher themselves. The appeal to their professional skills becomes somehow unidirectional, because of the focus on only one student. The teacher loses the diversity and the variety that are optimal requirements for inclusive practice. They also lose the diversity and variety needed for professional satisfaction related to the tasks reflecting all their students.

In The Netherlands and Flanders, two strategies for differentiation are used: (1) "convergent differentiation" and (2) "divergent differentiation". Convergent differentiation means that standard educational objectives, set for the entire group of students, are applied in a three-tiered structure (20%-60%-20%), based on Bell curve distribution. According to Meijer (2004), it acknowledges the natural differences between the characteristics of individual students. Divergent differentiation means that the response to a student's educational needs is as individualised as possible. It aims to create a personalised educational continuous process, in which each student has an individualised exit profile (Mol-Lous, 2011; Oostdam, 2009).

The concept of convergent differentiation is open to four points of criticism. Firstly, in reality, a Bell curve rarely applies to any classroom population. The idea of a normal distribution of abilities applies only if large numbers are involved. In a group of only 23-25 students, there is a decreased chance that a normal distribution curve can be observed. As a result of this, teachers do not experience the Bell curve in the reality of their daily practice. In fact, they may experience a relatively large group of students that have different needs than can be met within the three-tiered structure.

The second point of critique is that, in reality, even the subgroups based on the Bell curve are not as homogeneous as that way of grouping might suggest (Lo et al., 2019; Ziegler et al., 2017). A closer look at Tier 3 shows differences that may have been missed before. Gagné (2010a) states that gifted students lie in the top 10% of all students. Hence, it could be concluded, that gifted students lie in the top half of the group of students that have the potential to function at least at Tier 3. The problem is that Tier 3 has a ceiling based on the average abilities in Tier 3 and not on the abilities of the gifted students. Consequently, the educational goals and opportunities offered at Tier 3 are only a partial match to the abilities of gifted students.

The third point of criticism, regarding the concept of convergent differentiation, is that average and below average students gain more from this concept than gifted students. In a system of convergent differentiation, age peers are unlikely to act as role models or be relevant sparring partners when it comes to academic achievement. Gifted students can, however, gain from a homogeneous grouping that is the result of divergent differentiation, which provides them with the opportunity to learn from other able students (Kitsantas et al., 2017).

The fourth and final point of critique regarding the concept of convergent differentiation is that this kind of differentiation does not offer many opportunities to acknowledge natural differences, compared to the opportunities offered by divergent differentiation. Based on theories of Stevens (1997) and

Doornbos (1991), Oostdam (2009) suggests that it is important to do justice to mutual differences between students in order to be adaptive to the different needs they have. The construction of tiered goals, used for the concept of convergent differentiation, ignores the surplus of natural abilities of gifted children. How can one strive for excellence in education if the way education is organised crops the opportunities to excel?
Based on these points of criticism it seems rational to opt for divergent differentiation. Divergent differentiation provides opportunities for an optimal match between a student's needs and the education offered to stimulate the ongoing continuous progress of a child's development (Mol-Lous, 2011). The glass ceiling that exists in the approach of convergent differentiation does not exist in this approach. So, it provides for the opportunity of the optimal match between the student's needs and the education offered, to stimulate the ongoing continuous progress in the child's development (Tomlinson et al., 2009; Winstanley, 2010). Hence, this way of differentiation seems to offer more opportunities for the education of gifted students, not only in full-time segregated education for the gifted but also in a regular classroom (Janson, 2017; Lo et al., 2019).
However, a strong objection to divergent differentiation is that it stretches the teacher's capacity for classroom management above and beyond what might be seen as fair and reasonable (Kuipers, 2017; Mol-Lous, 2011). Secondly, advocates of convergent differentiation point out that, to be effective, the administrative workload of divergent differentiation requires too much time, which could be better invested in "real" teaching.

2.3.2 The System of Response to Educational Needs

In The Netherlands and Flanders, educational needs are ranked based on the complexity for teachers to respond to those needs. This ranking can be graphically represented in a pyramid (Armstrong et al., 2010). Every level in the pyramid represents another level of intensity in the way that the educational system responds to a student's needs. The intensity of support can be seen as a reversed continuum (Armstrong et al., 2010). The more complex a student's needs are, the more intense the support must be to sort, affect, and stimulate individual development. The required skills to recognise and understand a student's needs become more complex if the needs increase to the next level of complexity (Peters & Oveross, 2020). Without the ability to recognise, to understand, and to interpret a student's needs, a teacher cannot respond effectively.
This construct of achievement, the identification of needs, and a matching, meaningful response in support apply both in a heterogeneous setting (regular schools) and a homogeneous setting (full-time education and pull-out programmes for the gifted). Each student has a unique profile based on personal traits and natural capacity. The student's ecological system influences this profile (Lo et al., 2019; Mol-Lous, 2011; van Meersbergen & de Vries, 2017; van Meersbergen & Jeninga, 2012). This interaction results in a unique educational profile (van Gerven & Hoogenberg, 2011; Ziegler et al., 2017). Therefore, the

idea should be dismissed that all students within a homogeneous setting have the same educational needs.

Figure 4

Continuum of Educational Needs and Support

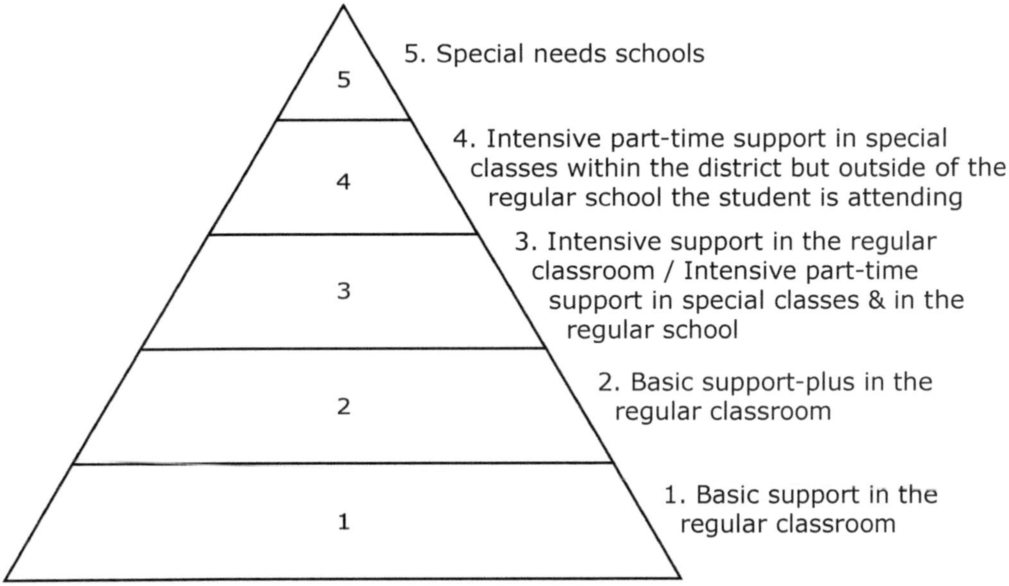

Note. From *Knapzak Praktijkgidsen: Uitdagend Onderwijs*, by E. van Gerven, 2014, Leuker.nu. Copyright 2014 by E. van Gerven.

Figure 4 shows a pyramid of support regarding gifted education in The Netherlands and Flanders (van Gerven, 2014). This pyramid is almost identical to the way overall educational support in both countries is organised (Prodia, 2012; van Gerven, 2014). The two bottom levels of the pyramid of support represent the basic support provided in every classroom. Every school is allowed to design its unique concept of basic support (levels 1 & 2). Level 1 represents the regular curriculum. Level 2 is the variation in the regular curriculum allowed to be applied within certain limits as the teacher sees fit for the student. The extent to which school management has confidence in their teaching staff determines how strictly they set these limits. Level 3 represents the support provided by the school's special educational needs coordinator (SENCo) when adjustments at level 2 are not sufficient and intensified support is needed. This support is additional or can be an alternative to what is offered at level 2. Support at this level is the adjustment towards a far-reaching kind of differentiation. Interventions at level 3 can be split into two kinds of support: support that helps the teacher to see and understand educational needs, and support that helps the teacher to provide an effective and meaningful response towards those needs. Both kinds of support aim to help the teacher, focus on what the teacher needs to become self-sufficient, and what the teacher needs to support or guide the student.

Additional support provided by level 4 is taken care of by district SENCo's. These district SENCo's are often teachers who specialise in a particular field of learning and behavioural problems. Interventions at this level can be aimed at supporting the in-school SENCo, the teacher or the student. Everything that applies for level 3 support applies for level 4 support as well. The major difference between the tasks of both types of educational advisors lies within the complexity of the cases, or the unicity of a problem, that is, the combination of giftedness with a specific learning or behavioural disorder. Level 5 refers to placement in special needs classes and is mostly reserved for students with learning and/or behavioural disabilities.

The application of the pyramid of support varies for each school. With reference to the education of gifted students, the stability of this pyramid of support is still weak in The Netherlands and Flanders. The characteristics of each school determines the shape and stability of their pyramid of support. The stability of the pyramid increases if the number of students who need support at the next level of intensity decreases per layer. The more competent teachers become, the wider the possible layer that they can serve. In 2012, Segers and Hoogeveen concluded that, within the context of the Dutch educational system, the competences that are needed by teachers to serve gifted students effectively were almost undiscovered ground. In 2013, de Boer et al. claimed that "in most cases, teachers lack the knowledge and understanding of how to differentiate and align their curriculum to the pedagogical and didactic needs of gifted students." In 2014, Haenen and Mol-Lous stated that, due to a curricular gap in basic Dutch teacher education, the education of the gifted was still unexplored ground. They all recommended that an inventory of necessary teacher competences for educating the gifted should be made. In *Slim beleid*, van Gerven (2008) published a preliminary list of teacher competences to get school policy on educating the gifted on track. In 2011 a revised list was published in *Begaafd begeleiden* (van Gerven & Hoogenberg, 2011). This list was placed in a matrix and combined with basic teacher competences as described by the SBL (Stichting Beroepskwalificaties Leraren – Foundation for Professional Teaching Standards) (Ministerie Onderwijs en Wetenschappen, 2006).

The adaptability of the pyramid of support is also affected by the school's student population. If the basic needs of the student population already call highly on a teacher, it is less likely that this teacher can respond individually to the needs of a large group of students. This is where the effect of the disturbance of the equilateral triangle takes its toll. Therefore, defining whether interventions belong to level 2 or level 3 is different in every school. When, for some reason, interventions for gifted students at level 2 are insufficient or difficult to organise, the call on level 3 to become more actively involved becomes more intense. Figure 5 visualises how this affects the stability of the pyramid.

Figure 5

Continuum of Educational Needs and Support Out of Balance

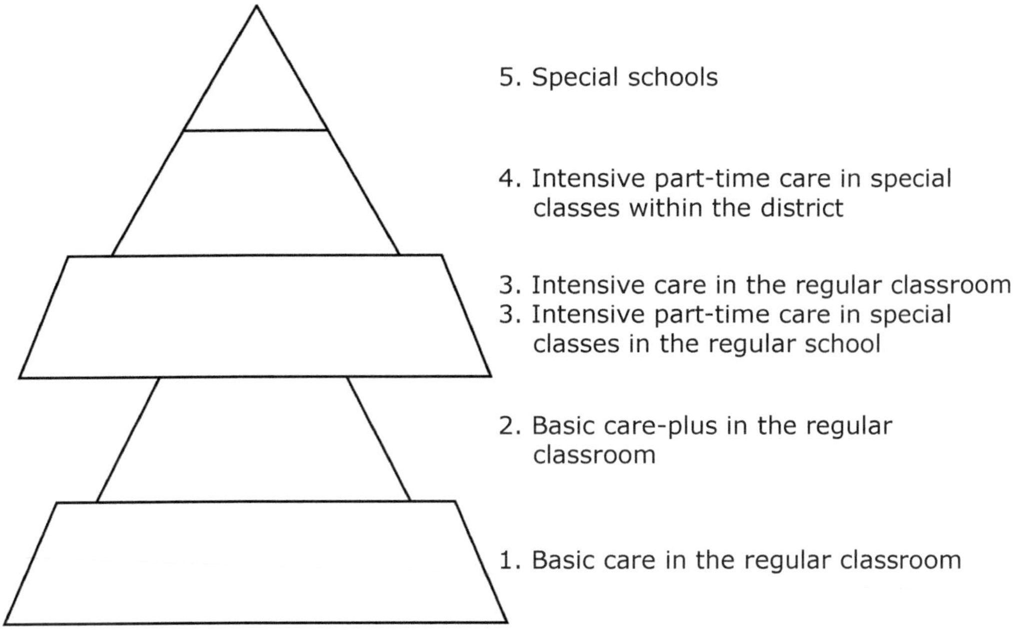

In more advanced teacher education courses on educating the gifted, participants express their discontent about the instability of the pyramid of support at their schools. They stress that, often, half of the students in their pull-out programmes could be served in the regular classroom, provided that the teacher has developed the necessary competences. Instead, their caseload as specialist teachers becomes heavier and more complex, resulting in a decreasing quality of care provided by these special programmes.
There are no statistics available, regarding the education of the gifted, that reflect the stability of the pyramid of support for Dutch and Flemish schools. However, the assumption that this pyramid is likely to be more stable in Dutch schools than in Flemish schools can be made when taking into account the following facts:

- The Dutch government invests in the education of the gifted (Slob, 2018a, 2018b).
- Education for the gifted is explicitly mentioned in Dutch education laws.
- Dutch schools invest a serious amount of money in the professionalisation of their teachers around the topic of giftedness.
- There are several intensive teacher education courses on educating the gifted in The Netherlands. Most of them require a time investment of 100-280 hours; one requires an investment of 450 hours.

- There is a Dutch professional society for formally accredited "specialists in gifted education" that stratifies their members by different degrees of professionalism, based on their previous education (LBBO, n.d.-b).
- Up until 2019, the Flemish government has not invested explicitly in education for the gifted.
- In Flemish educational law, there is no explicit reference to gifted students, but there is to students with disabilities (Decreet betreffende maatregelen voor leerlingen met specifieke onderwijsbehoeften, 2014). Starting at the beginning of the school year 2021-2022, the "Begeleidingsdecreet" will be introduced. In this decree, gifted learners will be mentioned as a group of students having additional educational needs to be addressed by schools.
- Until recently, there was hardly any budget for Flemish schools to invest in the professionalisation of their teachers (Hovius & Van Kessel, 2010). Therefore, it is likely that ongoing professionalisation regarding the topic of giftedness was not highly prioritised. However, this also seems to have changed as a result of the current coalition agreement. In the coalition agreement the necessity for teacher education and continuous professionalisation is stressed as conditional to the government's new plans.
- There are no educational programmes for teachers in Flanders to become specialists in gifted education that match the intensity and in-depth level of those that are available in The Netherlands (Vreijs et al., 2018).

2.4 The Third Paradigm Shift: New Perspectives on Giftedness

In 2005, in *Conceptions of Giftedness* (Sternberg & Davidson), James Borland published a chapter titled "Gifted education without gifted children: The case for no conception of giftedness". Borland successfully raised the discussion about the relation between the concept of inclusion in modern education and the definition of giftedness. He argues that it is no longer valid in modern education to use a conception of giftedness to underpin selection procedures, to determine which student can or cannot participate in alternative educational arrangements (i.e., enrichment classes, pull-out programmes, etc.). He refers to the fact that there is no universal conception of giftedness, and that this means that selection procedures are inequitable by nature.

Borland's argument was not about ignoring the educational needs of gifted students. It was, that the definition of giftedness becomes more or less a non-issue when goal-differentiated curricula and instructions are organised. According to Borland, goal-differentiated curricula and instruction could create commodious opportunities for all students to fulfil their potential. In an article in 2017, Janson describes several opportunities to create an educational environment where this becomes possible. In these settings, the need to define separate groups of students by their potential in order to respond to their educational needs is no longer necessary. Borland (2005) states, "Were we to set as our goal the creation of schools in which curricula and instruction mirrored

the diversity of students found in classrooms, and were we to achieve this goal, the only legitimate aim of gifted education would be achieved" (p. 13).
Borland was clearly not the only one who noticed the need for change in perspectives on giftedness and gifted education. It was only a year later that D. Matthews and Foster (2006) published their book *Being Smart About Gifted Education*. In it, they presented an analysis of a paradigm shift that was taking place regarding the conceptions of giftedness, and described the earlier mentioned shift from a mystery-perspective on giftedness towards a mastery-perspective on giftedness. In this chapter it is important to explain both ends of this theoretical continuum, before going into the consequences of this paradigm shift in relation to the context of educating gifted students in The Netherlands and Flanders.[2] The continuum is considered from an overall perspective and without going into details.

2.4.1 Three Paradigms

The question of "what is giftedness" is discussed worldwide. A wide range of different opinions can be noted (Ambrose & Sternberg, 2016; Balchin et al., 2009; Collangelo & Davis, 2003; Dai & Chen, 2014; Freeman et al., 2010; Kreger-Silverman, 2013; Lo & Porath, 2017; Pfeiffer, 2008; Shavinina, 2009; Subotnik et al., 2011). The fact that Sternberg, at the 15th ECHA conference in Vienna (March, 2016), explicitly mentioned that the identification of gifted students was still as much a hot topic at conferences about giftedness as it was thirty years ago, demonstrates that a united answer to the question "what is giftedness" still lies beyond the horizon. In 2012, Cohen gave a plausible and not to be misunderstood explanation for the lack of a joint conceptualisation:

> There is also dogmatic insularity in hanging on to one's interpretation of the term. Each definition is 'gospel to St. Guru' perpetuated by proponents whom each stay in their own silo, claiming theirs is the correct one. Their definition is learned and touted by new learners (their students), who defend their teachers' definition, or create their own, perpetuating the lack of coherence in defining their terms. (p. 26)

In 1996, Morelock distinguished two different groups of perspectives on giftedness; she refers to them as The Talent Development Movement and The Columbus Group Movement. Dai and Chen (2014) refer to both perspectives as different paradigms: the gifted child paradigm and the talent development paradigm. If we make a comparison between D. Matthews and Folsom's and Morelock's dichotomy, several similarities can be perceived. In this comparison, the perspective of gifted child paradigm is the equivalent of D. Matthews and Folsom's mystery perspective. The talent development paradigm represents the mastery perspective.

[2] At this point, I would like to stress that it is not my place, nor my intention, to determine which perspective on giftedness is "the best perspective". This short sketch of perspectives is merely to show that there is a relation between the notion of what teacher competences need to be developed and the different perspectives of giftedness. A different perspective requires different competences.

The Columbus Group describes giftedness as "an asynchronous development in which advanced cognitive abilities and heightened intensity combine to create inner experiences and awareness that are qualitatively different from the norm. This asynchrony increases with higher intellectual capacity" (Columbus Group, 1991 as cited in Morelock, 1996). Their threshold to label a person as gifted is 2 SD above average. This sets the standard at an IQ of 130. This threshold is based on a normal distribution of intelligence, and it implies that two-three percent of the population might be called gifted. The higher the IQ, the more this asynchronous development will manifest itself (Kreger-Silverman, 2013). In the gifted child paradigm, giftedness can best be seen as a "state of being". A person is born gifted, and signs of giftedness can be noticed from a very early age onwards. Gifted children can be recognised by specific personality traits combined with an advanced intellectual development. The Columbus Group's notion of giftedness is highly influenced by Dabrowski's theory of overexcitabilities. Overexcitabilities can best be described as an intense reaction to external and internal stimuli (Ackerman, 2009). The strong relationship between overexcitabilities and IQ in this concept of giftedness affects what are to be seen as indicators for giftedness (Kreger-Silverman, 2013). In this perspective on giftedness, insufficient justice is given to the complexity of being gifted if the focus is only on high achievement (Dai & Chen, 2014). Therefore, Kreger-Silverman (2013) states critically about her opponents in the Talent Development Movement that "The equation of giftedness with achievement seems to resonate with educators more than with parents" (p. 25). She states that the behavioural indicators are more important than isolated academic achievements. With regard to teacher competency, this means that the ability to observe and correctly interpret behavioural patterns prevails over selection through academic achievements (Haenen & Mol-Lous, 2014). This also implies that to make education successful for gifted children, a fine-tuned response to individual strengths and weaknesses within the context of this asynchronous development is of the highest importance.

Exponents of the talent development paradigm define giftedness as a concept of constructivism, a result of a developmental process where an individual's developmental potential as a key variable interplays with the pedagogic-didactic environment and with personal effort, in a way that demonstrates exceptional achievements (Cohen, 2012; Dai & Chen, 2014; Gagné, 2010a; Heller, 2010; Lo et al., 2019; Renzulli, 1985; Sternberg, 2002; Subotnik et al., 2011).

The threshold to be labelled gifted is set at 1.3 SD above average (IQ 120). This implies that about 10% of people can be called gifted (Gagné, 2010a), but IQ is not to be seen as the sole predictor of giftedness. Other indicators of giftedness mentioned by exponents of the talent development paradigm are creativity, motivation and perseverance, strong task orientation, and goal-oriented behaviour. However, IQ and previously observed high achievements are seen as the most reliable indicators of giftedness. Compared to the exponents of the gifted child paradigm, exponents of the talent development paradigm have more reservations about indicating behaviour. They state that, from the majority of indicators mentioned in the gifted child paradigm, it cannot be said beyond any doubt that this behaviour can only be the result of a high IQ:

> Too often, behaviours like maladaptive perfectionism, feelings of being different, or extreme sensitivity and intensity have been put forward as defining characteristics of giftedness, whereas these behaviors may, in fact, be outcomes of the interaction between gifted children and their home, community and school environment as a result of or independent of the 'gifted' label. (Subotnik et al., 2011, p. 10)

Although both paradigms represent opposing sides of the mystery vs mastery continuum, there is a third and relatively new paradigm on the rise: the differentiation paradigm. The international strive for inclusive education would appear to favour the differentiation paradigm as a way of thinking. Dividing the world into educational "haves and have-nots" is no longer appropriate (Borland, 2005; Dai & Chen, 2014; Lo & Porath, 2017; Lo et al., 2019). Therefore, in this paradigm, the identification of giftedness as a state of being or as a selection criterion for alternative educational arrangements is no longer relevant. The differentiation paradigm focuses on optimal interplay between child and the educational environment, to provide meaningful educational responses that encourage talent development. Within this context, D. Matthews and Foster (2006) presented an education-based definition of giftedness: "Giftedness is exceptionally advanced subject-specific ability at a particular point in time such that a student's learning needs cannot be well met without adaptations to the curriculum" (p. 28). In this educational needs-based concept of giftedness, it is only logical that education has the highest results if it is conceptualised as providing a dynamically responsive curriculum match for the educationally advanced (Lo et al., 2019).

With regard to educational interventions, all three paradigms stress the need to respond to individual differences between gifted students and to tune into their individual profiles of strengths and weaknesses (Dai & Chen, 2014; Kreger-Silverman, 2013; Subotnik et al., 2011). This implies that they all feel that the educational needs of gifted students are likely better served in an educational environment based on divergent differentiation. Therefore, competence in seeing and understanding educational needs and in creating meaningful educational responses goes hand in hand with the ability to acknowledge and recognise the individual needs of gifted students (Lo et al., 2019). Responding to those needs requires an ability in effective classroom management, and the ability to adjust the curriculum in a way that best serves the individual developmental opportunities of students (Calahan & Hertberg-Davis, 2018; Johnsen et al., 2016; Rogers, 2007).

In The Netherlands and Flanders, teachers are aware of the fact that there are different definitions. However, they are unaware of the different theoretical fundamentals which underpin these definitions. It is almost unknown that each perspective on giftedness has its own separate indicators, separate educational goals and consequently separate educational strategies (Dai & Chen, 2014). Consequently, communication and processes of decision making are often blurred. Every school is allowed to define giftedness differently (van Gerven, 2017). In day-to-day practice, even within one team of teachers, several interpretations of a chosen definition can exist without the team being aware of it. This can result in a seemingly arbitrary selection of who is to be labelled as

gifted or not (de Boer et al., 2013; Doolaard & Oudbier, 2010; Haenen & Mol-Lous, 2014).
If asked, teachers will summarise the personal traits of the gifted based on the gifted child paradigm. However, when it comes to selecting students for enrichment activities, they base their decisions on achievements, leave the earlier mentioned personal traits outside of their scope and limit the concept of educational needs of gifted students solely to curricular interventions (Slim! Educatief, Postgraduate Teacher Education Course Specialist in Gifted Education, personal communication/class discussions during modules 3 and 4, January-March, 2017, January-March 2018; January-March 2019; January-March 2020). Where they strive for an educational approach based on the concept of a differentiated response to individual educational needs for most students in their classroom, they move away from that approach when dealing with gifted students. Their pedagogical and didactical responses are mostly based on the concept of a "one-size-fits-all-gifted-students-curriculum". Standard pre-set coursework for students who are identified as gifted has become increasingly popular, leaving ideological convictions regarding optimal differentiation behind on the shelf (Kuipers, 2017).
The somehow unconscious ambivalence towards what it is to be defined as gifted often results in a Babylonian confusion between teachers within a team; teachers and parents; and even between teachers and specialists in gifted education (van Gerven, 2017). Available observation lists do not always lead to clarity. Dutch toolkits for teachers to identify gifted children all use a different theoretical model to underpin their tools (de Bruin-de Boer & Kuipers, 2011; Houkema, 2009; van Gerven & Drent, 2012). They all use one or more of the models based on the talent development paradigm, but they leave the chosen perspective after explaining it in the manual of their toolkits. It is possible to ascertain this by analysing the observation lists. The items on these lists are mostly consistent with behaviour as described within the gifted child paradigm on giftedness (Kreger-Silverman, 2013), and divided into categories as distinguished in the talent development models. In Flanders, Prodia (2012) published a protocol for teachers to help identify gifted students. In a similar way to the Dutch toolkits, it starts from the perspective of the talent development paradigm and, although the observation list is organised in a way that connects with Heller's theory on giftedness (2010), the behaviour described in the list originates from a more eclectic perspective. It incorporates characteristics described by the talent development paradigm, and personal traits described by the gifted child paradigm.
The use of these observation lists brings with it the risk of a circular argument about who is to be labelled as gifted. Students displaying these characteristics are labelled as gifted based on the toolkits guidelines, and the toolkits guidelines in how to label these students as gifted are based on the display of these traits. Successfully using any of the observation lists, requires the awareness of personal preference for a specific paradigm. Within this theoretical context, it requires an accurate knowledge of the overall characteristics on these lists and their association with the resulting educational needs. It also requires the skill to interpret student behaviour displayed in a meaningful educational environment,

within the context of these characteristics, and it requires the competency to create a stimulating learning environment that creates the opportunity for students to actualise these characteristics in a way that makes it possible to observe this behaviour. Whereas the paradigm shift worldwide is moving from the more mystery-oriented perspective towards a more mastery-oriented perspective and the differentiation paradigm is definitely on the rise, in The Netherlands and Flanders, mystery and mastery perspectives are becoming more and more intertwined. In this light, it is interesting to see how popular Dweck's theory (2006, 2017) on mindset is in both countries, because it creates an opportunity for those who adapted the gifted child paradigm to shift to a more mastery-oriented perspective, without letting go of the origins of the gifted child paradigm. After all, Dweck's theory is the perfect example of a mastery-oriented perception. Individual responsibility regarding one's personal development and the linked attitude to internally attributing successes and failure can be seen as the perfect steering mechanism for self-actualisation. An integrated perspective like this can be recognised in Kieboom and Venderickx's (2017) theory of giftedness. By integrating both ends of the continuum of perspectives on giftedness, they might become the new leaders in the Dutch and Flemish educational field regarding the definition and description of the notion of giftedness, because their perspective seems to appeal to parents as well as teachers. It is important to note, as a healthy criticism at this time, that the personal traits, as they are mentioned in this concept of giftedness, have not yet been objectively proven as being solely the result of a high IQ. A counter-argument may be that the traits are recognised in all the dossiers in their practice and that we might assume that all ravens are black until proven otherwise.

The educational effects of the rise of the differentiation paradigm cause serious concern amongst parents, teachers and Dutch and Flemish specialists in gifted education. The increasing number of segregated schools for full-time education for the gifted as discussed in paragraph 2.2.1.1 can be seen as a counter-reaction towards this democratisation of the concept of giftedness.

Inclusion versus segregation, responding to educational needs, the call for differentiation and the changing perspective on giftedness have to be seen within the context of another major paradigm shift in education: the movement towards social-constructivism educational strategies.

2.5 The Fourth Paradigm Shift: Social-Constructivism and 21st Century Skills

The fourth paradigm shift is demonstrated by a fast-emerging need to adjust education to the development of 21st century skills. In The Netherlands and Flanders, a traditional approach to education is no longer seen as sufficient (Onderwijsraad, 2014; Vlaams Ministerie van Onderwijs en Vorming, n.d.; Unesco, 2015). The high-speed development of knowledge and skills, and the increasing emphasis on technology, requires that people should be able to adjust themselves at the same pace (Onderwijsraad, 2014; Vlaams Ministerie van Onderwijs en Vorming, n.d.). Therefore, education should be modified by adjusting the traditional didactics to modern requirements (de Bie, 2016). This means a shift from a knowledge-transfer based didactic towards a knowledge-construction based didactic, also referred to as constructivism. This shift is a complex process. Knowledge-transfer can be a straightforward process. The knowledge and skills to be taught are known, as is how the results of the learning process can be assessed. Knowledge-construction, however, is not that straightforward. Process and results can vary enormously. Educational objectives are not always easy to stipulate because often new knowledge is constructed, and qualitative criteria to assess this new knowledge are difficult to predict (Marzano & Kendall, 2008).

While learning, this new didactic requires the application of different skills from those that were required by the traditional way of teaching. Marzano and Heflebower (2012) distinguish cognitive and conative skills that need to be developed by students. The Onderwijsraad, the Dutch National Board for Educational Advice (2014) points out three domains of skills: thinking skills, social skills, and metacognitive skills. In its report *Van lerensbelang* (the importance of learning) the Flemish Ministry of Education describes 14 different categories of skills that can be clustered as thinking skills, social skills, metacognitive skills and IT-skills (Bambs et al., 2016). Although there is a difference in terminology, all approaches focus on the same behaviour. For daily practice in education, it means that teachers have to teach children how to become self-regulating, creative and critical thinkers, able to solve relatively complex problems in a social context that requires them to be active participants and effective team players, with the ability to reflect on their behaviour as the trigger for processes and results (Vermeulen & Vrieling, n.d.). Due to the digitalisation of information and communication, the development of ICT skills should also be part of this modern curriculum (SLO, n.d.).

In paragraph 2.3.2, the concept of a differentiated response to educational needs was introduced. With the shift towards a social-constructivist approach to education, the necessity for differentiation becomes more compelling than ever. Not only does curriculum content have to be taught, but all students have to be equipped with a broad set of self-regulating, social and metacognitive skills (Ylvisaker & Feeney, 2014; van der Donk et al., 2015). Hence, the individualised teacher's response is conditional to its success (Kroesbergen et al., 2017). These skills have to be developed and internalised over a long period. A condition of developing these skills, is the taking on of all different kinds of assignments and projects in various educational contexts (Meltzer, 2018; Stoeger & Ziegler, 2011). Each student has their own strengths and weaknesses in these skills (Anderson et al., 2014; Barkley, 2012). This implies that this new didactic depends not only on a student's capacity and efforts but also very much on the teacher's capacity and attitude (Onderwijscoöperatie, 2014; Vlaamse Onderwijsraad, 2017). The teacher's ability to model preferred behaviour is crucial.

The so-called 21st century skills are not new in education (Thijs et al., 2014), nor are the necessary teacher competences (Mol-Lous, 2011). What is new is the fact that these 21st century skills are valued as important for everybody, and not only for the higher educated (Unesco, 2015). Following international politics, the Dutch and the Flemish governments, therefore, concluded that these skills should be developed by all students and that this learning process should be started at a very early age (Onderwijsraad, 2014; Vlaams Ministerie van Onderwijs en Vorming, n.d.). This conclusion affects the profession of teaching. To adjust their didactics to what is required by a social-constructivist approach to education, teachers first have to develop these 21st century skills themselves (Vermeulen, 2016; Vlaams Ministerie van Onderwijs en Vorming, n.d.; Voogt & Pareja Roblin, 2010).

Many similarities can be observed between the social-constructivist approach and the mastery-oriented perspective on the development of giftedness. Taking these similarities into account, it might be concluded that, provided there is sufficient teacher support, the skills aimed for could be a proper match for the way a lot of gifted students learn (Schrover, 2015). However, no paradigm is introduced without healthy criticism. Although this approach might appear to be a good fit for a lot of gifted students, based on the individual educational profiles of gifted students, there is not enough evidence that it is good for all gifted students. This conclusion brings the focus onto the last paradigm shift.

2.6 The Fifth Paradigm Shift: Evidence-Informed Teaching

In 2006, the Onderwijsraad, the Dutch National Board for Educational Advice wrote, "Research can provide a more reliable judgement on the appropriateness of methods and approaches, and can prevent the continuing of ideological discussions and 'trial and error'" (pp. 9-10). The advice to assign ideology and personal professional experiences (based on N=1) as secondary to generic proven principles, showed the direction of future discussions for new pedagogical and didactical approaches. Objectified knowledge prevails above the professional experiences of teachers. Although personal professional experiences were not completely excluded when it comes to determining "what works" (Onderwijsraad, 2006), they were marginalized. Evidence-based teaching became dominant in the discourse on the quality of education (Biesta, 2007; Diery et al., 2020).

In The Netherlands and Flanders, the discussion on strategies for evidence-based teaching is dominated by the research of John Hattie and Robert Marzano (Hattie, 2013, 2014; Hattie & Yates, 2015; Marzano, 2007, 2011). Their perspective includes a strong emphasis on datadriven education. That implies that students' results on standardised tests are considered to be the most relevant information for underpinning planned educational strategies (Hattie, 2013; Marzano, 2007). Data walls based on standardised test results are being used to help teachers to differentiate between their students by dividing their test results into three tiers (convergent differentiation). Educational goals and instructional strategies are different for each tier. The data wall is considered to be the step between the database of test results in the student assessment system and the teacher's action plans for clustering students based on the educational needs that can be deduced from this information (Kappen & Förren, n.d.).

Figure 6

Continuum of Evidence

Note. From *What Makes a Practice Best? Evidence Based Recommendations in Gifted Education*, by A. Robinson, 2015, World Council of Gifted and Talented Children, 21st World Conference, Odense, Denmark. Copyright 2015 by A. Robinson.

Although student assessment and standardised testing are without a doubt a solid starting point for planning educational goals (Chandra Handa, 2009; Roberts & Inman, 2015), this does not necessarily mean that standardised tests should be the only information teachers are using to underpin their educational strategies (van Meersbergen & de Vries, 2017). Voicing can be an effective strategy to gather additional information and the results of that process might be considered data as well (Pameijer, 2017; Mayes et al., 2019). In teacher education courses teachers report that the use of portfolios is becoming a more often taken approach as well (Slim! Educatief, Postgraduate Teacher Education Course Specialist in Gifted Education, personal communication/class discussions during modules 3 and 4, January-March 2017, January-March 2018; January-March 2019; January-March 2020). For teachers, the fast-growing emphasis on the concept of evidence-based teaching has had a major impact (Van Braak et al., 2008; Vandenberghe, 2005). They needed to be enabled to use empirical evidence productively (European Commission, 2007), and although they were expected to integrate empirical evidence with their teaching experience, their personal (individual) professional experiences seemed to be less valued than before (Biesta, 2007). They had to learn how to accept that *good for one* does not automatically mean that it is *good for all*, or even *good for most*. Research by Diery et al. (2020) suggests that students in teacher education are more likely to accept information derived from their daily practice than research-based sources of knowledge, and that they are more motivated to learn from these practical resources. This information tells us how teachers weigh the importance of their personal experiences and how complex it is for teacher educators to teach them to accept research-based sources of knowledge.

The process of learning how to use and value (empirical) evidence is widely discussed amongst teacher-educators (Diery et al., 2020; Simpson, 2020). In 2015, at the biannual conference of the World Council for Gifted and Talented Children, Robinson presented a hierarchy of evidence. She distinguished three separate levels of evidence. The first level is referred to as *soft evidence* or wisdom. At this level, the evidence is shaped by the personal experience of one professional with one student. Advice obtained at this level should be seen as a personal recommendation. The evidence for advice at this level lacks theoretical underpinning and is mostly shaped by trial and error (Robinson, 2015). The second level of evidence is referred to by Robinson as *best practice*. Best practice differs from common practice. In this context, best practice refers to a theoretically underpinned approach, applied within clear conditions but with the flexibility to adjust as seen fit. At level three, *hard evidence* forms the top level Robinson's hierarchy and is referred to as evidence-based. The level of stringency in the application of conditions and set rules is at its highest on this top level. This stringency has its roots in the field of medicine. Biesta (2007) states that, at this level, effective interventions are those that prove to have a positive causal relationship between the intervention and the response; he argues that a singular focus on effectiveness leaves the most important question of "effective for what" undiscussed.

So in light of using this hierarchy, it is therefore a valid question to ask, what can be considered as reliable evidence and whose evidence counts (Nelson & Campbell, 2017)? To what extend does the evidence gathered from overarching research prevail over evidence gathered by teachers in their daily practice? The complexity of evidence-based education leaves additional factors out of the scope, suggesting that what has been proven effective for most, is also effective in individual cases. It leaves undiscussed whether *effective* is a synonym for *good*; it is just assumed that it is a synonym. However, that also should be discussed (Biesta, 2012; Korthagen, 2017; Mol-Lous, 2011). Teacher-educators stress the importance of teachers also having to learn how to weigh when, from a change and solution-focused pedagogy, *good for one* should prevail over *good for all* (Biesta, 2007, 2012; Mol-Lous, 2011). Moreover, as a parallel process, teachers have to be enabled to distinguish when an approach is based on prejudice towards a group of students and to dismiss such an approach even when they feel personally attracted to it based on personal experience (Biesta, 2007).

Instead of evidence-based teaching, a better concept might be evidence-informed teaching. This last concept leaves room for teachers to make professional judgements about individual cases, based on different types of evidence and the different contexts in which this evidence needs to be applied (Nelson & Campbell, 2017). Evidence-informed teaching does not provide a recipe book. Sharples (2013) interprets it as being about integrating professional expertise with the best external evidence from research, to improve the quality of practice. Professional expertise does refer the teacher's professional ability "to integrate knowledge and handle complexity, and formulate judgements with incomplete or limited information, but that includes reflecting on social and ethical responsibilities linked to the application of their knowledge and judgements" (Bologna Working Group on Qualifications Frameworks, 2005, pp. 68).

The paradigm shift towards evidence-informed teaching brings two issues to light that are influencing the position of gifted education in The Netherlands and Flanders. The first issue has to do with the quality of postgraduate teacher training regarding gifted education. In order to meet current European standards in teacher education (Bologna Working Group on Qualifications Frameworks, 2005), the quality of teacher education is crucial (European Commission, 2007, 2013). The ability to make well-judged evidence-informed decisions, requires that teachers should be taught how to do this responsibly while staying true to their professional values (Biesta, 2012; Diery et al., 2020; Campbell et al., 2017). An important aspect affecting the quality of teacher education is the use of an evidence-informed approach by teacher educators in their courses (Diery et al., 2020; Pillen, 2018). This calls for qualified teacher educators who are more experienced in research and teaching and who are able to implement the use of reliable evidence (Diery et al., 2020, Lunenberg et al., 2014). This is where the paradigm shift towards evidence-informed teaching influences the position of gifted education in The Netherlands and Flanders. As not all specialists in giftedness involved in continuous professional development of teachers are qualified teacher educators, it raises questions about how, in the first place, that affects the quality of evidence-informed teacher education in courses for continuous professional development and, in the second place, how that affects the quality of gifted education. In other words, how do we ensure that gifted education is not based on quicksand?

The second issue regarding the paradigm shift towards evidence-informed teaching, is raised by the question of whether there is sufficient evidence to accept that proposed strategies for gifted education match the Dutch and Flemish social-economical and educational context. When it comes to gifted education, existing research about what works and what does not work is mostly based on projects in countries other than The Netherlands and Flanders (de Boer et al., 2013; Doolaard & Oudbier, 2010; Haenen & Mol-Lous, 2014; Hoogeveen et al., 2004; Mooij et al., 2007; Segers & Hoogeveen, 2012). Although that does not mean that this knowledge cannot be adopted, it prevents a one-on-one transferral of ideas. For every proven strategy, one has to take into account that there might be differences between the educational and social context within which the strategy is applied (Biesta, 2007, 2012). There may also be differences in the underlying concept of giftedness, cultural assumptions as to what should be set as educational goals, and in what is defined as good *in* education and good *for* education. Not only is the development of giftedness systemically influenced, but education is as well (van Meersbergen & Jeninga, 2012). Evidence of what works and what does not is highly dependent on the context within which this proof is gathered, applied and valued. Thus, the most interesting question is not whether a strategy for gifted education works or not, but is about the conditions for its application and the underlying normative concept of what a good education is, and in what context these students can best be served.

2.7 Summarising

In this chapter, Dutch and Flemish education of the gifted is positioned within the context of five interplaying significant paradigm shifts, all affecting Dutch and Flemish education. The interaction between the five paradigms, relating to the call on the expertise that each paradigm places on teachers, demonstrates the complexity of teaching in general and of teaching the gifted especially. Dutch and Flemish school administrations are free to decide how they think these students are served best.

As a result of the inclusive approach, the education of gifted students should be integrated into every school's practice (Dekker, 2014a, 2014b; Slob, 2018a; Vlaams Parlement, 2019). This calls highly on the teacher's ability to differentiate in pedagogics and didactics and requires well-developed managerial skills (Janson, 2017; Kuipers, 2017). The diversity of their students' educational profiles also calls on the teacher's creative problem-solving skills. To determine what strategies will work, or what the best intervention is under specific conditions, a further required competency is the teacher's ability to critically weigh their options (Biesta, 2012; Korthagen, 2017). Hence, a condition of competence is not just a declarative knowledge of giftedness but also a declarative knowledge of different educational strategies (de Boer et al., 2013; Haenen & Mol-Lous, 2014; Hoogeveen et al., 2004; van Gerven & Hoogenberg, 2011). To be able to put this knowledge into use, teachers need to develop perspectives on giftedness and align their actions with regard to their chosen perspectives (Dai & Chen, 2014).

The way educational needs call on a teacher's handling capacity, taking into account how other students also call on this capacity, determines the level of support they choose as a response to the needs of the gifted student. The width of the existing support-pyramid depends, therefore, on a school's population, the level of professionalism of the teachers, and the managerial options school administrators have. As a result of this, the level of support used to respond to the educational needs of gifted students may differ for all schools. Even within a teaching team, there will be differences between what each teacher can offer. It requires managerial responsibility to develop a continuum of support with the highest balance to circumvent possible inequalities in the care of students due to the different capabilities of a teaching team.

From the interdependency of the educational paradigms, five main domains of teacher competencies can be inferred as explicitly influencing the education of gifted students. (1) Concepts of giftedness: the need for in-depth knowledge of the theoretical perspectives on giftedness and gifted education and the ability to apply this knowledge contextually in educational practice. (2) Seeing educational needs: the change and solution-focused pedagogy for seeing and profiling the educational needs of gifted students. (3) Understanding educational needs: it is imperative to interpret and understand the educational needs of gifted students within their ecological context, within an interactive system of the student, the teacher, the curriculum, physical learning environment, the family, and peers. (4) Responding to educational needs: effective education requires meaningful responses to the educational needs of gifted students from a contextual and developmental perspective. In so doing, (learning) process and (learning) results are equally important. (5) Assessing the responses to interventions: tuning into a student's needs requires the competence to monitor the process of responding meaningfully to a student's educational needs and the ability to interpret the student's responses to these interventions contextually.

Taking into account the fact that basic teacher competencies for The Netherlands and Flanders are already prescribed by law, competence in the domains mentioned above should always be described in relation to those basic teacher competencies. It is recommended, therefore, that a competency matrix should be constructed which combines both sets of competencies.

3 A Theoretical Framework for a Professional Standard in Gifted Education

3.1 Introduction

In The Netherlands and Flanders, there are several perceptions of the notion of *specialist in gifted education*. These perceptions differ in the level of in-depth knowledge that has to be developed, the kind of knowledge that has to be developed, the theoretical underpinning of this knowledge, the terminology that is used, and the professional background of the specialist. Placing the growing awareness of the needs of gifted students within an historical context explains this phenomenon. In the mid-eighties of the last century, parental self-help organisations carefully started to raise awareness of their gifted children. In 1985, Bekina (n.d.), the association for parents of gifted children was founded in Flanders. In 1987, Pharos, a Dutch equivalent association was founded. In 1994 Hint, and in 1999 Choochem, both parental groups, were also established (van den Heuvel, 2007). Parents of gifted children noticed that their children did not get an education that stimulated them to actualise their intellectual potential. As parents, they developed practically based insights into "what worked" in education for their children and "what did not work" for them. Joining parenting groups gave them a voice in society that could no longer be ignored.

During the same period, giftedness became a topic for Dutch university scholars. Prof Dr Pieter Span and Prof Dr Franz Mönks focused their respective researches on the notion of giftedness and its influence on education and upbringing. Both founded centres of expertise that were supported by their respective universities.[3] Parents and schools could ask for their advice in raising or educating a gifted child. The first efforts for teacher education were, therefore, initiated by expert-parents and psychologists.

In the mid-nineties, an increasing number of parents and teachers approached local and regional special educational needs coordinators (SENCo's) with their questions about what to do with gifted students in education. In response, these consulted professionals enhanced their knowledge, for example, by joining conferences of international organisations such as the European Council of High Ability and the World Council for Gifted and Talented Children. Based on this expertise, these professionals started to deliver workshops and training courses for teachers. This movement in the Dutch professional field caused a snowball effect. Giftedness was placed on the educational agenda.

In 1997, The Dutch Centre for Gifted Studies (CBO) joined forces with Antwerp University to open the Flemish Centre for Gifted Studies, coordinated by dr Tessa Kieboom, who became its Director in 2001. The centre fulfilled a similar need to its Dutch counterpart. It played an important role in the education of Flemish teachers in gifted education and in the support of parents bringing up their gifted children.

[3] Prof Dr Span started PABU – currently Pieter Span Expertise Centre (Utrecht University). Prof Dr Mönks started the Centre for Gifted Studies-CBO (Radboud University).

Since 2001, on more than one occasion, the Dutch government has funded gifted education with large sums of money, simultaneously encouraging specialists in giftedness to become involved in both gifted education and in the continuing professional development of teachers on this subject (Dekker, 2014a, 2014b; Dijksma, 2008; Keuchenius, 2001; Slob, 2018b). In the following seventeen years, an increasing number of people have developed expertise in giftedness, either based on personal experience or based on a solid theoretical and practical underpinning. Currently, the level of expertise amongst these specialists on giftedness strongly fluctuates between being self-taught, and having a university degree in psychology, pedagogy or educational sciences. With or without a professional background in education, this group of self-acclaimed specialists in giftedness have responded to the growing demands of schools to professionalise them in gifted education. A broad spectrum of continuous professional development courses in giftedness and gifted education has become available. However, the response of Dutch and Flemish professional teacher educators to the professional dilemma's teachers were experiencing with their gifted students was weak. Therefore, continuous professional development on giftedness was often arranged without the involvement of teacher education institutes or even professional teacher educators. In Flanders, giftedness is not a special topic for continuous professional development within teacher education institutes, while in The Netherlands, teacher education institutes offer only very limited options for continuous professional development regarding gifted education. Consequently, this influences the quality of continuous professional development in the highly specialised domain of gifted education.

The absence of professional teacher educators in the field of continuous professional development in gifted education is striking, since aiming for quality in the areas of teacher education and teachers' continuous professional development is a serious issue that is openly discussed in both Flemish and Dutch society.

As a combined specialism of teacher education and gifted education is still rare, the chances are that few continuous professional development courses in gifted education are underpinned by both specialisms. This has consequences for the quality of gifted education itself. The complexity of teaching as a profession requires not only an in-depth knowledge of education but also the development of skills to apply this knowledge contextually. However, being a teacher educator is completely different from being a teacher (Murray & Male, 2005). Being a teacher educator is a profession in its own right. The call for professional teacher educators at both bachelor's and master's levels and in continuous professional development courses has increased (Lunenberg et al., 2014). The Netherlands and Flanders both have societies (Velon/Velov) for professional teacher education that are striving for a formal teacher educator register. Teacher educators need support in developing their professional behaviour. A solid knowledge base should underpin this professional behaviour. Lunenberg et al. (2014) state: "Teacher educators should be able to make this behaviour explicit. Implicit knowledge and 'practical wisdom' are in our view not a sufficient foundation of professional behaviour" (p. 6). They suggest that there is "a general consensus that, to a large degree, teacher educators determine the quality of teachers and that these teachers are a crucial factor in the quality of primary and secondary education" (Lunenberg et al., 2014, p. 1). Based on that suggestion, a high percentage of Dutch and Flemish current teacher education courses in gifted education do not qualify, because the specialists who deliver this training do not meet this standard. 10% of all children are gifted or high potential learners. Not only should their education, but also the education of their teachers, be taken as seriously as is seen in any other focus group in education (Slob, 2018b). The social and professional relevance of a professional standard for gifted education is, therefore, that this standard should aim to ensure as high a quality of gifted education as is ensured for all other students with complex educational needs.

In this chapter, we, therefore, focus on the conditions required to develop a professional standard for specialists in gifted education. Being a specialist in gifted education implies, first and foremost, that one has to be a specialist in education, that is, a fully qualified teacher. Consequently, one can only become a specialist in gifted education after undertaking a teaching degree. This approach has consequences for the perception of what defines a specialist in gifted education. In the previous chapter, we introduced the five paradigm shifts that influence teaching behaviour regarding gifted education in the Dutch and Flemish educational system: an inclusive approach; response to intervention; paradigms on giftedness; the influences of twenty-first-century skills; evidence-based intervention. Based on these paradigm shifts we deduced that five domains of teacher competences were conditional for gifted education in the current Dutch and Flemish educational system: theoretical approaches to the notion of giftedness; seeing the educational needs of gifted students; understanding the educational needs of gifted students; responding to the educational needs of gifted students; and assessing the results of meaningful responses to the educational needs of gifted students. To connect general education with gifted education, we combine the domains distinguished, with the competencies that every teacher has to acquire in their initial teacher education. This integration of general educational competence and specific educational competence provides the necessary insight into how a specialist in gifted education can be distinguished from other teachers. It also ensures that gifted education stays anchored in general education and does not become a differentiation in its own right, miles away from the reality of education (D. Matthews & Foster, 2009). One can recognise a similar intention in the "National Gifted Education Standards for Teacher Preparation". "Standards, therefore, assist candidates and practising teachers in having specific content knowledge and pedagogical skills in gifted education so that they are able to develop gifted learners' talents for high-level functioning in the 21st century" (Johnsen et al., 2016, p. 3).

To understand the underpinning theoretical framework of the competency matrix (presented in part 2, Chapters 4-10) that consequently follows this integrated approach, where we connect the competencies of general education with the competences of gifted education, we must explore four underpinning elements. Firstly, we briefly explore the fundaments of Dutch and Flemish initial teacher education. We describe, in short, the Dutch and Flemish qualifications for a general teaching degree in primary education. Secondly, we explore how the so-called "Dublin Descriptors" for higher education at bachelor's and master's degree levels underpin general teacher competences. Based on these descriptors, teacher education aims not just for high quality and content-based knowledge, but has a primary aim of developing a professional approach that reflects problem-solving skills, critical thinking, the ability of educational judgement, self-reflection and self-regulation and a love for lifelong learning (Biesta, 2012; King & Lau-Smith, 2013; Korthagen, 2017). Thirdly, we explore the notion of "competence" that includes the position deduced from the Dublin Descriptors. This will lead us to the criteria for a coherent framework of competence in gifted education as is intended by the construction of a matrix. Fourthly, we deduce how a model of intrapersonal adaptation explains the fundamental requirements of the modern teacher's attitude towards continuous professionalisation. Based on the requirements of the Dublin Descriptors and the notion of competence, this model explains the need for lifelong learning. These so defined behavioural aspects mirror a professional attitude which is requisite for the specialist in gifted education.

3.2 General Teaching Competences

As stated before, a specialist in gifted education is, first and foremost, a qualified teacher. A qualified teacher meets the professional demands as described in the professional teaching standard of their country.

> Teaching competences are thus complex combinations of knowledge, skills, understanding, values and attitudes, leading to effective action in situation. Since teaching is much more than a task, and involves values or assumptions concerning education, learning and society, the concept of teacher competences may resonate differently in different national contexts. (European Commission, 2013, p. 8)

In Flanders and The Netherlands, a teaching degree at bachelor's level is proof of this qualification.

The current description of the professional standard in Flanders dates from 2007. The description of the professional standard for Dutch teachers dates from 2004; this description dates from 2014; a revised description was implemented in 2017. This revision merely led to a reshuffle rather than to complete renewal. The recently stipulated professional standard is inferred from the so-called SBL competences reflecting the legislation enforced in 2006 (Onderwijscoöperatie, 2014a).

Although in Flanders one discerns kindergarten teachers from primary school teachers, and Dutch legislation does not make this distinction, there are hardly any major differences in the general professional standards for teachers in both countries. Observed differences are more of a semantic nature than content-based. In Flanders, one refers to this professional standard as "basic competences". In The Netherlands in the revised standards, the notion of competence is dismissed. Instead, one uses the phrase "professional demands" in combination with "professional qualification". The competency-based approach is now reserved for set educational objectives in teacher education courses

(Onderwijscoöperatie, 2014b). It is no longer seen fit to describe the professional standards that must lead to the general teaching qualification. We revisit the underpinning discussion in paragraph 4.
Flemish law divides teacher competences into three different types of responsibility: responsibility towards the learner, responsibility towards the school and/or the educational community, and responsibility towards society. One can describe the responsibility of each, in their essential roles, as follows (Struyven & Ceulemans, 2012).

- Towards the learner:
    - The teacher as a mentor in learning and developmental processes
    - The teacher as educator
    - The teacher as a curricular expert
    - The teacher as a manager
    - The teacher as an innovator and researcher.
- Towards the school:
    - The teacher as a partner for parents/carers
    - The teacher as a member of a teacher-team
    - The teacher as a partner for external contacts
    - The teacher as a member of the educational community.
- Towards society:
    - The teacher as a cultural participant.

These responsibilities are consistent with the roles as described by the European Commission in 2013. The commission sees teachers as:

- reflective agents
- knowledgeable experts
- skilful experts
- classroom actors
- social agents
- lifelong learners.

Based on these roles, appropriate teacher behaviour, and therefore, relevant teacher competences can be deduced.The professional demands, as described in current Dutch educational law, do not refer explicitly to these roles, although the roles can be recognised easily in the professional demands. In The Netherlands, professional demands leading to a professional qualification are divided into three areas: curricular expertise, didactical expertise and pedagogical expertise (Onderwijscoöperatie, 2014a). The construction of these professional demands as a triad shows their coherency and interdependency. A professional basis embeds this triad (see Figure 7).
The Onderwijscoöperatie (2014a), the "Teachers' Cooperative" states:

> In education, professional behaviour involves teamwork. Being a good teacher supposes the ability to organise, plan, collaborate and prioritise (...) A teacher has an investigative attitude, is focused both on a student's results and on their developmental process. A teacher values continuous professional development in interaction with colleagues. (p. 6)

Figure 7

Fundaments of professionalism

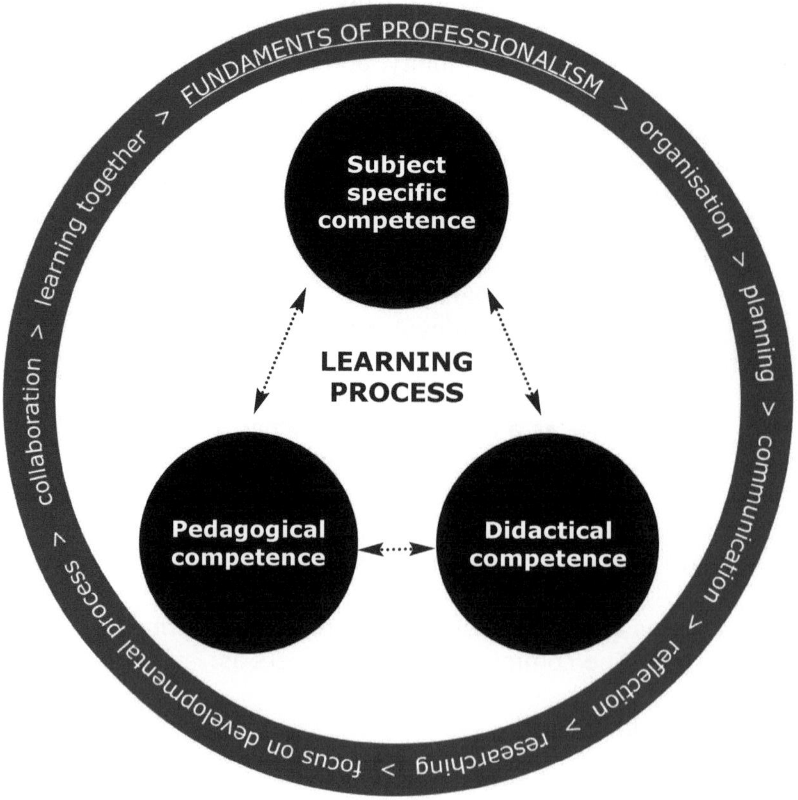

Note. From *Bekwaamheidseisen* [Professional standards for teachers]. Onderwijscoöperatie, 2014. Effective date August 1, 2017.

In this professional profile, one can recognise most aspects of twenty-first century skills. A call on the conscious application of twenty-first century skills becomes preeminent. The paradigm shift towards this educational approach, as it is described in the previous chapter, now formally affects the professional standard and therefore also affects the education of teachers. The fact that there is a generation of teachers active in schools who are not educated based on the requirements of this paradigm may complicate the implementation of these new standards.

Each aspect of the educational triad of professional demands has its definition. Curricular expertise for primary school teachers is to be interpreted as content-based expertise at a level above and beyond the curricular objectives which are set for students in primary education. Based on this expertise, the teacher must be able to compose a curriculum in a manner that will allow their students to master the content. Didactical expertise means that the teacher, in collaboration with their colleagues, transfers curricular objectives in a way that they become teachable in the context of a primary school, matching school policy. Pedagogical expertise means that the teacher creates a safe and stimulating learning environment for every student, tuned into their individual educational needs. This requires the assessment of a student's development in learning and behaviour. The teacher contributes to the social-emotional and moral development of their students in a way that enables them to become independent and responsible adults (Onderwijscoöperatie, 2014a, pp. 10-13). Both Flemish and Dutch laws on education describe the professional demands on teachers in a manner that leaves ample room for a more specific interpretation. Both laws refer to the individual responsibility of teacher education institutes to translate the professional demands into a curriculum as they see fit for the profession (Onderwijscoöperatie, 2014b; Vlaamse Regering, 2008). This responsibility applies not only to initial teacher education but also to postgraduate teacher education. The Dutch "Teachers' Cooperative" stated in their re-evaluation of professional demands and qualifications that "with a competence-oriented approach, professional knowledge and skills can be combined with professional actions and the effects of those actions, in different situations, in a specific educational context. That instrumental choice is to be made by the teachers' education institute" (Onderwijscoöperatie, 2014b, p. 5).

3.3 Dublin Descriptors

The interaction between the "Big Five" paradigms in gifted education (Figure 1 Interdependency of paradigms, discussed in Chapter 2) shows the complexity of teaching as a profession. This complexity requires highly educated and well-trained professionals in education.

As an aim of education is the development of a student's ability to participate in modern society, teaching standards and consequently teaching competences and qualifications cannot stand on their own. They are inferred from social and economic demands (Biesta, 2012; European Commission Education and Training, 2013; Kelchtermans, 2012; Korthagen, 2004; Onderwijscoöperatie, 2014b; Vlaamse Regering, 2008). These demands, therefore, must reflect a general attitude towards what society requires from its members, and in particular from higher education. The Bologna Declaration reflects this general attitude. In June 1999, 29 countries agreed on the Bologna Declaration. With this joint declaration, the European Ministers of Higher Education aimed for a system of "easily readable and comparable degrees (...) in order to promote European citizens employability and the international competitiveness of the European higher education system" (European Higher Education Area, 1999, p. 3). Initially, the agreement focused on two cycles: an undergraduate (bachelor's) level and a postgraduate (master's) level. Later a third cycle at a doctoral level and a fourth, short cycle at an associate level were added. Other objectives of this agreement were a universal system of credits, the promotion of student mobility in the European Higher Educational System, and the structural promotion of the concept of lifelong learning.

As might be expected, the system to be developed had to be acceptable to the joint partners. "We hereby undertake to attain these objectives – within the framework of our institutional competences and taking full respect of the diversity of cultures, languages, national education systems and of University autonomy – to consolidate the European area of higher education" (European Higher Education Area, 1999, p. 4). The Bologna declaration formed the basis of the later developed Dublin Descriptors. These descriptors are general statements about the ordinary outcomes that are achieved by students in higher education after the completion of a curriculum. These descriptors are not prescriptive, nor do they set a benchmark for those achievements. They are merely an overall description of the nature of a qualification. One must bear in mind that the Dublin Descriptors do not primarily focus on teacher education. They consider five generic domains as educational objectives for higher education in general (Bologna Working Group on Qualifications Frameworks, 2005).

- Knowledge and understanding
- Applying knowledge and understanding
- Making judgements
- Communication skills
- Learning skills.

The descriptors are set up in a tiered structure, slightly differently described and applied at each level. Each level requires that behaviours described for the previous levels are acquired (Table 1).

> Within such frameworks, degrees should have different defined outcomes. First and second cycle degrees should have different orientations and various profiles in order to accommodate a diversity of individual, academic and labour market needs. First cycle degrees should give access, in the sense of the Lisbon Recognition Convention, to second cycle programmes. Second cycle degrees should give access to doctoral studies. (Berlin Communiqué, 2003, p. 4)

In The Netherlands and Flanders, initial teacher education takes place at a bachelors' level (Onderwijscoöperatie, 2014a; Vlaamse Regering, 2008). Consequently, in their professional behaviour, graduate teachers should reflect the application of the Dublin Descriptors at this level. In their daily classroom practice, they independently provide education at level one, a "basic-arrangement", fit for most students. At level two, they provide a "basic-arrangement-plus" approach with additional interventions as necessary, for some students (de Bruïne et al., 2015; Houkema et al., 2018).

Table 1

Dublin Descriptors (Bologna Working Group on Qualifications Frameworks, 2005, pp. 67-68)

	Bachelor's level	Master's level
Knowledge and understanding	"have demonstrated knowledge and understanding in a field of study that builds upon and supersedes their general secondary education, and is typically at a level that, whilst supported by advanced textbooks, includes some aspects that will be informed by knowledge of the forefront of their field of study."	"have demonstrated knowledge and understanding that is founded upon and extends and/or enhances that typically associated with Bachelor's level, and that provides a basis or opportunity for originality in developing and/or applying ideas, often within a research context."
Applying knowledge and understanding	"can apply their knowledge and understanding in a manner that indicates a professional approach to their work or vocation, and have competences typically demonstrated through devising and sustaining arguments and solving problems within their field of study."	"can apply their knowledge and understanding, and problem-solving abilities in new or unfamiliar environments within broader (or multidisciplinary) contexts related to their field of study."
Making judgements	"have the ability to gather and interpret relevant data (usually within their field of study) to inform judgements that include reflection on relevant social, scientific or ethical issues."	"have the ability to integrate knowledge and handle complexity, and formulate judgements with incomplete or limited information, but that includes reflecting on social and ethical responsibilities linked to the application of their knowledge and judgements."
Communication skills	"can communicate information, ideas, problems and solutions to both specialist and nonspecialist audiences."	"can communicate their conclusions, and the knowledge and rationale underpinning these, to specialist and non-specialist audiences clearly and unambiguously."
Learning skills	"have developed those learning skills that are necessary for them to continue to undertake further study with a high degree of autonomy."	"have the learning skills to allow them to continue to study in a manner that may be largely self-directed or autonomous."

Teachers do not have a choice as to whether or not a gifted student takes part in their class. Having a gifted student in their class is a fact. Educational law demands that teachers provide good education for all students and that includes the gifted student. *Good education* demands the acknowledgement of individual differences between students and therefore, the acknowledgement of different educational needs (Florian & Black-Hawkins, 2011). Being a professional and being a good teacher means that, even without a law that sets this demand, teachers want to respond meaningfully to their students' educational needs. The educational needs of gifted students may differ from the needs of most of the students in their classroom (Ambrose & Sternberg, 2016; Dai & Chen, 2014; D. Matthews & Foster, 2009; Plucker et al., 2017; Subotnik et al., 2011; Vreys et al., 2017). Tuning into their educational needs is likely to start at level two, the basic-arrangement-plus-approach. However, in their initial teacher training, knowledge of gifted education is offered only marginally and with little training of specific pedagogical and didactical skills relating to gifted education (de Boer et al., 2013; Haenen & Mol-Lous, 2014; Hoogeveen et al., 2004; Vreys et al., 2017). As a result of this, teachers do not always feel capable and confident about developing arrangements at level two for this specific group of students (Houkema et al., 2018). They lack the necessary knowledge and skills. Moreover, teachers, in general, seem to find it difficult to judge professionally how interventions may not only lead to change but also how they may lead to improvement of the student's (educational) situation (Biesta, 2012). It is necessary to develop a mental framework of appropriate theoretical underlayers to create an in-depth understanding and meaningful response to the educational needs of this specific group of students (Haenen & Mol-Lous, 2014; Hoogeveen et al., 2004; Johnsen et al., 2016; Vreys et al., 2017). When it comes to other students with very specific educational needs, such as students with exceptional behavioural problems or students with specific learning disabilities, continuous professional development is supported. However, while giftedness should never be defined as a "learning disability" or a "behavioural problem", a gifted student may confront a teacher with as equally complex a professional dilemma (D. Matthews & Foster, 2009) as may a gifted student who is held back by a learning disability or behavioural problems. For continuous professional development in gifted education, an adequate response to these professional concerns is that, it should at least aim to develop a professional approach towards the subject at bachelor's degree level. That way it matches the overall professional standard based on the approach as described in the Dublin Descriptors (Registerleraar.nl, 2017).

There are even more complicated cases, for example, when a student's needs exceed the interventions that a teacher can offer at level two, or when there has been a long period in which a gifted student's educational needs were not effectively met at level two. The educational needs of this student may have become so complicated that specialised student and teacher support is needed. These cases exceed general teacher competencies and, therefore, require a specialist (Houkema et al., 2018). Again, when it comes to other groups of students with specific educational needs that exceed the curriculum as offered in initial teacher education, these specialised teachers, that is, school counsellors and (special) educational needs coordinators, take on a master's study as continuous professional development. Their knowledge and skills are not only broadened, but they develop a more in-depth professional attitude (see Table 1). Therefore, it is arguable that, if specialised student or teacher support at this level is needed for a gifted student, the specialist should receive similar in-depth training at a master's degree level. However, there are some reservations about this argument, as the current Dutch and Flemish postgraduate teacher training courses at master's level (as already seen at bachelor's level), have hardly any mandatory training on this topic. In Flanders, schools for higher education do not even offer masters studies in education as an option for continuous professional development. Although Flemish universities offer masters courses in education, they are hardly practically oriented (Amkreutz, 2017). At some Dutch schools for higher education, their educational masters' courses offer optional modules in gifted education. It is up to a student to select these modules as part of their personalised educational programme. It is regarded as a student's personal professional responsibility to make choices. One may argue that teacher education can only be effective if it departs from the teacher's personal professional concerns (Korthagen, 2017). Ownership is a modern adage. From this perspective, training on gifted education should only be provided to those who express their interest in this topic as a professional concern and feel authentic ownership towards the topic. However, as an inclusive approach is mandatory in Dutch and Flemish education, one could argue that it is morally wrong to provide for a student's education in an environment that cannot respond sufficiently to the student's needs because an educational needs coordinator did not consider giftedness as one of their professional concerns. Educational needs coordinators should, therefore, take ownership and make giftedness their professional concern. The difference between a teacher with the ability to respond meaningfully to a gifted student at a level two arrangement and an educational needs coordinator, lies not only in the problems that they have to solve but also in their specialised professional attitude. In a situation where a teacher has participated in a continuous professional development course to become a specialist in gifted education, the educational programme should match the Dublin Descriptors at master's level. In daily practice, acting as a specialist in gifted education should require the application of the Dublin Descriptors at master's level, otherwise there would be no distinction between a general teacher and a specialist teacher.

3.4 The Notion of Competence

In the context of the construction of a competency matrix, the use of the notion of "competence" in the Dublin Descriptors is relevant. In a working document of a Joint Quality Initiative Meeting, the members of this meeting stress that "the word "competence" is used in the descriptors in its broadest sense, allowing for gradation of abilities or skills. It is not used in the narrower sense identified solely on the basis of a "yes/no" assessment" (Joint Quality Initiative, 2004, p. 3). This means that one should never use a list of competences as a checklist as in 'to be or not to be'. However, the use of the notion of competence touches

a broader discussion regarding the use of a competence-oriented approach in teacher education. We briefly explore this discussion to clarify how, in this study on the competences of specialists in gifted education, the use of the notion of competence relates to this broader discussion.
Just as the Dublin Descriptors found their roots in the social-economic demands of modern society, the same applies to the notion of competence-oriented learning (Biesta, 2012; Korthagen, 2004; Merriënboer et al., 2002). Education and work were connected by the need to develop competences; this is no different for teacher education (Ceulemans et al., 2016). At the turn of this century, most Dutch and Flemish teacher educational institutes went along with the concept of competence-oriented teacher education (Ceulemans et al., 2016; Korthagen, 2004). The approach was seen as fit to improve the quality of teacher education. However, along with this approach came a problem; there is no agreement on a general definition of the notion of competence.
Dutch and Flemish teacher educators and educational scientists have been discussing the use of the notion of competence for a long time in relation to teacher education (Biesta, 2012; Ceulemans et al., 2016; Korthagen, 2004; Loughran, 2015; Struyven & Ceulemans, 2012). The discussion on the use and meaning of the concept shows similarities with the discussion on inclusion (Ainscow et al., 2006). There are almost as many definitions as there are users of the notion. There is no overall consensus on its meaning and, therefore, it is difficult to see the forest for the trees (Merriënboer et al., 2002; Mulder et al., 2007; Onderwijscoöperatie, 2014b).
Mulder et al. (2007) distinguish three main approaches to the notion of competence: a behaviourist approach, a generic approach and a cognitive approach. In the behaviourist approach, competence is defined as the "characteristics of a person that are related to superior performance in a job and can be common across situations" (Mulder et al., 2007, p. 5). The definition of competence from a general perspective is that competence is "more about framing an overall performance that is appropriate to a particular context. It is not about following simple recipes" (Hager, 1998, p. 533, as cited in Mulderet al., 2007). In the cognitive approach the definition of competence "includes all of the mental resources of individuals that are used to master tasks, acquire knowledge and achieve a good performance" (Weinert, 2001 as cited in Mulder et al., 2007).
In this study on a competency matrix for specialists in gifted education, the definition of competence by the European Commission on Education and Training is used as a starting point. This commission sees teaching competences as "complex combinations of knowledge, skills, understanding, values and attitudes, leading to effective action in situation" (European Commission, 2013, p. 8). This definition can be seen as an integration of all three approaches, with a strong accent on the cognitive approach.
Applying this definition to teaching as a profession and to gifted education, it becomes clear that the professional demands of a specialist in gifted education can only be stipulated as several different competences. In the way these competences are stipulated, they have to meet the criteria of the definition of the notion of competence. At the same time, the way each competence is

stipulated should be comparable with the stipulation of the other competences. To ensure this, the need for an operational approach to the definition becomes clear. Merriënboer et al. (2002) distinguished six dimensions which all contributed differently to the notion of competence. Competences are *contextually* bound. They are a *coherent*, inseparable cluster of knowledge, skills, attitudes, qualities and apprehensions, and they are *action-oriented*. Competences cannot be seen separately. They show an *interdependency* with each other. This interdependency empowers the individual meaning of each separate competence within a professional profile. Competences are *learnable*. Learning and professional development are conditional to the acquirement of competences. As competences find their roots in what society demands from a professional in a specific domain, indicators for competences will change over time, as the professional demands in a domain change over time as well. Therefore, competences themselves are *sustainable*, but the indicators for competent behaviour have limited sustainability. This stresses the need to revise examples for indicators when major changes in the professional domain take place.
Based on these six dimensions of the notion of competence, the construction of a competency matrix prevails over a competency list (Mulder et al., 2007; Merriënboer et al., 2002). The use of a matrix creates the opportunity to demonstrate the interdependency of the competences in a multi-dimensional perspective.
On its own, the combination of these six dimensions of competence and the Dublin Descriptors does not clarify the need for specific competences for the specialist in gifted education. Therefore, we must place the dimensions of competence in a more specific context, to clarify how these six dimensions relate to each other and the continuous professional development of teachers (Onderwijscoöperatie, 2014a). This is logical if one bears in mind, that the specialist in gifted education is first and foremost a qualified teacher and that becoming a specialist in gifted education is an additional development. Without the basic teacher qualification, the specialist in gifted education would not be allowed to take on tasks matching the teacher qualification standards without being supervised by a qualified teacher.
Although the professional roles of teachers are relatively stable, underlying indicators for competence change over time depending on paradigm shifts in education. New paradigms require an adjustment in knowledge, skills and attitudes. These paradigm shifts find their roots in the introduction of new normative concepts of what societies see as the core objectives of education.
If we want to describe what defines a specialist in gifted education, we should place the professional role of teachers within the context of the professional dilemmas that emerge when educating the gifted.

At this stage, we can conclude that a specialist in gifted education has a specific approach towards learning and professional behaviour. This approach is founded on generic teacher competences. Domain-specific applied skills and knowledge, and the ability to judge its use within a specific educational context, reflect this approach (Biesta, 2012; de Boer et al., 2013; European Commission, 2013;

Haenen & Mol-Lous, 2014; Johnsen et al., 2016; Onderwijscoöperatie, 2014a; Vreys et al., 2017). One can transfer this idea into a generic model that shows how this approach not only contributes to the lifelong development of teachers but also forms the theoretical basis for any pedagogical and didactical intervention in classrooms. The Model of Intrapersonal Adaptation (MIA), see Figure 8, reflects this process. In the next paragraph, we will explain the MIA in depth.

3.5 Learning as a Process of Intrapersonal Adaptation

The implementation of a competency matrix for specialists in gifted education can only be successful if, and when, a professional approach underpins the responsible use of these cohering competences. If this professional approach is lacking, a competency matrix will devalue, becoming a simple checklist answered with a simple "yes or no". This one-dimensional use of a competency matrix comes with the danger that the tool becomes a goal in itself, rather than a means of improving the quality of education (Biesta, 2012; Ceulemans et al., 2016; Korthagen, 2017). There is no reason to assume that this will be any different when it comes to the field of gifted education. Therefore, in this paragraph, we focus on the specific professional attitude which we describe as a process of intrapersonal adaptation (van Gerven & Hoogenberg, 2011).

Figure 8

The Model of Intrapersonal Adaptation

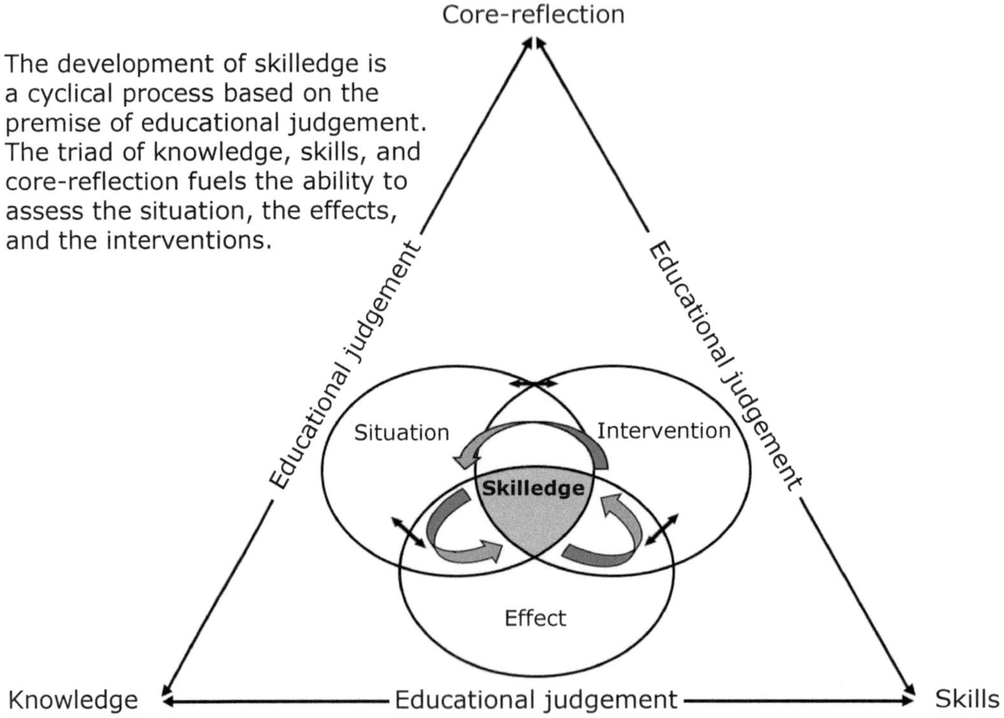

Researching within a professional context empowers a teacher's professionalism (Geerdink, 2018; Gerritsma et al., 2013; Kallenberg et al., 2007). An investigative approach is conditional to a teacher's success. It is an inseparable part of the professional identity of teachers (Holvoet et al., 2017; Onderwijscoöperatie, 2014a; Pols, 2017) and it leads to the lifelong learning of professionals (European Commission, 2013; van Gerven & Hoogenberg, 2011). Learning occurs when one experiences a process of transformation in thinking and acting, resulting in new (professional) behaviour (Maslow, 1954; Marzano & Kendall, 2008; Vygotsky, 1978). Personal concerns fuel this process (King & Lau-Smith, 2013; Korthagen, 2017; Loughran, 2002).

The Dublin Descriptors reflect the investigative reflective attitude needed during this process (see Table 1). If these descriptors underpin postgraduate teacher training, continuous professional development for teachers leads to intrapersonal adaptation. However, acting in a professional environment in which these Dublin Descriptors explicitly underpin professional behaviour is also a continuous learning process. Teachers not only learn through a formalised learning process but first and foremost through their daily practice. Being a teacher requires the ability to continuously consciously adapt one's professional actions (Biesta, 2012; European Commission, 2013; Gore et al., 2017; King & Lau-Smith, 2013; Korthagen, 2017). The teacher and the student find themselves in an ongoing process of developmental change. Therefore, working as a teacher requires a continuous intrapersonal adaptation (Kelchtermans, 2012).

The MIA is the revised version of the Intrapersonal Adaptation Model (IPA-Model), developed by van Gerven and Hoogenberg (2011). The MIA describes learning as an interactive process that starts with a teacher's professional dilemma. Professional dilemmas can be caused by several factors. When a teacher finds themself in a situation where the desired effect of their interventions lags behind, they may feel confused, dazzled or overwhelmed by the situation. If they have difficulty relating the situation to their experience as an educator, to their theoretical knowledge, to their core values or to their educational judgements, it is likely that they experience this contextual incompetence as a professional dilemma. In this way, a professional dilemma can explicate the core of a professional developmental need. Therefore, professional dilemmas should take a central position in the continuous professional development of teachers (Kelchtermans, 2012; King & Lau-Smith, 2013; Korthagen, 2017). The complexity of the five interacting paradigm shifts in gifted education (Figure 1) creates professional dilemmas for teachers on a daily basis. These dilemmas can focus on the question of what changes are needed in the curriculum for this student to ensure their further development. They can also focus on the more pedagogical or didactical question of how to help the student to develop an intrapersonal understanding of the consequences of being gifted. As expressed before, Dutch and Flemish teachers experience a gap between what gifted education requires and what they know about gifted education. They feel inadequately equipped to respond meaningfully to the needs of their gifted students (de Boer et al., 2013; Doolaard & Oudbier, 2010; Haenen & Mol-Lous, 2014; Janson, 2017; Kuipers, 2017; Vreys et al., 2017). That is how giftedness and gifted education become not just their personal

professional concerns, but also their personal professional dilemmas.
Although it is assumed that learning results, at least mentally, in new behaviour, it does not always lead to visibly new behaviour. It sometimes merely underpins current visible behaviour by creating an altered mental framework or by clarifying core values (King & Lau-Smith, 2013; Wilder et al., 2013). At times like this, it provides an appropriate language for the justification of one's actions. Institutionalised or formal learning is a process of deliberately influencing behaviour in a way that sets new mental boundaries for knowledge and skills, with the result of new or altered belief systems. These new boundaries explicate a higher level of the ability to apply knowledge, skills and educational judgements than before (Gore et al., 2017). This process shows how a zone of proximal development transfers into a zone of current development. As a result of this transfer, a new zone of proximal development becomes visible (Vygotsky, 1978). During this cumulative process, teachers develop knowledge and skills which are directly related to their professional concerns (Kelchtermans, 2012; Korthagen, 2017). The result is a new situation where the application of newly acquired knowledge and skills leads to new insights and reframed professional dilemmas; and to the process starting all over again (van Gerven & Hoogenberg, 2011). The MIA underpins this fundamental idea of the continuous development of the competences of the specialist in gifted education. The ability to adjust one's professional actions to what the situation - based on the five interdependent paradigm shifts on gifted education - demands, is a core condition of the intrapersonal adaptation of specialists in gifted education, and therefore a core condition for their professionalism.

3.5.1 Interaction as a Reason for and the Result of Learning

A teacher's continuous professional development aims for the evolution of their ability to employ well-balanced educational judgements towards the context-based use of knowledge and skills (Biesta, 2012; Kelchtermans, 2012; Wassink & Bakker, 2014). Interaction with significant others focuses and shapes this professional learning process through its influence on a teacher's professional behaviour and normative concepts (Jutte, 2008; King & Lau-Smith, 2013; Korthagen, 2017). This process equips them to deliberately influence their student's development while they aim for an improvement of the situation. Therefore, when a teacher is learning, this process always aims at their professional, interactive role as an agent of change in relation to their student (Kelchtermans, 2012). However, their interactions are not limited to the student. They also include colleagues, parents, school counsellors and so forth (de Bruïne et al., 2015; European Commission, 2013; Onderwijscoöperatie, 2014a; Pameijer & van Beukering, 2007; van Meersbergen & de Vries, 2017). One can state that the teacher's interaction always focuses on "the significant other", not only for change but most importantly to improve their student's situation (van Gerven & Hoogenberg, 2011).
A teacher's learning process is not only controlled by interaction but also aims for interaction. This interaction-based and focused process requires their active participation. It requires the ability to set personal learning objectives. It also

requires the ability to shape professional knowledge and skills by reflecting-in-action and reflection-on-action (Loughran, 2002, 2015; Onderwijscoöperatie, 2014a). Effective interaction requires not only the ability to communicate effectively but also the meaningful application of content-based knowledge and skills, and the ability to make proper judgements regarding content and the ethical employment of this knowledge and skills. If interaction forms the foundation for a process of intrapersonal adaptation, the employment of the Dublin Descriptors within the teacher's professional context is conditional.
In situations where it becomes clear that a student's educational needs differ from or are additional to the needs of most of the other students, the interaction between teacher and specific student intensifies. At the same time, interaction with the student's parents or carers is likely to intensify as well. These conditions also apply in gifted education. When educating gifted students, interaction with significant others is likely to be more frequent and intense. Due to the current educational system, their educational needs can often only be met by taking a different approach from that used to meet the needs of most students (Audenaert et al., 2015; Betts & Neihart, 2010; Calahan & Hertberg-Davis, 2018; Coleman & Johnsen, 2011; Davis et al., 2014; Doolaard & Oudbier, 2010; Johnsen et al., 2016; Kieboom & Venderickx, 2017b; Kuipers, 2017).
It is also likely that, as a situation becomes more complex, interaction may involve external school experts (de Bruin-de Boer & Kuipers, 2011; Drent & van Gerven, 2012). Context-relevant knowledge and skills contribute to high-level communication with the actors in a student's ecological system in order to be able to tune into that student's educational needs as seems fit for the situation (van Swet, 2009; P. Wolters, 2009).
Changes in education (i.e., applied didactics and pedagogics) can only be achieved when we approach schools from an ecological perspective. In using that perspective, schools are seen as a system in themselves and as part of a broader system (Engeström, 2001; Jutte, 2008; van Meersbergen & Jeninga, 2012). Every actor in this system influences the other actors. An approach which focuses on changing only one particular aspect or actor within the system is not likely to be effective (Caufman & van Dijk, 2009; Engeström, 2001; Jutte, 2008). This notion of systemic thinking contributes to the MIA. One should consider schools as learning organisations. Dublin Descriptor five, the application of learning skills, therefore, not only concerns individual professionals during a continuous professional development course, but also in their daily practice, as well as concerning the school as a professional organisation. A teacher's social-constructivist attitude becomes conditional for any further development of all the actors involved in the relation between "I" and "the significant others" (Kelchtermans, 2012; van Gerven & Hoogenberg, 2011). Using this principle as a point of deployment, means that a student's development can only be influenced positively if and when the teacher deliberately changes their professional behaviour.
This principle can be applied to situations that may occur during the education of a gifted student. If, for example, the curriculum on offer is based on the average ability level of age group, it is not likely that the gifted student needs to reach for a zone of proximal development (Bakx et al., 2016; Calahan & Hertberg-

Davis, 2018; Doolaard & Oudbier, 2010; Leavitt, 2017; D. Matthews & Foster, 2009). The teacher needs to alter their professional behaviour by adjusting the curriculum for this specific student to stimulate their further development. This adjustment requires interaction to determine the student's educational level. This interaction is a condition for change. However, if one expects a gifted student to reach a higher educational level or developmental stage, just by offering a more complex curriculum, this may lead to disappointment (Ahlers, 2017; Kuipers, 2017; Resing & Vogelaar, 2017). Additional interaction is needed. Although the adjustment to a more complex curriculum alters the student's situation, this alteration will not lead to an improvement of that situation if it does not come, hand-in-hand, with an alteration in the teacher's instructional attitude. In this case, the student needs an individualised instruction based on their needs and related to the offered curriculum. In order to do this, interaction between student and teacher is essential (Resing & Vogelaar, 2017; Vogelaar, 2017).

3.5.2 The Situation

As described in the previous paragraph, the interaction between the teacher and a student's significant others brings the need for change into focus. Interaction is the starting point for all interventions; then it steers the direction the intervention takes. Therefore, the perceptions of all involved parties in the current situation, determine the need for interventions and the directions that they take. The need to intervene in the current situation does not always mean that there is a problematic situation; the need for change often finds its roots in the student's regular developmental process. However, the need to intervene in a problematic situation or a problematic relationship can be part of the current situation.

In MIA, therefore, the current situation represents "the concrete experience" of Kolb's learning cycle (Hendriksen, 2005). In this current situation, all the actors involved experience a need for change. Without this need for change, the actors do not feel the need to learn. A second step is a reflective observation of the current situation. This reflective observation clarifies the need for change. Again, all actors involved take part in this reflective observation. By sharing their reflections, they can jointly steer towards the desired effect. This process of reflective observation is the process of profiling educational needs.

In educating gifted students, the process of profiling educational needs requires a clear perspective on theories of giftedness. Dai and Chen (2014) describe the three dominating paradigms of gifted education: the gifted child paradigm, the talent development paradigm and the differentiation paradigm. Each paradigm presents different traits and characteristics. One can observe these traits and characteristics in a student's behaviour. All traits lead to different educational needs (Borland, 2005; Dai & Chen, 2014; Kreger-Silverman, 2013; Subotnik et al., 2011). Consequently, the position taken about the notion of giftedness determines what behaviour will be recognised as a trait of giftedness and the subsequent educational needs. For schools, this implies the necessity of taking a stand in the discussion of the notion of giftedness. For the individual teacher, this implies that they have to be able to position their personal perspective in relation to the school's point of view. Perspective on the notion of giftedness has implications for the tools a specialist may want to use during the process of profiling educational needs (de Bruin-de Boer & Kuipers, 2011; Houkema, 2009; Prodia, 2012; van Gerven & Drent, 2012). Each tool has a slightly different focus. Therefore, each tool is of influence in how its user assesses the situation. Consequently, each tool requires different knowledge and skills from its users. Assessing a situation, and by so doing creating a student's educational profile and defining a student's educational needs, requires, therefore, not only a specific attitude but also specific competences related to gifted education.

3.5.3 The Effect

Setting educational objectives or setting goals, that is, describing the desired situation, precedes thinking about and the selection of possible interventions (Ajzen, 1991). There exists a pendular movement between assessing the current situation and determining the desired effect (van Gerven & Hoogenberg, 2011). Assessing what is needed depends highly on what is desired. Therefore, both situation and effect influence the choice to act, the moment to act and the chosen intervention.

Every action, every intervention, creates an effect. However, not all effects are as desirable as was previously estimated and not all changes are good changes (Biesta, 2012; King & Lau-Smith, 2013; Korthagen, 2017). Reflecting-in-action means that, while implementing interventions, the teacher, in relation to the actors involved, reviews the temporary results, the intended results and how they relate to their core values as a professional. By so doing, they can timely adjust their interventions if the situation requires them to do so. Reflecting-on-action takes place before one acts and after one has acted. Reflecting-on-action means that the teacher considers the quality, the efficacy and the "goodness" of their interventions in relation to the intended or the accomplished results. Based on these considerations, they decide whether the intended intervention is also the most desirable intervention, and also likely to be the most effective intervention. The ability to make contextual considerations like this requires an investigative attitude towards a teacher's existing knowledge and applied skills, an intrapersonal understanding of their core values, the ability of educational judgement and a highly developed ability in self-regulation.

The result of a teacher's learning process should be an accumulating ability to optimise the current situation for all the actors involved. Therefore, the opportunity to apply learned intervention strategies in their daily practice relies on the continuous professional development of teachers (Edinger, 2017; Gore et al., 2017; King & Lau-Smith, 2013; Korthagen, 2017; Kwakman, 2002). The completion of the learning cycle leads to a new current situation, and therefore to a starting point of a new learning process (Hendriksen, 2005).

When it comes to gifted education, educational objectives are easily subject to discussion. Depending on one's perspective on gifted education, different educational objectives become visible. Educational needs are intertwined with educational objectives. Each paradigm focuses on different educational objectives. In the gifted child paradigm, serving the gifted by developing thinking and leadership qualities is the primary goal of education (Dai & Chen, 2014, p. 49). In the talent development paradigm, supporting domain excellence and innovation by modelling after authentic professions and creativity is the primary goal of education (Dai & Chen, 2014, p. 49). In the differentiation paradigm, the diagnostic focus prevails in order to serve manifested individual educational needs within the confines of schooling (Dai & Chen, 2014, p. 49). As a result of these different main objectives, the desired effects on a specific student's situation can lead to emotional discussions among teachers, parents, external advisors or even within a teaching team. If the effect aimed for does not match one's perspective on giftedness or education, it may lead to a collision between the educational partners.

Personal core-values influence the level of comfort felt towards a paradigm. However, without an awareness that there are different paradigms, and with an assumption that the use of the words gifted and giftedness has the same meaning for everyone involved, miscommunication can fuel problematic interactions. As a result, the parties involved may not reach agreement on the objectives for interventions. Consequently, the student's development is not catered for as well as possible or the student's interest may be lost and out of sight. Therefore, in-depth knowledge of these paradigms as well as determining one's position about the paradigms, is conditional for the judgemental process, to decide if the effect aimed for is good and desirable for all the parties involved (Dai & Chen, 2014). This judgemental process should underpin any decision over whether or not to strive for the objective.

3.5.4 The Applied Intervention

Interventions lead to changes in the current situation. Changes can be valued as *good*, if and when they are desired by all the actors involved. This requires contextual knowledge and skills. Where one or more actors lack these skills, the teacher or in-school specialist has the responsibility to intermediate in this process of change. If and when interventions do not lead to the desired outcomes, a contextual review of the interventions chosen is justified.

The integration of knowledge, skills and reflection implies that during this process of change, the actors involved continuously develop a new perspective of the revised situation and the current opportunities for change. This revised perspective influences the actions of all the actors involved. Optimised arrangements and interventions replace previously ineffective ones. Effective interventions, leading to the desired effect continue. Educating is a continuous interactive process of estimating the situation, weighing possible effects and selecting the best possible interventions (Onderwijscoöperatie, 2014a; Vlaamse Regering, 2008). The change-oriented and solution-focused approach as mentioned in Chapter 2, has become conditional in modern education (Caufman & van Dijk, 2009; Pameijer & van Beukering, 2007; van Meersbergen & de Vries, 2017). During this process, all the actors involved can be defined as learners. Learning on its own merit can, therefore, be regarded as a natural intervention with short and long-term effects on daily educational practice (van Gerven &

Hoogenberg, 2011). For the professional involved, this requires the ability to apply a broad repertoire of meaningful educational responses and a flexible attitude towards the results of the changes in the current situation.
The fact that intentional interventions lead to changes in the current situation does not imply that unintentional influences on the current situation cannot play a part in the changes that take place. However, when applying meaningful educational responses, we assume that the teacher intervenes wilfully and consciously. We also assume that the teacher always bears in mind that unforeseen influences are part of day-to-day practice and therefore part of all aspects of education. This implies that the ability to adjust their behaviour to what a situation requires is an important aspect of a teacher's attitude (Biesta, 2012; Mol-Lous, 2011; Oostdam, 2009; Vlaamse Regering, 2008).
In gifted education, what is considered as educational needs depends on one's perspective of the notion of giftedness and on one's perspective of a pedagogical and didactical paradigm. This implies that interventions also depend on these perspectives. However, a competency matrix for specialists in gifted education, should not prescribe the perspective of the notion of giftedness. The matrix is supposed to describe competences at a generic level, as far as this can be expected in a specific field of expertise. The specialists' critical thinking skills are supposed to support their educational judgement in selecting and applying concrete interventions. Nevertheless, it is possible to distinguish different domains where the need for intervention becomes clear.
In her book on educating twice-exceptional children, Trail (2011) describes five domains of risk and resilience that can be affected by the clash between basic abilities due to giftedness and basic inabilities due to a learning or behavioural disability. The domains they distinguish are cognitive style, academic achievement, intrapersonal understanding, interpersonal functioning and self-actualisation. Taking a closer look at different educational paradigms and the different models and notions of giftedness, it can be concluded that these domains apply as risk and resilience domains for all (gifted) children. The concept of risk and resilience is, therefore, useful to map the educational needs of gifted students, to stipulate educational objectives for gifted students and to point out possible interventions in gifted education. The use of a so-called "risk and resilience disk" transforms the domains of abstract competences for the specialist in gifted education into domains with concrete options which respond meaningfully to educational needs.

3.5.5 Skilledge

Being confronted with professional dilemmas implies the chance of being confronted with personal incompetence in a specific situation. A professional reaction to help overcome this feeling of being consciously incompetent is the search for meaningful knowledge and useful skills. In their daily practice, a teacher with professional dilemmas focuses first on getting concrete answers to practical questions (Korthagen, 2017). There are no indications that this will be any different in gifted education (Johnsen et al., 2016). These problems can vary from "how can I teach this part of the curriculum in a way that all my students feel challenged?" to "how do I cope with the behavioural problems of one particular student?" These concrete dilemmas imply both the existence of knowledge and the need for knowledge (Johnsen et al., 2016). Without some knowledge, the teacher would not have defined their dilemma, nor would they have searched for answers. They can use their knowledge to both conceptualise the problem and the possible answers. New theoretical information connected to existing knowledge constructs a more in-depth knowledge. What follows is the implementation of the more in-depth knowledge towards new interventions. By doing so, the teacher meaningfully connects knowledge and skills. This connection in itself adds another layer to their in-depth knowledge and experiences. The immersion of a learning process like this contributes to the teacher's professionalism. Once successful in implementing interventions, the teacher feels consciously competent. The final step is a phase where the teacher internalises knowledge and skills, and where they transfer from being consciously competent to unconsciously competent. During this process, reflection-in-action and reflection-on-action are powerful intermediates to make that immersion of knowledge possible. This process of reflection empowers the interplay between knowledge and skills and thereby creates a higher level of professionalism. We can call this level of professionalism "competence", but we can also refer to it as "skilledge"[4]: the ultimate merge between knowledge and skills empowered by core-reflection. Developing skilledge is a never-ending concept. At each level of skilledge, the next level will emerge on the horizon. Knowledge is universal and can be taught. Based on this knowledge, skills can be developed. Core-reflection is a personal process referring to a specific context where knowledge and skills are applied based on personal, professional choices (King & Lau-Smith, 2013; Korthagen, 2017; van Gerven & Hoogenberg, 2011; Wassink et al., 2018). Therefore, as a result of this process, skilledge is non-transferable, always personal, always contextual, and its development never completes.

Critically reviewing self-evidential knowledge and skills via a process of reflection is a characteristic of professionalism. Statements such as "we have always done it like this" do not validly underpin one's actions. Simultaneously, interventions based on an assumption inherited from the past must also be explored for their validity in the current situation. This may lead to the conclusion that the validity of the assumption has expired or is at least doubtful. Beliefs are based on premises; they reflect the personal moral values of the professional in question. Developing skilledge means that the professional sincerely ponders if and how new information, new knowledge, and new skills should lead to a reassessment of these beliefs.

---

[4] The notion of skilledge is specifically designed for the purposes of the MIA and cannot be found in a dictionary.

3.6 Summarising

In this chapter, we described a theoretical framework that underpins a professional standard for specialists in gifted education. This professional standard describes specialists in gifted education first and foremost as teachers. This means that a specialist in gifted education is a qualified teacher. Undergraduate teacher training focuses on the development of competences deduced from professional standards for teachers. Although these standards may differ amongst the European countries, there is a high similarity between the Flemish and Dutch professional standards. Differences are merely semantic and not content-based. An initial teaching degree in Europe is based on the Dublin Descriptors at least to level 6, bachelor's level.

Being a teacher requires a specific attitude: skilledge. The process of developing and application of skilledge is visualised in a model of intrapersonal adaptation. In this model, knowledge and skills are strongly connected with reflection-in-action and reflection-on-action. The process of intrapersonal adaptation explains the contextual appliance of knowledge and skills in a process fuelled by and aiming for interaction. The entire process is based on the premise of educational judgement.

Specialists in education have received additional professionalisation at master's degree level (level 7). Their skilledge has more in-depth development. Their expertise is enhanced. They have acquired domain-specific knowledge and skills exceeding bachelor's level. The contextual application of this knowledge and skills shows in their high-level of professionalism. These specialists have developed the ability to support their colleagues and to intervene in complex educational situations. Consequently, the specialist in gifted education should be trained at master's level. Skilledge in this specialism combines the contextual knowledge and skills of giftedness with the contextual knowledge and skills of education.

The transformation of the theoretical framework, as described in this chapter, into a practical tool – a competency matrix for specialists in gifted education – takes place in Chapter 4. This matrix should reflect the six criteria that frame the notion of competence and should be expressed in terms that match the Dublin Descriptors. A meaningful application of each competence requires skilledge as the underpinning attitude.

Part 2

The Construction of the Competency Matrix

4 Methods Used for Literature Review and Empirical Research

This research uses a systematic review of literature and a questionnaire to better understand stakeholders' understanding of the competencies of the specialist in gifted education. This chapter describes the methodology used for both parts of the research.

4.1 Introduction to the Construction of a Matrix

Chapter 3 described how educational scientists and teacher educators worry that a competence-oriented teacher education programme might limit a more flexible approach to gifted education where a normative conceptualisation of the profession by the professional themselves is a crucial variable. Although a good teacher may be a competent teacher, the suggestion is raised that a competent teacher will not necessarily be a good teacher (Biesta, 2012; Ceulemans et al., 2016; Kelchtermans, 2012; Korthagen, 2017; Struyven & Ceulemans, 2012). The use of the notion of skilledge (see Chapter 3) as the conditional approach towards the use of a competency matrix, pre-empts the risk of the limitations mentioned previously. The Model of Intrapersonal Adaptation (MIA, see Chapter 3) offers teachers the opportunity to describe their competences meaningfully. Using the MIA to describe competences offers a flexibility that acknowledges the multidimensional perspective of teaching as a profession. It also acknowledges the importance of the teacher's personal normative concept related to their professional actions. Lastly, but not least, it acknowledges the fact that teaching is a profession empowered not only by the student's development but also by the development of the teacher.

A competency matrix should not be seen as the ultimate means of standardising the profession of the specialist in gifted education (Biesta, 2012; Ceulemans et al., 2016; Joint Quality Initiative, 2004) (see also Chapter 3). Dutch and Flemish teacher educators distance themselves explicitly from Likert-like scales as, all too often and too soon, they are applied as simple checklists (Ceulemans et al., 2016; Lunenberg et al., 2014). Therefore, there will be no suggestions made on how to transfer the matrix into a scale and checklist.

The competency matrix should be considered as a coherent series of recommended professional behaviour that is based on the analysis of a systematic literature review of combined scientific research and practically oriented publications as used by the professional community of teachers in gifted education. The matrix should be regarded as an opportunity to visualise how the different dimensions of professional behaviour represent themselves as strengths and weaknesses in the professional approach. Each competence can be presented on a continuum where the highest score, "achieved", merely represents the current situation in this specific context. This score does not represent the zone of proximal development for that competence in this specifically described context or any other context. To put it differently, even when a specialist scores "achieved" on a specific competence, it is not possible

to predict whether this same specialist would also achieve a similar score on the same competence if the circumstances were changed. As visualised by the MIA in the previous chapter, learning is a never-ending process. Hence, a professional standard as a way to measure a specialist's competencies is only reliable if it not only shows acquired competencies but also maps out what aspects of a competency can be developed in more depth. The indicators for competent behaviour offer that opportunity as each indicator focuses on different behavioural aspects of a competency. However, one should keep in mind that, due to the restricted amount of time available on teacher education courses, the number of specific contexts where taught knowledge and skills can be applied is limited. As teacher education is a never-ending story, successful teacher education should provide the teacher with the tools and approach for a process of lifelong learning (van Veen et al., 2010). The professional attitude the Dublin Descriptors aims to develop may offer the right skills for this process of lifelong learning (European Commission, 2013).
For each competency in the matrix, a coherent set of indicators describes how this competence can be recognised. These indicators are meant to be seen as examples. Teachers should be trusted in their ability to translate these examples to new contexts. They need the autonomy to decide if, when, and how their professional behaviour reflects these competencies, as they are highly capable professionals with a realistic understanding of their personal strengths and weaknesses (Biesta, 2012; Kelchtermans, 2012; D. Matthews & Foster, 2009).

4.2 A Brief History

Based on the topics for school policy on gifted education, in 2008, van Gerven described in a so-called "Knowledge and Skills List" the knowledge and skills recommended for the implementation of a – for Dutch standards – relatively new educational approach to gifted education (van Gerven, 2008). In 2009-2010 van Gerven, Weterings, Hoogenberg and P. Wolters developed the Master's SEN study in Specialist Gifted Education for Fontys School for Higher Education. They used this Knowledge and Skills List (2008) to set the educational objectives for the modules specific to gifted education. In 2020, Fontys School for Higher Education still uses a selection of these educational objectives to underpin their courses for gifted education.
van Gerven and Hoogenberg (2011) published a revised competency matrix for specialists in gifted education. Competencies for gifted education presented in this matrix were deduced from the Knowledge and Skills List (van Gerven, 2008). van Gerven and Hoogenberg combined seven domains relevant for gifted education with the seven competency domains of the generic professional standard for Dutch teachers. In so doing, they described 49 competencies for the specialist in gifted education. Since 2011, education in The Netherlands and Flanders has changed, as has the position of gifted education (see also Chapter 2). The need for a revised matrix emerges from the paradigm shifts influencing gifted education and in the changed perspective of (continuous) professional development for teachers. Therefore, the concept of the matrix to be constructed has its origin in this Knowledge and Skills List (van Gerven,

2008) and this first competency matrix (van Gerven & Hoogenberg, 2011); however, this concept matches the current demands of both gifted education and continuous professional development for teachers. The major difference between the current study and the previous matrix is that for the former competency matrix the competencies were based on what teachers who enrolled in different teacher education programmes for gifted education had expressed that they wanted to learn (see section 1.1).

4.3 The Construction of a Matrix

For the construction of the competency matrix, the results of the analyses presented in Chapter 2 were combined with an integrated approach to the Dutch and Flemish professional standards for teachers as they are presented in Chapter 3. It was examined to see whether the core domain descriptions and core competencies reflecting these domains were compliant with the six dimensions distinguished by Merriënboer et al. (2002).

4.3.1 Cross Domains: Generic Teacher Competencies

To create a new competency matrix for specialists in gifted education, generic teacher competencies as currently required for Flemish teachers (Vlaamse Regering, 2008) were examined and compared with the current professional demands for Dutch teachers (Onderwijscoöperatie, 2014a). For a short comparative description of both sets of professional standards, see Chapter 3, paragraph 2. These competencies were combined to represent the generic teaching competencies of both countries. This systematic review resulted in the selection of five domains. Due to the combination of two sets of professional standards, these five domains differ slightly from those that would be distinguished for each individual country. Using the Dutch standard, curricular expertise was integrated with didactical expertise. The overlay in both types of expertise was bigger than the difference between them. This approach matches the approach taken in van Gerven and Hoogenberg's competency matrix (2011). The two remaining domains of expertise (pedagogical and didactical expertise) of the Dutch standards were extended using three complementary domains: managerial expertise, communicational expertise, and expertise in research and innovations. In Flemish professional standards, these three domains are designated as teacher's responsibilities. These domains also reflect and make explicit the professional basis mentioned in the Dutch professional standard for teachers. These distinguished domains resemble five of the seven domains of the previous Dutch professional standards for teachers, the so-called SBL competencies (van Gerven & Hoogenberg, 2011). The generic description of these five domains is, however, a balanced mixture of the description of Flemish and Dutch current professional demands.

- **Pedagogical**
  The teacher creates a safe and stimulating learning environment for every student, tuned to their individual educational needs.

- **Didactical**
  The teacher transfers curricular objectives, in collaboration with their colleagues, in such a way that they become teachable in the context of a primary school, matching that school's policy. They must compose the curriculum in a way that enables their students to master the content.
- **Managerial**
  The teacher provides a structured, well-organised, task-oriented classroom atmosphere that supports the students' developmental process.
- **Communication**
  The teacher communicates professionally with all the educational partners who are involved with the student's education.
- **Research and innovation**
  The teacher adjusts their professional knowledge, perspectives and actions if and when required by relevant developments in education, society and science.

In the next step (Table 2), the generically described competencies in these domains were examined to determine if they are compliant with the six dimensions of the notion of competence (Merriënboer et al., 2002) described in Chapter 3, paragraph 4. Each of the five generally described competencies offers the opportunity to apply the competence in a personal, professional context. The meaningful connection between context and competency is to be made by the teacher (Ceulemans et al., 2016). Within this context, the teacher can apply the competencies coherently and interdependently. Each of the competencies solely contributes to the optimisation of 'the situation' but an integrated approach to the applied competencies as described in the MIA influences the extent to which a teacher's interventions will be effective. The described competencies are all action-oriented. The keywords for action that are used are respectively: creating, transferring, providing, communicating, and adjusting. The competencies describe skills that are learnable. Each competency is part of the general teacher education programmes in initial teacher education (Onderwijscoöperatie, 2014a; Vlaamse Regering, 2008). Finally, the competencies are described in a manner that shows their sustainability.

Table 2

Five Competencies Domains of General Education Compliance with the Six Dimensions of the Notion of Competence

	Contextual	Coherent	Interdependent	Action-oriented	Learnable	Sustainable
Pedagogical	x	x	x	x	x	x
Didactical	x	x	x	x	x	x
Managerial	x	x	x	x	x	x
Communication	x	x	x	x	x	x
Research and innovation	x	x	x	x	x	X

Although the teacher will always need these competencies when providing good education, there is also the flexibility to describe these competencies as seems fit for current standards in education. Hence, it can be concluded that they comply with all the dimensions.

4.3.2 Cross Domains: Specific Competencies for Giftedness

Running in parallel to the process of the examining the extent to which generic competencies are compliant with the six dimensions of the notion of competence, was the determination of if and how the competency domains specific for giftedness (as deduced in Chapter 2) met these dimensions. As identified in Chapter 2, the following descriptions of the competency domains were used:

- **Concepts of giftedness**
  The specialist has an in-depth knowledge of the theoretical perspectives of giftedness and gifted education and applies this knowledge contextually in their professional practice.
- **Seeing educational needs**
  The specialist takes a change and solution-focused approach to seeing and profiling the educational needs of gifted students. In this approach, all actors in the student's ecological system are seen as equal educational partners, including the student.
- **Understanding educational needs**
  The specialist interprets and understands the educational needs of gifted students in their ecological context, that is, a system that interacts between the student, the teacher, the curriculum, physical learning environment, the family, and peers.
- **Responding to educational needs**
  The specialist responds meaningfully to the educational needs of the gifted student from a contextual and developmental perspective. In so doing, (learning) process and (learning) results are equally important.
- **Assessing the responses to interventions**
  The specialist monitors the process of responding meaningfully to the gifted student's educational needs and monitors and interprets the gifted student's responses to these interventions contextually.

Each of the five competency domains described above can be applied and described contextually. Each competence refers specifically to this contextual application of knowledge and skills within this domain. The competencies are coherent and interdependent. For example, the perspective that a teacher has on giftedness influences actions and interventions in the other domains. However, the effects of actions and interventions in the other domains are mutually influential on the domain concepts of giftedness. All competencies are described in an action-oriented way. The keywords for action that are used are: applying, profiling, interpreting, responding, and monitoring. All competencies are described in a way that shows that they are learnable. They focus on knowledge, skills, and the use of educational judgement (Biesta, 2012;

Table 3

Five Competencies Domains of Gifted Education Compliance with the Six Dimensions of the Notion of Competence

	Contextual	Coherent	Interde-pendent	Action-oriented	Learnable	Sustainable
Concepts of giftedness	x	x	x	x	x	x
Seeing educational needs	x	x	x	x	x	x
Understanding educational needs	x	x	x	x	x	x
Responding to educational needs	x	x	x	x	x	x
Assessing the responses to interventions	x	x	x	x	x	x

Korthagen, 2017; van Veen et al., 2010). In the way these competencies are described they are also sustainable. The domains correspond with a general approach towards gifted education; indicating that behaviour can change over time without the risk that the competences themselves become out-dated. Hence, we can conclude that these competencies comply with all the dimensions of the notion of competence (see Table 3).

4.3.3 Indicators for Competent Professional Behaviour

The next step was to create a structure that allowed the translation of each competency into professional behaviour that can be observed. For each competency, six categories of indicators were selected. Like the competencies, the indicators met the five dimensions for the description of a competency described in Chapters 2 and 3 (contextuality, coherency, interdependency, action-oriented and learnability). The sixth dimension distinguished by Merriënboer et al. (2002), sustainability, is seen as a fluid dimension. Where the competencies are stipulated to meet also this sixth dimension, regarding the indicators, it is chosen for fluidity for the appliance of this sixth dimension to ensure that the indicators for competent professional behaviour will keep pace with any change of educational paradigms.
The indicators are divided into knowledge and skills indicators which match their specific competency (see Appendix B).[5] The categories A to E represent the domains of risk and resiliency that are likely to be influenced in education. This indexation is based on Trail's (2011) five categories of risk and resiliency: cognitive style, academic achievements, interpersonal relationships,

[5] Each indicator has an individual code. Whenever a context is set in the text that infers a specific indicator, a footnote is used. These footnotes refer to the indicator that can be inferred.

intrapersonal understanding and self-actualisation. How these categories could be seen as universal for all gifted students was argued in Chapter 3.
In The Netherlands and Flanders, a specialist in gifted education is likely to not only be a teacher but, moreover, is also likely to function at level 3 of the pyramid of support, similar to the role of a (Special) Educational Needs Coordinator[6] (S)ENCo (see also Chapter 2). Hence, a sixth category, referred to as category F, focusing on the responsibilities of a (S)ENCo was added. Indicators in this category express the knowledge and skills for this more generic role.

4.4 The Literature Review Study

An in-depth and semi-systematic literature review was undertaken, analysing the characteristics of each 'connection' in the matrix (Bryman, 2012; Franco-Santos & Bourne, 2005; Randolph, 2009). This was in addition to the narrative review completed for Chapters 2 and 3. A systematic literature study aims to collect and explore relevant literature (Franco-Santos & Bourne, 2005) in order to create a collective perspective of the theories involved; in this case, gifted education. Referring to research question 1 as presented in the next section, relevant literature implies a wider range of publications. This includes not only scientifically based publications, but also publications that are being used by the international professional community of specialists in gifted education themselves.
The following steps were taken (Bryman, 2012; Franco-Santos & Bourne, 2005; Randolph, 2009).

1. Focus of the research was determined.
2. Literature was selected.
3. Relevant literature was classified.
4. Literature was organised.
5. Data was analysed.
6. Data was presented.

4.4.1 Focus of the Literature Review

This research aimed to develop a knowledge and skills profile that, according to stakeholders, is the best representation of the competencies of the specialist in gifted education considering current educational paradigms. That way a qualitative instrument is created that matches the current perspectives on teacher education. It addressed the questions of:

[6] In The Netherlands and Flanders, the title of Special Educational Needs Coordinator is non-existent. The Dutch speak about the "Intern begeleider" being the in-service teacher advisor for all students at school. However, the Dutch society for "intern begeleiders", translates this title with Special Educational Needs Coordinator (SENCo). Most SENCo's have a master's degree, either in special educational needs or the currently used title for the degree as Master of Educational Needs (MEN). The underlying thought of transferring to MEN is that all students have educational needs. Defining needs as special is contradictory to current thoughts about inclusion and diversity (Florian & Black-Hawkins, 2011).

1. Which knowledge and skills for the specialist in gifted education are pointed out by the international professional community and international research on gifted education as essential for the optimal development of gifted students?
   a. Which advice is available in international literature for shaping gifted education?
   b. How does this advice fit into the Dutch and Flemish perspective on education in general?

2. How can advice in international research be translated to the Dutch and Flemish educational context?
   a. What knowledge and skills can be deduced from international research on gifted education, considering the international educational standards for specialised teachers?
   b. How can these indicators be stipulated to meet the Dublin Descriptors for professional behaviour?

4.4.2 Literature Selection

The process of the literature review study started with a systematic selection of literature (Randolph, 2009). This was done in a two-step process of global orientation and final selection.

4.4.2.1 Global Orientation

For the global orientation three main criteria were applied: type of resources, the use of up-to-date publications, and the theoretical underpinning. A flowchart has been used to stream the selection process (Figure 9).

A. **The Type of Resources.** Only books and articles aiming for a professional audience (or published in a peer reviewed journal) were included. Webpage publications that not included references, newspaper articles, magazine articles, and television reports were excluded.
B. **The Use of Up-to-Date Publications.** Aiming for a sustainable competency matrix means that giftedness should be regarded in the context of the paradigm shifts that take place currently in education. Therefore, the publications to be used should not be older than five years. Additional exceptions were made for specific publications if there was no equally relevant publication published in the selected time frame.
C. **Theoretical Underpinning.** (Meeting the criteria as might be expected at PhD level.) The level of this research immediately disclosed a problem for the selection process. Literature resources based on empirical research into giftedness and gifted education in The Netherlands and Flanders as well as Dutch and Flemish literature resources based on solid review studies are limited. The existing literature can be divided into three categories:

1. ***Reflecting Didactical and Pedagogical Interventions Based on the Author's Individual Experiences in Giftedness and/or Education.*** These publications show very little evidence of structural and solid literature review regarding the suggested interventions. In some publications, a list of resources is available. Reviewing the lists of resources and the use of these resources in these publications reveals four additional problems.
    - Quite often the used resources belong to the same level of proof as the publication itself. As a result, suggested interventions might be built on quicksand.
    - Statements are made, but the author does not always critically review and relate the used resources to different perspectives. Hence, suggested interventions might be based on a unidirectional perspective.
    - Often the used resources are older than five years and therefore not always still representative for the contemporary educational context.
    - In hardly any of these publications, are references implemented in the main text. Therefore, the value of the suggested strategies can neither be weighed in the context of the used sources nor in a broader educational perspective.

2. ***Reflecting Didactical and Pedagogical Interventions Based on a Solid (International) Literature Review of Gifted Education and General Education, Combined with Examples of Good Practices in The Netherlands and Flanders.*** These publications reflect a broad international theoretical underpinning for the deduced strategies. International research results are related to the contemporary Dutch and Flemish education context. Vignettes based on good practice and small case descriptions are used as showcases in how to apply deduced strategies.
   Reviewing these publications, one aspect required attention:
    - Analyses showed that studies at this level do not always follow the approach where recommendations from underpinning international research are reviewed in the educational context of the country of their origin before suggesting the adoption of them for the Dutch and Flemish educational system. This means that in some cases, recommendations based on a different cultural and educational context are suggested for implementation in the Dutch and Flemish educational system without checking whether there is a realistic chance of successful implementation.

3. ***Reflecting Didactical and Pedagogical Interventions Based on International Empiric Research in Gifted Education.*** These publications are peer-reviewed studies, presented in peer-reviewed journals or as a PhD-thesis.

Applying the selection criteria led to the exclusion of the first category (category C1) of literature resources. Regarding the pyramid of proof (Robinson, 2015), these publications belong to the base level of wisdom. Hence, their relevancy is at a personal level (see Chapter 2). Category C2 publications are included in this

Figure 9

Initial Literature Selection Criteria - Flowchart 1

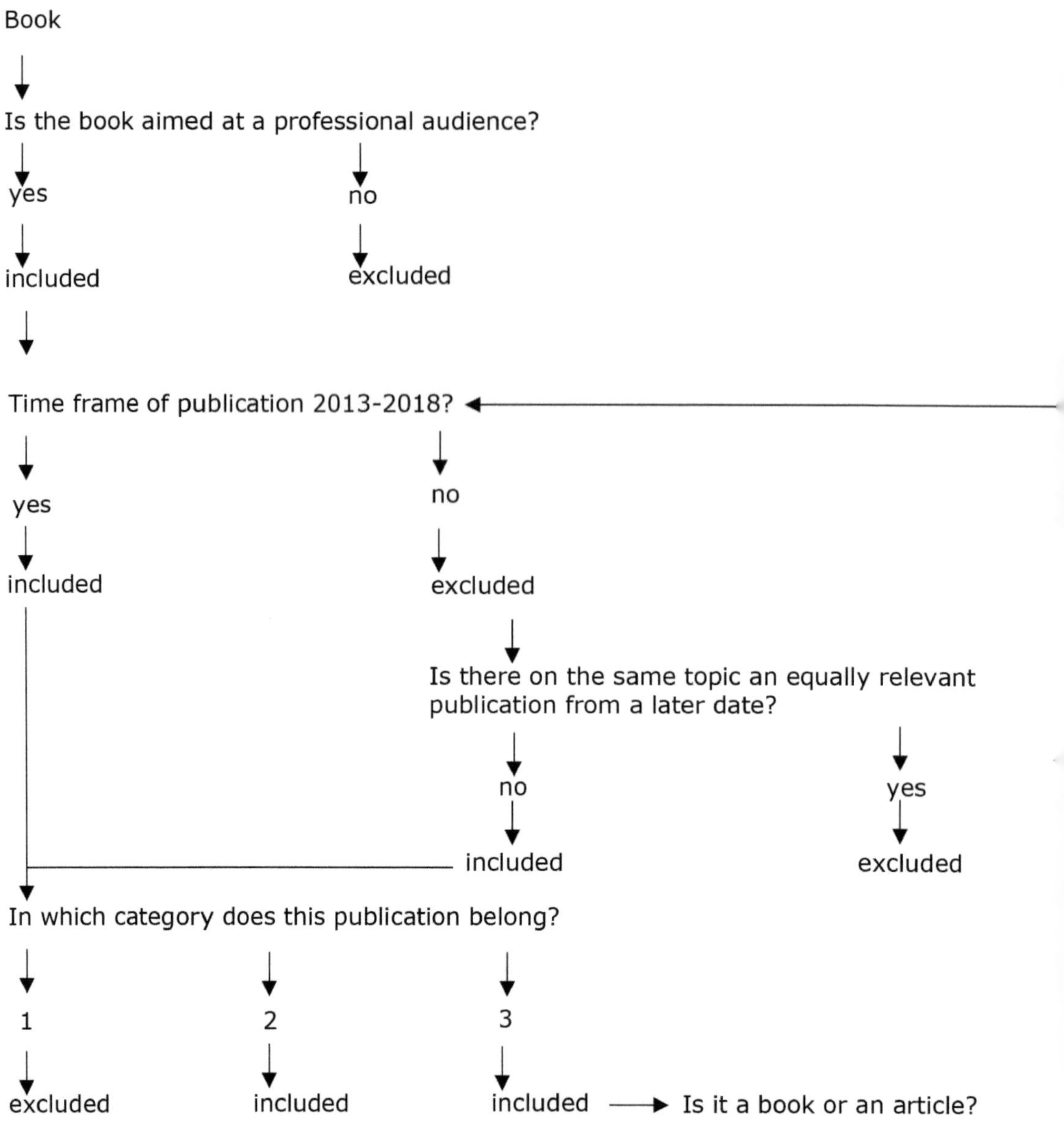

research project. Their underpinning is at level two of Robinson's pyramid of proof. Based on the critical use of theoretical underpinning referrals, publication belongs in either the lower half or the upper half of this level of proof. Category C3 publications are also included in this research project.

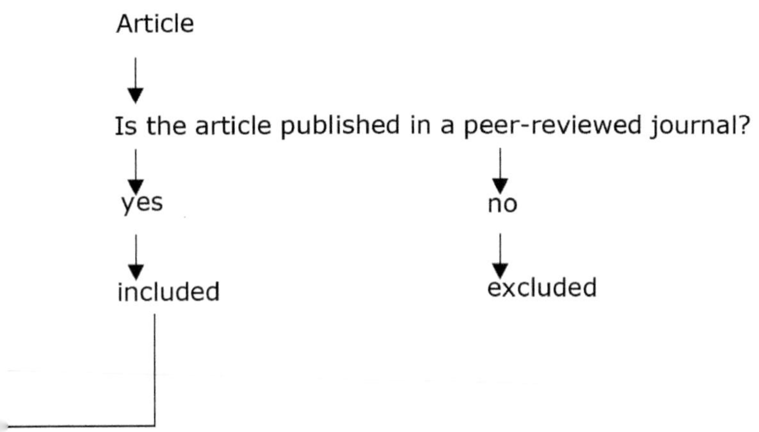

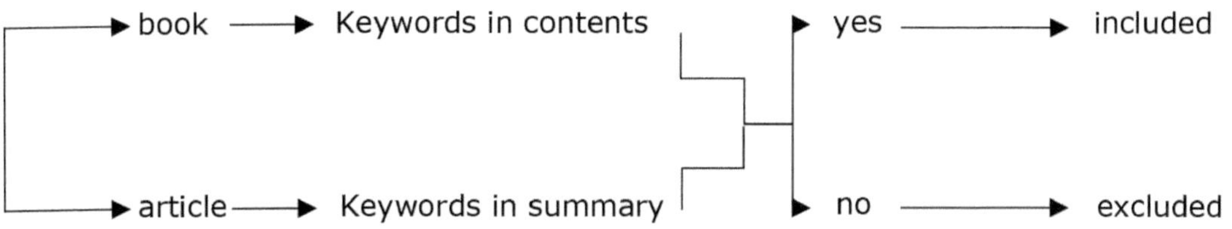

This selection process led to a limited selection of publications meeting all three selection criteria mentioned above. Hence, the underpinning of the described competencies would have been feeble. Criterion 2, the publication date between 2013 and 2018, was therefore extended to a period of ten years (2008-2018). The flowchart was adjusted (Figure 10).

Figure 10

Revised Literature Selection Criteria – Flowchart 2

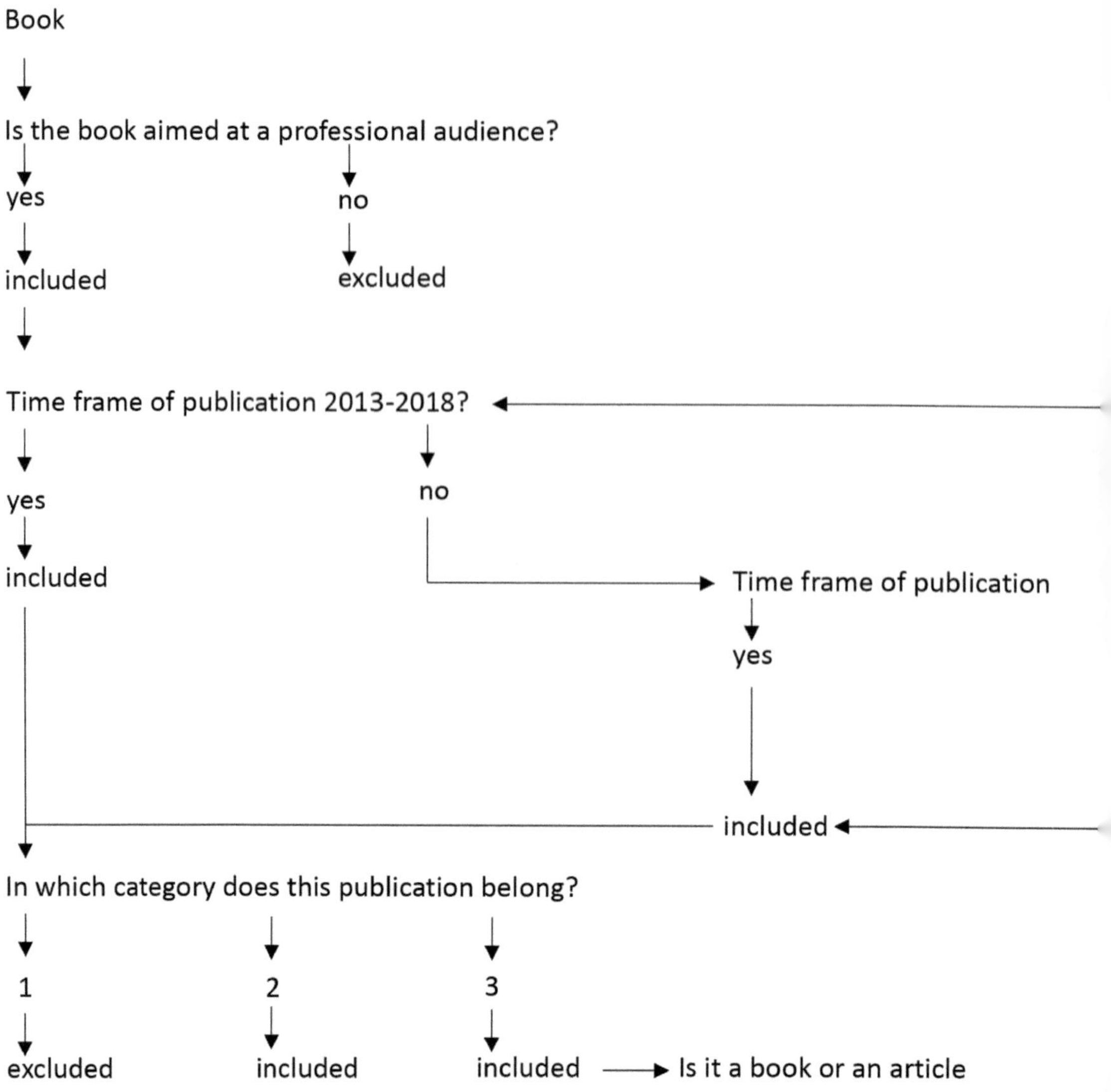

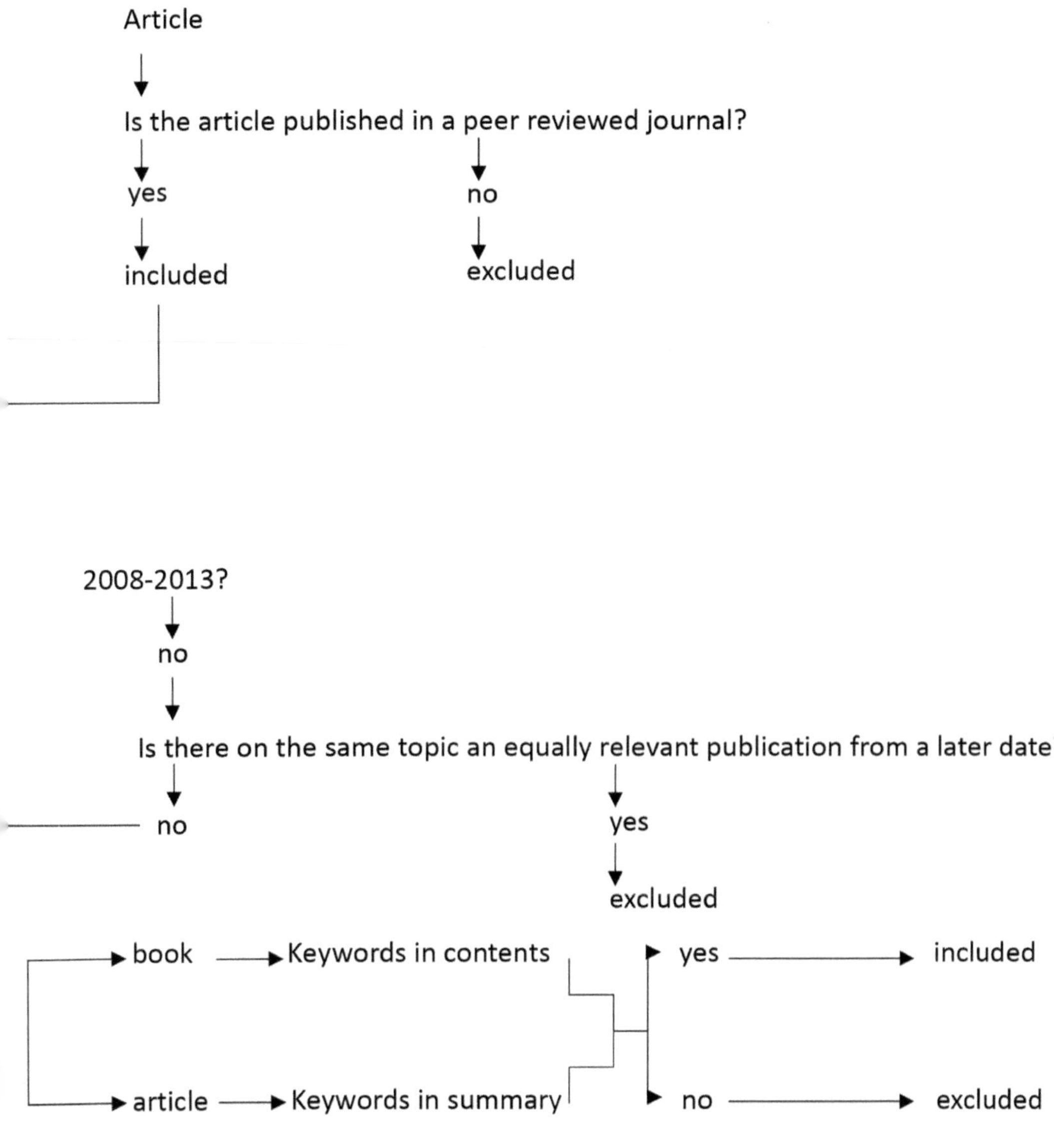
Article
Is the article published in a peer reviewed journal?
yes
no
included
excluded
2008-2013?
no
Is there on the same topic an equally relevant publication from a later date?
no
yes
excluded
book
Keywords in contents
article
Keywords in summary
yes
included
no
excluded

What followed was the extension of the existing list of resources with the inclusion of publications in international peer reviewed journals. The selection of these resources was based on the criteria used in flowchart 2 (Figure 10). The criteria were applied to four leading peer reviewed journals, in the field of gifted education:

- Gifted Child Quarterly
- High Ability Studies
- Roeper Review
- Journal for the Education of the Gifted.

In order to create a relevant selection, the summaries of all articles were read to determine if keywords used in the five domains of general teacher competencies and/or keywords used in the five domains of specific competences on giftedness were mentioned. Articles that did not match this selection criterion were excluded. Next, a group of English and American books that matched the criteria in Figure 10 was added to this selection. In a later stage, literature that was missed at the time of the global orientation, for instance because they were published in peer reviewed journals not specifically aimed at the field of gifted education, were added to the selection if they matched the criteria set up in flowchart 2 (Figure 10). This exception was made if, in the selection of literature, there was a reference to these resources that made inclusion of the original work relevant.

It was decided to include literature from category C2 as part of the systematic literature review. The rationale for this choice was not only that the number of resources at level C3 was limited but also the consideration that not only empirical studies do qualify to be included in teacher education. Research question 3 (see section 1.4) refers explicitly to the knowledge and skills for the specialist in gifted education as pointed out by the international professional community. As the international professional community on gifted education (e.g. teachers, educational specialists) often report about their approaches at level C2, this was a strong rationale to include this type of literature. The second rationale was that in day to day practice, if and when teachers underpin their didactical and pedagogical choices, they do so based on their personal experience, their level of expertise, and the information they have access to, which is – when it comes to gifted education – mostly literature at level C2. By including this category of literature, an opportunity is created not only to influence teacher education by empirical information, but also by the literature teachers use themselves.

For the interpretation of the results of the review study, this means that the selection of indicators will be based on a combination of both resources. As this research project aims to get a better understanding of stakeholders' understanding of gifted education, the choice to include resources from category C matches its purpose. Hence, level C2 and C3 resources are equally used.

4.4.2.2 Final Selection

Articles and books matching this first selection were included and were screened a second time. For this process, all the selected materials were studied. Studying these resources revealed another problem. A large number of the American publications refer for their underpinning to publications and research done before 2000-2008. The research done in those days is serious and reliable within the educational context of that time frame. However, education in The Netherlands and Flanders has changed significantly, as described in Chapters 2 and 3. It is questionable if and how data from before 2008 would still be applicable in the current Dutch and Flemish educational context and match the developments in Dutch and Flemish education. It is possible that changes in the educational context insufficiently take into account their effect on gifted education. However, excluding most of these publications was not an option, due to the fact that a huge percentage of them can be seen as standard works and that they are internationally well-reviewed. For the interpretation of these resources, a check regarding reforms in current education relating to the educational paradigm shifts described in Chapter 2, was undertaken. Criteria used are: the referral to constructivism as a didactical paradigm; a focus on the change and solution-focused approach instead of the so-called medical model approach (wherein classifying diagnoses were required); and the focus on interventions aiming for inclusive education. Publications that met these criteria were included when setting up the data set.

4.4.3 Organising Data and Data Analyses

The final data set of articles and books to be used, were first divided into five main categories, matching the five competency domains for gifted education. Part of the selected literature matched more than one domain. Hence, these articles and books were classified for all the domains they were addressing. Second, for each category/competency domain, the literature was flagged to identify the different categories of indicators they were addressing. Again, a lot of publications matched several categories and they were included for all the categories that they addressed. At this stage, literature that was selected at first, could still be excluded if the identified categories used for the competency indicators were not recognisably addressed. It was decided that only publications that are actually referred to in the final analyses were to end up in the references of the respective chapters.

4.4.4 Data Presentation

The results of the data analyses are presented in Chapters 5 to 9. Each chapter presents data of a specific domain of the competency matrix. As the construction of the matrix focuses on the creation of a coherent set of competencies, it was unavoidable that some issues were addressed more than once. To limit the presentation of data, issues were addressed only for a second or third time if the perspective of the domain the data was presented in, provided an additional perspective on indicating behaviour.

4.5 The Empirical Research: Stakeholders' Understandings

In a similar way to those steps taken for the literature review, the following steps were taken for the empirical research regarding the understandings of stakeholders (Bryman, 2012; Randolph, 2009).

1. The focus of the research was determined.
2. The notion of stakeholders was defined.
3. The research scenario was set up.
4. Data collection was undertaken.
5. Data management, reduction, and analyses took place.
6. The data was presented.

4.5.1 Focus of the Empirical Research

The empirical part of this research aims to get a better understanding of how Dutch and Flemish stakeholders in primary gifted education of gifted students understand the competencies of the specialist in gifted education. Research into the understandings of stakeholders of what they consider as competent professional behaviour for specialists in gifted education, has not been done before in The Netherlands and Flanders. That means that there is no national data available that can be used to compare current results. International research, including stakeholders' understandings from the USA by Johnsen et al. (2016), does not clarify how different stakeholders value different indicators for professional behaviour.

The actual research question was "What knowledge and skills indicators are selected by Dutch and Flemish stakeholders as the most relevant and important examples of competent professional behaviour of the specialist in gifted education?"

To streamline the process three subtopics were raised.

1. The results of the literature review on the competencies and indicators as to be presented in Chapters 5-9 suggested that there is no evidence that some indicators should be higher valued than others. Current theories do not provide a ranking of effective professional behaviour. Current educational paradigms in Dutch and Flemish education did raise the question of whether all indicators deduced from theory would be equally distributed amongst all stakeholders. Analysis of the responses of stakeholders may provide a better understanding of how these indicators are valued and to what extent these indicators of professional behaviour match the educational paradigms prevailing in The Netherlands and Flanders.
2. Stakeholders have different interests in gifted education and a different level of professionalisation. Hence, it was relevant to analyse if there were differences and similarities noticeable between the selection rates of indicators amongst the groups of stakeholders. As each group of stakeholders has a different role in the education and upbringing of gifted children, it was also interesting to analyse whether choices correlate with the role of different groups of stakeholders. If so, it may explain why stakeholders in their interactions might experience

differences in options that seem difficult to overcome. A proper understanding of what stakeholders consider important indicators of professional behaviour is necessary to anticipate the interplay of the different stakeholders' understandings of gifted education. It helps to prevent miscommunication once the parties involved can understand each other from their respective points of view. It can also help to select topics to be addressed in the curriculum of the continuous professional development of teachers and specialists in gifted education.

3. In Chapters 2 and 3, similarities and differences between the educational systems of The Netherlands and Flanders were described. This comparison showed that there are differences between both countries in the rendition of gifted education. These differences may affect the selection rates of stakeholders' of what is most important or most relevant. Therefore, we compared stakeholders' selection rates of indicators between both countries.

4.5.2 Who are Considered to be Stakeholders?

The notion of who should be considered as stakeholders in an organisation is a broad one (Freeman, 1987; Mitchel et al., 1997). In the context of education, the definition of stakeholders is "professionals and parents directly involved with the education of gifted children in schools". In the context of this specific research project, the notion of "schools" is limited to primary schools. According to Freeman (1987) and Mitchel et al. (1997), different groups of stakeholders have different roles, different responsibilities and different levels of influence on an organisation. Hence, regular classroom teachers, special project teachers, teachers in full-time classes for gifted education, school counsellors, specialists in gifted education, school administrators, school council officials and policymakers, teacher educators, psychologists and pedagogues involved in gifted education or in teacher education focusing on gifted education, and parents, are all to be seen as stakeholders. These different roles and responsibilities lead to different levels of contextual collaboration in educational partnership (de Wit, 2005). Broersen and Spreij (2009) distinguish four levels of collaboration: (1) displaying an interest in each other's experiences; (2) jointly exploring solutions; (3) collaboration for joint objectives; (4) making decisions together. Each level of collaboration has a different impact on how the understandings and opinions of stakeholders influence each other's behaviour (Smit et al., 2006; van Meersbergen & Jeninga, 2012).

Each group of stakeholders is an advocate for different interests (Freeman, 1987; Mitchel et al., 1997), and this is likely to be reflected in their selection of indicators. Although all stakeholders will have the interest of the gifted student at heart, additional interests depending on their role in the lives of gifted students, can result in different accents in the choices of what is seen as most valued, competent professional behaviour (Broersen & Spreij, 2009; de Vries, 2017; Pameijer, 2012). Hence, it was expected that selection rates of indicators would differ amongst the different groups of stakeholders. For example, general educational professionals and specialists in gifted education have to advocate for the interests of gifted students as well as for other students trusted in their care. Keeping the balance between the interest of the entire class or school and the interest of an individual student can be a delicate one (Broersen & Spreij, 2009). This delicate balance may occur in situations where responding to the needs of an individual student is relatively time-consuming and puts a strain on the teacher's classroom management (Kuipers, 2017; van Meersbergen & de Vries, 2017). Keeping that balance causes concerns to those who are involved. For parents, it is their responsibility to ensure that the educational interests of their child at school are served as well as possible (Broersen & Spreij, 2009). Although parents do understand that the interests of all the other students may differ from their child's interests, their child will often still be at the centre of their attention. From their perspective, it is understandable that they strive for a different focus on interventions and, therefore, that they might select different indicators for competent professional behaviour.

In gifted (primary) education, there are different groups of stakeholders, but one of these groups is not included in this research project. Purely from an educational perspective, gifted students should be considered as the most important group of stakeholders (Caufman & van Dijk, 2009; Pameijer & van Beukering, 2007). Teachers express that their students are capable of expressing their educational needs and their perspective of adequate professional teacher behaviour (den Otter, 2015; Mayes et al., 2019). Research by Bakx (2019) suggests that gifted students at primary schools often have a clear perspective of what they consider to be important skills for their teachers to acquire. Gifted students included in Bakx' research emphasised how the pedagogical relationship between themselves and their teachers trumps both the need for autonomy and the need for competency.

Even so, gifted students were not included as a group of stakeholders in their own right in this research. The choice to exclude them was determined by the role of the specialist in gifted education. The job description or professional role of the specialist in gifted education can be considerably different from the role of a classroom teacher (see also Chapters 2 and 3) (van Gerven & Hoogenberg, 2011). As a result of this, in most cases in The Netherlands and Flanders, the specialist in gifted education will hold a more distant relationship to the gifted student than a teacher of the gifted or a regular classroom teacher does (see also Chapters 2 and 3). This role can often be compared with the role of the (S)ENCo in Dutch schools (LBBO, 2019). Depending on the role the specialist in gifted education has at a school, there may be a broad range of competencies needed to fulfil that role, which remain completely out of the scope for gifted students (LBBO, 2019). Hence, although the voice of a student reviewing the competencies of their teacher is important and should be taken into account, for this research, they are excluded from participation.

Once different stakeholders were identified, calls for respondents were made. It was expected that most respondents would have a special interest in gifted education. For them, having the opportunity to express themselves about what they consider to be important professional behaviour, is likely to have high relevance. It was decided to make use of a wide range of focus group (social) media platforms and magazines to widen the call for respondents, involving stakeholders in education in general (who are not directly involved with gifted students) and offering them the opportunity to respond and participate in the research project. To reach parents, parental self-help organisations for parents of gifted children and magazines on the upbringing of children in general (not focusing on giftedness in particular) came in very helpful. The call was published in their magazines, newsletters and on their websites and additional social media platforms. Teachers of gifted students reached out to parents at their schools and forwarded the call for respondents as well. To reach respondents working in the field of primary education, professional magazines for teachers and/or school management and policymakers, educational organisations and institutes (i.e., a teachers union, the national society for remedial teachers, the national society for school counsellors and advisors, the national society for specialists in gifted education) were asked for their support. They published the call for respondents in their magazines, websites, and newsletters. Several commercial organisations who are known nationally for their teacher conferences and continuous professional development courses for teachers also published the call in their newsletters and on their websites. Finally, several calls via LinkedIn and Twitter were published.

By definition stakeholders in this research are persons that have an interest in gifted education. This does not necessarily mean that they have to be a teacher of a gifted student or a parent of a gifted child. Without realising it, teachers who are currently not educating a gifted learner can be considered as a stakeholder as well. The same applies to parents without a gifted child. The professional attitude of the specialist in gifted education may affect their student or their child as well. However, it seems logical that they don't consider themselves as stakeholders and therefore, that they did not reply to the call for respondents. Hence the current results may be generalised towards a population that consider themselves to be stakeholders, but it cannot be suggested that the results apply to 'all parents' or 'all educational specialists'. It may well be that not being a teacher of a gifted student means that such a professional may take a different position towards the competencies of specialists in gifted education.

4.5.3 The Research Scenario and Data Collection

As the topic of the research was complex, and the time-investment of respondents was considerable, it was decided to open up a website available at any time convenient for respondents (Brinkman, 2014). People could subscribe to participate in the research to get access to the website. On the website, the research process was divided into two parts.

4.5.3.1 Part One: Information About the Research

Part one aimed to inform people who were interested in the objectives of the research and the research process itself (Bryman, 2012; Dijkstra et al., 2014). General information was offered about the relevancy of the research, the structure and theoretical background of the competency matrix, and the structure of the questionnaire. This information was presented in three

PowerPoint presentations, each taking approximately 4-8 minutes. To process all the information took 20-30 minutes.

- PowerPoint 1 described the relevancy of the matrix and explained the role of the specialist in gifted education. The research question was presented, and respondents were informed of how their data was used.
- PowerPoint 2 described the professional standards for teachers and the competency domains deduced for the competency matrix. Additionally, a glossary (see Appendix C) was available, in case respondents would like to read a clarification of used terminology. Respondents were able to download the glossary from the website.
- PowerPoint 3 introduced the use of the online questionnaire.

There was also a trial questionnaire to introduce the process of answering the actual questionnaire. Respondents were free to decide whether or not to watch the PowerPoint presentations or to respond to the trial questionnaire.
Having processed the information, respondents could decide if they would continue their involvement and start answering the actual questionnaire, published as part two of the webpages. Having decided to continue their involvement, respondents who had subscribed for the website but had not answered the questionnaire received a reminder once a month that their questionnaire was still open. After three reminders, their name was removed from the list of respondents. The website opened up on November 1st 2018 and closed on March 31st 2019. 52 out of 406 respondents decided to withdraw from participating.

4.5.3.2 Part Two: The Actual Questionnaire

The construction of the questionnaire used for the research was based on the construction of the competency matrix (see Appendix A) and the indicators for competent behaviour (see Appendix B). The questionnaire had two sections. Section A collected general data regarding the role of the responding stakeholder. The addressed topics were:

- Country of residence
- The stakeholder's position and role in gifted education. Respondents could only select one position and/or role. They were asked to select the role that they thought relevant to their involvement in gifted education.
  - Teachers with gifted students in their classroom
  - Teachers without a gifted student in their classroom
  - Teachers of pull-out programmes
  - Teachers in full-time education/special schools for gifted students
  - Specialist in gifted education
  - Internal school counsellor/advisor
  - Director/headmaster
  - Decision-makers at school council level or member of a school board
  - Pedagogue/psychologist
  - Parent of a gifted child
  - None of the above

- The age group of students the stakeholder is primarily focused on
- The stakeholder's background regarding gifted education
- The existence of school policy at the school where the stakeholder is involved.

Table 4

Distribution of Stakeholders by Subgroups and Per Country

Subgroups	Percentages subgroups combined N=354	Percentages subgroups Netherlands N=299	Percentages subgroups Flanders N=55
Educational professionals (EP)	37.01 (N=131)	34.45 (N=103)	50.91 (N=28)
Specialists in gifted education (SGE)	29.10 (N=103)	33.44 (N=100)	5.45 (N=3)
Parents (Pa)	23.45 (N=83)	20.74 (N=62)	38.18 (N=21)
Rest	10.45 (N=37)	11.37 (N=34)	5.45 (N=3)

Note. This table is identical to Table D1.

Section B referred to the vertical domains of the competency matrix (see Appendix A).

- Concepts of giftedness
- Seeing educational needs
- Understanding educational needs
- Responding to educational needs
- Assessing the response to interventions

For each domain, all competencies were sequentially presented with their indicators. Additionally, for each domain, an open question was added. That question provided respondents with the opportunity to describe indicating professional behaviour that they had missed in the domain presented. The actual questionnaire is presented as The List of Indicators for Competent Behaviour in Appendix B.
In the online questionnaire, for each competency the indicators were presented randomly. This randomisation prevented respondents from selecting indicators that would answer a specific category of indicators based on the category itself rather than based on the described behaviour (Dijkstra et al., 2014). Additionally, this presentation randomly differed for each respondent. Respondents were unlikely to be able to identify in which category an indicator belonged as the identifying codes or the indicators (Figures 11 and 12) were not shown in the questionnaire.
The respondents had a repeating task on each question: select the three indicators that you consider the most relevant example for the described competency (Brinkman, 2014). The questionnaire was conditioned to prevent a respondent from answering the next question unless exactly three indicators were selected (Dijkstra et al., 2014). Once all the indicators were selected for a domain, the respondent had the opportunity to mention additional indicators

that they also considered important examples of competent behaviour. Respondents could decide to skip these questions for the domains if they thought them not to be relevant.
Answering the questionnaire took approximately one hour. It was not possible to set the questionnaire on hold, saving data and leaving the website and return at a later date or a more convenient time. This choice was made deliberately. It ensured that every respondent replied in full, and all questionnaires that were submitted were completed. Incomplete questionnaires could not be submitted. The glossary referred to in paragraph 3.5.3.1 stayed available for respondents during the process of answering the questionnaire.

Figure 11

Knows that the Interaction Between Aptitude and Development Influences a Student's Achievements

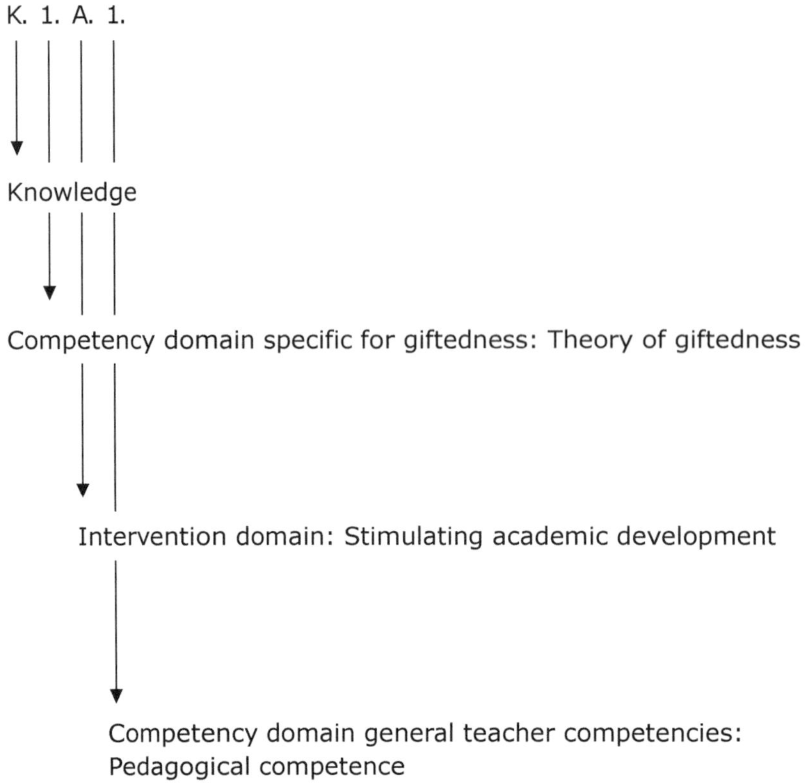

4.5.4 Data Management, Reduction and Analyses

The large corpus of collected data had to be reduced to a manageable size to be able to process the data and to make sense of it (Bryman, 2012). At the construction of the questionnaire, all indicators were given a four-digit code, that is, K.1.A.1 or V.1.A.1 (see Figures 11 and 12). Each digit in the code referred to a different aspect of the matrix and the domains of competencies. Using this code provided the opportunity to construct a strategy for data reduction that is best presented as a pyramid (see Figure 13).

Figure 12

Stimulates the Student's Understanding of Their Personal Ability

V. 1. A. 1.

Skills

Competency domain specific for giftedness: Theory of giftedness

Intervention domain: Stimulating academic development

Competency domain general teacher competencies:
Pedagogical competence

Each layer of the pyramid represents a more in-depth level of analysis for a more selective data set. The top level (level 3) was used to analyse data on a general level, to identify if differences between stakeholder selection rates per category of indicators would be noticeable. Firstly, descriptive statistics were used to identify differences. Secondly, One-Sample tests were employed to determine if observed differences means differed from the expected mean of 2.50. Paired Sample Statistics Tests were used to determine if there were significant differences in selection rates if the indicators were reorganised in two groups. Group ADE represents the indicators with the highest selection rates. Group BCF represents the indicators with the lowest selection rates[7] (see Appendix J, Table J26 & J27).

[7] Indicators from category A, D, and E form group ADE. Indicators from group B, C, and F form group BCF.

Figure 13

A Pyramid Representing a Strategy for Data Reduction

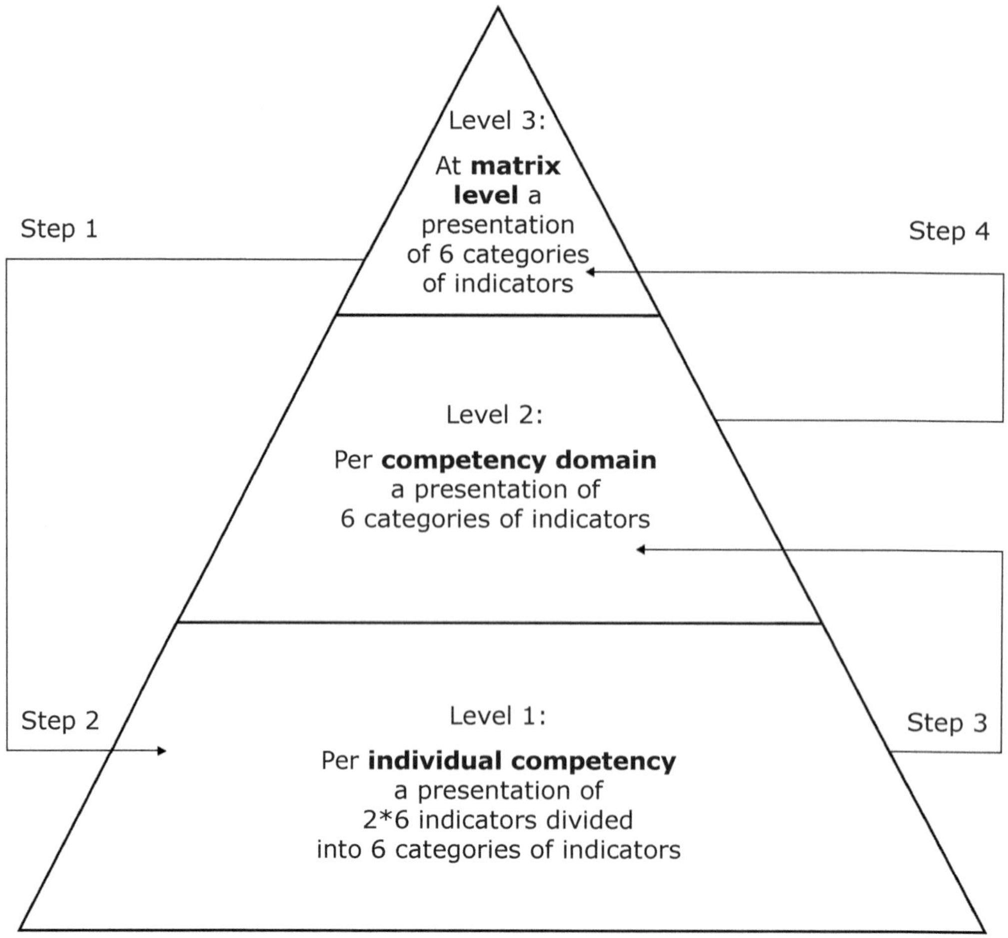

Observed differences led to the analysis of the data at the bottom level of the pyramid, level 1. This analysis allowed for the exploration of how stakeholder selection rates were divided amongst the indicator categories. The differences between the selection rates and the different groups of stakeholders were analysed and described by using frequency analyses. The selection rates for the presented indicators were analysed for each competency. The next step was to determine the selection rates per indicator and per group of stakeholders for each competency. The three indicators that had the highest selection rates per group of stakeholders and the one indicator that had the lowest selection rates per competency, was identified.[8] In order to do this, frequency analysis was used.

[8] At this in-depth level, possible differences between the countries were left out of the scope.

Whether different groups of stakeholders made different choices, as in what they thought to be the best indicators for competent professional behaviour was explored. Differences are described if they are relevant for the educational practice.
The following definition of relevant differences was used. Relevant differences are differences between two or more groups of stakeholders that suggest strongly that both groups of stakeholders have a different opinion but that this difference is not statistically significant. Relevant differences can only be identified if the data are presented and compared in the context in which they were presented to respondents. There are two criteria for relevant differences. The first criterion for relevant differences applies for indicators that have the highest selection rate within one group of stakeholders but that did not end up in the top three selection rates by one or more of the other groups of stakeholders. The second criterion applies for indicators that have the lowest selection rate within one group of stakeholders and did end up in the top three selection rates by one or more of the other groups of stakeholders.
These relevant differences are important, as they present the stakeholders' perspective in more detail, creating nuances that would otherwise be lost. These nuances create a better understanding between educational partners of what they consider important for the competencies of the specialist in gifted education and indirectly what they think to be important in gifted education.
Once the differences were identified, common ground between all groups of stakeholders was explored. Two criteria were defined as representing common ground:

- Common ground refers to situations where an indicator belonged to the three indicators that showed the highest selection rates amongst all groups of stakeholders.
- Common ground refers to all indicators that received the highest and lowest selection rates from all different groups of stakeholders.

Stakeholders had to select three indicators for each competency. Hence, 300 indicators were divided into selected/not selected. That means that each stakeholder selected 150 indicators. As a result, all 300 indicators could be ranked from least selected to most selected. This ranking is presented separately for each domain in Appendix J, where the statistical findings are included. An additional ranking for the entire set of indicators is also included in Appendix J.
What followed was data analyses at level 2. This is the level where indicators are organised per category and no longer per competency (see Figure 13). Per competency domain, selection rates of the indicators were compared amongst all four groups of stakeholders. The statistical significance of the findings was determined. To identify if there were categories of indicators showing significant differences in selection rates compared to other categories, One-Sample Statistics and One-Sample Tests were used. Paired-Samples Tests were also used to identify if there were differences when the six categories of indicators were divided into two different sets of indicators, similar to what was done at the start of the data analyses at level 3. The analysis to determine if different

groups of stakeholders had different understandings was done by comparing selection rates for categories of indicators. To identify significant differences for each comparison, an ANOVA test, with findings based on α 0.01 was used. To identify specific differences between groups, the Bonferroni's Test for Multiple Comparisons was employed. The effect size for ANOVA was calculated based on η^2. Traditionally, effect sizes between 0.2 and 0.5 are considered to be small, between 0.5 and 0.8 they are considered to be medium-sized and > 0.8 large. Based on Coe's concept of practical significance and the threshold he uses for practical significance, the effect sizes > 0.3 were classified as having practical significance. Referring to Glass et al. (1981), Coe (2002) describes that in education, the effectiveness of interventions can only be interpreted within the context in which they are implemented. Practical significance refers to situations in which the balance of the effort of an intervention and the effect of the intervention are described contextually. An intervention that requires only a small effort to apply but that raises the wellbeing or achievements of an entire group, may have an effect size statistically defined as low, but should be considered as practically significant (Hattie, 2013).
Thirdly, similarities and differences between The Netherlands and Flanders were analysed for levels 2 and 3. Stakeholder selection rates between stakeholders in the two countries were compared. The Independent Samples T-test was used to identify the significant differences of the means for the selection rates per category indicators between the Dutch and Flemish stakeholders. Levene's Test for Equality of Variances was applied, to determine the homogeneity of variances. It showed that the assumption of homogeneity of variances was confirmed. Hence, for all tests for equality of variances, the homogeneity of variances could be assumed. Then, the test for Equality of Means was used. The analyses of similarities and differences are only applied between the two countries at the generic levels (levels 2 and 3) and without dividing both groups into subgroups of stakeholders. That means that results for Dutch and Flemish stakeholders have been compared, but not to a level where, for example, Dutch parents were compared with Flemish parents or where Dutch educational specialists were compared with Flemish educational specialists.
Finally, returning to the top level of the data pyramid (level 3), observed differences at the lower levels were explored to see if they were still noticeable at this general level. An Independent Samples T-test was used to identify significant differences for the selection rates per category indicators between The Netherlands and Flanders. At this level, differences between groups of stakeholders were analysed by using ANOVA. First Levene's test of Homogeneity of Variances was applied, then analyses of variances to identify whether there were significant differences for each category of indicators between the group means of the selection rates of groups of stakeholders. By using Bonferroni's Multiple Comparison's test, it was possible to determine which groups of stakeholders differed significantly from each other in their selection rates.
In Chapters 5-9, once the theoretical underpinning for the selected competencies and indicators is presented, results of the empirical research are included per domain. This provides a direct understanding of the selection rates of stakeholders regarding the indicators for competent professional behaviour. In

Chapter 10, the domain results observed at level 2 of the data pyramid and results at level 3 are presented.

4.5.5 Respondents

The current data are based on N=354. There were 299 Dutch stakeholders (N=299) and 55 Flemish stakeholders (N=55). Both groups can be divided into four subgroups, as respondents who categorised themselves as a Specialist in Gifted Education; were separated from all other educational professionals (Appendix D, Tables D1 and D2) (1) Educational professionals; (2) Specialists in Gifted education; (3) Parents; and (4) Not Otherwise Specified. 23.45% of all respondents were parents; all other respondents were professionally involved in education.

Amongst all respondents, there was only one teacher who was not teaching a gifted student in their classroom. All other respondents were either involved in the education of gifted students or their upbringing. It is likely that other teachers without gifted students in their classroom either did not see themselves as stakeholders or have set different priorities for their daily practice.

Remarkably, there were no respondents who categorised themselves as "decision-makers at school council level of members of a school board". They were considered to be stakeholders as school boards in both countries have a legal obligation to provide education for all of their students (Decreet betreffende maatregelen voor leerlingen met specifieke onderwijsbehoeften, 2014; Wet passend onderwijs, 2012; Wet primair onderwijs BES, 2016) and to do this, they may depend on specialists in gifted education. In our call for respondents, we used several specific channels to reach out to these decision-makers and members of school boards (LinkedIn focus groups, page-wide calls in professional journals aiming for educational managers, websites of professional unions). There is no obvious explanation for the lack of responses from this group of stakeholders.

Although an attempt was made to involve school directors and administrators, there were very few respondents who categorised themselves as such. Responses via social media (i.e., LinkedIn, Facebook and Twitter) on calls to participate in the research show that these calls reached this group. It was noticed that these calls were often forwarded to designated staff members as could be noted in the social media responses.

Most respondents did not have a separate focus group but described themselves as involved with children in all year groups (Appendix D, Table D3). Respondents only involved with children in Year groups 1-2 (equivalent to kindergarten in Flanders) are relatively the least represented.

There are differences between the level of continuous professional development (CPD) between both subgroups of stakeholders and the two countries (Appendix D, Tables D4 and D5). Dutch respondents had a higher CPD level than respondents from Flanders. Among the Dutch group of educational professionals, 33.01% had a CPD level of 1-2 years. That is compatible with the CPD level of most specialists in gifted education. Educational professionals who have this CPD training level often categorised their role as being a special class

teacher working in a pull-out programme (N=17; 50% out of 34) or as a teacher in a segregated setting for full-time education for gifted students (N=7; 20.59%).
Respondents who categorised themselves as specialists in gifted education from both countries had a higher CPD level than other subgroups of stakeholders. 25% of the educational professionals from Flanders mentioned not having had any CPD training in the field of gifted education compared to 6.8% of the Dutch educational professionals. Although the CPD level of Flemish specialists in gifted education seems relatively high, is noteworthy that in this subgroup N=3 and the percentage merely means that two out of three specialists had intensive training and one respondent has a role of Specialist in Gifted Education without extensive training.

4.6 Summarising

This chapter first described how the results of the theoretical research presented in Chapters 2 and 3, leading to the construction of the matrix, has been analysed and evaluated. The cross domains for generic teacher competencies and specific competencies for the specialist in gifted education, match the criteria for a competence-oriented approach as described by Merriënboer et al. (2002).
Secondly, the methodology used for the literature review was described. A six-step process was used to select and analyse literature. The criteria for literature selection and how exceptions to that process were dealt with were both described.
Thirdly, the methodology for the empirical part of this project was described. Again, a six-step process was applied, similar to the process for the literature review. Special attention was paid to the definition of the notion of stakeholders. The chapter ended with a description of the respondents, based on the results of the first part of the empirical research. The most remarkable result was that there were no responses from policy makers and educational managers. The respondents were divided into four groups of educational professionals in various roles, specialists in gifted education who identified themselves as such, parents, and a relatively small NOS (not otherwise specified) group.
For both the literature review and the empirical research, the results will be presented in Chapters 5-9. A "vertical approach" is taken, describing the analyses and presenting the results clustered by the five domains influencing gifted education. Each domain is described in a separate chapter.

5 Domain 1: Theory of Giftedness

5.1 Introduction

Different notions of giftedness have different consequences for education (Dai & Chen, 2014). Therefore, a specialist in gifted education needs to have an in-depth knowledge of the theoretical perspectives of giftedness and gifted education.[9] Moreover, the specialist needs to be able to interpret their knowledge contextually in their professional practice (Borland, 2012; Dai & Chen, 2014; Kreger-Silverman, 2013; Leavitt, 2017; D. Matthews & Foster, 2009; Touron & Freeman, 2017). Insight into the perspective and consequences for education is necessary to plan one's actions.[10] In this first (vertical) domain of the competency matrix (see Appendix A), the five generic teacher competencies relate to the specific competencies regarding the theoretical concepts of giftedness.

5.2 How Gifted Paradigms Influence Gifted Education in General

In their analyses of the so-called mystery and mastery approach, D. Matthews and Foster (2009) point out how personal perspective on giftedness directly influences the identification of gifted students. Within the context of identification determining which students are selected for an adjusted educational programme, the specialist in gifted education should be aware of this relationship and its consequences.[11] The recognition of educational needs differs according to the paradigm the specialist holds as their point of departure (Dai & Chen, 2014).

The mystery approach (the gifted child paradigm) considers that a gifted child is born with high academic potential. In this paradigm, this potential leads to a score in the superior range of intelligence tests and the child's ability is perceived as relatively constant over time. It is also assumed, in this paradigm, that gifted students can operate at a high level in all academic areas. The meaningful educational response to this perspective is that gifted students should participate in a gifted programme (Kreger-Silverman, 2013). This gifted programme is based on a rather homogeneous perspective of the needs of gifted students (Dai & Chen, 2014; D. Matthews & Foster, 2009).

In the mastery approach (the talent development paradigm and the differentiation paradigm), giftedness is, according to D. Matthews and Foster, best defined as "an exceptionally advanced subject-specific ability at a particular point in time, such that the student's educational needs cannot be well met without significant adaptations to the curriculum or other learning experiences" (2009, p. 28). Based on this perspective, giftedness is recognised through focusing on the learning process and the subsequent results (Dai & Chen, 2014;

[9] Appendix B, Table B1, K.1.F.1.
[10] Appendix B, Table B5, K.1.B.5.
[11] Appendix B, Table B3, K.1.A.3.

Subotnik et al., 2011). It leads to the understanding that subject-specific interventions are needed. Gifted students are seen as a heterogeneous group with individual educational needs. These needs lead to the necessity of individual educational programming (Borland, 2012; Dai & Chen, 2014; Subotnik et al., 2011).

In *Paradigms of Gifted Education* Dai and Chen (2014) discuss how different conceptions of giftedness influence gifted education. They demonstrate the relationship between "what" (definition and concept), "why" (theoretical underpinning of the definition and concept), "who" (focus group for gifted education), and "how" (consequences for educational programming). Every perspective of giftedness stresses different aspects of education differently[12], and, as a result of this, curricular objectives can vary according to differences in these perspectives of giftedness.[13] An equilateral triangle between "who", "what", and "how" is conditional for success. Therefore, coherency between the concepts of giftedness, educational objectives, and educational programming is just as much a condition as knowledge about different opportunities for organisation and implementation of the curriculum (Borland, 2005; Dai & Chen, 2014; Kreger-Silverman, 2013; Leavitt, 2017; D. Matthews & Foster, 2009; Subotnik et al., 2011; Touron & Freeman, 2017).[14]

Chapter 2 discussed how, in the current scientific discourse, the paradigm shift from the gifted child paradigm via the talent development paradigm towards the differentiation paradigm takes place. The need for this paradigm shift is stressed by the interaction between modern educational perspectives and concepts of giftedness (Dai & Chen, 2014, Stack & Sutherland, 2017; Ziegler et al., 2017).[15] In current educational perspectives, educational partnership and open communication between educational partners are important (de Vries, 2017; de Wit, 2005).[16] Broersen and Spreij (2009) describe how this partnership could function. They see education as a joint venture between several educational partners. Teachers, parents and carers, students, counsellors, and (S)ENCo's have different, but equally important roles in this venture. Each partner has its influence on the student's development. Interaction between the partners often adds an unnoticed dimension to this development (Garces-Bacsal, 2012; Ziegler et al., 2017).[17] Therefore, in a solution-focused and ecological or systemic approach, collaboration between educational partners is essential (Caufman & van Dijk, 2009; de Shazer et al., 2012; Pameijer, 2012; van Meersbergen & de Vries, 2017).[18]

It is important for educational partners to know how their perspective on giftedness influences a student's education (Morris Miller, 2008). Helping educational partners to come to this understanding is, therefore, a basic task for

---

[12] Appendix B, Table B2, K.1.E.2.
[13] Appendix B, Table B2, K.1.F.2.
[14] Appendix B, Table B3, K.1.F.3.
[15] Appendix B, Table B5, K.1.E.5.
[16] Appendix B, Table B4, V.1.C.4.
[17] Appendix B, Table B4, K.1.F.4.
[18] Appendix B, Table B4, V.1.B.4.

the specialist in gifted education.[19] Different perspectives on giftedness each have a focus on different educational needs (Day, 2010) and offer diverse approaches as to how these needs are best met. Different educational needs require different meaningful responses (Coleman & Johnsen, 2011; Pameijer & van Beukering, 2007; van Meersbergen & de Vries, 2017). Through the lens of the educational partners, education can only be successful if there is a match between the student's educational needs and the educational programme provided. School teams need to realise that their perspective on gifted education affects their perspective of relevant educational objectives and educational programming (Morris Miller, 2008).[20] Educational partnership requires common ground (Pameijer, 2012). To avoid misunderstanding, communication between schools and parents is essential. Openly discussing where they both stand and what their perspective is of giftedness and gifted education is a prerequisite. Broersen and Spreij (2009) describe the four different levels of educational partnership (see also Chapter 3). The highest level of educational partnership is co-deciding. In The Netherlands and Flanders, schools are at liberty to decide what they set as educational objectives and how they organise educational programming in a way that is likely to reach these goals. Although parents may influence these decisions at the third level of educational partnership (thinking about and discussing together), the ultimate decision is for the school to make. In order to be able to make well-informed decisions for educational programming, the specialist in gifted education needs to know what expertise is available amongst their team members and how this expertise is used.[21] This way they will be able to support their team members to optimise gifted education[22] in a manner that is realistic and within reach of their daily practice (Wood, 2018). Approaches that are desirable, such as daily teacher-student conversations about the student's learning process (den Otter, 2015; Mayes et al., 2019)[23], can be placed within the perspective of what is doable in often hectic educational practice. In this context, D. Matthews and Foster (2009) describe their so-called "DSM": Dynamic Scaffolding Model of teacher development. This approach towards team professionalisation shows several resemblances to the approach van Gerven suggests in *Slim Beleid* (2008). It is consistent with the role of the specialist in gifted education at support level 3 and the (S)ENCo in the pyramid of support as described in Chapter 2. This specialist does not always work directly with the gifted student but might develop a support structure that enables teachers to work with the students themselves while calling on the teacher's professional knowledge and skills in a needs-based context. If the teacher requires the specialist's support, this is provided based on the teacher's own professional needs, that is, what the teacher needs to meet the needs of the gifted student.

5

[19] Appendix B, Table B1, V.1.F.1.
[20] Appendix B, Table B2, V.1.F.2; Table B4, V.1.F.4, V.1.A.4, K.1.A.4.
[21] Appendix B, Table B5, K.1.F.5.
[22] Appendix B, Table B5, V.1.F.5.
[23] Appendix B, Table B3, V.1.F.3, K.1.D.3.

5.3 Matching Paradigms with a Social-Constructivist Approach to Education

Current education in The Netherlands and Flanders stresses the concept of the social constructivist approach to education (Marzano & Kendall, 2007). The development of 21st century skills serve this approach (Bambs et al., 2016; de Bie, 2016; Marzano & Heflebower, 2012). In this didactical concept, children should learn how to become self-regulating, creative and critical thinkers who are able to solve relatively complex problems in a social context that requires them to be active participants and effective team players, with the ability to reflect on their own behaviour as the trigger for processes and results (Vermeulen & Vrieling, n.d.).

There is a discrepancy between one of the axioms of the social constructivist approach and the approach of gifted education based on the gifted child paradigm. For the social constructivist approach to education, the desire to fulfil one's potential (self-actualisation) is conditional for learning. In this approach, a growth mindset (Dweck, 2006, 2017) is needed for educational success. The gifted child paradigm, however, is, according to Dai and Chen (2014), based on a fixed mindset where innate abilities are the main requirements for educational success. In the gifted child paradigm, self-actualisation is seen as a result of education (Kreger-Silverman, 2013). Educational success is defined by the person's ability for self-actualisation. This makes the match between a social constructivist approach and gifted education based on the gifted child paradigm a more complex one.

Connecting the social constructivist approach to education to the mastery-oriented approach is less complex. There is a match between the social-constructivist approach towards education and the constructivist approach towards giftedness, as in the talent development paradigm (Sternberg, 2009; Subotnik et al., 2011). The idea that (outstanding) achievements are the result of a combination of potential and effort (Dweck, 2006, 2017), combined with taking ownership of one's development, makes it likely that a concept of giftedness based on the talent development paradigm can be implemented successfully as a school strategy for educating the gifted (Schrover, 2015).

5.3.1 How Giftedness Might Interfere with the Learning Process

Creating an optimal fit between the social constructivist educational approach and gifted students requires that the specialist in gifted education be aware of how the learning process of gifted students is perceived in the different theoretical views of giftedness (Morris Miller, 2008).[24] Knowledge and understanding of how a gifted student's personal traits influence this learning process are required to meet the needs of gifted students (Heacox & Cash, 2019; Houkema, 2017).[25] In this paragraph, we highlight specific skills that are

[24] Appendix B, Table B4, K.1.B.4.

[25] Appendix B, Table B1, K.1.B.1, K.1.A.1.

needed to be an effective learner, in a context where social-constructivism is the leading concept in the didactics that are used.
Due to the importance of executive skills in the 21st century approach, specialists in gifted education need to understand how education can interfere with the development of the executive skills of their gifted students (Scott et al., 2008; Stoeger & Ziegler, 2011).[26] All three paradigms of giftedness take into account that gifted children might develop difficulties with their executive skills during their school career. All three paradigms focus on different causes of conceivable problems. In the gifted child paradigm, asynchronous development and so-called overexcitabilities are often referred to as an explanation for a student's behaviour (Kreger-Silverman, 2013; Piechowski, 2014). In this perspective on giftedness, having difficulties in staying focused, staying task-oriented, and inhibiting dominant reactions may be caused by the so-called overexcitabilities combined with too much or too little stimuli to regulate these overexcitabilities. Consequently, an educational approach that calls highly on these executive functions may cause difficulties for the learning process, especially when there is a large intrapersonal discrepancy between intelligence and the developmental pace of executive skills (van Gerven, 2020).
In a more social-constructivist-oriented approach to giftedness, such as the talent development paradigm and the differentiation paradigm, regulation skills – which include executive skills – are considered to be the result of interaction between nature and nurture. Kroesbergen et al. (2017) argue that, although gifted children are born with sufficient executive functions, the transfer between their executive functions and the development of their executive skills might be influenced negatively by education. This transfer is only stimulated if students are challenged to work in their zone of proximal development (Vygotsky, 1978). If the curriculum offered does not call on the use of these skills, they will not develop them at the correct level for adulthood. When the curriculum offered does not match a student's abilities, but the student still achieves to a high level, these high achievements may suggest that the student's executive skills are well developed, whereas, in fact, the student hardly needed these skills to achieve at this level. This may lead to misconceptions about the student's self-regulation skills, ending up with the expectation that the student is capable of regulating their learning process whilst, in reality, they have not yet developed the required skills. These misconceptions may result in a mismatch between the student, the curriculum offered next, and the teacher's expectations of the student's ability to self-regulate the learning process.[27]
Monitoring the student's learning process and monitoring how the student responds to intervention provides the necessary insight to tune into the student's needs.[28] From this point of view, 21st century skills not only call on executive skills, but also offer the opportunity to teach a student how to take on personal responsibility for their actions and how to develop appropriate

[26] Appendix B, Table B3, K.1.B.3.
[27] Appendix B, Table B3, V.1.B.3.
[28] Appendix B, Table B5, V.1.B.5.

executive skills (Scott et al., 2008).[29] This requires frequent interaction with the student over what they think contributes to their learning and achievements.[30] Success depends on keeping in mind that the level of responsibility handed over to the student and the way that the student is coached in exercising this responsibility, should differ for each age group (Jolles, 2011). Therefore, assessment of a student's progress and achievements is not only necessary for administrative objectives but, first and foremost, necessary in helping a student to develop a personal understanding of their abilities.[31] The concept of feed forward is very helpful.[32] Feed forward shows the student what their developmental level is regarding the set objectives and what their next developmental step should be. It also provides clear suggestions of what the student should actually do to reach the set objectives (Hattie, 2013; Hattie & Yates, 2015; Wiliam, 2011).

5.3.2 Giftedness and Educational Programming

To create coherency between the student's educational needs, selecting educational objectives, and educational programming, it is essential that the specialist in gifted education can make a long-term projection about the educational programme that is to be provided (Johnsen et al., 2016).[33] Both The Netherlands and Flanders have set objectives for the end terms of primary education. These objectives can best be compared with the outline for a national curriculum. Educational programming in The Netherlands and Flanders is, therefore, primarily based on these set objectives and implemented by use of predefined coursework for all school subjects. However, schools are at liberty to implement the curriculum in whatever way they see fit for their population. Perspectives on giftedness do not influence the end terms of primary education as set by the respective governments, but might influence the strategies used to match the pacing and content of the curriculum with the educational needs of every individual gifted student (see also Chapter 7).
The current educational approach requires the ability to differentiate between the needs of individual students (Chandra Handa, 2009; Heacox & Cash, 2019; Mol-Lous, 2011; Oostdam, 2009; Roberts & Inman, 2015; van Meersbergen & de Vries, 2017). An in-depth understanding of the curriculum and the opportunities for differentiation are needed in order to tune the curriculum to the individual needs of the student.[34] There are didactical consequences to a more individual approach to providing a curriculum (Florian & Black-Hawkins, 2011; Roberts & Inman, 2015). Adjusting the curriculum means that adjustments to the supporting didactical approach will become necessary.[35] The specialist in gifted education should be aware of these consequences and should be aware of

[29] Appendix B, Table B5, V.1.E.5.
[30] Appendix B, Table B3, V.1.D.3.
[31] Appendix B, Table B3, K.1.E.3.
[32] Appendix B, Table B3, V.1.E.3.
[33] Appendix B, Table B2, K.1.A.2.
[34] Appendix B, Table B2, V.1.A.2.
[35] Appendix B, Table B2, K.1.D.2.

how these consequences may affect interventions in the education of the gifted.[36] Just like for every other student, gifted students depend on their teacher's ability to provide appropriate instructions and matching scaffolding to master learning content (Johnsen et al., 2016; Kieboom & Venderickx, 2017b; Roberts & Inman, 2015).[37] The more the regular programme and didactics match the needs of different students, the easier it will be for a teacher to respond meaningfully (Janson, 2017).

In the talent development and the differentiation paradigms, differentiation in pedagogical and didactical instructional strategies is essential in transferring from the perspective of giftedness to the perspective of gifted education. Gifted students are seen as a heterogeneous group with individual educational needs. It is accepted that it is likely that these students may not excel in all subjects at school, that they all have different talents, and that their educational needs will differ. Schools, looking at education through this lens, are more likely to focus on a more individualised trajectory for all of their students. That way, a more individualised approach for gifted students becomes integrated in the didactical-pedagogical approach teachers use. In the gifted child paradigm, it is expected that the student is able to achieve highly in all academic domains. Even though it is acknowledged that gifted students are a heterogeneous group, it is also expected that all gifted students have similar educational needs. If a school adopts this paradigm, it is likely that all gifted students will be offered the same curricular adaptations (Dai & Chen, 2014).

In all paradigms, acceleration is often suggested as a solution to meeting the individual needs of gifted learners. Research shows that, in general, acceleration – as in grade skipping – might indeed be a proper solution, but it has not been said that this intervention would fit all gifted students under all circumstances (Hoogeveen, 2017a). A careful risk assessment, not only focusing on the student's personal traits and developmental stage, but also focusing on ecological influences on the students functioning, should provide essential information to make any final decisions (Hoogeveen, 2008).[38]

Acceleration, however, does not always mean that the needs of gifted students are truly met. If there are no adjustments made regarding instructional strategies, and if there are no adjustments towards the required thinking levels for learning, the student may be offered a curriculum at an appropriate achievement level, but the curriculum may not really be challenging (D. Matthews & Foster, 2009; Morris Miller, 2008; Winstanley, 2010).

The ability to oversee long-term curricular objectives offers the opportunity to consider if and when a gifted student might be better served at a higher grade or in a different educational context (Borland, 2012; Peters et al., 2014; D. Matthews & Foster, 2009). One does not have to exclude the other. Sometimes, to meet the needs of individual students, a special programme, a part-time class for gifted education, and acceleration, should go hand in hand.

[36] Appendix B, Table B2, K.1.B.2.
[37] Appendix B, Table B2, V.1.B.2, V.1.E.2.
[38] Appendix B, Table B5, K.1.A.5, V.1.A.5.

However, under different circumstances, a different decision may be made (Borland, 2005; D. Matthews & Foster, 2009) (see also Chapter 8).

5.3.3 Giftedness, a Social Constructivist Didactic Approach and the Call on Interpersonal Skills

Teaching based on didactical strategies to develop 21st century skills, means an increasing focus on opportunities for cooperation and collaboration amongst students during the learning process (Marzano & Heflebower, 2012). Cooperative and collaborative learning in itself is nothing new in education. However, the explicit call made by the social constructivist approach to being a team player, shows that educational success no longer depends on the teacher alone, but also on the quality of cooperation and collaboration amongst students (Chandra Handa, 2009; Marzano & Heflebower, 2012; Voogt & Pareja Roblin, 2010). That quality is affected by the social climate in the classroom, interpersonal skills, and the opportunity to learn with like-minded students or students with equal levels of experience and achievements.

The social classroom climate influences a student's attitude toward learning (T. Cross, 2018; D. Matthews & Foster, 2009; Moon, 2009b). Understanding how age-peers and ability-peers can be of influence on an individual gifted student's developmental process is therefore essential (Riedl-Cross & Cross, 2017).[39] Social codes can enhance development or they can interfere negatively with a child's further development (D. Matthews & Foster, 2009). Unspoken social rules amongst students can be complicated to understand and even more complicated to live up to. Students need to develop an antenna for these rules as well as needing to learn how to balance when to live up to the rules and when to set them aside. The only way to learn this is in a community of peers. Hence, all children need peers to mirror their own behaviour as well as to stimulate development and achievement (Rinn, 2018).[40]

Provoking social interaction among students should, therefore, be a natural part of teaching (Janson, 2017). A stimulating learning environment based on social constructivism involves opportunities for students to learn from each other as much as it offers children space to work on their own (Marzano & Heflebower, 2012; D. Matthews & Foster, 2009; Riley, 2011; Tomlinson et al., 2009).[41] Interaction offers students the possibility of checking their ideas, and to match and stretch their knowledge and experiences with those of other students (D. Matthews & Foster, 2009). It is necessary for students to meet their academic equals if we want them to learn from each other (Heacox & Cash, 2019; Hertzog, 2017a).[42]

Intensive interaction during the learning process calls highly on a student's social skills. In an inclusive environment, the call on a student's socials skills

[39] Appendix B, Table B1, K.1.C.1.
[40] Appendix B, Table B1, V.1.C.1.
[41] Appendix B, Table B2, V.1.C.2, K.1.C.2.
[42] Appendix B, Table B3, V.1.C.3.

becomes even more demanding due to the heterogeneity of the group of students.
Research is inconclusive when it comes to the social-emotional development of gifted students (T. Cross, 2018; Lavrijssen & Verschueren, 2019; Rinn & Majority, 2018). Social relationships with age-peers might be complicated for some gifted students. However, there is not sufficient evidence that all gifted students are likely to experience more social and emotional problems than non-gifted children (T. Cross, 2018; Freeman, 2019; Lavrijssen & Verschueren, 2019; D. Matthews & Foster, 2009; Peterson, 2009; Rinn, 2018).[43] Any social and emotional issues gifted students might have to deal with are based on the interaction of the student with their ecological system (T. Cross, 2018; Rinn, 2018). Interplay between students and their ecological system influences their development. This emphasises the need for both a knowledge of risk and resiliency factors and an in-depth understanding of this specific interaction (van Meersbergen & de Vries, 2017; Ziegler et al., 2017).[44] It also emphasises the need for specialists in gifted education to be able to create a wide variety of educational experiences that require the explicit use of social skills.[45] van Gerven and Hoogenberg (2011) described various opportunities for a social constructivist approach, including in-classroom activities, pull-out programmes, and working with *hotspots* and *learning landscapes*. This wide range of opportunities offers the chance for students to group themselves based on like-minded interests or equal levels of experience and achievements (Little, 2017). In so doing, challenging learning opportunities become available for a wide range of students, including gifted learners (Janson, 2017).[46]

5.3.4 Gifted Education and Intrapersonal Understanding

Competency, autonomy, and relationship are basic pedagogical-psychological needs for every human being (Deci & Ryan, 2000). Gifted children form no exception when it comes to these needs.[47] These basic needs underpin a pedagogical approach where it is relevant for children to develop an intrapersonal understanding of their strengths and weaknesses.[48] This understanding does not develop spontaneously, it needs careful coaching, and it is an ongoing and complex process (Moon, 2009a, 2009b). The influence of all the (f)actors of the ecological system plays a big part in self-concept.[49] The specialist in gifted education should be able to analyse how, as an ecological factor, the curriculum in general and specific assignments in particular, influence the student's emotions during the learning process. These emotions influence the student's wellbeing at school.[50] Teaching the student how to cope with

[43] Appendix B, Table B5, K.1.C.5.
[44] Appendix B, Table B1, K.1.D.1.
[45] Appendix B, Table B3, K.1.C.3.
[46] Appendix B, Table B3, V.1.A.3.
[47] Appendix B, Table B1, K.1.E.1.
[48] Appendix B, Table B1, V.1.A.1; Table B4, V.1.D.4.
[49] Appendix B, Table B4, K.1.E.4.
[50] Appendix B, Table B5, K.1.D.5.

success and failure is part of their job. It is important for gifted children to learn how to control their emotions regarding success and failure and how to vent their emotions in a socially acceptable way (Dweck, 2006, 2017). There is nothing wrong with being disappointed when one is not successful. However, there is something wrong if disappointment leads to meltdowns or avoiding challenges entirely. Reflective conversations, where process and achievements are interrelatedly discussed, help gifted children to develop a realistic global-self-concept (Kaplan et al., 2019).[51]
Personal self-concept differs between domains (Siegle & McCoach, 2017). In a domain based on strengths, there is likely to be a positive self-concept. In a domain of relative weakness, there is likely to be a more nuanced self-perspective on personal abilities. For gifted children, it can be difficult to understand why their strength in one domain does not automatically mean that they have the same strength in other domains. This applies especially when the child is relatively late in being confronted with limitations to their capabilities due to the lack of realistic learning experiences (Dweck, 2006, 2017). A global-self-concept is the most important predictor of emotional wellbeing (D. Matthews & Foster, 2009). Both a student's wellbeing and a student's self-concept are influenced by interaction with others during the learning process. To acquire a well-balanced global self-concept means that students need to understand how they can manipulate their development personally (Dweck, 2006, 2017; Kaplan, 2019). This understanding can be stimulated through regular student-teacher conversations reflecting on the learning process, as much as through learning results and consideration of the student's wellbeing.[52] These teacher-student conversations offer great opportunities to explore the student's wellbeing[53] and to explore how the student responds to the intervention strategies chosen.[54] At the same time, stimulating a student's awareness of risk and resiliency comes almost naturally during these conversations, and this contributes to a well-balanced self-concept (Baum et al., 2017; Rinn & Majority, 2018; Trail, 2011).[55]
Having reflective conversations with students requires that the specialist in gifted education has well-trained communication skills (Caufman & van Dijk, 2009; de Shazer et al., 2012; Wood, 2018).[56] According to Wood (2018) a solution-focused strategy in communication with gifted students offers ample strategies for calling on their strengths. For example, in this approach, imagining the future by answering "the magic question" requires great creativity. Ownership and autonomy are handed to the student, and the process of reaching future goals is translated into learnable skills (Caufman & van Dijk, 2009; de Shazer et al., 2012). In the change and solution-focused approach, communication with all the actors in the student's ecology is seen as essential (Pameijer & van Beukering, 2007; van Meersbergen & de Vries, 2017). The

[51] Appendix B, Table B2, V.1.D.2.
[52] Appendix B, Table B1, V.1.B.1, V.1.C.1.
[53] Appendix B, Table B1, V.1.D.1.
[54] Appendix B, Table B5, V.1.D.5.
[55] Appendix B, Table B4, V.1.E.4.
[56] Appendix B, Table B4, K.1.D.4.

student's social-emotional wellbeing partially depends on a solid relationship between the ecological partners. Open communication is an important strategy in building this relationship (Rinn, 2018).[57]
Global self-concept is positively influenced when students learn how to take responsibility for their actions.[58] Internal attribution for success and failure stimulates an attitude towards learning based on resilience and the courage for well-calculated risk-taking. Whilst the gifted child paradigm adopts an approach where success in education is mainly due to natural abilities, both the talent development paradigm and the differentiation paradigm focus on the influence the child has on their own personal development (Rinn & Majority, 2018).

5.4 Five Competencies in Domain 1

In this chapter, we focused on interaction between theoretical knowledge of concepts of giftedness and the five generic teacher competencies. Systematic analysis of the literature, where both topics were addressed together, provided each *connection* to the matrix, with a specifically defined competency (see Table 5). Although the competencies are each stipulated in their own right, they comply with the six dimensions as defined by Merriënboer et al. (2002).
The theoretical review study underpins 60 indicators that were identified for domain 1, 12 for each of these competencies (see Appendix B). Each indicator should be regarded as the underpinning for the competency on the one hand, and on the other hand as an example of putting theory into practice. As the first domain of competencies for the specialist in gifted education merely sets the outline for the following domains, it shows how current knowledge and skills affect the indicators of the following domains. Five out of the six dimensions, as identified by Merriënboer et al. (2002), apply to these indicators, which show that these indicators cannot apply without each other.
The specialist in gifted education defines their professional perspective on giftedness. They should be able to translate this theoretical perspective into identifiable knowledge and skills that will underpin their actions. The indicators are influenced by the social constructs of inclusion and diversity, and strongly connected with current educational paradigms such as the change and solution-focused approach and social-constructivist didactical strategies.
The so-constructed theoretical framework was presented to stakeholders in gifted education. The next paragraph in this chapter presents the results of their response to this framework.

5.5 Selection Rates of Indicators for Competency Domain 1

In the literature, researchers stressed the need for specialists in gifted education to have up-to-date knowledge of the different theories of giftedness (van Gerven & Hoogenberg, 2011). This first competency domain is based on this suggestion.

[57] Appendix B, Table B4, K.1.C.4.
[58] Appendix B, Table B1, V.1.E.1.

Table 5

Domain 1, From Appendix A: Competency Matrix for Specialists in Gifted Education

Generic Teaching Competencies	1. Concepts of Giftedness
	The specialist has an in-depth knowledge of the theoretical perspectives of giftedness and gifted education and is able to interpret this knowledge contextually in their professional practice.
1. Pedagogical The teacher creates a safe and stimulating learning environment for every student, tuned to their individual educational needs.	***1.1.*** The specialist in gifted education understands and envisions how their perspective on giftedness influences their ability to see, understand and respond to the pedagogical needs of gifted children.
2. Didactical In collaboration with their colleagues, the teacher transfers curricular objectives in such a way that they become teachable in the context of a primary school, matching school policy. They must compose the curriculum in a manner that enables their student to master the content.	***1.2.*** The specialist creates a coherency between the concepts of giftedness, educational objectives and educational programming and the intervention strategies to be used.
3. Managerial The teacher provides a structured, well-organised, task-oriented classroom atmosphere that supports the students' developmental process.	***1.3.*** The specialist in gifted education understands how their perspective of giftedness influences the organisation of the educational programme provided.
4. Communication The teacher communicates professionally with all the educational partners who are involved with the student's education.	***1.4.*** Based on interaction between the educational partners, the specialist in gifted education develops a perspective of giftedness, and an approach to gifted students that matches their individual needs.
5. Research and innovation The teacher adjusts their professional knowledge, perspectives, and actions if and when relevant developments in education, society, and science require it.	***1.5.*** The specialist in gifted education relates their in-depth knowledge of giftedness to perspectives of current society in general and more specifically to the current perspectives on education.

The empirical research focused on what stakeholders consider to be the most representative indicators for competent professional behaviour in the specialist in gifted education.[59] These indicators are, as explained in Chapter 4, divided into 6 categories.

A. Stimulating academic development
B. Stimulating the use of different cognitive styles
C. Stimulating interpersonal functioning
D. Stimulating intrapersonal functioning
E. Stimulating self-actualisation
F. Generic counselling and advisory skills matching the skills of a (S)ENCo.

It is important to realise that the indicators were presented to stakeholders in the context of the matching competencies (see Appendix B). This means that stakeholders made a selection of indicators based within a mental framework of one specific competency at a time. Between the selected indicators, stakeholders were not asked to rank their selection. Any ranking that is presented is the result of the frequency of selection of an indicator by stakeholders.[60]
In this chapter, an analysis of differences in choices between the separate groups of stakeholders is presented first.[61] Relevant differences[62] between groups of stakeholders are identified per competency. This provides an understanding at a detailed level of what each group considers imperative knowledge and skills for specialists in gifted education, and indirectly what they think to be important in gifted education.
Secondly, significant differences between stakeholders in their selection rates of categories of indicators are identified and presented. Although all stakeholders take the education of gifted students to heart, it was assumed that they would differ to a certain extent in their opinion of the best way to care for gifted education.
Thirdly, the analyses identified common ground amongst the different groups of stakeholders. The identification of common ground provides ample opportunities for the underpinning of a solid curriculum for the continuous professional development of specialists in gifted education.

5.5.1 Selection Rates per Group of Stakeholders

As the empirical part of the research aimed to identify what was identified by stakeholders as relevant indicators for competent professional behaviour, results are presented for each group of stakeholders separately.

---

59 The specialist has an in-depth knowledge of the theoretical perspectives of giftedness and gifted education and is able to interpret this knowledge contextually in their professional practice.
60 In Chapter 4, the procedure and statistical test used for data analyses are described.
61 The structure of data presentation will be identical in Chapters 5-8.
62 These are differences between two or more groups of stakeholders that suggest strongly that both groups of stakeholders have a different opinion but that this difference is not statistically significant. There are two thresholds for relevant differences. The first one is set for indicators that have the highest selection rate within one group of stakeholders but that did not end up in the top three selection rates by one or more of the other groups of stakeholders. The second threshold is set for indicators that have the lowest selection rate within one group of stakeholders and did end up in the top three selection rates by one or more of the other groups of stakeholders (see Chapter 4).

5.5.1.1 Educational Professionals

Results per Competency

- Competency 1.1 aims for the specialist in gifted education to understand and envision how their perspective on giftedness influences their ability to see, understand, and respond to the pedagogical needs of gifted children. Educational professionals highly valued a specialist's knowledge and skills regarding risk factors influencing the student's wellbeing (23.41%, Table E1; 21.37%, Table E10).[63] Theory suggests that it is considered to be a risk factor if students have little understanding of their abilities (Dweck, 2006; D. Matthews & Foster, 2009; Moon, 2009b). Educational professionals showed a high selection rate for the indicator that the specialist in gifted education should have the skills to stimulate the gifted students' understanding of their abilities (23.41%).[64]
- Competency 1.2 aims for coherency between the concepts of giftedness, educational objectives and educational programming and the intervention strategies to be used. Dai and Chen (2014) suggest that knowledge of different perspectives of giftedness has implications for how each perspective focuses differently on the ultimate educational objectives. This knowledge indicator, however, showed a low selection rate amongst educational professionals (8.14%).[65] Morris Miller (2008) pointed out the relevancy of schools setting educational objectives that are fit for the school's perspective on giftedness.[66] Of all the skills indicators for competency 1.2, educational professionals valued this indicator the least (10.18%).
- Competency 1.3 focuses on understanding how a perspective on giftedness influences the organisation of the educational programme that is provided (Day, 2010; Little, 2017; Morris Miller, 2008). The indicator with the highest selection rate in this competency, K.1.B.3, was that aimed at the specialists' understanding of the importance of their students' capacity to use executive skills (25.7%).[67] According to Kroesbergen et al. (2017), this knowledge is relevant in helping educational professionals when setting up interventions that match the educational needs of gifted learners within the classroom context. The matching skills indicator for K.1.B.3 was V.1.B.3. This indicator showed the lowest selection rate in this domain[68] amongst educational professionals (12.47%). These stakeholders considered the skill to provide effective feed forward to be a better example of competent behaviour (18.32%).[69] This skill was more specifically articulated, and,

---

[63] K.1.D.1 - Table E1; V.1.D.5 - Table E10
[64] V.1.A.1 - Table E6
[65] K.1.F.2 - Table E2
[66] V.1.F.2 - Table E7
[67] K.1.B.3 - Table E3
[68] V.1.B.3 - Table E8
[69] V.1.E.3 - Table E8

therefore, likely to be a clearer example of the translation from knowledge to skill.

- Competency 1.4 focuses on the interaction of educational partners and the development of a joint perspective on giftedness. In the context of this competency, educational professionals thought it was important that specialists in gifted education use their knowledge of giftedness to set up open communication with educational partners. The results of this communication should substantiate the organisation of gifted education (25.95% / 26.97%).[70] This matches the general ideas on educational partnership in The Netherlands and Flanders. Parents are considered as relevant partners in education. Hence, schools value their input for the education of their children (Broersen & Spreij, 2009; Pameijer, 2012). This input can be used to select interventions that are a match with the student's needs (de Bruin-de Boer & Kuipers, 2011; Houkema, 2009; Pameijer & van Beukering, 2007; van Gerven & Drent, 2012; van Meersbergen & de Vries, 2017). The fact that for the same competency (1.4) the skills indicator V.1.F.4 "creates a joint perspective on giftedness between the educational partners"[71] showed the lowest selection rates (0.25%), does not imply per se that educational professionals contradict themselves when it comes to educational partnership. The indicator showing the highest selection rate, V.1.C.4,[72] is more explicit than indicator V.1.F.4, and likely to be considered as a better example.
- Competency 1.5 aims for an in-depth development of knowledge and understanding to create meaningful connections between different perspectives of giftedness and current educational and social paradigms. This connection ensures that the educational approach of gifted learners is embedded into general education. Educational professionals articulated its relevancy for the specialist investing in personal continuous professional development (25.19%).[73] What was remarkable was that educational professionals showed that their highest selection rate for skills indicators matching this competency aimed for a strong focus on the student's wellbeing (21.37%)[74] but that their knowledge of different theoretical perspectives on the social functioning of gifted students showed a relatively low selection rate (11.96%).[75] In this case, a more generic indicator was found to be more important than a very specific indicator.

[70] K.1.A.4 - Table E4; V.1.C.4 - Table E9
[71] V.1.F.4 - Table E9
[72] V.1.C.4 - Table E9
[73] K.1.B.5 - Table E5
[74] V.1.D.5 - Table E10
[75] K.1.C.5 - Table E5

Overall Results for Educational Professionals for Domain 1

Comparing the selection rates for all individual indicators showed that educational professionals considered it important for the specialist in gifted education to know what different didactical strategies there are to challenge gifted students in their zone of proximal development.[76] This indicator has the highest selection rates of all indicators that they selected in competency domain 1 (28.5% of all Educational Professionals selected this indicator, Table E11). There were two indicators that showed an unexpectedly low selection rate. These indicators had to do with the organisation of peer group education.

- V.1.C.1 "Stimulates gifted students to interact with gifted peers" (7.63%).[77]
- K.1.C.3 "Knows that there are different strategies for cooperative and collaborative learning for peer group education" (9.92%).[78]

Given the huge number of publications in professional magazines for teachers that stress the need for peer group education as extremely relevant for the wellbeing of gifted learners, these low selection rates were unexpected.[79] Skills indicator V.1.C.3 "Deliberately organises interaction between gifted students of an equal level to stimulate their learning process" showed a far higher selection rate of 17.81%. As a result of this discrepancy, special attention was paid to the selection rates for the same indicators amongst other groups of stakeholders.

5.5.1.2 Specialists in Gifted Education

The selection of indicators by the specialists in gifted education and educational professionals showed several similarities as can be identified in Appendix E Tables E1-E11 and Tables E36-E37.

Results per Competency

- For competency 1.1, the two indicators showing the highest selection rates amongst specialists in gifted education are "Knows which risk factors might influence a student's wellbeing" (25.57%)[80] and "Makes gifted students responsible for their development, matching their individual ability to take upon this responsibility" (27.51%)[81]. These selection rates match the general teacher competencies (pedagogical

---

[76] K.1.D.2 - Table E2
[77] V.1.C.1 - Table E6
[78] K.1.C.3 - Table E3
[79] Between 2010-2020 The Dutch magazines on giftedness published on average 1 article regarding peer group education in each issue. The Dutch professional magazines on education in general published on average 5 articles on gifted education per year. At least one of these articles focuses on peer group education. In 2018 at council level, 44.9% of all school councils organised peer group education at council level (Houkema et al., 2018). There are no current numbers about the amount of individual schools that organise peer group education.
[80] K.1.D.1 - Table E1
[81] V.1.E.1 - Table E6

knowledge and skills) that are set by the Dutch and Flemish governments in their professional standards (see Chapter 3). Identically to the selection rates of educational professionals, indicators aiming for knowledge and skills regarding the necessity of peer group education showed the lowest selection rates.[82]

- Competency 1.2 combines didactical knowledge and skills with theories of giftedness. Two indicators showed high selection rates. Knowledge indicator K.1.D.2, aiming for the knowledge of different didactical strategies to challenge gifted students (28.48%)[83], and skills indicator V.1.E.2, aiming for the specialist to develop the skills to select interventions matching to their students' needs (28.16%).[84]
  For this competency, the lowest selection rate amongst specialists in gifted education was for the more generic indicators in category F, focusing on the understanding of the relationship between perspectives on giftedness and educational objectives[85], and in category C, the organisation of cooperative and collaborative learning (7.44%).[86]
- Competency 1.3 focuses on the relationship between theories of giftedness and classroom management and organisation. According to Kuipers (2017), teachers should provide a structured, well-organised, and task-oriented classroom atmosphere that supports the students' developmental process. The knowledge of how a personal perspective of giftedness influences the selection of students who are offered an adjusted curriculum (D. Matthews & Foster, 2009), was found to be important by specialists in gifted education (24.60%).[87] Including the daily provision of feed forward for gifted students on their learning process was considered as valuable as well (20.06%).[88]
- Competency 1.4 considers communication with educational partners in combination with theories on giftedness. Identically to the other groups of stakeholders, specialists in gifted education showed a high selection rate in aiming for a strong educational partnership based on open communication (28.16%).[89] Theoretical perspectives on the learning process (3.88%)[90] were not as highly valued as a joint perspective by the educational partners (27.51%).[91] The extremely low selection rate for indicator V.1.F.4 (0.97%)[92] is likely to be caused by stakeholders having already selected a knowledge indicator for this same competency with an identical focus.

[82] K.1.C.1 - 8.74%; V.1.C.1 - 4.21%, Table E1.
[83] K.1.D.2 - Table E2
[84] V.1.E.2 - Table E7
[85] K.1.F.2 - Table E2
[86] V.1.C.2 - Table E7
[87] K.1.A.3 - Table E3
[88] V.1.E.3 - Table E8
[89] V.1.C.4 - Table E9
[90] K.1.B.4 - Table E4
[91] K.1.A.4 - Table E4
[92] V.1.F.4 - Table E9

- Competency 1.5 combines professional development with theories on giftedness. Like the educational professionals, specialists in gifted education showed the highest selection rate for the indicator regarding personal continuous professional development regarding theories of giftedness (28.48%).[93] Interestingly specialists in gifted education considered it important that they should be able to offer feedback to their colleagues to improve the quality of gifted education (25.89%).[94] For this competency, they thought that the knowledge of different theoretical perspectives on the social functioning of gifted students was less relevant. This was also the case for the ability to create a risk analysis considering the pros and cons of acceleration of individual students.

Overall Results for Specialists in Gifted Education for Domain 1

Two indicators both showed the highest selection rate in this domain. Identically to educational professionals, specialists in gifted education highly valued the knowledge of didactical strategies to challenge gifted students in their zone of proximal development, with an identical selection rate (28.48%) for the indicator about aiming at investing in personal continuous professional development around theories of giftedness.[95]
The previously mentioned low selection rates amongst educational professionals regarding the knowledge and skills related to peer group education were unexpected. Their selection rates were compared with those amongst specialists in gifted education. These rates were found to be relatively low (K.1.C.3 7.12%[96]; V.1.C.1 4.21%[97]; V.1.C.3 17.15%[98]). These rates suggest that the first two indicators were again not considered to be the most relevant indicators for the competency with which they were matched.
There was one relevant difference observed in the selection of indicators between specialists in gifted education and educational professionals. For competency 1.1, the highest selection rate for an indicator for educational professionals was V.1.A.1: Stimulates the student's understanding of their personal ability. Amongst specialists in gifted education, this indicator did not end up in the top three selected indicators for this competency. There was a slightly higher selection rate for indicator V.1.F.1, aiming for the ability to explain to educational partners how their perspective of giftedness influences their choices for the education of a gifted child. This choice matches the role of the specialist in gifted education working at level 3.

---

[93] K.1.B.5 - Table E5
[94] V.1.F.5 - Table E10
[95] K.1.D.2, K.1.B.5 - Table E11
[96] K.1.C.3 - Table E3
[97] V.1.C.1 - Table E6
[98] V.1.C.3 - Table E8

5.5.1.3 Parents

The third group of stakeholders whose selection rates for domain 1 were analysed are parents.

Results per Competency

- For competency 1.1, both the highest and lowest selection rates of parents matched the choices for indicators of educational professionals. There was a relevant difference between selection rates of parents and the specialist in gifted education. This difference was identical to the difference found between educational professionals and specialists in gifted education.
  As seen with the educational professionals, parents thought the indicator aiming for students' understanding of their personal abilities to be the best example of professional behaviour. Indicator V.1.A.1 showed a selection rate of 24.10%.[99]
- For competency 1.2 the highest selection rates amongst parents were for the same indicators as both educational professionals and specialists in gifted education.[100] The lowest selection rate for the knowledge indicator was also identical to the selection of educational professionals and specialists in gifted education.[101] The lowest selection rate for the skills indicator was that which focused on setting educational objectives that fit the school's perspective of giftedness (4.82%).[102] Considering the high value for indicators that were aimed at educational partnership, these selection rates were not unexpected. After all, this indicator may suggest that in the end educational partnership is a farce.
- Analysis of selection rates for competency 1.3 showed that parents highly valued the specialist paying attention to the gifted student's executive skills (22.49%).[103] "Learning how to learn" is an important topic amongst parents and teachers in both The Netherlands and Flanders (see also Chapters 2, 6, 7, and 8). Within this context, the parents' high selection rates of indicator V.1.D.3 (25.70%)[104], aiming for the skills to discuss with students what they think might contribute positively to the effectiveness of their learning process, is understandable (Baum et al., 2017; Caufman & van Dijk, 2009; Rinn & Majority, 2018; Trail, 2011).[105] This selection also matches the strive for voicing as a pedagogical strategy as discussed in Chapter 3. Indicators that were least selected by parents focused on the specialists' understanding of how a specialist's perspective on giftedness influences

[99] K.1.D.1 – Table E1; V.1.A.1 – Table E6
[100] K.1.D.2 – Table E2; V.1.E.2 – Table E7
[101] K.1.F.2 – Table E2
[102] V.1.F.2 – Table E7
[103] K.1.B.3 – Table E3
[104] V.1.D.3 – Table E8
[105] V.1.D.3 – Table E8

the selection criterion for students being offered an adjusted curriculum.[106] The indicator to ensure that the curriculum challenges all students in the classroom showed equally low selection rates.

- For competency 1.4, the prioritising of the highest and lowest rates of knowledge indicators were identical for all groups (see Appendix E, Table E4). Considering the skills indicators, we see small differences. Parents valued the opportunity to contribute to their child's education highly. This matches the Dutch and Flemish perspective on educational partnership (Pameijer, 2012). This indicator got their highest selection rate (26.51%).[107] In line with the parents' choice of the indicators on voicing and intrapersonal understanding, indicator V.1.D.4 (20.48%), aiming for a dialogue with gifted students to stimulate the development of a realistic self-concept was often selected. This reveals a relevant difference between parents and other groups of stakeholders. Amongst educational professionals and specialists in gifted education Indicator V.1.D.4 was not selected often enough to end up in their top 3 for this competency.
- For competency 1.5, the highest selection rate amongst parents was for the indicator where specialists in gifted education know how to monitor the gifted student's wellbeing (22.89%).[108] For parents, their child's wellbeing is often the signal of whether education still matches their child's needs. Hence, their focus matches their role as parents. The highest scoring skills indicator selected aims for a schoolwide high quality of gifted education (22.49%).[109] The lowest selection rates of the knowledge indicators focused on the specialists' awareness of the different theoretical perspectives of the social functioning of gifted students (9.24%).[110] Skills indicator V.1.B.5 about monitoring the effect how didactical and pedagogical strategies influence the students' learning process may have been too specific for parents (12.85%).[111] Parents do value high quality in education, but again, due to their role as parents, they may not focus on the details of which the impact on their child may not be directly clear to them.

Overall Results for Parents for Domain 1

As seen with educational professionals, the overall highest selection rate by parents, in this first domain, was for the specialist's competency to use different instructional strategies to challenge gifted students in their zone of proximal development (30.12%).[112]

[106] K.1.A.3 (12.45%) – Table E3; V.1.A.3 (6.45%) – Table E8
[107] V.1.B.4 – Table E9
[108] K.1.D.5 – Table E5
[109] V.1.F.5 – Table E10
[110] K.1.C.5 – Table E5
[111] V.1.B.5 – Table E10
[112] K.1.D.2 – Table E11

Identically to the selection rates amongst educational professionals and specialists in gifted education, the indicator aiming for knowledge and skills about peer group education, did not end up in the top three rates for parents. Although the selection rate for indicator V.1.C.3 was almost identical (17.67%)[113], this selection rate was still not high enough to end up amongst the three most selected indicators for parents for competency 3. These results contradict educational practice, where, during conferences and teacher education courses, teachers report that parents consider the opportunity to participate in special programmes and peer group education very important. There were relevant differences noticed between parents and educational professionals and parents and specialists in gifted education regarding competency 1.3 (see Appendix E, Table E3). Parents showed the lowest selection rate for indicator K.1.A.3 (Knows how their perspective on giftedness influences the selection of students who are offered an adjusted curriculum). The same indicator ended up amongst the top three most selected indicators for educational professionals and specialists in gifted education. Instead of this indicator, parents selected for this competency an indicator aiming for communication strategies with gifted students that are helpful in coaching processes (19.68%).[114]

A second relevant difference was observed for skills indicator V.1.D.3. Amongst parents, this indicator showed the highest selection rate while the same indicator did not end up amongst the top three selection rates for specialists in gifted education. Instead, the specialists selected an indicator that aimed to develop classroom interventions within realistic limits in their daily practice. This is where it becomes clear that different groups of stakeholders have different interests that run concurrently with the role the stakeholders play in the education and upbringing of gifted students (see also Chapter 4) (Broersen & Spreij, 2009; Freeman, 1987; Mitchel et al., 1997).

5.5.2 Significant Differences Between Groups of Stakeholders in Domain 1

The ANOVA test and Bonferroni's Multiple Comparisons Test were used for the analyses of differences between the groups of stakeholders. Significant differences between the groups of stakeholders were found when selection rates per category of indicators matching domain 1 were identified. These differences apply to knowledge indicators in the categories A ($p = .009$), B ($p = .002$), C ($p = .007$) and D ($p = < .001$), (see Appendix E, Table E12; see also Figure 14).[115]

Parents made significantly different choices compared to the other groups of stakeholders (see Appendix E, Tables E12-E14). They were more focused on knowledge indicators regarding the stimulation of interpersonal and

[113] V.1.C.3 – Table E8

[114] K.1.D.3 – Table E3

[115] A: Stimulating academic development | B: Stimulating the use of different cognitive styles | C: Stimulating interpersonal functioning | D: Stimulating intrapersonal functioning.

Figure 14

Comparison of Selection Rates of the Knowledge Indicators in Domain 1 Between Groups of Stakeholders - (see also Table E12)

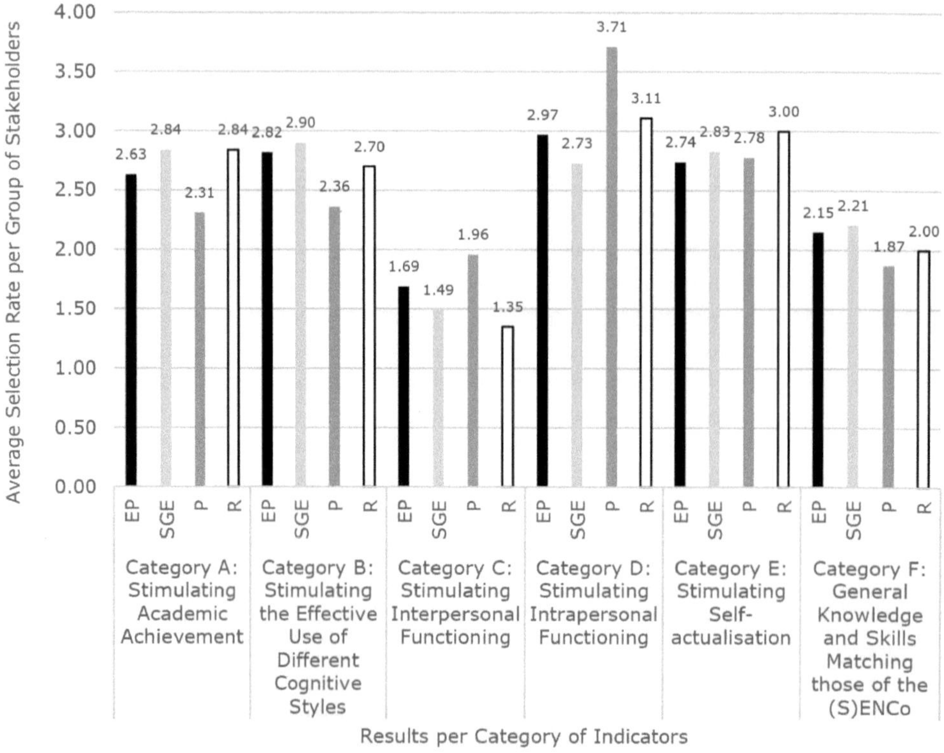

Note. N = 354 divided into four groups of stakeholders: EP = Educational Professionals, SGE = Specialists in Gifted Education, P = Parents, R = Rest. EP: N = 131, SGE: N = 103, P: N = 83, R: N = 37.

Test value = 2.5

intrapersonal functioning. Their selection rate for C-indicators showed a significant difference compared to the selection rate of specialists in gifted education ($p = .017$). The parents' selection rates of D-indicators differed significantly from both educational professionals ($p < .001$) and specialists in gifted education ($p < .001$). Compared to specialists in gifted education, parents were less focused on selecting knowledge indicators that stimulated academic development ($p = .008$). This was also the case for E-indicators[116], with the aim of effective use of different cognitive styles in the learning process. Both observed differences are significant ($p = .002$; $p = .009$).
Differences between the stakeholders were also found in the skills indicators.

[116] E: Stimulating self-actualisation

Figure 15

Comparison of Selection Rates of the Skills Indicators in Domain 1 Between Groups of Stakeholders - (see also Table E15)

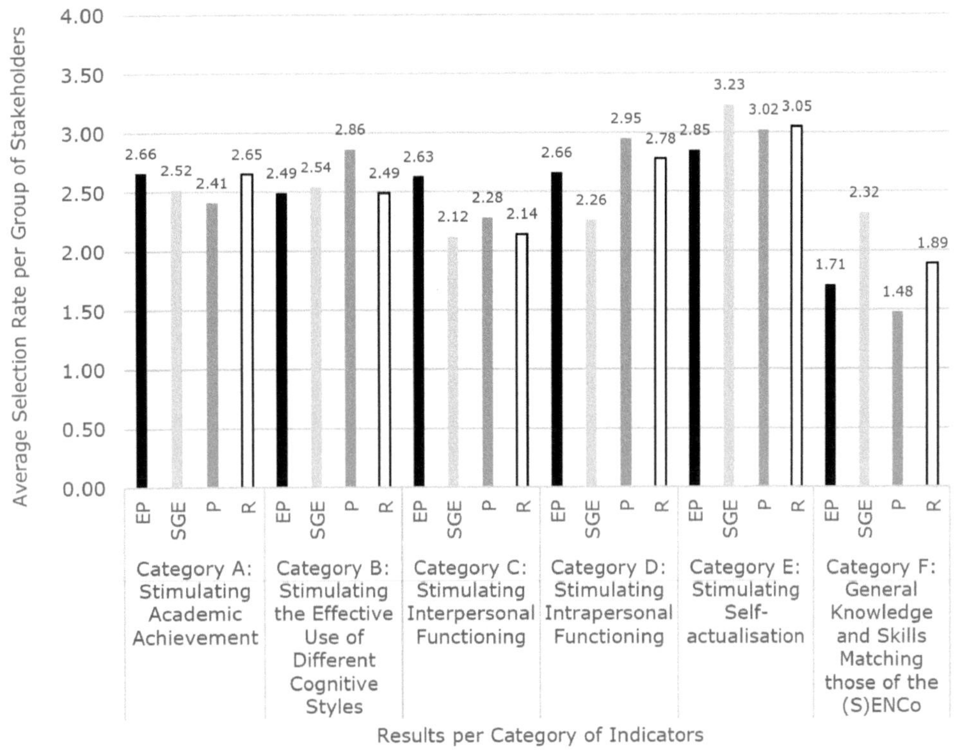

Note. N = 354 divided into four groups of stakeholders: EP = Educational Professionals, SGE = Specialists in Gifted Education, P = Parents, R = Rest. EP: N = 131, SGE: N = 103, P: N = 83, R: N = 37.

Test value = 2.5

Differences were identified for categories C ($p = .004$), D ($p = .001$), E ($p = .045$) and F[117] ($p < .001$) (see Appendix E, Tables E15-17; see also Figure 15). In category C, specialists in gifted education selected less skills indicators aimed at stimulating their students' functioning in interpersonal relations, than did educational professionals ($p = .004$). In category E, focusing on the stimulation of their students' ability for self-actualisation, specialists in gifted education had a higher selection rate than educational professionals ($p = .028$). Specialists in gifted education selected more indicators in category F, the generic counselling and advisory skills matching the skills of a (S)ENCo, than parents did ($p < .001$).

[117] F: Generic counselling and advisory skills matching the skills of a (S)ENCo

5.5.3 Common Ground Between Groups of Stakeholders in Domain 1 Common Ground Amongst all Groups

Initial analysis of common ground aimed to identify the common ground between all groups of stakeholders. Two frequency analyses were used. The primary frequencies analysis showed that there were 10 knowledge indicators and 8 skills indicators that ended up amongst the three most selected indicators between all groups of stakeholders (see Appendix E, Tables E18-E19).[118] This selection of 18 indicators represented five different categories:

A. Stimulating academic development (3)
B. Stimulating the use of different cognitive styles (6)
D. Stimulating intrapersonal understanding (3)
E. Stimulating self-actualisation (5)
F. Generic counselling and advisory skills matching the skills of a (S)ENCo (1).

The secondary frequencies analysis was aimed at determining if, amongst these 18 indicators, there were indicators that had the highest selection rates between all groups. There were no skills indicators found meeting that criterion. However, three knowledge indicators that had the highest selection rate amongst all groups were identified (see Appendix E, Table E29). These indicators aimed for knowledge to stimulate gifted students' intrapersonal understanding and academic development. The indicators matched:

- competency 1.1
    - Knows which risk factors might influence a student's wellbeing.[119]
- competency 1.2
    - Knows different didactical strategies to challenge gifted students in their zone of proximal development.[120]
- competency 1.4
    - Knows why it is important to organise gifted education from the joint perspective of the educational partners.[121]

Remarkably, indicators regarding interpersonal functioning were never high enough to end up amongst the indicators with the highest selection rates for all stakeholders for the same competency. Using a One-Samples Statistics test and a One Sample test, it was found that, as a category, these so-called C-indicators were selected significantly less for the entire domain once there was no differentiation between the separate groups of stakeholders. The knowledge indicators C had a selection rate of $\overline{X}$ 1.66. The difference, -0.842, to the expected rate ($\mu = 2.5$) is significant, $t(353) = -14.458$, $p < .001$, the effect

[118] The analysis at competency and indicator level was used to explore common ground between stakeholders. Per competency, it was determined which three indicators belonged to the three most frequently selected indicators by all groups of stakeholders. This has been done separately for the knowledge and the skills indicators.
[119] K.1.D.1 – Table E29
[120] K.1.D.2 – Table E29
[121] K.1.A.4 – Table E29

size represents a practical significance, d = .77 (see Appendix E, Tables E20-E21; see also Figure 16). The skills indicators C had a selection rate of $\bar{X}$ 2.34. The difference, -0.155, to the expected rate (μ = 2.5) is significant, t(353) = -5.533, p = .012, representing a small effect size, d = .14 (see Appendix E, Tables E22-E23).
Although as a category, F-indicators are included in the list with indicators that received the highest selection rates regarding one competency and amongst all

Figure 16

Mean Selection Rate of the Knowledge Indicators in Domain 1 - (See Table E20)

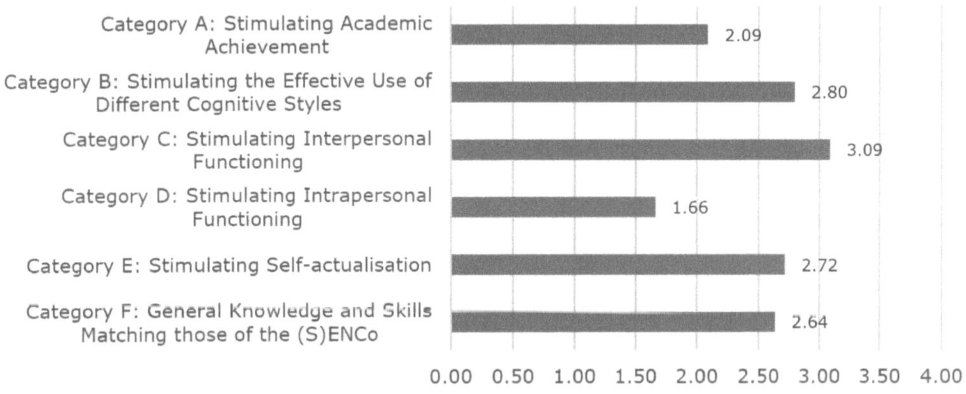

Note. N = 354. Test value = 2.5

Figure 17

Mean Selection Rate of the Skills Indicators in Domain 1 - (see Table E22)

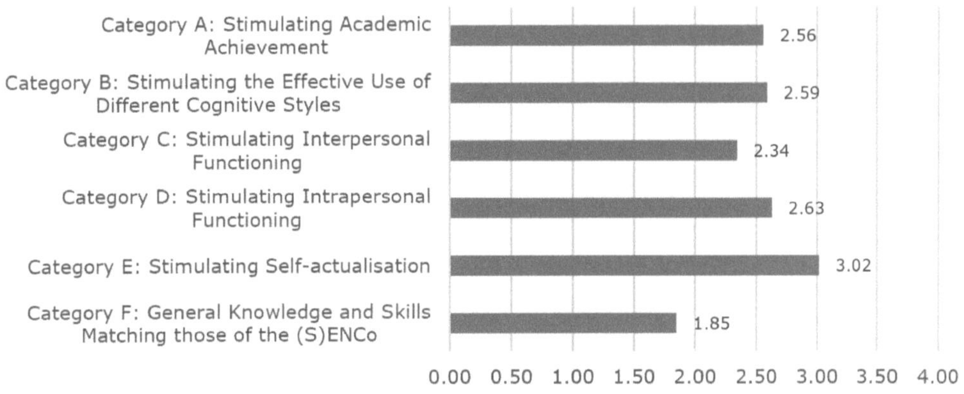

Note. N = 354. Test value = 2.5

groups of stakeholders, it is relevant to mention that only one skills indicator from this category could be included in the list. As a category, the skills indicators F showed low selection rates, $\overline{X} = 1.85$ compared to the expected rate ($\mu = 2.5$). The difference, -0.647, is significant, $t(353) = -10.947$, $p < .001$, the effect size represents a practical significance, $d = .58$ (see Appendix E, Tables E22-E23; see also Figure 17).
All groups of stakeholders showed high selection rates for the skills indicators aiming for self-actualisation. As a category, the skills indicators E had a selection rate of $\overline{X}$ 3.02. The difference, 0.523, to the expected rate ($\mu = 2.5$) is significant, $t(353) = 9.479$, $p < .001$, the effect size represents a practical significance, $d = .50$ (see Appendix E, Tables E22-E23).
The overall results for domain 1 suggested that some categories of indicators were more selected than others (see Appendix E, Tables 25-28). Analysis suggested that they could be divided into two combined groups: a category clustering ADE-indicators and a category clustering the BCF-indicators. A Paired Sample Statistics Test showed that clustered knowledge indicators in the categories ADE had a significantly higher selection rate than clustered indicators in the categories BCF ($p < .001$). Clustered skills indicators in the categories ADE had a significantly higher selection rate than clustered indicators in the categories BCF ($p < .001$). This suggests that, overall, the ADE indicators were considered to be more relevant examples for the competencies they were supposed to match.

5.5.4 Comparison Between The Netherlands and Flanders

The final analysis for this domain considered similarities and differences between results for The Netherlands and Flanders. Using group statistics and an Independent Samples test, the following results were found.
Compared to the other categories of indicators, indicators belonging to category C (stimulating interpersonal functioning) were significantly less selected by all groups of stakeholders; this applied to both countries (see Appendix E, Tables E31-E32; see also Figure 18). However, there is a noticeable difference between both countries. Amongst the Flemish stakeholders, the C-indicators for knowledge showed significantly higher selection rates than amongst the Dutch stakeholders; the difference, -0.362, is significant, $t(352) = -2.26$, $p = .024$, representing a small effect size, $d = .33$. A similar difference is noticeable regarding the C-indicators for skills. For Flanders the selection rate was slightly higher, the difference, -0.345, is significant, $t(352) = -2.05$, $p = .041$, the effect size meets the threshold for practical significance, $d = .30$ (see Appendix E, Tables E31-E32).
The Dutch stakeholders selected knowledge indicators from category B more often than Flemish stakeholders, the difference, 0.490, is significant $t(352) = 3.29$, $p = .001$, the effect size represents a practical significance, $d = .48$. Dutch stakeholders also selected skills indicators from category F significantly more often than Flemish stakeholders; the difference, 0.343, is significant $t(352) = 2.11$, $p = .035$, the threshold for practical significance is met, $d = .31$ (see Appendix E, Tables E33-E34; see also Figure 19).

Figure 18

Comparison of Mean Selection Rates in Domain 1 per Category of Knowledge Indicators Between Dutch and Flemish Stakeholders - (See Table E31)

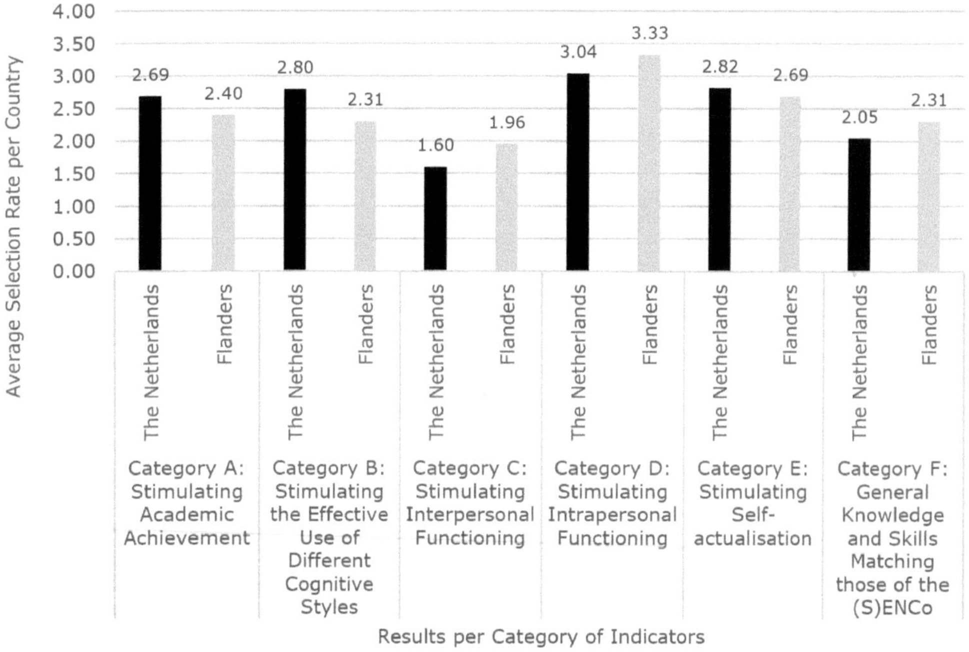

Note. N = 354. The Netherlands N = 299, Flanders N = 55. Test value = 2.5

Figure 19

Comparison of Mean Selection Rates in Domain 1 per Category of Skills Indicators Between Dutch and Flemish Stakeholders - (See Table E33)

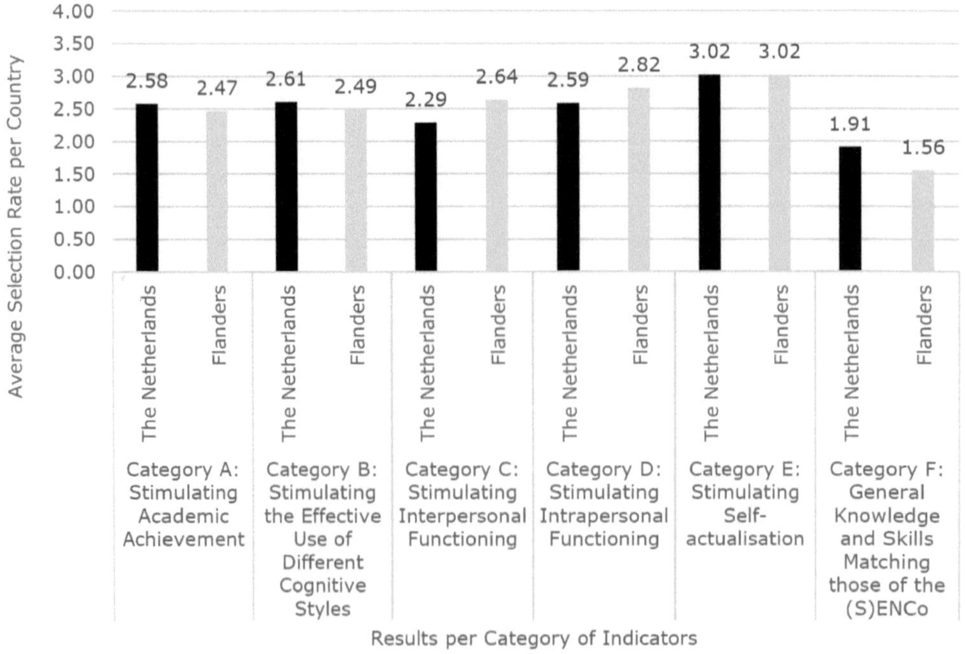

Note. N = 354. The Netherlands N = 299, Flanders N = 55. Test value = 2.5

5.6 Summarising

This chapter focused on the combination of general teaching standards and domain 1, general theories of giftedness. The competencies that are described for this domain represent five coherent topics in gifted education.

- 1.1 The specialist in gifted education understands and envisions how their perspective on giftedness influences their ability to see, understand, and respond to the pedagogical needs of gifted children.
- 1.2 The specialist creates a coherency between the concepts of giftedness, educational objectives and educational programming and the intervention strategies to be used.
- 1.3 The specialist in gifted education understands how their perspective on giftedness influences the organisation of the educational programme provided.
- 1.4 Based on interaction between the educational partners, the specialist in gifted education develops a perspective on giftedness in how to approach gifted students and match their individual needs.

- 1.5 The specialist in gifted education relates their in-depth knowledge of giftedness to the perspectives on current society in general and more specifically to the current perspectives on education.

For each competency, stakeholders selected the indicators that they considered as being the most relevant and best examples of competent professional behaviour (see Appendix E, Table E37). The observed differences between them matched the respective roles of the different groups of stakeholders (see Appendix E, Table E35-E36). These differences should be taken into account in CPD courses, as the explanation of these differences to specialists in gifted education may create a better understanding of their educational partners. The selection rates of indicators were not equally distributed. Some categories of indicators showed significantly higher selection rates than others. All groups of stakeholders selected skills indicators in category E (stimulating the student's strategies for self-actualisation) most often (see Appendix E, Table E24). Educational professionals and parents had the highest selection rates for knowledge indicators in category D (stimulating the student's intrapersonal understanding). At this point, specialists in gifted education showed a different selection pattern; they chose indicators in category B (stimulating the effective use of different cognitive styles) most often. The overall low selection rates for indicators regarding the stimulation of gifted students' interpersonal functioning (C-indicators) were unexpected. The low selection rates for indicators aiming for peer group education were also unexpected. Analysing if both trends are also noticeable for the other four competency domains would be relevant.
The differences in selection rates per category indicators between both countries were relatively small. Indicators in the categories C and F scored relatively low amongst stakeholders from both countries. There were differences observed between both countries in selection rates for indicators in these categories. Although these differences are significant, against the overall picture of what stakeholders think to be important indicators for competent behaviour of specialists in gifted education, they are not likely to result in entirely different curriculums for CPD courses in both countries. However, it may have implications for small nuances in topics that teacher educators may want to address regarding general theories of giftedness.

6 Domain 2: Seeing Educational Needs

6.1 Introduction

The second domain of the competency matrix concerns the interplay of giftedness with generic teacher competencies and focuses on *Seeing Educational Needs*. Gathering information concerning gifted students fits into the role of the specialist of gifted education. The knowledge and skills needed in this role show similarities to the professional repertoire of SENCo's (Coleman & Johnsen, 2011; Pameijer & van Beukering, 2007; van Meersbergen & de Vries, 2017; Wood, 2018).
Competencies in this domain are not only influenced by the domain itself or by generic teacher competencies, but also by perspectives on giftedness. A professional's perspective on giftedness influences how that professional sees and interprets educational needs (Dai & Chen, 2014; Godor, 2019).
The current direction of the profiling of educational needs in Dutch and Flemish education is more towards a change and solution-focused approach (Caufman & van Dijk, 2009; de Bruin-de Boer & Kuipers, 2011; Houkema, 2009; Pameijer & van Beukering, 2007; van Gerven & Drent, 2012; van Meersbergen & de Vries, 2017; van Swet, 2009). This process aims for the optimisation of education for all students. In this approach, it is understood that *all students*, not just *special students*, have educational needs. Therefore, the process is not focused on labelling students or dividing them into separate categories. Educational needs are described as teachable objectives needed for a student's optimal development. Based on these needs, classroom teachers design their interventions to respond meaningfully to those needs in the regular classroom environment. The abbreviation SEN, previously used for the term Special Educational Needs, could now stand for the term Seeing Educational Needs. According to van Meersbergen and de Vries (2017), the process of seeing educational needs is focused on describing the student's strengths and weaknesses in an educational context, while taking the student's ecological system into account.[122] In this approach actors in the ecological system are equal educational partners. This matches a systemic approach as described by Ziegler et al. (2017) and Pameijer and van Beukering (2007). Based on this perspective, in this chapter we will analyse competency domain 2, seeing educational needs. This analysis leads to the second set of five competencies with 60 matching indicators. The results for domain 2 of the empirical part of the research are presented as a separate paragraph in an identical structure to the results of domain 1, at the end of this chapter.

[122] Appendix B, Table B6, K.2.C.1, V.2.C.1.

6.2 Including the Student's Ecological System

The process of seeing educational needs focuses both on the student's personal traits and on mapping the influences of the student's ecological system on their behaviour (Johnsen et al., 2016; Ziegler et al., 2017).[123] By taking this systemic approach, the specialist in gifted education ensures that their actions are based not only on their own professional perspective of giftedness, but also on other current perspectives on the process of seeing educational needs. In a change and solution-focused approach, explaining the importance of seeing a student's educational needs to the educational partners precedes the actual process of seeing educational needs. This explanation helps them to understand both the process itself and the effects that this process may have on the student's development and wellbeing (Peters & Oveross, 2020; van Meersbergen & de Vries, 2017).[124] It is important to bear in mind that the notion of educational partners refers to all the persons involved in the student's education. It refers to the relationship between parents and teachers but also to the relationship between classroom teachers and special project teachers, the relationship between administrators and classroom teachers and the relationship between the teacher/specialist in gifted education and the student. They are all actors in the student's ecological system (van Meersbergen & Jeninga, 2012; Ziegler et al., 2017).

The process of seeing educational needs starts with the description of everything that can be observed: behaviour, development, personal characteristics, achievements. As giftedness in all paradigms is described as more than mere achievements, this broad perspective on everything that can be observed is relevant (Gagné, 2010a; Houkema, 2017; Kieboom & Venderickx, 2017b; Kreger-Silverman, 2013; Subotnik et al., 2011).[125] The second step describes the ecological influences on this behaviour. The third step is the analysis of educational needs: what exactly does this student have to learn to empower their ability to self-actualise? The previously mentioned teachable objectives exceed the cognitive domain (the academically oriented curriculum), trying also to encompass the conative domain (cognitive style, interpersonal functioning, intrapersonal functioning) (Marzano & Heflebower, 2012). The process concludes with the specialist writing a final report that provides suggestions and approaches that will adjust instructional strategies and interventions in a way that optimises the student's education and wellbeing.[126]

6.3 Using a School Toolkit for Seeing the Educational Needs of Gifted Students

There exists a subtle distinction between education for gifted students and gifted education (Borland, 2012; Plucker et al., 2017). In the first approach, educational interventions are reserved for gifted students, whatever the school's

---

[123] Appendix B, Table B7, K.2.C.2; Table B8, K.2.F.3; V.2.F.3.
[124] Appendix B, Table B9, K.2.F.4, V.2.F.4.
[125] Appendix B, Table B7, V.2.F.2.
[126] Appendix B, Table B9, V.2.A.4.

definition of giftedness may be. In the second approach, educational interventions are needs-based, and the label of being gifted is therefore not conditional (Frumau et al., 2011). In both The Netherlands and in Flanders, there exists a contradiction in daily practice between what is said to be leading and what is leading in reality. From a political point of view, teachers should strive for gifted education (Slob, 2018a, 2018b). However, in daily practice, although often unconsciously, the concept of education for the gifted still leads. As a result of this, it is still good practice to assess the likelihood of the student being gifted before profiling the student's educational needs. This assessment is undertaken through identification based on characteristics of the gifted. Consequently, the process of profiling educational needs is influenced by this routine (Subotnik et al., 2017).

The assessment of the educational needs of gifted students is mainly done with the help of school and teacher toolkits. These toolkits were developed to estimate the likelihood of a student being gifted (de Bruin-de Boer & Kuipers, 2011; Houkema, 2009; Prodia, 2012; van Gerven & Drent, 2012). These toolkits can help to profile the educational needs of gifted students who fit into the *gifted child profile* as set out in the toolkit. Students who do not match this profile are not likely to be selected unless they show extremely high academic achievements. Whether it is concluded that a student is gifted or not gifted therefore depends on the toolkit that is used (see also Chapters 2 and 3). For example, when the identification of a student as being gifted takes place simultaneously using two toolkits, it may lead to two contradictory conclusions. The exploration of a student's strengths and weaknesses with these toolkits takes place based on the assumption that the student may be gifted. As seen with the varying definitions of giftedness, strengths and weaknesses are defined differently within the separate toolkits. Each toolkit focuses on different kinds of strengths and weaknesses. None of these toolkits are validated, and, therefore, none of these toolkits can lead to a formal classifying diagnosis. They are designed to be used by teachers or educational specialists, and they do not replace psychological assessment, that is, intelligence tests. The use of these toolkits takes place in what can be seen as a pre-diagnostic phase. Most often, this is not followed up by a diagnostic phase. As most schools in The Netherlands and Flanders cannot afford psychological assessment for students who are thought to be gifted, the toolkits are seen as a reasonable alternative for the identification of gifted students (Bakx et al., 2016; de Bruin-de Boer & Kuipers, 2011; Houkema, 2009; van Gerven & Drent, 2012). Psychological assessment is reserved for cases that are considered to be too complex to be handled by the educational specialist or the specialist in gifted education (de Bruin-de Boer & Kuipers, 2011; van Gerven & Drent, 2012). Bearing all this in mind, we can conclude that, regardless of whether the perspective taken is that of education for the gifted or gifted education, only highly skilled professionals are able to use these toolkits responsibly. The success of using these toolkits depends on the approach that is taken to the assessment. This context of using these toolkits shapes an outline for specific competencies in seeing the educational needs of gifted students for the specialist in gifted education.

For a successful approach, a clear identification question regarding the situation that needs a meaningful educational response is required.[127] Based on the MIA (see Chapter 3) the process starts with the description of the current situation that is being considered for change. What issues are hindering or stimulating the student's development or functioning? This is discussed between the educational team at school, parents and the student (Caufman & van Dijk, 2009; Pameijer & van Beukering, 2007; van Meersbergen & de Vries, 2017; van Swet, 2009). The second step is to set the educational objective, as in describing what the preferred situation should look like. In other words: after consulting with the educational partners, the specialist should be able to define, from a joint perspective, what they all want to be different in the future and what are the concrete effects that they are aiming for (Caufman & van Dijk, 2009; Coleman & Johnsen, 2011; Pameijer & van Beukering, 2007; Peters et al., 2017; van Meersbergen & de Vries, 2017; Wood, 2018). Effects may differ according to the way a curriculum is adjusted for the development of positive behaviour, a growth mindset, self-regulation skills, and so forth (Johnsen et al., 2016; Pameijer & van Beukering, 2007). It is also relevant to include a student's perspective on how effects and interventions could best be prioritised. In the change and solution-focused approach, students taking ownership of their developmental process is an important condition for success.[128] The specialist's ability to estimate if and how the student is ready to oversee the (long-term) consequences of their choices, influences the level of ownership transferred to the student.[129] The third step is making an inventory of the information needed to select a matching intervention.[130] Making this inventory is influenced by the paradigm of giftedness that underpins the professional's actions. In The Netherlands, in 2018, the Minister for Primary and Secondary Education and Media announced additional funding for gifted education (Slob, 2018b). However, he did not describe a conditional definition of the concept of giftedness. The conditions for funding only refer to students showing characteristics of giftedness without describing what these characteristics are. This places a huge responsibility on the schools and therefore on the specialists in gifted education who are likely to explore the needs of these students. Acting from the differentiation paradigm, the specialist is likely to focus on academic achievements and the way the educational setting is able to meet the student's need from this perspective (Peters et al., 2017). Acting from the gifted child paradigm, the specialist may be looking for concrete behavioural indicators that the student is gifted. This specialist may be focusing on personal traits and matching profiles (Houkema, 2017). Acting from a talent development paradigm, the specialist may want to uncover the special talents a student has (Subotnik et al., 2011). The ability to take a position next to the different paradigms is, therefore, conditional to seeing educational needs. The fourth step is to determine what information is already known and what remains to be

[127] Appendix B, Table B10, V.2.F.5.
[128] Appendix B, Table B9, V.2.E.4.
[129] Appendix B, Table B9, K.2.E.4.
[130] Appendix B, Table B9, K.2.A.4.

explored and to infer from this what question requires investigation. Finally, this explorative question can be operationalised, and at that point, the selection of tools becomes relevant.
Although the previously described process is the preferred protocol in the change and solution-focused approach, the Dutch and Flemish toolkits that are described at the start of paragraph 6.3 are created in such a way that users could ignore this protocol. These toolkits each have their own protocols. These protocols do not match the change and solution-focused approach and may even work against the change and solution-focused approach. Explorations with these toolkits are focused on the student and leave a systematic exploration of the influence of the ecological system out of the picture. In current practice, the results of the assessment obtained with these toolkits forms the basis of the decision about whether or not a student can benefit from interventions designed for gifted education. The unspoken assumption is that if a teacher knows whether or not a student is gifted, then the teacher also knows what the student's educational needs are (Dai & Chen, 2014). This assumption can be traced back to the perspective of the gifted child paradigm. Therefore, if the school or the specialist in gifted education has a different perspective on giftedness, that is, the talent development perspective or the differentiation perspective, the use of these toolkits is not sufficient. The current Dutch and Flemish toolkits are therefore useful if the specialist in gifted education has a direct focus on what they need to know to profile the student's needs without focusing on the question of whether the student can be labelled as gifted or not gifted. Within this context, it is necessary that a specialist in gifted education is able to set up a personal toolkit. When the specialist in gifted education sets up their personal toolkit and selects the most helpful assessment instruments on a case by case basis, personalised assessment of the needs of individual students becomes possible without the risk that a specific perspective on giftedness will unconsciously direct the exploration (Kroesbergen et al., 2015; Wood, 2018).[131]

6.4 Seeing Academic Needs

In all three paradigms of giftedness, the academic development of the student is seen as a leading trait for interventions. In all paradigms, the mismatch between the student and the curriculum offered is described as an important trigger for the development of problems around the student's development and wellbeing (Heacox & Cash, 2017; Hoogeveen, 2017b; Siegle et al., 2017).[132] Within this context, it is important that the specialist in gifted education is able to determine a student's academic level (Coleman & Johnsen, 2011; Hoogeveen, 2017a, 2017b; Johnsen et al., 2016).[133] Tuning to educational needs starts with the ability to monitor a student's academic development.[134]

[131] Appendix B, Table B6, V.2.F.1; Table B10, K.2.F.5.
[132] Appendix B, Table B6, K.2.A.1, V.2.A.1.
[133] Appendix B, Table B10, K.2.A.5.
[134] Appendix B, Table B8, K.2.A.3, V.2.A.3.

Gifted students are often able to function at an advanced didactical level compared to their age group (Gagné, 2010a; Johnsen et al., 2016; Kreger-Silverman, 2013; Peters et al., 2017; Subotnik et al., 2011). This influences any assessment of the student's development. Several problems may occur. In the case of the specialist who is only interested in whether a student can meet the set objectives for their age group, it is likely that they will not be able to recognise the student's full ability. A 100% score on a standardised test measuring whether or not the student has met the set objectives provides only a clear view of the bottom level of the student's didactical level (Hollenberg et al., 2014). The student is able to achieve at this level, but the results still do not inform about whether the student is able to achieve at a higher level. Therefore, the optimal estimation of a student's didactical level should show not only the mastered objectives but also the objectives that are not yet mastered.
Underachieving students may score below the set standards for a pass (Siegle et al., 2017). They may not feel challenged by the test and may even understand that this test does not provide a correct representation of their capabilities (Heacox & Cash, 2017). Another problem that can occur is the interference of a learning disability (Webb et al., 2016). The intrapersonal discrepancies between intelligence and school results that occur with twice-exceptional students, often affect the conclusions of an assessment (Reis et al., 2014; Weterings, 2017). If the learning disability is not recognised or not taken into account, the results may not match the student's real abilities. A different way of assessment may provide a more realistic understanding of the student's abilities (Baumet al., 2014; Trail, 2011).[135] Therefore it is necessary to have the correct knowledge of measurement principles, to have practices that differentiate assessments and to be able to interpret the results (Johnsen et al., 2016). Assessment should not be focused on the question of whether the student meets standards and levels set for their peer group, but on what the student's zone of actual development is and what the zone of proximal development will be (Peters et al., 2017).
Formative, dynamic assessment of the student's academic development shows which educational objectives match the student's zone of proximal development (Hattie, 2013; Vogelaar, 2017). In addition to assessment via standardised tests, the student's perspective on the curriculum should also be taken into account. This provides for an opportunity to understand how the curriculum influences the student's wellbeing and behaviour. Discussing this in teacher-student meetings can be a useful strategy to aid this understanding (den Otter, 2015; Mayes et al., 2019; Siegle et al., 2017; Wood, 2018).
Roberts and Inman (2013) describe how differentiation should not only be visible in assignments but also in assessments. If the product that is used to assess the student's mastery level matches the student's personal preference when producing work, then it is likely that the student may show a higher mastery level than they would in an assessment where the student does not have a choice in how to show their level of understanding.[136] Students who are

[135] Appendix B, Table B7, K.2.F.2.
[136] Appendix B, Table B10, V.2.A.5; Table B8, K.2.B.3.

encouraged to try different methods to illustrate their mastery level are likely to develop a better personal understanding of their own learning preferences and how they interact with the objective to be mastered.[137] This encouragement will be even more successful if teachers set up opportunities to reflect on the learning process as both learning products and achievements (Roberts & Inman, 2015).

6.5 Seeing the Needs Regarding the Student's Cognitive Style

The notion of cognitive style refers to the way a student processes and produces information. Cognitive style can be described through four interplaying concepts: learning style, multiple intelligence, levels of thinking skills and executive skills. Although the topics of learning style and multiple intelligence are highly discussed and more or less labelled as non-issues in education (Hattie & Yates, 2015; De Bruyckere et al., 2016) they are relevant to the discussion at this point as many books on gifted education still refer to these theories and see them as the starting point for matching instructional strategies (Houkema, 2009; Renzulli, 2012; Roberts & Inman, 2013). According to Hattie (2013) the main myth is that teaching in a student's preferred learning style results in higher achievements.

A specialist in gifted education should be aware of the theoretical discussion that goes on about the concepts of learning styles.[138] The current idea in general theories of education is that education based on a student's preferred learning style was hyped up in the previous century and should be put behind us as soon as possible (De Bruyckere et al., 2016; Dai, 2008; Hattie, 2013). De Bruyckere et al. (2016) refer to research done by Coffield et al. in 2004, where they analysed 71 different learning styles. Thirteen of them were considered as highly influential. Their research suggests that there is insufficient evidence to assume that people differ in preferences for a specific learning style and that being taught in this preferred learning style would lead to better educational results. Meta-analytical research shows that a person's learning style is not fixed but depends on the situation and what is being learned. Under different circumstances, students may apply different learning strategies and learning styles.[139] Successful students know when and how to switch styles in order to get the best achievements (Sternberg, 2009). To estimate a student's chances of successful learning, it is therefore relevant to ask the educational partners involved how the student uses different strategies under different circumstances.[140]

The same meta-analytical research also shows that there is no evidence that knowing a student's learning style preferences provides instructional designs of a higher quality leading to higher achievements (Pashler et al., 2008). What does lead to higher achievements is the actual effort it takes to learn new

[137] Appendix B, Table B8, V.2.B.3.
[138] Appendix B, Table B7, K.2.B.2.
[139] Appendix B, Table B9, K.2.B.4.
[140] Appendix B, Table B9, V.2.B.4.

knowledge and skills (Willingham, 2009). A complex task in a student's zone of proximal development requiring the use of higher order thinking skills, calls intensely on a student's learning process (Marzano & Kendall, 2007). Intense thinking processes lead to more long-term results (Willingham, 2009). Based on this idea, it may be assumed that teaching in a so-called preferred learning style makes the learning process easier and therefore is not likely to be as fruitful as it could have been were a student to have stepped out of their comfort zone. Hence, it is relevant that the specialist in gifted education is able to observe how a student responds when offered assignments that call on various learning styles.[141] For a student, the complexity of a task may increase when offered outside the comfort zone of their preferred learning style. In these situations, students may respond differently when compared to situations where assignments are offered within their preferred learning style. Being confronted with an assignment outside the comfort zone of the preferred learning style may confront them with the notion of being consciously incompetent and may even set emotions free that they find difficult to handle. This may influence the motivation needed to finish the assignment. In these cases, listening to the student and helping them in the struggle for the discovery of self-efficacy by having reflective teacher-student conversations, provides the necessary insight for seeing educational needs (Siegle et al., 2017).[142]

As stated, in gifted education, curricular adjustments are seen as the main strategy to optimise the student's education and wellbeing. However, these adjustments should not only match a student's academic levels but also their skills for self-regulation. Self-regulation skills interplay with talent development (Portenga, 2017). Marzano and Kendall (2007) explain how the complexity of assignments interplays with the complexity of regulation skills. The higher the level of required thinking skills, the greater the call on regulation skills.[143] Although some gifted students may prefer the use of higher order thinking skills as an approach to discovering new knowledge and learning new skills (Schrover, 2015), this preference does not apply to all gifted students and will vary under different circumstances. As gifted students are a heterogeneous group, they have individual profiles regarding their educational preferences and individual profiles of strengths and weaknesses with regard to the use of different thinking skills. Knowing how to map these strengths and weaknesses and how to translate them for teachers into clear instructional strategies, increases the success of curricular adjustments (Dawson & Guare, 2009; Johnsen et al., 2016; Kroesbergen et al., 2017).[144]

According to Portenga (2017), self-regulated learners take ownership of their developmental goals and learning strategies. Their ability in self-actualisation depends highly on their effective use of self-regulation skills. They compare their current knowledge and skills with the knowledge and skills that will be developed by taking on the assignment. Teachers need to identify any support needed for

---

[141] Appendix B, Table B7, V.2.B.2.
[142] Appendix B, Table B7, V.2.D.2.
[143] Appendix B, Table B7, K.2.E.2.
[144] Appendix B, Table B10, V.2.B.5.

the student to span gaps between the level of self-regulation that is required for the student *to learn*, and the student's *actual* already developed self-regulation skills (Dawson & Guare, 2009; Kroesbergen et al., 2017; Stoeger & Ziegler, 2011). A specialist in gifted education should therefore be able to assess the level of thinking skills an assignment calls on, as this provides a first indication of the level of self-regulation that is required during the learning process. Then the specialist needs to assess the student's ability for self-regulation and what type and intensity of support the student needs to complete the assignment successfully (Kroesbergen et al., 2017). This all requires a knowledge of theories of thinking skills and how these skills are used in current educational paradigms such as a more social-constructivist approach to education.[145] At the same time, it requires the skill to observe the student's ability to use self-regulation skills in different educational contexts.[146] In situations where a student's wellbeing decreases, self-regulation may be difficult as emotions may interfere with the ability to set realistic goals (van Gerven, 2020).[147] Underachievement, for example, affects a student's self-concept. Research shows that underachievement also affects the use of regulation skills (Obergriesser & Stoeger, 2015). Gifted students with limited self-regulating learning skills are likely to be less successful compared to their peers who have well-developed self-regulating learning skills. This affects their feeling of self-efficacy. At the same time, environmental responses to their achievements may also affect their self-concept (Siegle et al., 2017). This shows the necessity to explore if and how a student's self-regulation skills interplay with the student's feeling of self-efficacy and self-concept.[148]

The result of these observations should be a student's profile of strengths and weaknesses in regulation skills providing information to a teacher to help determine which instructional strategies and interventions may be of most benefit to the student.[149]

6.6 Seeing Interpersonal and Intrapersonal Needs

In the current pedagogical and didactical approach, intrapersonal understanding and interpersonal functioning have become increasingly relevant (Biesta, 2018b). The concept of 21st century skills requires a reflective attitude and the ability to be an active team player alongside fellow students (Bambs et al., 2016). Students need an in-depth intrapersonal understanding in order to reflect on their personal role with regard to the learning processes. The value of cooperative and collaborative learning in this educational paradigm stresses the capability of high-level functioning in the interpersonal domain (Marzano & Heflebower, 2012).[150] For some gifted students, this may not be a difficulty, while for others it may create problematic experiences (Freeman, 2019). Some

[145] Appendix B, Table B10, K.2.B.5.
[146] Appendix B, Table B8, K.2.E.3, V.2.E.3.
[147] Appendix B, Table B10, K.2.E.5.
[148] Appendix B, Table B10, V.2.E.5.
[149] Appendix B, Table B7, V.2.E.2.
[150] Appendix B, Table B8, K.2.C.3.

gifted students develop asynchronously, and their social and emotional skills lag behind their cognitive skills. These students may experience problems in connecting with other students and may also experience problems developing a realistic self-concept (Kreger-Silverman, 2013; Wood, 2018).

The notion of self-concept is best described as the general perception individuals have about themselves in achievement situations (Obergriesser & Stoeger, 2015). Self-concept relies strongly on comparative social information (Obergriesser & Stoeger, 2015). A mismatch between the student and their social environment may affect the student's self-concept. If, over time, there is a consistently negative response to a student's achievements or a student's learning behaviour within their social environment, the student may develop a negative self-concept. Negative responses towards the gifted student, stereotyping them as "the little professor" or "nerd", are as devastating as the proverbial Homeric laughter by classmates when the gifted student finally makes a mistake while doing a relatively easy task. Eventually, in future learning situations, the gifted student may withdraw from participation or adjust to the same level as their classmates.[151] As self-concept is thought to be relatively stable over time (Obergriesser & Stoeger, 2015), signs that indicate the development of a negative self-concept are therefore important to spot as early as possible.[152]

Other gifted students may be well ahead of their age group and may not experience the previously mentioned problems at all (Neihart et al., 2016; Rinn & Majority, 2018). Peterson (2009) states that there is insufficient evidence to establish whether gifted students are more or less likely to develop social-emotional problems. However, the fact that giftedness may not necessarily be the trigger for social-emotional problems does not mean that a student's functioning or development cannot be at risk (Rinn, 2018). Therefore, if it is necessary for a gifted student to be supported by a teacher in developing their interpersonal skills, it is important that these skills can be assessed within a suitable context, which is where the specialist in gifted education comes in (Johnsen et al., 2016).[153] Knowing how to do this and knowing how to use instruments such as observation lists or specific communication techniques for diagnostic talks with students, is conditional for a good assessment.[154]

This assessment not only builds on the proper use of instruments and techniques, but the results and the reliability of these results are influenced by the match between the topic being explored, the instrument and technique used for the exploration, and the student's capabilities (Kallenberg et al., 2007).[155] That is to say, a student who has difficulties with written expression may find it difficult to answer a written questionnaire about their wellbeing or interests. In this case, although there may be a match between the topic being explored and the questionnaire, a mismatch between the student and the instrument (writting)

[151] Appendix B, Table B7, V.2.C.2.
[152] Appendix B, Table B6, K.2.D.1.
[153] Appendix B, Table B8, V.2.C.3; Table B10, K.2.C.5, V.2.C.5.
[154] Appendix B, Table B6, K.2.F.1; Table B8, K.2.D.3.
[155] Appendix B, Table B6, V.2.D.1; Table B8, V.2.D.3.

makes the results less reliable. Therefore, in this situation, the results of the exploration may be more reliable if the specialist uses the approach of observing the student, combined with a teacher-student and teacher-parent conversation (Wood, 2018).[156] This observation gives them the chance to explore the student's behaviour in a specific context. It also creates the opportunity for a contextual interpretation of the results of the exploration (Wood, 2018). Suppose that a teacher notices that the gifted student has very few interpersonal relationships with other students. Suppose that they find this situation worrying and that they think that the student is socially not as well-developed as may be expected in relation to their age group. They may even conclude that the student is socially challenged or that the student's social behaviour is the result of an unidentified disability. Observation by a specialist in gifted education could offer a different perspective, may even provide more accurate information, and may exclude the chance of a misdiagnosis or a missed diagnosis (Webb et al., 2016).[157] For example, the gifted student in our example may have difficulty finding like-minded peers in a classroom where students are grouped based on their age. The gifted student may use language that is too complex for their age group. Their experience may be that very few of the other students share the same interests or knowledge and, therefore, they tend to be ignored if they want to contribute to a group discussion, which may hurt their feelings. This could have a negative effect on the student's wellbeing.[158] Another interaction problem may be raised by the student's cognitive style. Some gifted students may prefer to explore learning content top-down while their peers may prefer a more step-by-step approach (Bakx et al., 2016; Callahan & Hertberg-Davis, 2018; Danielian et al., 2018; Wallace et al., 2019).[159] If communication is already strained due to huge differences in the language used by the students, different approaches may lead to a clash where both parties find it too complex to overcome their differences (de Bruin-de Boer & van Gerven, 2009). A diagnostic and solution-focused conversation with the students involved provides additional contextual information (Wood, 2018). Focusing on solutions and opportunities for change, by describing weaknesses in terms of opportunities for learning new skills, can be helpful during these conversations (de Shazer et al., 2012). It may help the teacher to see the needs of all the students involved in the collaborative work. Identifying those needs may empower all of them.[160] Therefore, a meeting with the student is not only helpful in the process of seeing the educational need. Simultaneously it offers scaffolding that helps understand their educational needs and to explore possible meaningful responses (Caufman & van Dijk, 2009; de Shazer et al., 2012). In the previous example, the student's coping mechanism may be to exclude themselves by choice, rather than feel excluded, that is, "If I choose not to participate, they cannot exclude me." An additional diagnostic meeting with their parents could bring a wider

[156] Appendix B, Table B10, K.2.D.5.
[157] Appendix B, Table B10, V.2.C.5.
[158] Appendix B, Table B6, K.2.E.1.
[159] Appendix B, Table B6, K.2.B.1.
[160] Appendix B, Table B6, V.2.B.1, V.2.E.1.

ecological context into perspective (Ziegler et al., 2017) and may lead to the following information.[161] At home, this child has three special friends who share their interests and with whom they can be themselves, and with whom they have no problems with social functioning. By comparing the different social settings described by parents, student and the teacher, a new light is shed on the student's functioning.[162] This example of a systemic and solution-focused approach stresses that the specialist in gifted education needs both well-developed communication skills and well-developed skills to analyse, evaluate and generalise gathered information.
A student's wellbeing may be affected by the student's self-concept. Underachieving students tend more often to have a negative self-concept than students who are achieving commensurate to their natural abilities (Figg et al., 2012).[163] Therefore, it is important that the specialist is able to explore if and how the student's self-concept matches the perception of the educational partners and to what extent the student's self-concept is realistic. In cases of a big gap between the student's perception and the perception of the educational partners, it becomes important to explore the origin of the gap.[164]
In the previous paragraphs, conditions for self-actualisation are described, without being pointed out as such explicitly. Self-actualisation is the drive to realise one's potential. From an educational point of view, this drive is influenced by factors such as academic opportunities, the development of a cognitive style that supports the use of those academic opportunities, interpersonal and intrapersonal functioning, and by the opportunities provided by the student's ecological system. Therefore, the process of seeing a student's needs concerning self-actualisation can only be done from a holistic point of view, taking into account all the developmental aspects that influence a student's development.

6.7 Five Competencies in Domain 2

In this chapter, we have focused on the relationship between the notion of seeing the educational needs of gifted students and the five generic teacher competencies outlined in Table 6. The way educational needs of gifted students are seen and recognised as such, depends on the perspective on giftedness held by the specialist in gifted education (Dai & Chen, 2014; D. Matthews & Foster, 2009). At the same time perspective on these needs is defined by the current constructive approach to education. The use of higher order thinking skills, the increasing need for the development of strong intrapersonal and interpersonal skills as well as the importance of the ability for self-regulation, affect the focus of the process of seeing educational needs. This process of seeing educational needs rests solidly on the change and solution-focused approach, providing ownership to all of the educational partners involved. From student to teacher to parents, they are all included in this process. Seeing educational needs is

[161] Appendix B, Table B9, K.2.C.4.
[162] Appendix B, Table B9, V.2.C.4.
[163] Appendix B, Table B9, K.2.D.2.
[164] Appendix B, Table B9, K.2.D.4.

Table 6

Domain 2, From Appendix A: Competency Matrix for Specialists in Gifted Education

Generic Teaching Competencies	2. Seeing Educational Needs
	The specialist takes a change and solution-focused approach to profiling the educational needs of gifted students. In this approach, all actors in the student's ecological system are seen as equal educational partners.
1. Pedagogical The teacher creates a safe and stimulating learning environment for every student, tuned to their individual educational needs.	**2.1.** The specialist explores the strengths and weaknesses of the gifted student within the intrapersonal and interpersonal developmental domain.
2. Didactical In collaboration with their colleagues, the teacher transfers curricular objectives in such a way that they become teachable in the context of a primary school, matching school policy. They must compose the curriculum in a manner that enables their student to master the content.	**2.2.** The specialist assesses the student's developmental levels in all academic domains offered, considering the student's cognitive style.
3. Managerial The teacher provides a structured, well-organised, task-oriented classroom atmosphere that supports the students' developmental process.	**2.3.** The specialist takes on a systematic approach to the process of profiling educational needs and manages their time and actions in a way that contributes to the optimal results of this process.
4. Communication The teacher communicates professionally with all educational partners who are involved with the student's education.	**2.4.** The specialist approaches all actors in the student's ecological system as equal partners and values their information about the student's educational needs as such.
5. Research and innovation The teacher adjusts their professional knowledge, perspectives, and actions if and when relevant developments in education, society, and science require it.	**2.5.** The specialist is able to develop a personal toolkit for profiling the educational needs of gifted students flexibly and appropriately, using different instruments in different contexts.

therefore no longer limited to the exploration of a gifted student's personal traits, but includes the way the ecological system influences the translation of the student's gifted capital into becoming a successful, intelligent person (Freeman, 2008; Freeman, 2017; Sternberg, 2009).
As the competency matrix is to be constructed as a set of coherent, interdependent, action-oriented, sustainable and learnable knowledge and skills that the specialist should be able to apply contextually, five competencies have been inferred for domain 2 - Seeing Educational Needs (see Table 5). They are preceded by the competencies as described for domain 1 (see Appendix A). 60 new indicators were also deduced as examples of competent behaviour (see Appendix B). The next paragraphs of this chapter explore what stakeholders consider to be the best indicators of professional behaviour matching the competencies of the matrix.

6.8 Selection Rates of Indicators for Competency Domain 2

In this competency domain, it became clear that seeing educational needs differs from identifying students as gifted. The shift towards an educational paradigm based on the notions of inclusion and diversity, emphasises that all students have educational needs and that these needs should be the starting point for educational interventions. Hence, seeing educational needs is conditional to the setting up of educational programmes that best meet the needs of students. In the context of educating gifted learners, the process of seeing educational needs is covered by the six themes used to categorise the indicators for professional knowledge and skills.
The analyses of the data gathered in the research amongst stakeholders are presented in a similar way to that in Domain 1. All tables referred to are included in Appendix F.

6.8.1 Selection Rates per Group of Stakeholders

6.8.1.1 Educational Professionals

Results per Competency

- Competency 2.1 holds as its aim that specialists in gifted education are able to identify individual strengths and weaknesses of gifted students, especially where it concerns their interpersonal and intrapersonal development. The knowledge indicator with the highest selection rate amongst educational professionals for this competency is indicator K.2.D.1, "Knows what types of behaviour by the gifted student may be a sign for the development of negative self-concept" (27.23%). The matching skills indicator ("Selects appropriate techniques to explore and describe the student's self-concept") shows the lowest selection rate (10.18%) among the skills indicators.
  Educational professionals value other indicators for this competency higher. For instance, the indicator focusing on communication with

students in how they experience and value the curriculum is valued with a selection rate of 22.90%.[165] This suggests that educators see their gifted students as relevant partners in the educational process.

- Competency 2.2 is about the ability to assess students' academic levels in relation to their cognitive style. Educational professionals show high selection rates for the knowledge indicator that focuses on the understanding of how executive skills affect the student's development (24.43%).[166] In line with these selection rates is the percentage of educational professionals that selected indicator V.2.F.2 (27.74%). This indicator emphasises the importance of not only including academic achievements in assessing the student's academic level but other information as well. Results for educational professionals show them to be aware that academic achievements are not the only indicator of whether a student may be considered gifted. However, knowledge about learning styles[167] (11.45%) and skills concerning how classmates can affect gifted students' development[168] (4.33%) were not considered as important examples in competency 2.2.
- Competency 2.3 focuses explicitly on the ability to take a systematic approach to profiling educational needs. One aspect of this process is identifying a student's achievement levels. Within this context, educational professionals consider it to be important that the specialist in gifted education knows how to offer the student different opportunities in a choice of products to show the results of the learning process (20.87%).[169]
  For the skills indicators, the three most selected indicators for this competency have selection rates that are relatively close to each other.[170] Educational professionals consider the assessment of learning, the student's self-concept and self-regulation skills as the best indicators for this competency. A low selection rate for the skills indicator V.2.F.3 looking at how (f)actors in the student's ecological system influence the student's academic achievements, stands out[171] (8.91%).
- Competency 2.4 aims for educational partnership with the actors of the student's ecological system. In domain 1, results suggested that this partnership is highly valued by educational professionals. For the current domain, actors in the ecological system are considered mainly as a source of information about their students and not so much as an actor influencing the student's behaviour (compare the selection rates between indicator V.2.F.3 and indicators K.2.A.4 and V.2.B.4). This can be explained, if we take into account that toolkits for identifying the needs of gifted learners leave the influence of the ecological system out

6

[165] V.2.A.1 – Table F6
[166] K.2.E.2 – Table F2
[167] K.2.B.1 – Table F2
[168] V.2.C.2 – Table F7
[169] K.2.B.3 – Table F3
[170] V.2.E.3 – Table F8 – 20.10%; V.2.B.3 – Table F8 – 18.58%; V.2.D.3 – Table F8 – 18.32%
[171] V.2.F.3 – Table F8

of their scope (de Bruin-de Boer & Kuipers, 2011; Houkema, 2009; van Gerven & Drent, 2012).
An understanding of why it is important to explain the objectives of the assessment of educational needs to the educational partners[172] has relatively low selection rates (7.38%). This is also the case for the indicator that stresses that concrete suggestions of how interventions should take place should be included with the reported results of the assessment[173] (8.40%). Both indicators may be considered as too obvious and, therefore, less relevant to select than skills indicators aiming for the strategy of voicing[174] (23.66%) and the solution-focused approach (26.72%)[175].

- Competency 2.5 stresses the importance of specialists in gifted education constructing their own toolkits to assess students' educational needs. The highest selection rate for a knowledge indicator matching this competency is about the understanding of why it is important to consider different perspectives when exploring a gifted student's problematic social-emotional behaviour (19.08%).[176] However, three other indicators showed almost similar selection rates, suggesting that, as a group of stakeholders, educational professionals found most indicators equally important. The relatively low selection rates for indicator K.2.E.5 pointed out for teachers the knowledge that self-regulation skills may decrease when wellbeing decreases is less relevant in relation to the specific competency this indicator was combined with[177] (12.47%). The exact same selection rate is observed for indicator K.2.F.5, regarding the specialist's competency to create custom made assessment strategies to identify the educational needs of each gifted student.

Overall Results of Educational Professionals for Domain 2

The highest selection rate in this domain was for indicator V.2.F.2, "Takes other indicators besides school achievements into consideration when estimating if the student may be gifted" (27.74%, Appendix F, Table F11). This shows that educational professionals are aware that not every gifted learner can be profiled as a high achiever. They view educational partners as important sources for seeing educational needs and often select indicators that focus on including their students' voice. Based on the ranking of selection rates, indicators focusing on self-regulation skills and the impact of these skills on learning are considered to be relevant.

---

[172] K.2.F.4 - Table F4
[173] V.2.A.4 - Table F9
[174] V.2.E.4 - Table F9
[175] V.2.D.4 - Table F9
[176] K.2.C.5 - Table F5
[177] K.2.E.5 - Table F5

6.8.1.2 Specialists in Gifted Education

Results per Competency

There are similarities between the selection rates of specialists in gifted education and educational professionals. These can be found in Appendix F, Tables F1-F10.

- For competency 2.1, specialists in gifted education show the highest selection rate for the indicator that recognises signs of the development of negative self-concept[178] (26.21%). This high selection rate matches the emphasis of educational professionals on this indicator. The lowest selection rate amongst the knowledge indicators was for indicator K.2.B.1, aiming for the knowledge of how a student's cognitive style affects their interaction with classmates (11.00%).
  The selection rates for the three skills indicators that received the highest selection rates differ only slightly (see Appendix F, Table F6). This suggests that these three indicators are valued equally by this group of stakeholders. These indicators focus on (1) listening to the students' perception of the curriculum (22.01%)[179]; (2) grouping students in a way that stimulates them to complement and support each other (17.15%)[180]; (3) observing how students use strengths and weaknesses in interaction with classmates (22.33%)[181].
- For competency 2.2, both the highest selection rate[182] and the lowest selection rate[183] amongst the knowledge indicators were similar to the selection rates of educational professionals. This similarity was also observed for the skills indicators. Obviously, for this competency, specialists in gifted education and educational professionals had a common perspective.
  The selection rate of indicator V.2.C.2 (2.27%) was unexpected but can be explained. The matching knowledge indicator K.2.C.2 was already selected by 21.04% of the specialists in gifted education. Being forced to make a choice of three out of six indicators, it seems likely that the effect of peers on a student's achievements (V.2.C.2) was thought to be covered by the choice for indicator K.2.C.2, "knows how actors in the student's ecological system affect a student's achievements".
  The selection rate (12.62%) for indicator K.2.D.2 (how under-achievement affects a student's self-concept) was unexpected but this score can be explained as competency 2.2 focuses on the assessment of academic development. However, an identical selection rate for indicator K.2.A.2, aiming for the skill to determine a gifted learner's zone of proximal development, was also unexpected. The context of the covering competency offers no explanation in this case.

[178] K.2.D.1 - Table F1
[179] V.2.A.1 - Table F6
[180] V.2.B.1 - Table F6
[181] V.2.C.1 - Table F6
[182] K.2.E.2 - Table F2 - 22.98%
[183] K.2.B.2 - Table F2 - 9.06%

- Competency 2.3 aims for a systematic approach when profiling educational needs. Including actors of the ecological system for this process has become common practice in The Netherlands and Flanders (Pameijer & van Beukering, 2007; Prodia, 2012; van Meersbergen & de Vries, 2017; van Swet, 2009). Indicator K.2.F.3 and indicator K.2.C.2 are identical. Again, this indicator had a high selection rate amongst specialists in gifted education; 22.98% of all specialists in gifted education selected the indicator K.2.F.3, suggesting that it is important to understand the influence of the ecological system on the student's development. The matching skills indicator[184] showed only a selection rate of 13.27%.
- Competency 2.4 is about the use of effective communication skills in the process of seeing educational needs. Again, the selection rates of specialists in gifted education match the selection rates of educational professionals.
  Knowledge indicator K.2.E.4 is about the specialist's ability to assess to what extent their student is capable of seeing the consequences of their actions. Understanding the consequences of one's actions is relevant considering the increasing emphasis that is now being placed on making students own their education (Caufman & van Dijk, 2009). Interestingly, this indicator showed the lowest selection rate (9.06%) while the matching skills indicator (V.2.E.4, "consulting the student when prioritising intervention strategies") showed the second highest selection rate (22.65%).
- Competency 2.5 aims to integrate skills for research and innovation into the process of seeing educational needs. For specialists in gifted education, involving educational partners in the process of assessing the educational needs of their gifted students proved to be relevant.[185] This matches earlier selection rates for similar items. Similarly to the educational professionals, the relationship between wellbeing and the ability to self-regulate is the least selected knowledge indicator.[186] Although this seemingly contradicts the high selection rates of an indicator with similar intentions for competency 2.1, we have to bear in mind that respondents were asked to select indicators that were the best example for the presented competency.
  The skills indicator with the highest selection rate V.2.C.5 (24.27%) points out that social and emotional problems can not only be related to giftedness, but also to learning or developmental problems.[187] This competency which explores and explains problems from different perspectives is highly valued by specialists in gifted education.

---

[184] V.2.F.3 - Table F8
[185] K.2.D.5 - Table F5 - 21.68%
[186] K.2.E.5 - Table F5 - 13.59%
[187] V.2.C.5 - Table F10

Overall Results of Specialists in Gifted Education for Domain 2

Interestingly, specialists in gifted education for all competencies in this domain included the C-indicators in their selection of knowledge indicators. This contradicts their selection in domain 1 where indicators from the C-category showed relatively low selection rates.
For domain 2, for three out of five competencies, the skills indicators in category C were also selected as good examples of their matching competency. The two skills indicators that were not selected showed the lowest selection rates.
In six out of ten cases, the specialists in gifted education chose for their highest selection rates the same indicators as those of the educational professionals. Between both groups of stakeholders, no relevant differences were observed. Combined with the previous conclusion regarding the high similarities in selection rates, this suggests compatibility between the interests of both groups of stakeholders.
Identically to domain 1, there was an unexpectedly low selection rate of 11.33% for indicator K.2.E.1 ("Knows how a peer group can influence a gifted student's wellbeing"). Although this indicator does not specifically aim for peer-group education as defined as a pull-out programme or a special class, it was not expected to have a selection rate at this level. However, if we consider the question of what indicators were thought to be the best match for a specific competency, the selection rates are understandable.

6.8.1.3 Parents

Results per Competency

- For competency 2.1, parents selected knowledge indicators that were similar to those selected by the educational professionals and specialists in gifted education. Identically to both other groups, the highest selection rate was for K.2.D.1, the knowledge indicator of what behaviour could be a sign for the development of a negative self-concept (26.51%). Their highest selection rate for a skills indicator in competency 2.1 matches the selection of the educational professionals for competency 2.1.[188] The parents considered inclusion of their child's opinion of the curriculum to be important to the process of seeing their educational needs.
  Their lowest selection rate is for the knowledge indicator K.2.E.1, "Knows how a peer group can influence a gifted student's wellbeing" (10.44%). As explained in the previous chapter, this was an unexpectedly low selection rate.
- The highest selection rate for a knowledge indicator for competency 2.2 (22.49%) is for indicator K.2.D.2. This indicator ("knows how underachievement affects a student's self-concept") is not selected often

[188] V.2.A.1 - Table F6 - 23.69%

enough by specialists in gifted education to meet the criteria for the three highest scores. Hence, a relevant difference between parents and specialists in gifted education becomes clear.
Indicator V.2.B.2 (the skill to observe how differentiation in learning style affects learning behaviour) was not highly selected by educational professionals nor by specialists in gifted education. However, amongst parents, this indicator shows a selection rate of 20.08%. Even though the difference in selection rates for this item does not meet the criteria to be considered a relevant difference, it is an interesting difference. It suggests that parents have the idea that if their child can use a preferred learning style, their child's learning behaviour could be more effective.

- For competency 2.3, parents show the lowest selection rate for knowledge indicator K.2.F.3 (12.45%). The understanding of how the ecological system of the student influences their achievements seems to be less relevant to them than the specialists' knowledge of how to observe the student's ability for self-regulation.[189]
  Skills indicator V.2.C.3 is differently phrased but has the same objective as indicator K.2.F.3. Interestingly, this indicator got the highest selection rate of the skills indicators (19.28%).
  By far the most interesting selection rate for this competency is the low rate for indicator V.2.A.3 ("systematically acquires information on a student's academic achievements" –12.45%). Apparently, parents consider academic achievements less important than interpersonal functioning and intrapersonal understanding.
- For competency 2.4, the selection pattern of parents and the other groups of stakeholders show similarities. Differences are mostly small and only occur for the skills indicators. 16.6% of the parents selected indicator V.2.A.4. This indicator stipulates that the specialist in gifted education should provide concrete suggestions for curricular adjustments for their gifted learners. Although this selection rate is almost double the selection rate of educational professionals, it does not meet the threshold for what has been defined as a relevant difference.
- For competency 2.5, parents have a selection rate of 12.05% for indicator K.2.D.5 while specialists in gifted education have the highest selection rate for this indicator (21.68%). This suggests that for this competency, educational partnership between parents and the school was not considered to be as important for parents as it was for specialists in gifted education. However, vice versa, indicator K.2.E.5 (the relation between self-regulation skills and wellbeing) showed the lowest selection rate amongst educational professionals and a relatively high selection rate amongst parents. The difference between both was, however, not large enough to be considered relevant.
  The skills indicator that showed the highest selection rate for parents was indicator V.2.B.5 (21.69%). This indicator focuses on the specialist's

[189] K.2.E.3 – Table F3 – 22.49%

ability to profile strengths and weaknesses in the student's use of thinking skills.

Overall Results of Parents for Domain 2

For parents, the specialist's ability to pick up on signs that their child may develop a negative self-concept was very important. This indicator was the most frequently selected indicator in domain 2. This selection rate was in line with the selection rates of the other groups of stakeholders.
For four competencies, parents selected knowledge indicators concerning academic development. For three competencies they also selected the skills indicators. Competency 2.3 is the only competency where indicators from category A were left out of the scope.
In the competencies in domain 2, there are several relevant differences observed between the different groups of stakeholders. For competency 2.2 (see Appendix F, Table F2) parents selected the indicator that stressed the knowledge of how underachievement affects a student's self-concept. For the specialists in gifted education that specific indicator did not reach the top three of selected examples of professional behaviour.
For competency 2.3, we see more differences between selection patterns of the groups of stakeholders (see Appendix F, Table F3). Educational professionals find it important to offer a student different opportunities for choosing a product that shows the results of the learning process. However, for parents, this indicator seemed not to be as important as other indicators did; their highest selection rate was for the indicator K.2.E.3 aiming for the specialist to know how to use observation techniques as a way to explore the student's ability for self-regulation.
For this same competency, specialists in gifted education have the highest selection rate for indicator K.2.F.3, knowing how (f)actors of the student's ecological system affect a student's academic achievements. Amongst the parents, this specific indicator has the lowest selection rate of the six presented knowledge indicators.
Parents show the highest selection rate for the skills indicator aiming for exploration of how the student's social context affects the learning process,[190] whilst amongst educational professionals this item did not end up in their top three (see Appendix F, Table F8).
Competency 2.4 aims for educational partnership between the actors in the student's ecological system (see Appendix F, Table F4). Another difference can be noticed between the selection rates of parents and specialists in gifted education. Parents selected indicator K.2.B.4, focusing on the knowledge that students may show different task-orientation strategies in different circumstances. For this indicator, the selection rates of specialists in gifted education are not high enough to be included in their top three. However, the difference between the selection rates of indicator K.2.B.4 and K.2.D.4 ("knows

[190] V.2.C.3 – Table F8

how solution-focused conversation techniques can contribute to the development of a student's self-concept") is only marginal.
Competency 2.5 is about the specialists' ability to use instruments to identify and profile the educational needs of gifted students. The difference between the selection rates of parents and specialists in gifted education for the knowledge indicators is relatively large (see Appendix F, Table F5). Specialists in gifted education selected indicator K.2.D.5 the most. In this indicator, educational partners are presented as important sources when exploring a student's strengths and weaknesses. Parents, however, selected this indicator and the indicator K.2.B.5 the least.
A similar difference, although reversed, between parents and specialists in gifted education applies for the selection of indicator K.2.A.5. Parents selected this indicator, regarding the assessment of an underachieving student's academic level, more often than specialists in gifted education.
For the skills indicators, again we see a difference between parents and specialists in gifted education (see Appendix F, Table F10). Specialists in gifted education had high selection rates for indicator V.2.C.5, valuing the necessity of exploring gifted students' social-emotional problems from both the perspective of giftedness and the perspective of possible learning problems. In this competency domain on seeing educational needs of gifted students and the competency of the use of relevant instruments and tools for this assessment, parents had a higher selection rate for competency V.2.A.5 (assessing academic levels in order to provide the optimal educational challenges in the curriculum) than for competency V.2.C.5.
These differences of selection rates amongst the different groups of stakeholders are similar to the observed differences for domain 1.

6.8.2 Significant Differences between Groups of Stakeholders in Domain 2

In domain 2, differences between the selection rates amongst separate groups of stakeholders are significant for the knowledge indicators in the categories B ($p < .001$), C ($p = .034$) and D ($p < .001$) (See Appendix F, Table F12; see also Figure 20).
Parents made significantly different choices when we compare their selection rates to other groups of stakeholders (see Appendix F, Tables F13-F14). They selected less often the indicators concerning the effective use of different cognitive styles, the so-called B-indicators. Compared to specialists in gifted education, parents selected significantly more indicators that were aimed at stimulating interpersonal functioning (C-indicators).[191] They also selected significantly more often the indicators that were aimed at intrapersonal understanding (D-indicators) than the other groups of stakeholders.

[191] Bonferroni's multiple comparisons test is a more conservative approach than the ANOVA. That explains why, when using the ANOVA, a significant difference is reported while this is not the case using Bonferroni's test. As in this case the difference is so small it has been decided to mark the observed difference as significant.

Figure 20

Comparison of Selection Rates of the Knowledge Indicators in Domain 2 Between Groups of Stakeholders - (see also Table F12)

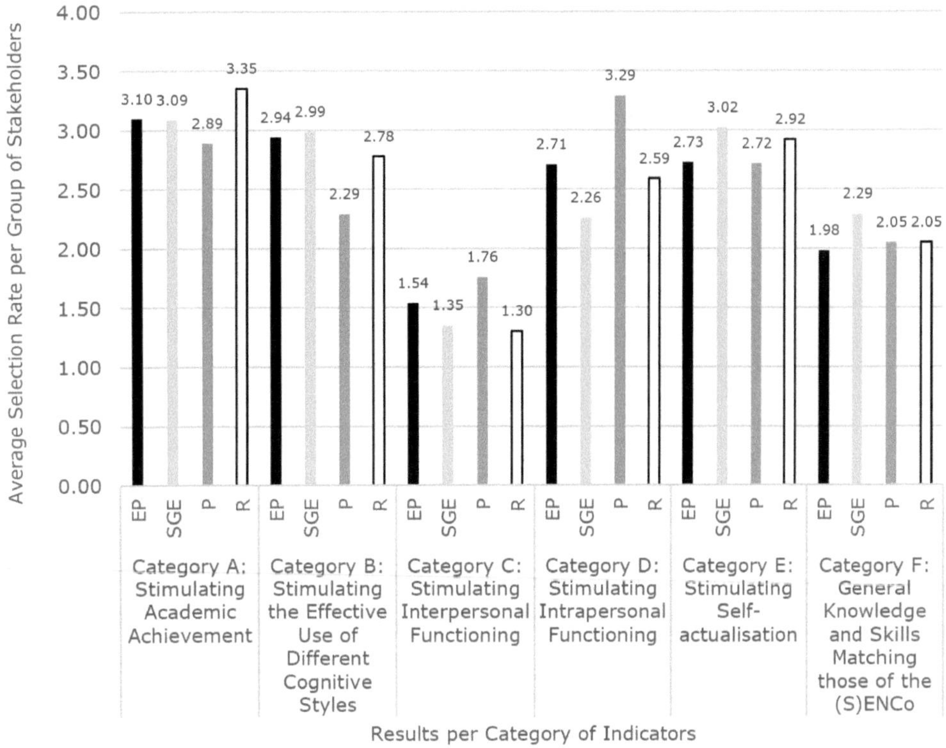

Note. N = 354 divided into four groups of stakeholders: EP = Educational Professionals, SGE = Specialists in Gifted Education, P = Parents, R = Rest. EP: N = 131, SGE: N = 103, P: N = 83, R: N = 37.

Test value = 2.5

Although educational professionals selected significantly fewer D-indicators than parents, they still selected these indicators significantly more often than specialists in gifted education.
Parents selected more skills indicators regarding the effective use of different cognitive styles, compared to those selected by the other groups of stakeholders (see Appendix F, Table F15; see also Figure 21). They also selected more indicators in category D than specialists in gifted education. These findings are significant for skills indicators B ($p < .001$), C ($p = .025$), D ($p = .007$) and F ($p < .001$) (see Appendix F, Tables F16-F17).
Educational professionals selected more indicators aiming for the stimulation of interpersonal functioning (C-indicators) than specialists in gifted education.

Finally, specialists in gifted education selected significantly more indicators that focused on generic counselling and advisory skills, matching the skills of a (S)ENCo (F-indicators) than both other groups of stakeholders.

Figure 21

Comparison of Selection Rates of the Skills Indicators in Domain 2 Between Groups of Stakeholders - (see also Table F15)

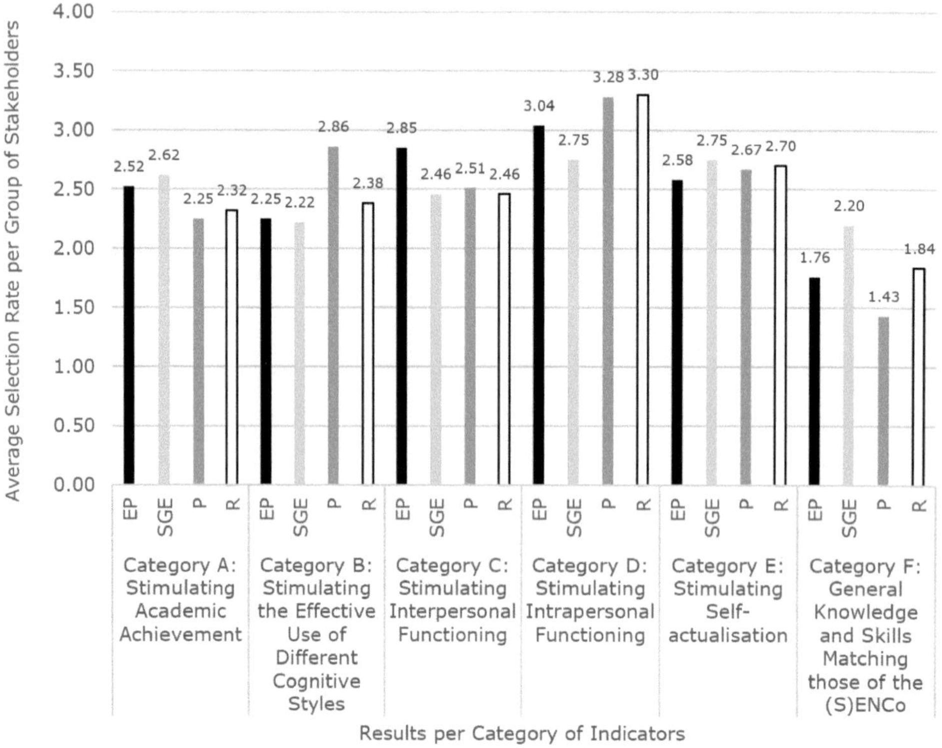

Note. N = 354 divided into four groups of stakeholders: EP = Educational Professionals, SGE = Specialists in Gifted Education, P = Parents, R = Rest. EP: N = 131, SGE: N = 103, P: N = 83, R: N = 37.

Test value = 2.5

6.8.3 Common Ground between Groups of Stakeholders in Domain 2

Common Ground Amongst all Groups

The initial frequencies analysis showed that there were 8 knowledge indicators and 9 skills indicators that ended up amongst the three most selected indicators between all groups of stakeholders (see Appendix F, Tables F18-F19). Indicators included all 6 categories.

A. Stimulating academic development (3)
B. Stimulating the use of different cognitive styles (2)
C. Stimulating interpersonal functioning (3)
D. Stimulating intrapersonal understanding (2)
E. Stimulating self-actualisation (5)
F. Generic counselling and advisory skills matching the skills of a (S)ENCo (2).

6

The second analysis aimed at determining if, amongst the 17 indicators, there were indicators that had the highest selection rates amongst all groups for one competency. One knowledge indicator and two skills indicators met that criterion (see Appendix F, Tables F30-F31). Two indicators aimed to develop the students' intrapersonal understanding, and one criterion could be considered a more generic counselling and advisory skill from category F.

- Competency 2.1
  - Knows what types of behaviour of the gifted student may be a sign for the development of negative self-concept.[192]
- Competency 2.2
  - Takes other indicators besides school achievements into consideration when estimating if the student may be gifted.[193]
- Competency 2.4
  - Is focused on solutions and learnable skills instead of being focused on problems when talking with a student.[194]

In Appendix F, Tables F20, F21 and F22 show that as separate categories, the knowledge indicators A ($\overline{X} = 3.07$), B ($\overline{X} = 2.79$, D ($\overline{X} = 2.70$) and E ($\overline{X} = 2.83$) were significantly more selected than expected ($\mu = 2.5$). This suggests that for seeing educational needs, indicators from these categories were thought to offer better examples of professional behaviour. As a category, the knowledge indicators C had a selection rate of $\overline{X} = 1.51$. The difference, -0.989, to the expected rate ($\mu = 2.5$) is significant, $t(353) = -17.598$, $p < .001$, this effect size represents a practical significance, $d = .94$. Indicators from category F also showed low selection rates, $\overline{X} = 2.10$. The difference to the expected rate ($\mu = 2.5$) is also significant, $t(353) = -6.945$, $p < .001$, the effect size represents a practical significance, $d = .37$ (see Figure 22).

[192] K.2.D.1 - Table F29
[193] V.2.F.2 - Table F30
[194] V.2.D.4 - Table F30

Figure 22

Mean Selection Rate of the Knowledge Indicators in Domain 2 - (See Table F20)

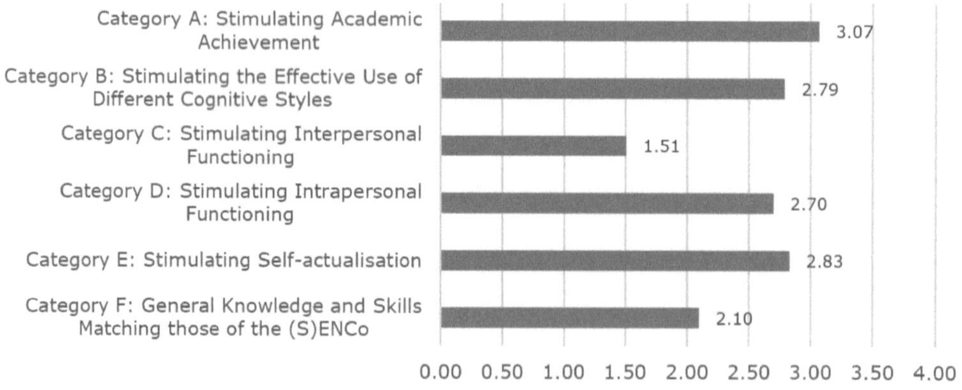

Note. N = 354. Test value = 2.5

Figure 23

Mean Selection Rate of the Skills Indicators in Domain 2 - (See Table F22)

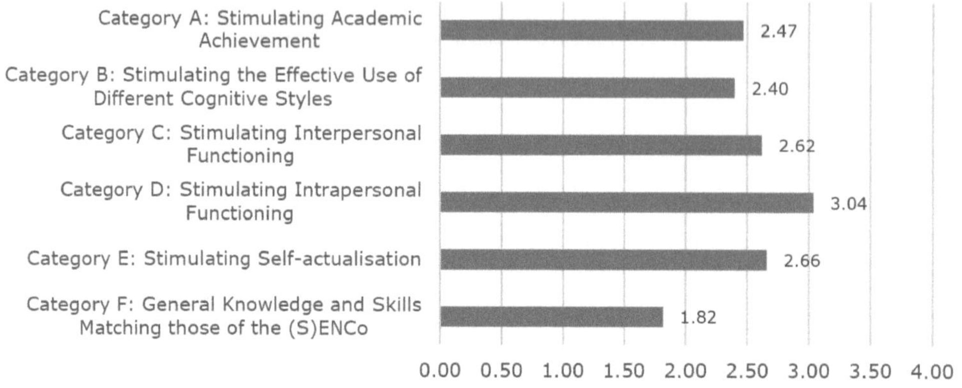

Note. N = 354. Test value = 2.5

Analysis of the selection rates of the skills indicators (Appendix F, Tables F23-F24; see also Figure 23) showed that indicators in category D ($\overline{X}$ = 3.04), E ($\overline{X}$ = 2.66) were selected significantly more often. Indicators in category F ($\overline{X}$ = 1.82) were selected significantly less than indicators in the other categories. Identically to domain 1, a Paired Samples Statistics test was undertaken to determine if differences between the grouped indicators ADE vs BCF would be significant (see Appendix F, Tables F25 to F28). Viewing the knowledge indicators as a clustered group, indicators in the cluster ADE $\bar{X} = 2.87$ were more often selected than indicators in the cluster BCF $\bar{X} = 2.13$. The difference, 0.738, is significant, $t(353) = 15.232$, $p < .001$, $d = .81$, the effect size is of practical significance. Viewing the skills indicators as a clustered group, the cluster ADE also showed higher selection rates, $\bar{X} = 2.72$ compared to the cluster BCF $\bar{X} = 2.28$. The difference, 0.444, is significant, $t(353) = 9.110$, $p < .001$, $d = .48$, the effect size represents a practical significance. The findings are similar to those of domain 1, so again, this suggests that indicators from the categories ADE are considered better examples of the competencies that they match.

6.8.4 Comparison between The Netherlands and Flanders

As with the previous competency domain, the final analyses for domain 2 focused on similarities and differences between The Netherlands and Flanders. Both countries showed significantly different selection rates for four indicators. Compared to the other categories of indicators, in domain 2, knowledge

Figure 24

Comparison of Mean Selection Rates in Domain 2 per Category of Knowledge Indicators Between Dutch and Flemish Stakeholders - (See Table F32)

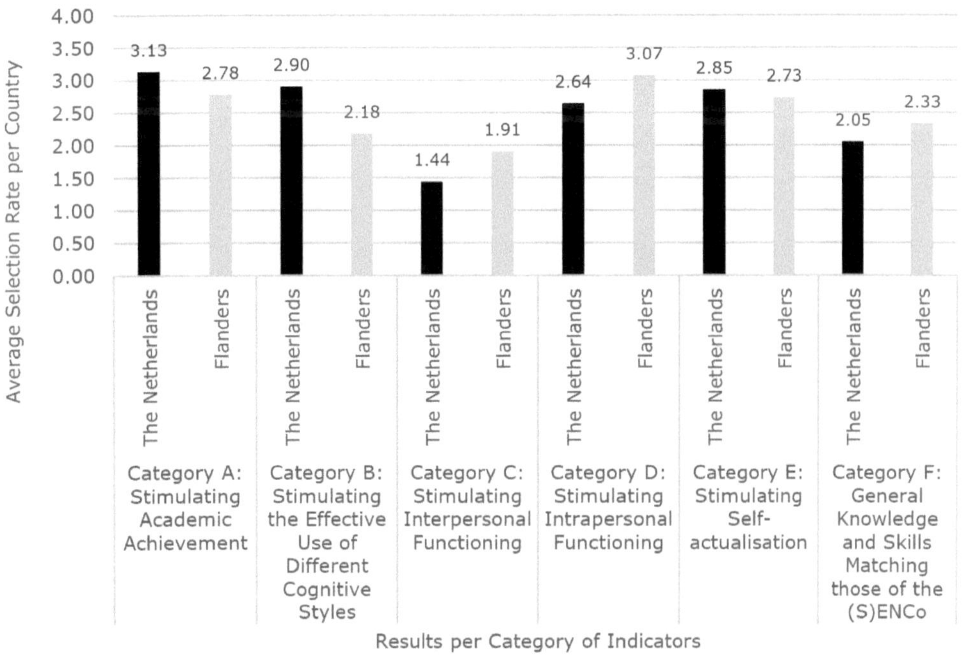

Note. N = 354. The Netherlands N = 299, Flanders N = 55. Test value = 2.5

indicators focusing on stimulating interpersonal functioning (C-indicators) showed the lowest selection rate ($\bar{X}$ = 1.51) (see Appendix F, Tables F20-F21). However, a comparison between both countries (see Appendix F, Tables F32-F33; see also Figure 24) showed that similarly to domain 1, Flemish stakeholders again had a significantly higher selection rate ($\bar{X}$ = 1.91) for this category of knowledge indicators than the Dutch stakeholders ($\bar{X}$ = 1.44). The difference, 0.471, is significant, $t(352) = -3.07$, $p = .002$, the effect size represents a practical significance, $d = .45$.
The so-called D-indicators showed a higher selection rate for both countries. A comparison between Dutch and Flemish stakeholders showed that stakeholders from Flanders selected the D-indicators more often ($\bar{X}$ = 3.07) than stakeholders from The Netherlands ($\bar{X}$ =2.64). The difference, 0.437, is significant, $t(352) = -2.42$, $p = .016$, the effect size meets the threshold for practical significance, $d = .36$.
In both countries, indicators in category A had the highest overall selection rates in domain 2. However, once again, there are differences between both countries. For the indicators in category A, the Dutch stakeholders showed a higher selection rate ($\bar{X}$ = 3.13) than stakeholders from Flanders ($\bar{X}$ = 2.78). The

Figure 25

Comparison of Mean Selection Rates in Domain 2 per Category of Skills Indicators Between Dutch and Flemish Stakeholders - (See Table F34)

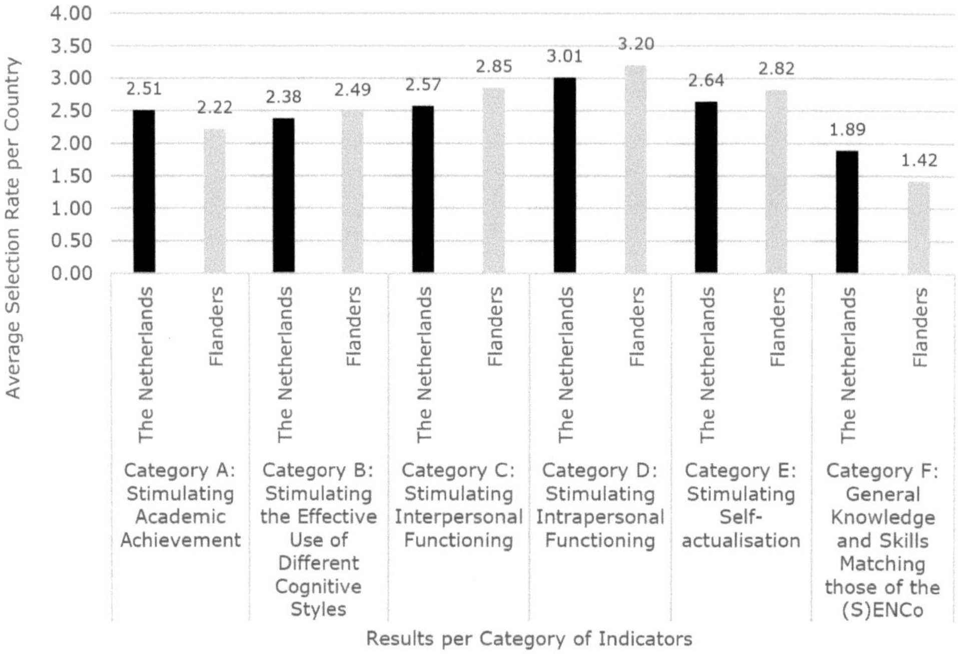

Note. N = 354. The Netherlands N = 299, Flanders N = 55. Test value = 2.5

difference, 0.345, is significant, t(352) = 2.23, p = .026, this effect size represents a practical significance, d = .33.
For the indicators in category B, the Dutch stakeholders also showed a higher selection rate ($\bar{X}$ = 2.90) than stakeholders from Flanders ($\bar{X}$ = 2.18). The difference, 0.715, is significant, t(352) = 4.926, p < .001, the effect size meets the threshold for practical significance, d = .72.
Regarding the skills indicators (see Appendix F, Tables F34-F35; see also Figure 25), the only difference in selection rates between both countries concerned indicators in category F. Skills Indicators in this category had the lowest selection rates ($\bar{X}$ = 1.82), but they showed higher selection rates amongst the Dutch stakeholders ($\bar{X}$ = 1.89) than amongst Flemish stakeholders ($\bar{X}$ = 1.42). The difference, 0.475, is significant, t(352) = 3.00 p = .003, the effect size represents a practical significance, d = .44.

6.9 Summarising

This chapter described domain 2 of the competency matrix: seeing the educational needs of gifted learners. The way specialists in gifted education apply the process of seeing the educational needs of gifted learners depends on their perspective on gifted education in general. At the same time, the process of seeing educational needs is influenced by current perspectives on how to identify educational needs in general. Both in The Netherlands and Flanders, a change and solution-focused approach with an explicit emphasis on the pre-diagnostic stage can be observed (Caufman & van Dijk, 2009; Pameijer & van Beukering, 2007; Prodia, 2012; van Meersbergen & de Vries, 2017).
The competences that are deduced from the literature analyses (see also Appendix A) are:

- 2.1 The specialist explores the strengths and weaknesses of the gifted student within the intrapersonal and interpersonal developmental domain.
- 2.2 The specialist assesses the student's developmental levels in all academic domains offered, considering the student's cognitive style.
- 2.3 The specialist takes on a systematic approach to the process of profiling educational needs and manages their time and actions in a way that contributes to the optimal results of this process.
- 2.4 The specialist approaches all actors in the student's ecological system as equal partners and values their information about the student's educational needs as such.
- 2.5 The specialist is able to develop a personal toolkit for profiling the educational needs of gifted students flexibly and appropriately using different instruments in different contexts.

In domain 2, as a joint group, stakeholders selected indicators aimed at stimulating academic development, the effective use of cognitive styles, intrapersonal understanding, and self-actualisation significantly more often than indicators for interpersonal functioning and the more generic counselling and advisory skills of specialists in gifted education. Educational professionals and specialists in gifted education had the highest selection rates for indicators A and D (see Appendix F, Tables F36-F37). Specialists in gifted education had an identical selection rate for knowledge indicators D and E, suggesting that for the competency of seeing educational needs, these indicators are equally important to them. The highest selection rates from parents were for indicators aiming at intrapersonal understanding.

If it is assumed that the selection of indicators is an indirect reference to what each group of stakeholders considers to be the most important in gifted education, a trend can be observed. In a similar way to that seen in domain 1, the different groups of stakeholders select indicators that are in line with the role they each have in the education and upbringing of gifted children. Parents have a strong focus on social and emotional needs. Educational professionals and specialists in gifted education include seeing a child's academic needs as part of the relevant skillset for this domain.

The differences between the stakeholders of the two countries are interesting. Although they have in common that the same indicators have relatively low or high selection rates compared to the selection rates of the other categories of indicators, there are small but significant differences. Stakeholders in Flanders place a greater emphasis on social and emotional indicators, while Dutch stakeholders accentuate in their selection the indicators that influence academic achievements. A partial explanation may be that the percentage of parents amongst the Flemish stakeholders was higher than the percentage of parents amongst Dutch stakeholders (see Appendix D, Tables D1-D2). As a group, parents showed higher selection rates for the C-indicators and D-indicators. This may have influenced the analysis of the differences between the selection rates for both countries.

The fact that Dutch stakeholders show a higher selection rate for the so-called F-indicators, may be explained by the fact that only 5.45% respondents from Flanders labelled themselves as specialists in gifted education, compared to 33.44% of specialists in gifted education amongst the Dutch stakeholders. Hence, these generic counselling and advisory skills may not be as relevant for the entire group of participating Flemish stakeholders.

7 Domain 3: Understanding the Educational Needs of Gifted Students

7.1 Introduction

In the third domain of the competency matrix, the knowledge and skills necessary to understand the educational needs of gifted students are explored. Again, the focus is on the interplay between generic teacher competences and the five domains affected by the notion of giftedness.
In previous chapters, a systemic approach towards gifted education and educational needs was advocated. Actors in the ecological system are not only partners in education, but also influence a student's development and, therefore, influence the student's educational needs. In this chapter it becomes even clearer that educational needs can only be understood within the context in which they appear (van Meersbergen & Jeninga, 2012; Zhang et al., 2012; Ziegler et al., 2017).[195] As understanding educational needs precedes the ability to respond meaningfully towards those needs, acquiring the knowledge, skills and attitude to interpret the educational needs of gifted students contextually is fundamental to the competencies of the specialist in gifted education.

7.2 The Necessity for a Multi-Focused Perspective

To stress the importance of the contextual interpretation of observed behaviour and educational needs, consider this vignette that could occur in any Dutch or Flemish school or classroom. Theories of gifted education are used to analyse the vignette.

> ***The Case.*** A gifted student confronted with a complex classroom assignment, complains that they find the task (still) boring. The student keeps postponing working on the task and finally misses the deadline. When asked, they refer to the task as being irrelevant, unimportant and moreover far below their capabilities. The teacher interprets these comments and the matching behaviour as a sign of the student being a so-called "non-consumer" (Figg et al., 2012; Galbraith & Delisle, 2015; Heacox & Cash, 2017; Siegle & McCoach, 2017). The teacher decides that they want to address this type of underachievement and adjust the student's curriculum by offering more complex tasks. After a while, they find themselves bending over backwards to find assignments that match this student's interests and capabilities. However, the desired result lags behind. The student still complains that the curriculum offered is boring and below their capabilities, and they still do not hand in any of the scheduled assignments.
> ***A Different Perspective.*** Boredom may not be the only factor that is influencing this student's behaviour. Something else may be the trigger

[195] Appendix B, Table B11, K.3.C.1.

for this response as well. Previously, this student may always have been successful; for the first time in their educational career they feel consciously incompetent, which may trigger unpleasant emotions that the student has never experienced before (Siegle & McCoach, 2020; Venderickx & Kieboom, 2017). The student may consider this new situation incompatible with their social status, which is, in their perception, defined by their success rate. Therefore, they may avoid being upfront about their feelings of incompetence. They may have the idea that feeling incompetent means that they are no longer gifted or that they were not gifted after all (Dweck, 2006, 2017).[196] Perhaps the student may think that their incompetence will make other students bad-mouth them. Over the years, within their social context, it may have become socially acceptable for them to use the expression "I am bored". Hence, they may see the use of this expression as the perfect emergency escape from doing the assignments. As soon as they inform the teacher that the task is boring, the teacher starts searching for a new task, and the process starts all over again.

Educational Needs. Analysis of this vignette stresses the importance of understanding educational needs contextually. The student in this vignette has several needs that can be described as educational objectives[197]: learning how to deal with a feeling of incompetence, learning how to adjust their estimation of self-efficacy and learning how to ask for support effectively. This student needs to be provided with additional scaffolds and instructional support in order to master the set objectives, and maybe even some extrinsic motivation to keep on trying when they are not successful at the first attempt. A student who is bored because assignments are below their abilities needs a different educational response. Their educational need is to be challenged at the right level and with sufficient support. The specialist in gifted education, therefore, should never interpret the student's educational needs without having explored the educational context and without having consulted with the student.[198]

The skills to contextually understand a gifted student's behaviour are underpinned by theoretical knowledge (Bronfenbrenner, 1979; Pameijer & van Beukering, 2007; van Meersbergen & de Vries, 2017; van Meersbergen & Jeninga, 2012; Ziegler et al., 2017).[199] Considering that at Dublin Descriptor level 7, the attitude towards critical thinking is seen as an important sign of professional behaviour, the specialist in gifted education should not only take notice of theories of giftedness that match their personal perspective, but should also take alternative perspectives into account when they try to understand the educational needs of their gifted students.[200] The ability to see behaviour from a

---

[196] Appendix B, Table B12, K.3.D.2.
[197] Appendix B, Table B15, V.3.F.5.
[198] Appendix B, Table B13, K.3.D.3.
[199] Appendix B, Table B11, K.3.E.1; Table B15, K.3.C.5.
[200] Appendix B, Table B15, K.3.B.5, V.3.B.5.

different perspective will create a better understanding of the needs to be inferred.

As explained in previous chapters, different theories of giftedness offer different explanations for behaviour that may be characterised as a personal trait in gifted students (Dai & Chen, 2014).[201] The skill of identifying how giftedness may influence a student's behaviour is, therefore, not an easy one to develop.[202] The notion of overexcitabilities in relation to giftedness, for example, is just one of the topics that have caused serious discussions during recent years (Pfeiffer & Muniz Prado, 2018; Vuyk et al., 2016). The interpretation of behaviour corresponding with the characteristics of overexcitabilities as being typical for gifted students is correct from one perspective but has also come in for criticism from a different perspective on giftedness. In theories based on the gifted child paradigm, Dabrowski's overexitabilities are seen as a logical explanation for certain behaviour (Kreger-Silverman, 2013; Piechowski, 2014), while in theories based on the talent development paradigm, scholars may not agree with this explanation (Freeman, 2019; Gubbels, 2016; Subotnik et al., 2011). However, that does not mean that an individual gifted student may not be struggling to cope with overexcitabilities (Rinn, 2018). The fact that overexcitabilities cannot merely be explained by a high IQ does not mean that the specialist in gifted education can ignore the possibility that a student indeed has to cope with overexcitabilities.

Another example demonstrating the importance of a multidimensional perspective concerns twice-exceptional students. Webb et al. (2016) described the risk of misdiagnoses and missed diagnoses of twice-exceptional students. If a student's results lag behind what is expected, underachievement is often the first thing that comes to mind (Weterings, 2017). However, just as it can be for any other student, gifted students can be hindered by learning difficulties and learning disabilities (Reis et al., 2014). Lack of a multidimensional perspective on giftedness may prevent understanding of why this student is failing to achieve as expected (Baum et al., 2017; Burger-Veltmeijer & Minnaert, 2016). This student may display serious behavioural problems as a result of the collision between their natural abilities and their natural inabilities (Trail, 2011). In this case, focusing on behavioural problems without taking their intellectual abilities into account does not help the student; things may become worse because the real problem remains unaddressed. To understand the student's behaviour and their future educational needs, a multi-focus approach is necessary (Baum et al., 2017). In cases like these, it might be helpful for teachers to have instruments, at a pre-diagnostic stage, that determine a perspective which may provide fruitful intervention strategies, while also offering a combined perspective (Amend et al., 2009; McCallum et al., 2013).

The use of these instruments rightfully requires the involvement of different actors from the student's ecological system. This is similar to the involvement of

[201] Appendix B, Table B15, K.3.E.5.
[202] Appendix B, Table B15, V.3.E.5.

the educational partners for the identification of gifted and talented students, as described in Chapter 5.
Based on observations of the student's behaviour in different situations and under different circumstances, the instruments for profiling a student's educational needs provide the specialist in gifted education with a *snapshot* for a *specific* time frame. Changes in conditions may lead to different observations (McBee, 2010). Meeting educational partners to discuss how to interpret their observations provides a perspective for understanding (Pameijer & van Beukering, 2007; Subotnik et al., 2017; Ziegler et al., 2017). Setting up regular meetings between the specialist in gifted education and these partners provides opportunities to spot changes in the student's ecological system and the effects on their education over a relatively short term (Coleman & Johnsen, 2011; van Swet, 2009).[203] That creates an in-depth understanding of a student's case.
Parents and carers not only have the right to be involved in the process of understanding educational needs; they often express the importance and the necessity of becoming involved. Pameijer and van Beukering (2007) found that including parents and carers offers them the opportunity to explain how they experience the student in the home environment. That information may make a difference in the search for the best possible educational strategies (Peters & Oveross, 2020; van Swet, 2009).[204]
Working from premises based on the change and solution-focused approach, the specialist in gifted education focuses on chances and opportunities (Cauffman & van Dijk, 2009; de Shazer et al., 2012; van Meersbergen & de Vries, 2017). Working with the educational partners to determine when a problem is not as intense as thought or is non-existent, is an important part of this approach.[205] It creates an understanding of how strengths and weaknesses can be translated into educational needs.[206] This approach also creates the opportunity to explore, with the educational partners, how their educational strategies and responses influence the student's behaviour and needs (Ziegler et al., 2017).[207] Exploration of the exception in an educational situation brings to light the circumstances under which the educational environment is most successful (Cauffman & van Dijk, 2009).[208] That information provides clues for a more effective approach to those situations that need optimisation (de Shazer et al., 2012; van Swet, 2009).

7.3 Understanding Academic Needs from an Ecological Perspective

Gagné (2010a) describes how talent development can only take place if the student takes part in an educational process that systematically strives to develop certain skills, aiming for a specific goal over a long period and without disruptions to the process. As in other talent development theories, Gagné

[203] Appendix B, Table B14, V.3.F.4.
[204] Appendix B, Table B14, K.3.F.4.
[205] Appendix B, Table B11, K.3.F.1, V.3.F.1.
[206] Appendix B, Table B15, K.3.F.5.
[207] Appendix B, Table B15, V.3.A.5.
[208] Appendix B, Table B15, V.3.C.5.

(2010a) specifically refers to the interplay between the ecological system and the student (Gubbels, 2016; Kieboom & Venderickx, 2017a, 2017b; D. Matthews & Foster, 2009; Renzulli, 2012; Sternberg, 2009; Subotnik et al., 2017). Based on Bronfenbrenner's (1979) concept of systemic thinking, van Meersbergen and de Vries (2017) describe behaviour to be the result of interaction between personal traits under different circumstances and in multiple contexts. Viewed from this perspective, the development of a gifted student's behaviour is no different to that of any other student. Based on talent development theories it can be stated that the student's personal traits influence the actors in the ecological system and vice versa.[209]

The ecological system includes the physical learning environment. Hence, the physical learning environment affects the student's development (Hertzog, 2017a, 2017b; van Meersbergen & Jeninga, 2012). A talent unsupported by opportunities in the physical learning environment, cannot be developed. However, in the regular classroom, interplay between personal traits such as learning behaviour and the physical educational environment, although certainly there, may not be so obvious at first sight. As a result of this interplay, expected results may lag behind in reality. For example, highly creative students with great problem-solving skills need the opportunity to give their thoughts free flow (Kroesbergen, 2017). Without the opportunity to wander off in thought, these students may lack the chance to explore all of the options to solve the problem that confronts them. Some of these students need physical space to move around in, helping them to order their thoughts. If that space is not available, the opportunity is not offered, or the teacher experiences their need to move whilst thinking as a disturbance for the daily classroom routine, these students may not be able to use their creative ability fully. Other students may prefer an educational environment that is optimally structured to keep their minds going. If these students have to function in a classroom with lots of stimuli, not structured at all, it calls highly on their executive skills and may even prevent them from achieving highly.[210] The specialist in gifted education, should, therefore, be capable of understanding how the physical environment affects a student's behaviour. This understanding will offer them clues on how to optimise interventions but also acknowledge the limitations of possible interventions.

In the NAGC standard for professional behaviour of the specialist in gifted education, there is a strong emphasis on knowledge and skills around the curriculum, and educational and instructional strategies (Johnsen et al., 2016). Next to enrichment, strategies for curriculum compacting is thought to be one of the most successful interventions in gifted education (Ahlers, 2017; Coleman & Johnsen, 2011; Heacox & Cash, 2017; Johnsen et al., 2016; Leavitt, 2017; Renzulli, 2012). The main reason for its success is that compacting responds to the student's abilities as it affects the pacing of the curriculum. Enlarging the gaps between the educational stepping stones in the curriculum is a relatively natural way to make the learning content more complex to grasp, and therefore

[209] Appendix B, Table B15, K.3.A.5.
[210] Appendix B, Table B11, K.3.A.1; Table B13, K.3.B.3.

more challenging for gifted learners, without losing the high achievement level that may be expected of gifted students (Henshon, 2018). Understanding that a student's motivation to practise the same knowledge and skills decreases once the knowledge and skills are mastered, paves the way for compacting as a successful intervention (Bakx et al., 2016; Heacox & Cash, 2017; Troxclair, 2010). To interpret to what extent the pacing in the required curriculum matches both the student's ability and their existing knowledge and skills, the specialist should be aware of the curricular constructs within the different academic domains (Ahlers, 2017; Drent, 2009).[211] At this point, exploration of the balance between the most desirable and most realistic interventions affects the pedagogical and didactical strategies that the specialist may choose.[212] Ideally, an individualised curriculum may offer the best opportunities for the student's developmental process. In reality, this may not be achievable in the context of the regular classroom, but may have a better chance if supported via pull-out programmes aimed at supporting the classroom teacher through personalised learning. Drent (2009) describes how compacting can be based not only on short-term results through skipping what the student has proven to have mastered but can be even more successful from a classroom management perspective, when using their long-term strategy. This strategy is based on the assumption that there are similarities between the needs of high ability learners in the way that they can process curriculum content: the ability to take big learning steps and little need for repetitive exercise. These similarities in needs can be addressed in small groups.[213] Understanding these needs leads to a strategy where the specialist in gifted education selects what assignments can be missed in the curriculum without the student previously having to prove that this content has already been mastered and without the loss of the student's high achievement levels. The point of deployment for this strategy is to be sure that the student's achievement levels are known, that these levels have been stable over a long period, and that the latest achievement test results show the student's zone of proximal development in the domain to be compacted. In this strategy, the specialist in gifted education preselects elements of the curriculum that are new to the student. Only this new learning content is offered and taught. Of what is left, a limited selection is made to ensure that the student gets enough practice to internalise strategies and develop the ability to apply them within an acceptable amount of time. Conditional to this approach is the understanding that being gifted or being an able learner in a specific domain does not mean that the student can manage new content without instruction, once the pacing has stretched to the student's abilities (Drent, 2009). An implication of the compacting strategy is the creation of a situation where stretched pacing in the curriculum allows the student to initially experience being consciously incompetent. Instruction provides the student with sufficient knowledge and skills to develop a positive feeling of self-efficacy which then motivates them to reach their zone of proximal development (Bakx et al., 2016;

---

[211] Appendix B, Table B12, K.3.A.2; V.3.A.2.
[212] Appendix B, Table B13, K.3.F.3.
[213] Appendix B, Table B13, V.3.F.3.

Maslow, 1954; Vygotsky, 1978).[214] Van Gerven and Drent (2012) applied their compacting strategy to the majority of the Dutch curricular methods. They show how teachers can adjust their classroom management effectively to create the opportunity for gifted students to join group instruction when needed.[215] Communication with the student is necessary to understand how much instruction, and of what kind, the student needs. Gifted students are very well able to describe what they need to make instruction effective (Bakx, 2019).[216] Involving them in the process of making decisions about outlines of an individual educational plan calls on their feelings of being competent, autonomous learners acting in a safe pedagogical relationship with their teachers.[217] This approach is in line with the change and solution-focused strategies as suggested by Pameijer and van Beukering (2007) and van Meersbergen and de Vries (2017). It also matches the concept of ownership as discussed in Chapter 5, and as will be discussed further in paragraph 7.4.

Regarding communication between the specialist in gifted education and their gifted students, catering for the needs of the gifted student implies that the specialist in gifted education modifies their use of language to the student's level of understanding (Peterson, 2008; Wood, 2018).[218] Although gifted students are often ahead in their language abilities, that does not automatically imply that they have the same level of understanding where their emotions are involved (Freeman, 2019; Kieboom & Venderickx, 2017a; Kreger-Silverman, 2013; Neihart et al., 2016). To understand a student's educational needs, the specialist in gifted education needs highly developed communication skills that integrate verbal and non-verbal messages.

Successfully challenging gifted students comes with conditions. To understand the student's response to the curriculum offered, the teacher's attitude also needs to be examined. How a big part of understanding the student's educational needs is influenced by the way ecological influences on the student's behaviour are taken into account, was detailed at the start of this chapter. Siegle and McCoach (2017) describe, through their achievement orientation model, how the student's motivation for learning is influenced by the way and to what extent they feel supported by their educational environment.[219] Vygotsky's description of the zone of proximal development can best be explained as a process wherein the student, supported by the teacher, closes the gap between what they have mastered and the next level of knowledge and skills (Vygotsky, 1978). During this process, the teacher aims to support the student to use existing knowledge and skills effectively. Successful teachers, therefore, understand that scaffolding the learning content and the application of differentiated instructional strategies are conditional to the learning process of all students, including gifted students (Kroesbergen et al., 2015; Leavitt, 2017;

[214] Appendix B, Table B13, K.3.A.3; Table B11, K.3.D.1.
[215] Appendix B, Table B13, V.3.A.3.
[216] Appendix B, Table B11, V.3.D.1.
[217] Appendix B, Table B13, V.3.E.3.
[218] Appendix B, Table B14, K.3.D.4, V.3.D.4.
[219] Appendix B, Table B12, K.3.F.2.

Roberts & Inman, 2013).[220] Myths about gifted education still exist (Moon, 2009b; Sisk, 2009). As a result, some teachers still think that gifted students should be able to master new objectives without support. Thus, it is important that the specialist in gifted education can communicate to their colleagues what would be helpful for the student's learning processes. Showing their colleagues how a teacher's attitude and support contributes not only to the gifted student's wellbeing but also to the gifted student's achievements is relevant to the process of understanding educational needs.[221]

In their behavioural model, Marzano and Kendall (2007) describe how a student's perception of their ability regarding the requirements of a task influences the learning process. The behavioural model is based on Marzano and Kendall's taxonomy of thinking skills. This taxonomy includes thinking skills at the level of the cognitive system (retrieval, comprehension, analysis, and knowledge utilisation), skills based on the metacognitive system (specifying goals, process monitoring, monitoring clarity and monitoring accuracy) and skills regarding the self-system (examining importance, examining efficacy, examining emotional responses and examining motivation).

In this taxonomy, examination of the importance of a task refers to the student's personal goal valuation and the social meaning and context that this knowledge and skills may have for a student. The affective component of learning is not only determined by the student's goal valuation, but also by the goal valuation of their peers. Learning takes place in a social context (Wallace & Adams, 2019). In a learning environment where classmates express a negative judgement towards curricular objectives related to the gifted student's assignments, or where classmates respond negatively to a student's achievements based on jealousy or social codes, gifted students may adjust their original goal valuation (Freeman, 2019). Understanding this impact means that it may be easier for the specialist in gifted education to understand how the social environment affects the gifted student's choices.[222]

Integration of the top levels of thinking skills with thinking skills based on the cognitive system, as presented in this taxonomy by Marzano and Kendall (2008), offers insight into a circular process that takes place during the entire learning process, for the behavioural model does not end after the first cycle. The cyclical process of this model stays active as long as the student is involved in a task. This behavioural model provides the specialist in gifted education with the opportunity to understand which factors during the learning process may cause the student to give up working on an assignment, even when the topic has the student's interest and when the student was enthusiastic at the start of the project. It shows that the process and results are equally important. For gifted students, teachers and parents often focus on their achievements, and, when these achievements do not match their expectations, the learning process is often reviewed with a central focus on the student's efforts, motivation, and willingness to learn. The mindset theory is, although very useful, a clear

[220] Appendix B, Table B12, K.3.F.2.
[221] Appendix B, Table B12, V.3.F.2.
[222] Appendix B, Table B12, K.3.C.2, V.3.C.2.

example of that focus (Dweck, 2006, 2017). Widening that perspective to the influence of all actors in the ecological system, not only to the student's mindset but also to the educational context in which it is provided, is relevant. With the focus on learning and the learning process, teachers sometimes forget that reflecting on the learning process is only relevant within the context of the aimed objectives and the results of the process (see paragraph 6.4). Creating a learning environment and taking a didactical approach that aims for equal importance for both process and results, provides a solid context for academic development.[223]

In Marzano and Kendall's taxonomy (2008), each level of thinking skills calls on different teacher support strategies. To what extent the intensity of these support strategies should be applied, is different for each student.[224] Different thinking skills call on different approaches to applying these skills. Having a preference for specific thinking activities may be helpful in one task, but may be limiting for another task. The ability to switch between tactics and approaches leads to higher achievements (Coffield et al., 2004; Hattie, 2013).[225] Understanding what instructional strategies should be recommended, depends on analysis of which thinking skills are being called on by the curriculum offered, combined with the student's profile of strengths and weaknesses in the application of these thinking skills.[226] Marzano and Kendall's theory is not focused on gifted education. However, their theory offers a combination of regulation and motivational theories, relating the notions offered in these theories to the actual level of mental activities needed in the current social-constructivist approach to education.

7.4 Understanding of Cognitive Needs from an Ecological Perspective

The understanding of cognitive needs, within the educational paradigm of social-constructivism, focuses on several aspects of ownership (Jolles, 2011; Marzano & Heflebower, 2012; Oostdam et al., 2007; Thijs et al., 2014). Ownership in education is about *how* to take responsibility for your actions during the learning process. Students have to learn how to take on this responsibility within the institutional context of their education, as well as in the context of intrapersonal understanding.[227] Cognitive needs are, therefore, intertwined with intrapersonal needs. Peterson and Lorimer (2011) discuss how some gifted students experience difficulties in intrapersonal and interpersonal understanding. As a result of this, these students may interpret giftedness as the obligation to be completely self-sufficient, as in: having the responsibility of solving their own problems, maintaining a facade of invincibility or being reluctant to ask for assistance. Based on this perspective, ownership is limited to taking care of oneself without the help of others. From a similar perspective, some teachers may interpret being gifted as the student's ability to learn without the daily

[223] Appendix B, Table B11, V.3.A.1.
[224] Appendix B, Table B14, K.3.A.4.
[225] Appendix B, Table B12, K.3.B.2.
[226] Appendix B, Table B14, V.3.A.4, V.3.B.2.
[227] Appendix B, Table B13, K.3.E.3.

support of a teacher (Peterson & Lorimer, 2011). They may see giftedness as a signal for transferring their responsibility for the education of the student to the student themselves and may forget that, however gifted the child, they still need adult support and adult responsibility (Frumau-van Pinxten, 2012). Jolles (2011) describes the complexity of the transfer of responsibilities attached to ownership between a teacher and their students. Teachers cannot expect students to take ownership without teaching them how to do this and without regard to what should be expected developmentally considering their age.[228] Gifted students are no exception to the rule. The younger the student is, the more limitations there are for transferring responsibility and ownership. Hence, "ownership" is a modern but complex adagio.

Ownership requires commitment as much as it can result in a commitment (Cauffman & van Dijk, 2009; de Shazer et al., 2012). Students who do not feel a commitment to a task or educational objective are not likely to take ownership of it. With the lack of ownership, chances for success decrease (Cauffman & van Dijk, 2009; de Shazer et al., 2012).[229] Commitment requires motivation. Siegle and McCoach (2017) showed through their Achievement Orientation Model how the interplay between different factors influences motivation. Although the original model dates from 2005, in the current literature on underachievement and motivation, this model is still referred to as relevant (Heacox & Cash, 2017; Hoogeveen, 2017b). The balance between a student's goal-valuation, environmental perspective and self-efficacy is crucial. This balance leads to self-regulated behaviour, as in having realistic expectations and implementing the appropriate strategies to complete tasks successfully (Siegle & McCoach, 2017). The reverse argument would be that positive goal-valuation, a positive environmental perspective and a high level of self-efficacy contribute to the motivation to show commitment to a task. If the commitment is there, the student is likely to take ownership of their developmental process. Understanding of this mechanism helps the specialist in gifted education to recognise why some students do not achieve to their potential.[230] Based on this understanding, it is easier to recognise which factors form possible risks in the student's developmental process.[231]

Metacognition is conditional for the use of self-regulation skills (Snyder et al., 2011). Consequently, the development of metacognitive skills within the educational paradigm of social-constructivism is highly valued. Metacognition is the ability to understand, regulate and use one's cognitive processes in a constructive manner (Figg et al., 2012; Oppong et al., 2018; Snyder et al., 2011). Snyder et al. (2011) describe the different skills students should have in order to be able to assess their performances and achievements. The knowledge of cognition and the regulation of cognition are considered as key components. The knowledge of cognition refers to the actual knowledge a student has about different cognitive processes, relevant strategies, and their own abilities. To

[228] Appendix B, Table B14, V.3.E.4.
[229] Appendix B, Table B11, K.3.B.1.
[230] Appendix B, Table B15, V.3.D.5.
[231] Appendix B, Table B15, K.3.D.5.

transfer this knowledge of cognition to the regulation of cognition, students need to develop a sufficient level of self-understanding.[232] The regulation of cognition refers to the ability to effectively implement cognitive strategies and monitor one's performances and achievements. Research shows that the development of metacognitive skills takes effort (Kroesbergen et al., 2017; Obergriesser & Stoeger, 2015). Students develop metacognitive skills and self-regulated learning in an educational context that calls on these skills explicitly. Teachers have to invest in the development of these skills (Hertzog, 2017a, 2017b). Through this perspective, it becomes clear that the specialist in gifted education benefits from the knowledge of how to stimulate the development of metacognition and self-regulated learning in gifted students.[233]

Self-reflection is needed for the development of metacognition (Anderson et al., 2014; Dawson & Guare, 2009; Ylvisaker & Feeney, 2014; Zelazo & Carlson, 2012). The actual knowledge of different reflection techniques provides strategies both for the specialist to explore their own choices and actions, and strategies that help students to develop their ability for self-reflection (Biesta, 2012; Korthagen & Nuijten, 2018; Loughran, 2002, 2015).[234]

Self-regulated learners are able to think contextually about their knowledge and actions in relation to the effects of their actions and beliefs (Oppong et al., 2018). Stimulating students to think about a sequence of consecutive statements helps them to develop strategies for self-reflection. An example of a series of statements is: "This is what I knew; based on my knowledge, this is what I thought; based on my thoughts, these were my actions; and as a result of my actions I got... Now I know that...; based on my current knowledge, I think that...; based on my thoughts my next action will be to...; I aim for the following result; and I expect that..." Teaching students to explicate these sequences of thoughts provides the student with the necessary self-understanding. It makes clear how and where they can regulate their behaviour during a learning process. At the same time, it provides the specialist in gifted education with the necessary understanding of the student's educational needs. It gives them the opportunity to explore what their students expect of themselves during the learning process and what they experience as successful learning behaviour.[235] The student's explicated use of self-regulated learning strategies helps the specialist in gifted education to gain a better understanding of the student's educational needs (Oppong et al., 2018). At the same time, examining the student's strengths and weaknesses in their ability to regulate their thinking and actions, in relation to their overall development, adds to the level of understanding. From this perspective, metacognitive skills not only keep the student on track, but also help the specialist in gifted education to stay on track while guiding this student. This knowledge creates an opportunity for the specialist to discuss with the student what they experience as helpful and stimulating.[236]

---

[232] Appendix B, Table B13, V.3.D.3.
[233] Appendix B, Table B14, K.3.E.4.
[234] Appendix B, Table B14, K.3.B.4, V.3.B.4.
[235] Appendix B, Table B12, V.3.D.2; Table B11, V.3.B.1.
[236] Appendix B, Table B11, V.3.E.1.

As stated, Marzano and Kendall's new taxonomy (2007, 2008) is a clear example of a didactical perspective on how to translate the 21st century skills into educational objectives. This taxonomy also emphasises the call on regulation skills. The more complex the mental task is, the higher the call on the student's executive skills. In their taxonomy, researching and experimenting, together with decision making and problem solving, are higher order thinking skills. These tasks require, amongst others, the ability to structure information, organise procedures and processes, and to regulate attention. In an educational perspective where gifted students should be offered authentic learning experiences to develop domain-related expertise, learning calls on existing executive skills and strengthens these skills at the same time (Kroesbergen et al., 2017; Sternberg, 2009).[237] Understanding this reciprocal process may influence the specialist in gifted education when selecting students for special programmes and group work.[238]
As explained in Chapter 1, curricular changes in The Netherlands and Flanders are usually provided within the regular classroom. Gifted children spend a relatively small part of their time in ability groups outside of their classrooms. As a result, most enrichment activities are offered within the physical environment of the regular classroom. Teachers and specialists in gifted education provide challenging and complex tasks to respond to the academic needs of gifted students. However, with a large number of students in their classroom, the growing complexity of the educational needs of students with mixed abilities in the regular classroom, limited resources for additional classroom assistance, and misconceptions about what giftedness means, this sometimes leads to the unintended educational neglect of students working on their enrichment projects (Bakx et al., 2016; Frumau-van Pinxten, 2012). These enrichment projects aim to challenge gifted students. The challenge may call on their executive skills in a way which exceeds their current capabilities, just as happened to the student in the vignette used at the start of this chapter. The teacher's expectations of a student's executive skills, in a situation where this student is consciously incompetent towards the learning content, influences how they support that student (van Gerven, 2020). To understand the required level of attention and support that the student needs, the teacher needs to explore to what extent the student meets expectations regarding their executive skills.[239]
The understanding of the academic and cognitive needs of students cannot be seen as separate from their intrapersonal and interpersonal needs. It is a well-accepted understanding that development requires a safe pedagogical environment (Buyse et al., 2009; Leeman & Wardekker, 2004; Nieuwboer, 2015). At school, that environment not only depends on the teacher, the curriculum, and the physical environment; it also depends on the opportunity to meet like-minded peers with similar intellectual abilities (Venderickx & Kieboom, 2017).

---

[237] Appendix B, Table B13, V.3.B.3.
[238] Appendix B, Table B12, K.3.E.2.
[239] Appendix B, Table B12, V.3.E.2.

7.5 Understanding Intrapersonal and Interpersonal Needs from an Ecological Perspective

Peterson and Lorimer (2011) describe how, from a perspective of prevention, gifted students can benefit from a small-group affective curriculum for gifted children.[240] Although they stress that there is no evidence that gifted students are more likely to develop social-emotional problems than their non-gifted peers, they do point out that these students may be more at risk. Research undertaken by Lee et al. (2012) is helpful in understanding these risks, because these authors provide a context within which these risks may occur.[241] Referring to research done by experts like Austin and Draper, Gross, Kreger-Silverman, Davis and Rimm, between 1981 and 2002, the three authors explain that gifted students who are not able to find like-minded peers may opt-out of social activities. They explain how being ahead of their age group may lead to situations of misunderstanding and miscommunication between the gifted student and their non-gifted classmates. Dissimilarities between interests and abilities on the one hand and, on the other, the wish to belong to a peer group, may lead to compromises that feel good in the short term but may feel uncomfortable in the long run. Students who find it difficult to compromise or who are having difficulties with the consequences of their compromises may prefer to work alone. Another risk factor mentioned for some gifted students may be their wish to belong and to conform, which can lead to a masking of their giftedness and the development of social identities that they experience to be more socially acceptable. For The Netherlands and Flanders, there is no current research that either confirms or contradicts these risk factors, nor is there any current research that estimates how many Dutch and Flemish gifted students in the age group of 4-12 years are affected by these risks in reality. Considering the risk mentioned, the specialist in gifted education should understand why it is important for gifted students to have the opportunity to work with like-minded peers (Oppong et al., 2018).[242] Gifted students who find it difficult to fit in socially may benefit from the opportunity to meet other gifted students. Sharing the same interests and being able to communicate on an equal level creates a sense of belonging (Kitsantas et al., 2017; Oppong et al., 2018). It offers the opportunity to engage in social relationships from a similar perspective (Neihart & Yeo, 2018). Friendship is important for all children and often develops among children sharing the same interests, having similar traits, behaviours and attitudes. The chance that gifted students will find like-minded peers in a regular group is considered to be smaller than for non-gifted students (Neihart & Yeo, 2018; Rinn, 2018). Next to participating in a mixed ability group, having the opportunity to meet like-minded peers at school may contribute to the wellbeing of gifted students (J. Cross, 2012; T. Cross, 2018).
Learning and working with other students of equal ability has positive effects on gifted academic and social development (Kitsantas et al., 2017; Lee et al.,

[240] Appendix B, Table B13, V.3.C.3.
[241] Appendix B, Table B11, V.3.C.1.
[242] Appendix B, Table B13, K.3.C.3; Table B14, K.3.C.4.

2012).[243] It teaches them to listen to each other, take on different perspectives and reflect on their own knowledge, skills, and behaviour (Oppong et al., 2018). When gifted students experience difficulties in their interpersonal functioning, it is advisable to explore to what extent these students have the opportunity to meet gifted or like-minded peers (Neihart & Yeo, 2018; Rinn, 2018).[244] It could well be that students have difficulties in their interpersonal functioning due to the lack of opportunities to practise social skills or to discuss how they can best respond to social situations with non-gifted students (Peterson & Lorimer, 2011). However, as pointed out before, there is no prevalence to the idea that problems in interpersonal relationships can only be caused by a student's high IQ (J. Cross, 2012; Freeman, 2019). Specialists in gifted education should also keep in mind that there are various other reasons why a gifted student may experience challenges in interpersonal relationships (Freeman, 2017, 2019). Freeman for instance refers to the influence of families on the social development of gifted students, while being twice-exceptional often also has an impact on the interpersonal relationships of gifted students (Baum et al., 2017; Reis et al., 2014; Trail, 2011).

7.6 Five Competencies in Domain 3

This chapter stressed the importance of understanding the educational needs of gifted students from an ecological perspective. This approach makes it possible to let go of a somewhat one-dimensional understanding of giftedness that may isolate gifted education from other developments in the educational field (Florian & Black-Hawkins, 2011; van Meersbergen & de Vries, 2017; Ziegler et al., 2017). By taking the influence of the actors of the ecological system into account, student behaviour can be seen as the result of being an active participant in an environment where the interplay between the different actors leads to different outcomes for different students. This understanding creates opportunities for a broad spectrum of interventions. The connection of these interventions to what is known about giftedness and gifted education, results in a better understanding of how to possibly respond to the student in the actual classroom.
Interplay between the curriculum offered and the student's cognitive style shows which instructional needs the individual student may have. In the current educational paradigm of a social-constructivist approach to education, gifted students may experience a better fit with the applied instructional strategies.

The accent on self-regulated learning is supposed to be a better match with the learning characteristics of gifted students (Oppong et al., 2018; Schrover, 2015). However, this does not depend only on the educational paradigm or the student being gifted; also of influence are the quality of instruction, the amount of teacher support during the learning process, as well as a well-balanced physical

---

[243] Appendix B, Table B14, K.3.C.4.
[244] Appendix B, Table B14, V.3.C.4.

Table 7

Domain 3, From Appendix A: Competency Matrix for Specialists in Gifted Education

Generic Teaching Competencies	3. Understanding Educational Needs
	The specialist interprets and understands the educational needs of gifted students in their ecological context, that is, an interactive system of the student, the teacher, the curriculum, physical learning environment, the family, and peers.
1. Pedagogical The teacher creates a safe and stimulating learning environment for every student, tuned to their individual educational needs.	**3.1.** The specialist develops an understanding of how the actors in the student's ecology influence the student's strengths and weaknesses.
2. Didactical In collaboration with their colleagues, the teacher transfers curricular objectives in such a way that they become teachable in the context of school and/or district policy. They must compose the curriculum in a manner that enables their student to master the content.	**3.2.** The specialist explains how the interplay between the curriculum and the student's cognitive style influences their development and instructional needs.
3. Managerial The teacher provides a structured, well-organised, task-oriented classroom atmosphere that supports the students' developmental process.	**3.3.** The specialist understands how the individual student's needs are of consequence for their classroom management.
4. Communication The teacher communicates professionally with all educational partners who are involved with the student's education.	**3.4.** Based on the results of their exploration of the student's educational needs, the specialist explains to the actors in the student's ecological system how their behaviour influences the student's development.
5. Research and Innovation The teacher adjusts their professional knowledge, perspectives, and actions if and when relevant developments in education, society, and science require it.	**3.5.** The specialist uses theory on giftedness and RtI to interpret the results of the exploration of the student's educational needs in the context of educational development.

learning environment. As gifted students in The Netherlands and Flanders are often catered for in the regular classroom, their needs have consequences for the teacher's classroom management. For the specialist in gifted education, this means that they need to adjust their classroom management strategies based on the balance between what is most desirable and most doable. In The Netherlands and Flanders the specialist in gifted education not only acts as a teacher, but very often at the same time as a gifted education coordinator, and sometimes even the equivalent to a school counsellor (van Gerven & Hoogenberg, 2011). Communicating with the different actors in the student's ecological system is important. Their ability to explain the student's educational needs to the other actors depends on their level of understanding of these needs. Being this spider in a web, their interactions with the student are the most logical starting point for their communications. A theoretical knowledge of giftedness, gifted education and general educational strategies underpins their competence in taking a broad perspective of the student's educational needs. The change-oriented and solution-focused approach provides the specialist in gifted education with a primary strategy for applying successful interventions. For the competency matrix, this brings us to the competencies described in Table 7.

7.7 Selection Rates of Indicators for Competency Domain 3

Understanding educational needs precedes the design of meaningful responses to observed behaviour and achievements. It requires a high interaction level between all educational partners as behaviour and achievements can only be understood within the context in which they are exhibited. Identically to the previous domains, the process of understanding educational needs is covered by the six themes used to categorise the indicators for professional knowledge and skills. Analyses of the data gathered in the stakeholders' research is presented in a similar way to that in Chapters 5 and 6. All tables referred to are included in Appendix G.

7.7.1 Selection Rates per Group of Stakeholders

7.7.1.1 Educational Professionals

Results per Competency

- Competency 3.1 focuses on a specialist's understanding of how the actors in the student's ecology influence the student's strengths and weaknesses. Understanding this interplay creates opportunities at the next stage to organise effective student support. Educational professionals showed the highest selection rate for indicator K.3.B.1, understanding the relevance of ownership for all parties involved in finding solutions during the process of mentoring a student (22.39%).[245]

---

[245] Table G1

The highest selection rate for the skills indicator amongst them aimed for communication with the student, discussing what they see as helpful learning strategies and interventions (24.68%).[246] Both indicators show great concern about involving the student. This suggests that voicing is seen as a relevant strategy.

- For competency 3.2 the focus was on the specialist's ability to explain how a student's development was affected by the interplay between the curriculum and the student's cognitive style. The knowledge indicator with the highest selection rate aimed at understanding the impact of the student's mindset on their development (29.26%).[247] Interestingly the skills indicator with the highest selection rate, V.3.E.2, supports the necessity of evaluating whether a teacher's expectations towards a student's executive skills are realistic (21.12%). The lowest scores for knowledge and skills for this competency are interesting, as they both belong to category C: social environment. The social relevance of learning objectives and the impact peers have on the student's perspective of their curriculum were selected by only a few educational professionals as good translations for competency 3.2 into daily practice.[248]
- Competency 3.3 stresses the relationship of effective classroom management and the ability to tune into the individual student's educational needs. The two indicators with the highest selection rates focused on the specialist stimulating the development of effective executive skills (22.90%)[249] and the specialist teaching their gifted learners to take responsibility for their actions (22.39%)[250]. Both indicators can be connected directly with the role most educational professionals will have. Students with limited executive skills and/or limited feeling of responsibility for their own actions (and learning) require more support from the teacher. This can put a strain on the teacher's classroom management. As teachers must divide their time to match the needs of all of their students, both items are considered relevant to proper balance in a teacher's time management. A remarkably low selection rate for this competency was for skills indicator V.3.F.3, aimed at clustering students with identical educational needs into small groups. As this is often already in practice (Pameijer & van Beukering, 2007; van Meersbergen & de Vries, 2017), it may not have been seen as a necessary skill to emphasise.
- The fourth competency in this domain, 3.4, stresses the necessity of good communication skills. The specialist creates an understanding amongst the actors in the student's ecological system of how their behaviour influences the student's development. The knowledge

[246] V.3.E.1 – Table G6
[247] K.3.D.2 – Table G2
[248] K.3.C.2 – 6.62% - Table G2; V.3.C.2 – 5.34% - Table G7
[249] V.3.B.3 – Table G8
[250] K.3.E.3 – Table G3

7

indicator with the highest selection rate, K.3.A.4, considered how different levels of thinking skills may require different support strategies (23.66%). This suggests that educational professionals are aware that tasks requiring higher order thinking skills may require something different of their pedagogical-didactical strategies than lower order thinking skills. For the skills indicators, the highest selection rate was for the indicator V.3.B.4, about helping students to develop a variety of reflection strategies (24.17%).[251] For the other competencies in this domain, teaching students to take responsibility for their learning and achievements was already stressed. The high selection rate of indicator V.3.B.4 confirms this observed emphasis.

- Competency 3.5 focuses on the specialist's ability to interpret their student's behaviour and achievements from the joint perspective of theory of giftedness and RtI. Amongst educational professionals, there were two knowledge indicators with an equally high selection rate of 22.65%: knowing the risk factors contributing to underachievement[252] and the understanding of how strengths and weaknesses can be translated into educational needs[253]. The lowest selection rates for this competency were for a knowledge indicator aimed at understanding how the student's ecological system affects the student's development (10.69%)[254] and the specialist's skills in selecting educational strategies based on their perspective of giftedness (8.40%)[255].

Overall Results of Educational Professionals for Domain 3

The highest selection rate for this domain, indicator K.3.D.2, focused on the effect of a student's mindset on their educational process (29.26%, Appendix G, Table G11). It suggests that educational professionals value this notion highly. This matches growing attention to the notion of mindset in Dutch and Flemish schools. It also suggests that educational professionals emphasise their student's role in their development, not only attributing success and failure to their teacher, but also attributing it to the student's attitude towards learning (Dweck, 2006, 2017).

In this third domain, educational professionals gave low selection rates to the so-called C-indicators, which focused on stimulation of interpersonal functioning (Appendix G, Tables G1-G10). In five out of ten times, the C-indicators had the lowest selection rates. Only three times did the C-indicators end up amongst the selected indicators with the second-highest selection rate, suggesting that most times, indicators from different categories were thought to better represent the specific competency.

---

[251] V.3.B.4 – Table G9
[252] K.3.D.5 – Table G5
[253] K.3.F.5 – Table G5
[254] K.3.C.5 – Table G5
[255] V.3.B.5 – Table G10

Although 19.85% of the educational professionals selected K.3.C.3, "Understands the relevance for gifted students to work with like-minded peers for their social development"[256], as a good example of professional behaviour, the matching skills indicator V.3.C.3, "Creates the opportunity for gifted students to work with like-minded peers"[257], was only selected by 13.49% of all educational professionals. A similar difference can be noticed for indicators K.3.C.4 and V.3.C.4 (Appendix G, Tables G4 and Table G7). The knowledge indicator states "Knows that students need like-minded peers in order for well-balanced social development" and is selected by 16.28%. The skills indicator aims for good practice and states "Determines if the gifted student has ample opportunity to meet and work with like-minded peers". This skills indicator gets a selection rate of 7.63%.

7.7.1.2 Specialists in Gifted Education

Results per Competency

There are similarities between the selection rates of specialists in gifted education and those of educational professionals. These can be found in Appendix G, Tables G1-G10.

- For competency 3.1, the selection rates of specialists in gifted education were similar to those of the educational professionals. Ownership regarding interventions and voicing as a pedagogical strategy were highly valued. Their lowest selection rate for knowledge indicators was for indicator K.3.C.1, "Understands how in the change-oriented approach the process of data interpretation takes place", focusing on the relevancy of how to interpret collected data from educational partners during an exploratory assessment regarding the student's educational needs (7.12%). Their lowest skills indicator selection rate (V.3.D.1, "Discusses with the student their need for instruction when new elements in the curriculum are offered", 10.36%) seemingly contradicts their highest selection rate (V.3.E.1, 25.24%). However, it is more likely that specialists in gifted education thought that this skill was already covered by selecting indicator V.3.E.1 and, therefore, chose another indicator.
- For competency 3.2, again, a similar selection pattern to that of the educational professionals was found. Although the selection rates were slightly different, they all stayed within the same range. No relevant differences were found between educational professionals and specialists in gifted education, suggesting mostly common ground between these two groups of stakeholders.
- Competency 3.3 showed a similar pattern. Although there were slight differences between the height of the selection rates, specialists in gifted education stayed close to the choices and selection rates of educational professionals.

[256] K.3.C.3 – Table G6
[257] V.3.C.3 – Table G8

The knowledge of how the physical learning environment influences the student's development of executive skills got the lowest selection rate (8.41%).[258] The counterpart to this knowledge indicator, stipulated as a skills indicator, got their highest selection rates (22.98%).[259]

- For competency 3.4, 20.71% of the specialists in gifted education included knowledge indicator K.3.F.4[260] in their selection for this competency. This indicator focuses on specialists of gifted education understanding why it is important for educational partners to be included in the assessment process of a student's educational needs. This selection rate will be discussed in the next paragraph.
  Regarding the skills indicators, similar selection patterns between specialists in gifted education and educational professionals can be observed.
- For competency 3.5, a high similarity can again be observed in selection rates between educational professionals and specialists in gifted education. There are no selection rates that stand out or suggest relevant differences in understandings.

Overall Results of Specialists in Gifted Education for Domain 3

Specialists in gifted education mostly showed a similar selection pattern to educational professionals (Appendix G, Tables G1-G10). Often, indicators with high selection rates amongst educational professionals also showed high selection rates amongst specialists in gifted education. The same applied to indicators showing low selection rates. For instance, nine out of ten times an indicator that had low selection rates amongst specialists in gifted education had also low selection rates amongst educational professionals. There was only one relevant difference between these two groups of stakeholders. 23.66% of the educational professionals selected indicator K.3.A.4[261], aiming at specialists knowing how the application of different levels of thinking skills require different support strategies. This indicator only had a selection rate of 15.86% amongst the specialists in gifted education.
The selection rate for specialists in gifted education of 20.71% for indicator K.3.F.4[262] was relatively high compared to the selection rates for the other groups of stakeholders. Only 13.23% of educational professionals selected this indicator, and it had the lowest selection rate for this competency of only 14.06% amongst parents, highlighting a second relevant difference between specialist and parent. This difference once more points out that the role stakeholders have in the education and upbringing of a student influences their choices for indicators for competent behaviour. Most educational professionals have to offer direct support to their students during the learning process.

---

[258] K.3.B.3 – Table G3
[259] V.3.B.3 – Table G8
[260] Table G4
[261] Table G4
[262] Table G4

Specialists in gifted education often have (as described in previous chapters) a more distanced role towards gifted students. As a result of this role, their focus may be more on the generic skills that match the role of the (S)ENCo. Identically to the result in domain 1, one of the indicators focusing on peer-group education had very low selection rates amongst the specialists in gifted education. Skills indicator V.3.C.4 ("Determines if the gifted student has ample opportunity to meet and work with like-minded peers") received a selection rate of a meagre 6.80% from specialists in gifted education.[263] The counterpart to this indicator (K.3.C.4) received a selection rate of 17.48%.[264] Therefore, although specialists in gifted education indicated the importance of having knowledge and understanding of the need for gifted students to have like-minded peers for a well-balanced social-emotional development, if they have to make a choice, putting theory into practice seems to become less relevant.

7.7.1.3 Parents

Results per Competency

- For competency 3.1 parents have knowledge indicator selection rates that are similar to those of educational professionals and specialists in gifted education. 21.29% of them found it important for specialists in gifted education to understand the relevance of providing their students with instruction when new elements in the curriculum are offered.[265] Although the other groups of stakeholders had other indicators with a higher selection rate, these differences were only marginal.
  For the skills indicators, the highest selection rate amongst parents was for indicator V.3.E.1[266], "Discusses with the student what they see as helpful learning strategies and interventions". Interestingly 19.68% of all parents selected indicator V.3.B.1[267] which focuses on the development of a growth mindset and how the student's own personal behaviour contributes to success and failure. The other groups of stakeholders did not select this indicator often enough to include it in the top 3 of either group. However, the knowledge counterpart to this indicator (K.3.B.1) was most often selected by them.
- Competency 3.2 focuses on how the interaction between the curriculum and the student's cognitive style influences their development and instructional needs. Indicator K.3.E.2 ("Knows how the learning process itself contributes to the development of a student's executive skills") was selected by 16.06% of all the parents.[268] That selection rate was not high enough to include this indicator amongst their three most selected indicators. The third indicator parents selected most in this competency

[263] Table G9
[264] Table G4
[265] K.3.D.1 - Table G1
[266] Table G6 - 22.49%
[267] Table G6
[268] Table G2

was indicator K.3.B.2 (17.27%), stressing the understanding of how differentiation between learning styles contributes to the student's development.
The skills indicators for this competency are interesting. The selection rates for the three indicators that were selected most often are ranked exactly the same amongst parents as amongst educational professionals and specialists in gifted education.

- In competency 3.3, parents had an identical selection rate for two indicators. Indicator K.3.A.3 and indicator K.3.C.3 both had a selection rate of 19.68%.[269] This represents the second-highest selection rate for this competency. As a result, a fourth indicator was selected to present three different selection rates. Indicator K.3.D.3 was selected by 16.47% of all the parents. This indicator aims at understanding of how interaction with a gifted student can provide insight for specialists, enabling them to adjust their actions to meet the student's needs. Parents value the input of their child in the process of tuning to their educational needs.
For this same competency, two skills indicators also had an identical selection rate. An equal percentage of parents (17.27%) selected indicators V.3.D.3 and V.3.E.3.[270] Both indicators aim at including the student's voice when designing interventions. The strategy of voicing to create ownership for gifted learners reoccurs as a point of focus for parents in all competency domains.
- The selection rates for knowledge indicators for competency 3.4 were relatively indecisive, as two indicators got an identically low selection rate of 14.06% (K.3.B.4 and K.3.F.4), and two indicators (K.3.D.4 and K.3.E.4) showed a selection rate that was only 1.2% higher (15.26%).[271] The highest selection rate for this competency was indicator K.3.A.4 (21.29%), which matched the choice of the educational professionals and which was focused on specialists knowing that different thinking skills require different support strategies. For the skills indicators, parents made similar choices to the other groups of stakeholders.
- For competency 3.5 the selection rate of indicator K.3.E.5 (27.31%) stands out.[272] This indicator focuses on the knowledge that giftedness may affect a student's behaviour. Although the difference in selection rates between the three groups of stakeholders cannot be identified as relevant, when conforming to the threshold for relevant differences (see also Chapter 4), the difference between the percentage of parents and the other stakeholders who selected this indicator was 8.2%, showing that this indicator was enormously important for parents.
The selection rates for the skills indicators showed minor differences but were overall similar to those of the other groups of stakeholders.

[269] Table G3
[270] Table G8
[271] Table G4
[272] Table G5

Overall Results of Parents for Domain 3

The highest selection rate in this domain (21.29%) was for K.3.A.4 ("Knows how different levels of thinking skills may require different support strategies"). This indicator also ended up in the top ten of selected indicators amongst educational professionals, but did not end up in the top ten of specialists in gifted education. At competency level, it can be noticed that the difference between the selection rate of parents and specialists in gifted education is relevant, just as this difference was noticeable between educational professionals and specialists in gifted education. Other relevant differences are mentioned in the overall results of specialists in gifted education in paragraph 7.7.1.2.
In this domain, the highest selection rates by parents were for indicators that aim for the stimulation of the student's ability for self-actualisation (indicators in category E) as 6/10 indicators in their top ten belong to this category.
Indicators in category F (both for knowledge and skills) were the least selected indicators by parents. Understanding their child's educational needs can be better achieved by indicators from the other categories.

7

7.7.2 Significant Differences Between Groups of Stakeholders

For analysis of the differences between the groups of stakeholders, the ANOVA test and Bonferroni's Multiple Comparisons Test were used. In domain 3, differences between the selection rates amongst separate groups of stakeholders are significant for the knowledge indicators in the categories B ($p = .005$), D ($p < .001$), and F ($p = .028$) (see Appendix G, Tables G12-G14; see also Figure 26).
In a similar way to the previous competency domains, parents made significantly different choices compared to the other groups of stakeholders. These differences are observed for categories B and D. Their selection rate for B-indicators ($\overline{X} = 2.33$) was significantly lower than the selection rate of the educational professionals ($p = .005$) and specialists in gifted education ($p = .037$). Their selection rate for D-indicators ($\overline{X} = 2.96$) was significantly higher compared to educational professionals ($p = .002$) and specialists in gifted education ($p < .001$). This suggests that, for parents, the indicators focusing on intrapersonal understanding were better examples of professional behaviour than indicators in category B, focusing on stimulating the gifted student to use different cognitive styles.
Significant differences were also found between specialists in gifted education and educational professionals. Although both their selection rates for the D-indicators were not as high as the selection rates of parents, specialists in gifted education selected these indicators even less than did the educational professionals ($p < .001$). However, specialists in gifted education selected the F-indicators significantly more often ($\overline{X} = 2.47$) than educational professionals ($\overline{X} = 2.05$, $p = .048$).
For the skills indicators, significant differences were observed for the same categories, B ($p = .007$), D ($p < .001$), F ($p < .001$) (see Appendix G, Tables G15-G17; see also Figure 27).

Figure 26

Comparison of Selection Rates of the Knowledge Indicators in Domain 3 Between Groups of Stakeholders - (see also Table G12)

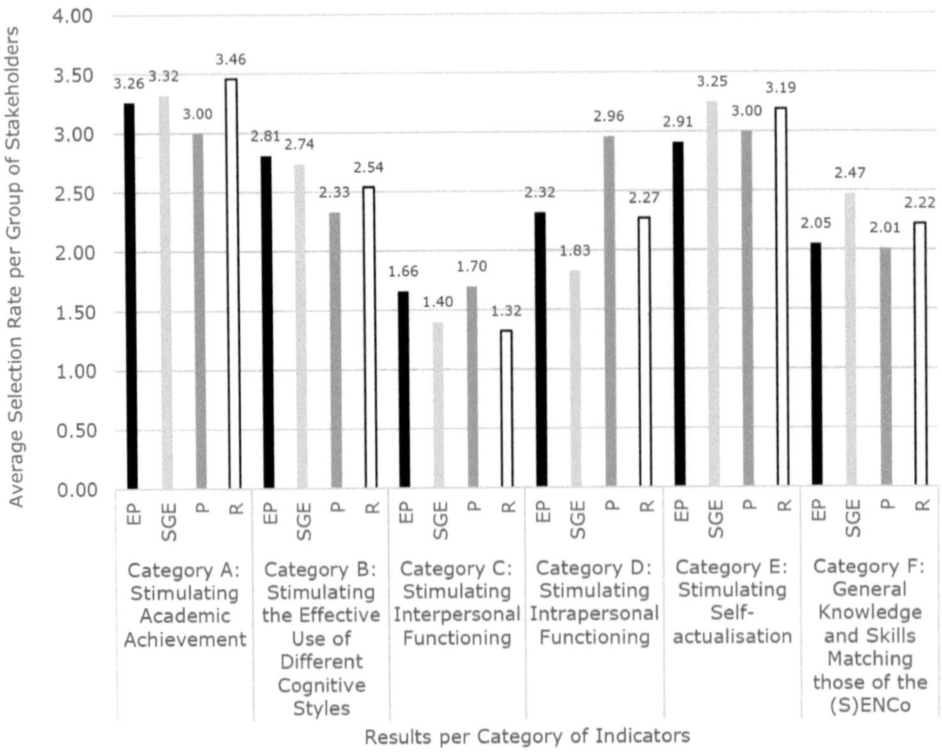

Note. N = 354 divided into four groups of stakeholders: EP = Educational Professionals, SGE = Specialists in Gifted Education, P = Parents, R = Rest. EP: N = 131, SGE: N = 103, P: N = 83, R: N = 37.

Test value = 2.5

Contrary to that seen with the knowledge indicators, for the skills indicators parents selected B-indicators significantly more often than both educational professionals ($p = .048$) and specialists in gifted education ($p = .006$). Specialists in gifted education showed significantly lower selection rates for the D-indicators compared to parents ($p = .026$) and educational professionals ($p = .001$). However, once again, the specialists in gifted education placed a stronger emphasis on the F-indicators compared to both other groups of stakeholders, with selection rates that were significantly higher (educational professionals $p = .005$; parents $p < .001$).

The differences in selection rates, especially the differences regarding the F-indicators, keep indicating that the role of stakeholders in a student's education and upbringing governs with their emphasis on what they consider to be relevant indicators.

Figure 27

Comparison of Selection Rates of the Skills Indicators in Domain 3 Between Groups of Stakeholders - (see also Table G15)

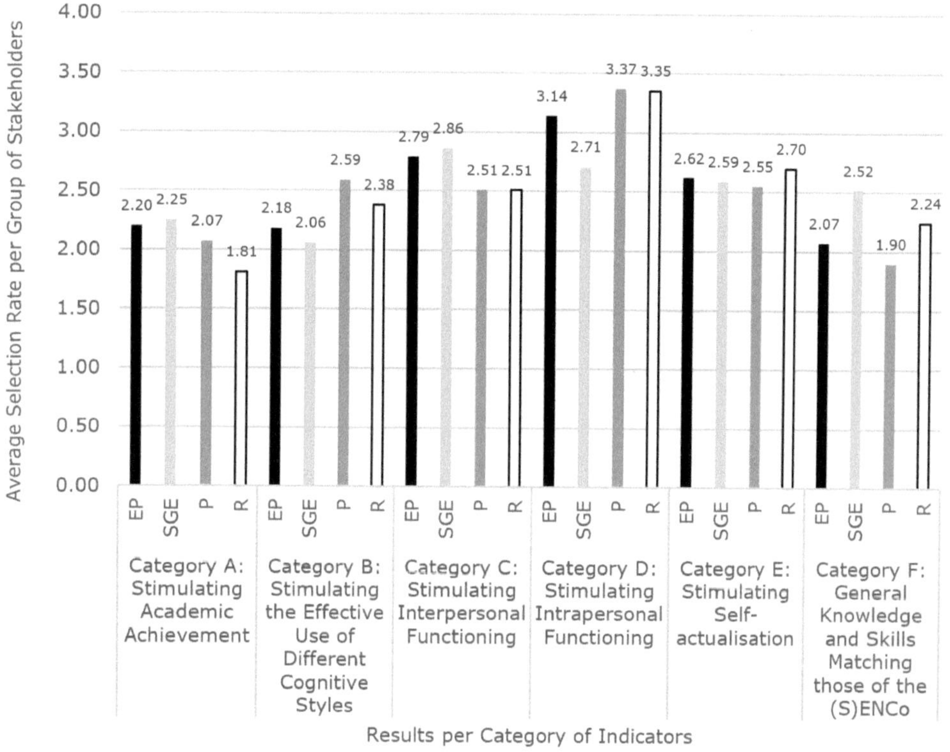

Note. N = 354 divided into four groups of stakeholders: EP = Educational Professionals, SGE = Specialists in Gifted Education, P = Parents, R = Rest. EP: N = 131, SGE: N = 103, P: N = 83, R: N = 37.

Test value = 2.5

7.7.3 Common Ground Between Groups of Stakeholders

Common Ground Amongst All Groups

Initial analysis of common ground was aimed at identifying common ground between all groups of stakeholders. Two frequency analyses were applied. A primary frequency analysis showed 12 knowledge indicators and 13 skills indicators distributed between the three most selected indicators amongst all groups of stakeholders (see Appendix G, Tables G18-G19). Indicators included all 6 categories.

G. Stimulating academic development (4)
H. Stimulating the use of different cognitive styles (3)
I. Stimulating interpersonal functioning (1)

J. Stimulating intrapersonal understanding (5)
K. Stimulating self-actualisation (8)
L. Generic counselling and advisory skills matching the skills of a (S)ENCo (4)

The secondary frequency analysis was aimed at determining whether, amongst these 25 indicators, there were four indicators that had the highest selection rates in all three groups. One knowledge indicator and three skills indicators met that criterion (see Appendix G, Tables G28-G29).

- Competency 3.1
    - Discusses with the student what they see as helpful learning strategies and interventions.[273]
- Competency 3.2
    - Knows how mindset influences a student's learning process.[274]
    - Evaluates whether the teacher's expectations towards a student's executive skills are realistic.[275]
- Competency 3.3
    - Creates a learning environment where the use of executive skills is an intrinsic condition during the learning process.[276]

If not differentiated between the groups of stakeholders, the selection rates of all knowledge indicators in domain 3 differ significantly from the expected rate (see Appendix G, Tables G20-G21; see also Figure 28). Knowledge indicator A is the most frequently selected, $\overline{X} = 3.24$. The difference, 0.737, to the expected selection rate ($\mu = 2.5$) is significant, $t(353) = 13.476$, $p < .001$, representing an effect size for practical significance ($d = .72$). This suggests a strong emphasis on the attention of specialists of gifted education on the stimulation of their students' academic achievements as indicators for competent behaviour. Analysis per separate group of stakeholders shows that the indicators in category A had the highest selection rates, suggesting that the observed difference to the expected rate was not caused by a strong deviance in the selection rates of only one group of stakeholders.
Knowledge indicators in category C showed the lowest selection rate, $\overline{X} = 1.56$. This difference, -0.944, to the expected selection rate ($\mu = 2.5$) is significant, $t(353) = -16.134$, $p < .001$, representing an effect size for practical significance $d = .85$. If differentiated between the separate groups of stakeholders, knowledge indicators in domain C showed the lowest selection rate for all groups of stakeholders (see Appendix G, Table G12). This suggests that C-indicators were not seen as relevant examples of competent behaviour.
However, in the competencies in domain 3, there is a noticeable discrepancy between the selection rates for knowledge and skills indicators in category C (see Appendix G, Tables G20-G21 and G22-G23). For the skills indicators in

---

[273] V.3.E.1.
[274] K.3.D.2.
[275] V.3.E.2.
[276] V.3.B.3.

category C, the selection rate $\overline{X} = 2.72$. They were selected more often than was expected ($\mu = 2.5$). The difference, 0.218, is significant, $t(353) = 3.771$, $p < .001$, $d = .20$, representing a small effect size. The skills indicators were probably considered more relevant examples of competent behaviour than their counterparts in the knowledge indicators.

Figure 28

Mean Selection Rate of the Knowledge Indicators in Domain 3 - (See Table G20)

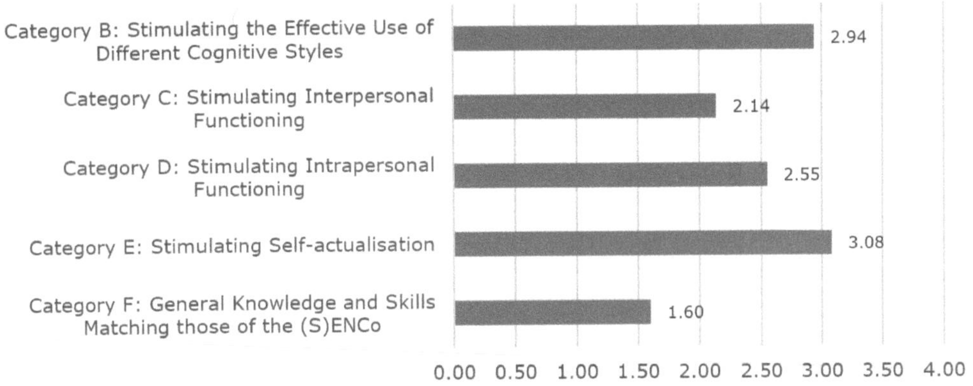

Note. N = 354. Test value = 2.5

Figure 29

Mean Selection Rate of the Skills Indicators in Domain 3 - (See Table G22)

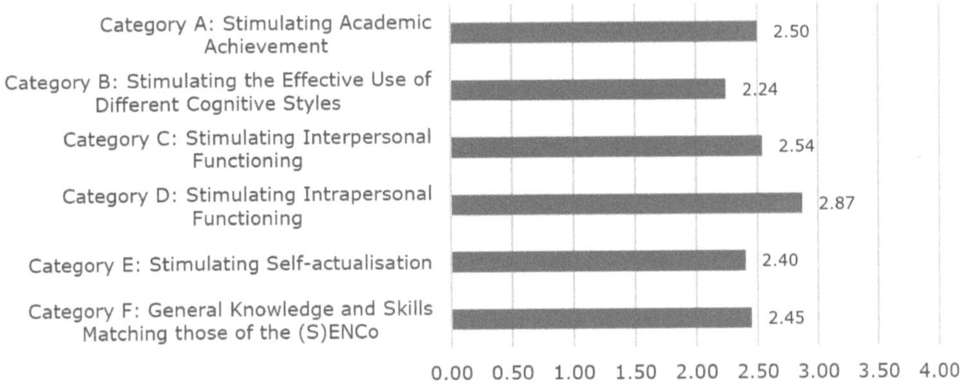

Note. N = 354. Test value = 2.5

A similar discrepancy is found regarding skills indicators in category A, showing significantly lower selection rates than was expected ($\overline{X}$ = 2.14, μ = 2.5, MD -0.356, t(353) = -5.757, p < .001, d = .31) (see Appendix G, Tables G22-G23; see also Figure 29). There is no direct explanation for these differences. Although it happens more often that an indicator category has high selection rates either for skills or knowledge, there is no pattern to be observed that may lead to an overall conclusion over whether when a category has high rates for knowledge it gets low rates for skills and vice versa.

In a similar way to competency domains 1 and 2, the overall results of domain 3 suggest that some categories of indicators were selected more often than others (Appendix G, Tables G25-G28). Again, a clustering in two categories (ADE vs BCF) was possible. The Paired Samples Statistics Test showed that the selection rates of clustered categories (both for knowledge and skills) differed significantly from the expected selection rate (p < .001). The clustered knowledge indicators ADE, $\overline{X}$ = 2.87, were selected significantly more often than the clustered knowledge indicators BCF, $\overline{X}$ = 2.13 (p < .001, d = .74). The clustered skills indicators ADE, $\overline{X}$ = 2.61, were selected significantly more than the clustered skills indicators BCF, $\overline{X}$ = 2.39 (p < .001, d = .25). This suggests that, overall, the indicators in category ADE were considered to be more relevant examples for the competencies they were supposed to match.

7.7.4 Comparison Between The Netherlands and Flanders

The final analyses for this domain focus on similarities and differences between The Netherlands and Flanders. Using Group statistics and an Independent Samples test, the following results were found that matched the pattern of results for competencies in domains 1 and 2 (see Appendix G, Tables G33-G36). Knowledge indicators in category A were selected more often by stakeholders from both countries than expected (Appendix G, Tables G33-G36; see also Figure 30). A difference between Dutch and Flemish stakeholders can be observed. Amongst Dutch stakeholders A-indicators had an average selection rate of $\overline{X}$ = 3.29, Flemish stakeholders showed a selection rate of $\overline{X}$ = 2.93. The difference between the selection rate of stakeholders from both countries, 0.367, was significant, p = .015, the effect size represents a practical significance, d = .36.

For the B-indicators, it can be observed that they are selected significantly more often by Dutch stakeholders ($\overline{X}$ = 2.76) than by Flemish stakeholders ($\overline{X}$ = 2.04). There is a significant difference, 0.723, p < .001, the effect size represents a practical significance, d = .72 (Appendix G, Tables G33-G34).

Knowledge indicators in category C are selected significantly less by stakeholders from both countries. However, comparing both countries shows that stakeholders from Flanders selected C-indicators more often ($\overline{X}$ = 1.96) than Dutch stakeholders ($\overline{X}$ = 1.48). The difference, 0.482, is significant, t(352) = -3.02, p = .003, the effect size meets the threshold for practical significance, d = .44 (Appendix G, Tables G33-G34).

For the knowledge indicators in category D, an almost identical pattern can be observed. Although the selection rates for both countries stayed closer to the

expected value (μ = 2.5), comparing both countries, the results showed a selection rate amongst Flemish stakeholders of $\overline{X}$ = 2.69, compared to the Dutch stakeholders of $\overline{X}$ = 2.25. The difference, -0.437, is significant t(352) = -2.23, p = .027, the threshold for practical significance is met, d = .33 (Appendix G, Tables G33-G34).
For the skills indicators, differences amongst Dutch and Flemish stakeholders are found for categories D and F (Appendix G, Tables G35-G36; see also Figure 31). D-indicators showed a higher selection rate amongst stakeholders from both countries, but were often selected even more by Flemish stakeholders ($\overline{X}$ = 3.42) than by Dutch stakeholders ($\overline{X}$ = 3.03). The difference, -.0388, is significant, t(352) = -2.29, p = .022, the effect size represents a practical significance, d = .34.
F-indicators showed lower selection rates for both countries but were more often selected by Dutch stakeholders ($\overline{X}$ = 2.23) compared to Flemish stakeholders ($\overline{X}$ = 1.89). The difference, 0.343, is significant, t(352) = 2.24, p = .026, the effect size represents a practical significance, d = .33.

7

Figure 30

Comparison of Mean Selection Rates in Domain 3 per Category of Knowledge Indicators Between Dutch and Flemish Stakeholders - (See Table G33)

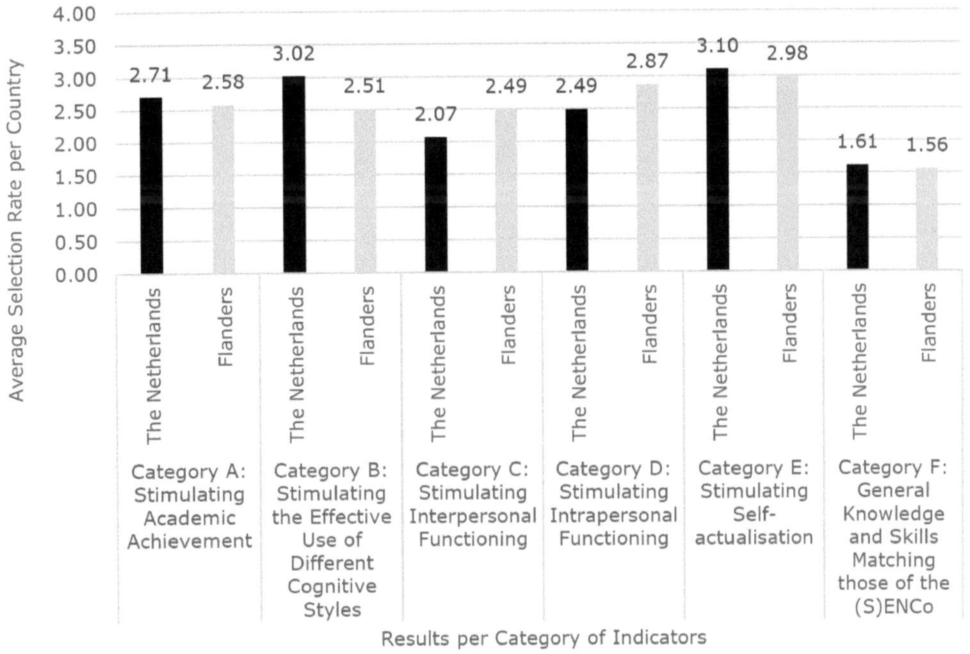

Note. N = 354. The Netherlands N = 299, Flanders N = 55. Test value = 2.5

Figure 31

Comparison of Mean Selection Rates in Domain 3 per Category of Skills Indicators Between Dutch and Flemish Stakeholders - (See Table G35)

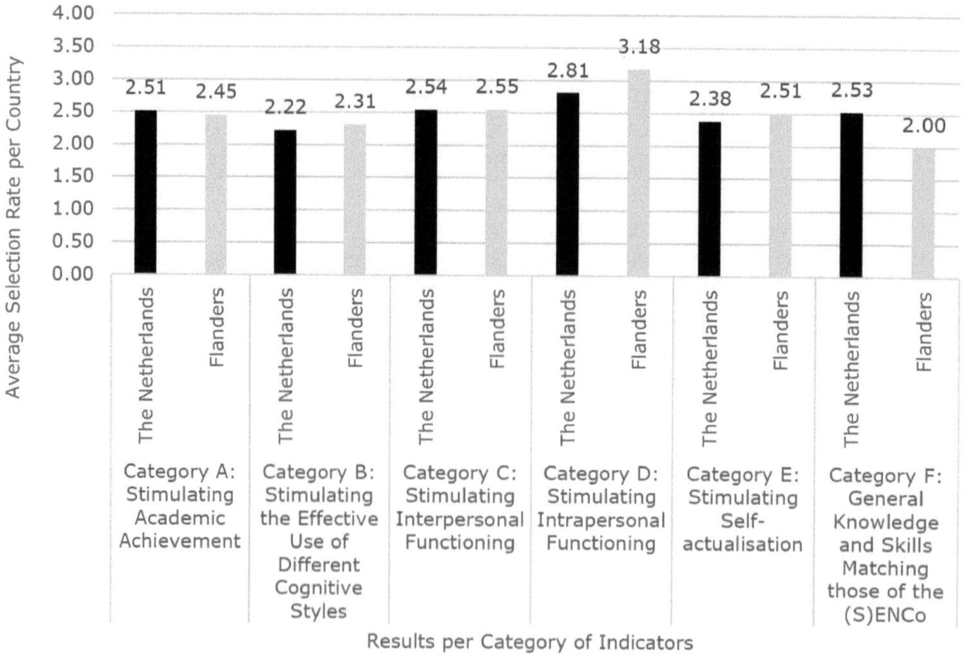

Note. N = 354. The Netherlands N = 299, Flanders N = 55. Test value = 2.5

Although the effect sizes of the observed differences are relatively small, they suggest that there is a trend where Flemish stakeholders emphasise different knowledge and skills in their specialists in gifted education than Dutch stakeholders. Previously, it was suggested that observed differences between groups of stakeholders may be caused by the role each group has in the education and upbringing of gifted students. As the number of specialists in gifted education amongst Dutch stakeholders (N = 100) is higher than amongst Flemish stakeholders (N = 3), this might have affected the selection rates for both countries.

7.8 Summarising

This chapter focused on the combination of general teaching standards and understanding the educational needs of gifted students. Understanding educational needs requires an educational partnership between all actors in the ecological system: teacher, student, parents and sometimes also peers. For specialists in gifted education, this partnership is not only necessary as a source of information but also as an opportunity for optimising education. Five competencies were deduced from the theory review presented in the first paragraphs of this chapter.

- 3.1 The specialist develops an understanding of how the actors in the student's ecology influence the student's strengths and weaknesses.
- 3.2 The specialist explains how the interplay between the curriculum and the student's cognitive style influences their development and instructional needs.
- 3.3 The specialist understands how the individual student's needs are of consequence for their classroom management.
- 3.4 Based on the results of their exploration of the student's educational needs, the specialist explains to the actors in the student's ecological system how their behaviour influences the student's development.
- 3.5 The specialist uses theory of giftedness and RtI to interpret the results of the exploration of the student's educational needs in the context of educational development.

In domain 3, as a joint group, indicators in category ADE were selected more often than indicators in category BCF. This suggests that stakeholders thought that the ADE-indicators offered more relevant examples for the individual competencies than BCF-indicators. Divided per category of indicators, for the knowledge indicators, those from category A had the highest selection rate amongst all groups of stakeholders. This suggests a relatively strong emphasis on understanding the educational needs of gifted students via their academic achievements and opportunities to stimulate these academic achievements.
For the skills indicators, parents and educational specialists showed highest selection rates for D-indicators, while specialists in gifted education had the highest selection rates for C-indicators. Interestingly although there are differences between the average selection rates of specialists in gifted education, the difference between their highest selection rate and the other five selection rates was not as extreme as the difference that can be observed at this point for the selection rates of parents and educational professionals. This suggests that, for both of these groups, the D-indicators were not only the most relevant examples but they also represent what should be emphasised the most in gifted education.
In domain 1, the overall low selection rates for the C-indicators were unexpected. However, having seen a similar pattern in domain 2 and now also in domain 3, these low rates may be seen as an identification of a more general trend. Specialists in gifted education showed higher selection rates for the skills indicators in this C-category, suggesting that most of them need these indicators to gain a better understanding of the educational needs of the gifted students at their school. However, at this point, it is too soon to draw conclusions or to suggest final explanations.

8 Domain 4: Meaningful Responses to the Educational Needs of Gifted Students

8.1 Introduction

A systemic approach to seeing and understanding the educational needs of gifted students clarifies the consequences of the five paradigm shifts in education that were described in Chapter 2. An inclusive perspective on education, transferred to a change and solution-focused approach and reflecting current perspectives of giftedness, leads us to a didactical paradigm for building 21st century skills. A strong emphasis on evidence-based instructional and pedagogical strategies is part of the current approach in education. This creates a framework for general teaching competencies. These competencies have five distinguished domains: pedagogical competencies, didactical competencies, classroom management competencies, communication competencies and competence in reflection and research.
In previous chapters it was concluded that the perspective on giftedness has changed over the years. It was also concluded that different perspectives on giftedness lead to different consequences for gifted education. However, what is stable across these different perspectives is that all experts describe them as having consequences for the curriculum of gifted students. Taking these consequences into consideration, it is clear that the competencies of the specialist in gifted education in this context are affected as well.
This chapter, which explores the fourth domain of the competency matrix, describes the interplay between general teaching competencies and pedagogical and didactical intervention strategies as they apply to gifted students. To emphasise the importance of tuning to the individual student's needs based on the in-depth understanding of those needs (Tomlinson et al., 2009), these interventions are called *meaningful responses*. In an identical format to previous chapters, the theory of meaningful responses is described first, followed by the results of the empirical research. The selections of knowledge and skills indicators by stakeholders are then analysed and presented.

8.2 Designing Meaningful Responses is a Joint Venture

Fundamental to a systemic or ecological approach to education is the participation of all the different actors of the student's ecological system (Bronfenbrenner, 1979; van Meersbergen & de Vries, 2017; van Swet, 2009). These actors are a source of information about the student. They have influenced the student's development in the past and they will be important contributors to future intervention plans (van Swet, 2009).
The importance of an educational partnership between parents or carers and the school has been described (Pameijer, 2012). However, parents and carers are not the only actors in the student's ecological system who should be taken into account; teachers who work together are educational partners as well. Taking care of the education of gifted students often includes time-sharing the

responsibility for a student between different teachers (Houkema et al., 2018). Besides participating in regular classroom learning, gifted students may take part in special classes, pull-out programmes and so forth (Bakx et al., 2016; Heacox & Cash, 2019; Janson, 2009; Winstanley, 2010). The specialist in gifted education may be the classroom teacher or the teacher who runs the pull-out programme, and could also be the person who has a more distant role in the gifted student's education, that is, being the gifted programme coordinator or acting as a school counsellor for gifted students (van Gerven & Hoogenberg, 2011). Hence, as seen with other educational specialists, it is important for the specialist in gifted education to identify educational partners in the individual cases in which they are involved and to decide how to communicate with them (Pameijer & van Beukering, 2007; van Meersbergen & de Vries, 2017; van Swet, 2009). Based on the exploration as described for the process of seeing and understanding educational needs, the specialist is able to recognise how individual (f)actors can positively influence a student's development.[277] This ensures that the role of all relevant (f)actors in the student's ecological system is tuned in such a way that they support each other (van Swet, 2009).[278] In order to do this, regular meetings between the specialist in gifted education and educational partners should be organised (Pameijer & van Beukering, 2007; Peters & Oveross, 2020; van Meersbergen & de Vries, 2017). During these meetings, the specialist in gifted education discusses educational programming with the educational partners.[279] In so doing, the specialist is including the educational partners as relevant actors in the plan and creating shared ownership to enhance the chances of a successful educational process.[280]
In The Netherlands and Flanders, teachers involved in pull-out programmes often voice their discontent that their efforts in pull-out programmes lag behind expectation as a result of the lack of communication between themselves, parents and regular classroom teachers (Houkema et al., 2018). They understand that continuous programming has a positive effect on the educational process for gifted students. They also understand that shared ownership regarding their responsibilities towards the gifted students in their care would be beneficial. Although they all see communication as a key factor for success, pull-out programmes often function in relative isolation.[281] Communication between teachers in these programmes and classroom teachers is often unidirectional and when discussion is reciprocal, classroom teachers often focus on the student's behaviour and results and forget to mention their personal professional influence on this behaviour and the results. Continuous lack of time for individual coaching of gifted students in the regular classroom and a lack of mutual understanding for each other's roles in the educational process, directly affect the success of these pull-out programmes. The opportunity to set educational objectives for individual students collaboratively

[277] Appendix B, Table B20, K.4.D.5.
[278] Appendix B, Table B20, V.4.D.5.
[279] Appendix B, Table B19, V.4.F.4.
[280] Appendix B, Table B19, K.4.F.4; Table B20, V.4.F.5.
[281] Appendix B, Table B19, K.4.A.4.

and to exchange activities focused on realising these objectives would be beneficial (Coleman & Johnsen, 2011; Kuipers, 2017; Pameijer & van Beukering, 2007; van Meersbergen & de Vries, 2017).[282]

8.3 Adjusting the Curriculum

Chapter 6 describes how researchers concluded that a mismatch between the gifted student and the curriculum offered often triggers a descent in their overall feelings of wellbeing (Heacox & Cash, 2014; Hoogeveen, 2017b; Siegle et al., 2017). Adjusting the curriculum and creating a better match for the educational needs of gifted students means providing sufficient educational challenges to help them prosper at school (Calahan & Hertberg-Davis, 2018; Hertzog, 2017b; Johnsen et al., 2016; D. Matthews & Foster, 2009; Renzulli, 2012).[283] Adjusting the curriculum starts not only with a perspective on giftedness, but also with a more general perspective on education. The fast-growing emphasis on a more social-constructivist approach to education through focusing on 21st century skills has been recognised worldwide (Marzano & Heflebower, 2012; Schleiger, 2012; Voogt & Pareja Roblin, 2010). This paradigm shift meets the shift in thinking about the notion of giftedness.

In their mega-model of talent development, Subotnik et al. (2011) outline four distinct stages of talent development: discovering potential, developing competency, developing expertise and developing eminence. The stage of discovering potential coincides with the timeframe of primary education. Based on the mega-model developed in 2011, in 2016 they once again stressed the importance of early exposure to a wide variety of domains to "ferret out interest and observe exceptional potential" (Olszewski-Kubilius et al., 2016, p. 141). They build a strong case for potential only being spotted in the context of students being exposed to situations that call on both their interests and their abilities. However, their theory lacks reflection on issues of equity, such as how to find gifted learners from a minority population if educators hold low expectations of these populations. According to Grantham (2012), the chance that teachers will discover these students through challenging levels of work, is limited.

Subotnik et al. (2011) acknowledge that finding a balance between focusing on discovered talent and exposing students to a broad range of activities in different talent domains may be contradictory and difficult to realise. They find a solution through defining education as no longer being a process that takes place exclusively at school. According to them, in our current society, education should be placed in a broader perspective than the limited construction of formal learning in a formal context. They describe numerous opportunities for informal learning and talent development. Here again, they don't consider that access to most types of informal programmes is limited to students who are relatively well off. Students from low-income families or living in an area where the variety of

[282] Appendix B, Table B19, V.4.A.4.
[283] Appendix B, Table B17, K.4.E.2.

extra-curricular activities is limited will be deprived of an education that meets their potential (Grantham, 2012).
According to Subotnik et al. (2011), the shift towards a new educational paradigm in institutionalised education is unavoidable. Therefore, changes in the curriculum are unavoidable, as are changes in the pedagogical and didactical strategies used.[284] By adopting this perspective, it once again becomes clear that there is a need for collaboration between educational partners and that the influences of the student's ecological system should be taken into account. Subotnik et al. (2011) see education through a lens that focuses on the fast changes of a society where new domains with new knowledge and skills can be identified. As a result, identification of talent as the cultivation of talent is deeply influenced. By taking this position, they have joined a fast-growing number of experts in gifted education who have assumed the same perspective (Ambrose & Sternberg, 2016; Hocket, 2009; Johnsen et al., 2016; Plucker et al., 2017; Renzulli, 2012; Sternberg, 2017; Tirri, 2017; Tomlinson et al., 2009). However, in Dutch and Flemish education, emphasis is placed on the academic development of students (Vlaams Ministerie van Onderwijs en Vorming, n.d.; Wet op het primair onderwijs (WPO), 2005). The extension of the key-objectives of primary education towards non-academic subjects seems unlikely.
In *The Parallel Curriculum,* Tomlinson et al. (2009) suggest that the changes as described previously, cause two major challenges for education:

- Students are changing and have learning characteristics that differ from learning characteristics in the past. Thus, our educational responses need a make-over. Reshaping teaching from the perspective of a student, shifting them from being a consumer to a co-creator in the curriculum, has become conditional for modern times.
- Different talents require different responses. At a time when an inclusive approach to education has begun to be a leading strategy, teachers should be prepared to deal with the demands resulting from these differences in a single classroom.

The two challenges for gifted education as described by Tomlinson et al. match three out of the five paradigm shifts that currently exist in the Dutch and Flemish perspectives on education: the development towards the social-constructivist approach, the need for increased differentiation as a result of a more inclusive approach to education, and the educational response as used in the change and solution-focused approach.
An effective curriculum respects the unique characteristics of the learner and ensures, at the same time, that the student is an active member of the classroom society (Robertson & Pfeiffer, 2016).[285] This requires an active approach to learning in a social context (Wallace et al., 2010). This matches with the effects that the inclusive approach taken in The Netherlands and Flanders has had on education. The increasing need for differentiation shows that a

---

[284] Appendix B, Table B20, K.4.B.5.
[285] Appendix B, Table B19, V.4.C.4.

curricular Procrustean bed is no longer seen as sufficient; no single curriculum can adequately address the needs of all learners in a classroom at the same time (Mol-Lous, 2011; Oostdam, 2009). However, this puts pressure on the teacher's ability for classroom management (Kuipers, 2017). The teacher must be capable of using different classroom managerial strategies.[286] Kuipers states, therefore, that responding meaningfully to the educational needs of gifted students requires not only an educational partnership with parents and other teachers but also a partnership with the school management.

Tomlinson et al. point out five opportunities for differentiation between students: their (a) cognitive and (b) academic abilities; (c) their social and emotional needs and characteristics; (d) their interests; and (e) their learning preferences. Similar opportunities to differentiate are mentioned by Roberts and Inman (2015) and Riley (2011). The five aspects referred to by Tomlinson et al. also match the categorisation Trail (2011) suggests: cognitive style, academic abilities, interpersonal and intrapersonal functioning and, as a result of a positive interplay between these four aspects, self-actualisation.

In Chapter 2, the position of gifted education in The Netherlands and Flanders was examined through reference to different levels of response, based on the more general system of response to educational needs that currently leads in The Netherlands and Flanders. This is similar to the tiered structure for RtI (Coleman & Johnsen, 2011; Robertson & Pfeiffer, 2016). Meaningful educational responses lean heavily on an effective curriculum (Hocket, 2009) and the use of different levels of curriculum provision in the educational system has been proven effective (Ainscow et al., 2006; Coleman & Johnsen, 2011; Robertson & Pfeiffer, 2016). Hu et al. (2016) refer to the three levels of learning described by Betts (2004). Levels 1 and 2 in the Dutch and Flemish system are similar to the first two levels Betts described.

Level 1 focuses on a set curriculum and objectives that are a fit for all students. At this level, it is important that the specialist in gifted education is able to estimate to what extent it would be relevant for gifted students to participate in the curriculum that is on offer (Heacox & Cash, 2019; Tomlinson et al., 2009; Winstanley, 2010). Curriculum compacting creates the opportunity to address the needs of these students based on their ability to digest the curriculum at a higher pace (Robertson & Pfeiffer, 2016). Depending on the individual needs of the student, the specialist chooses the strategy that offers the best match under the circumstances[287] (Ahlers, 2017; Drent, 2009).

Level 2 offers opportunities for differentiation to students based on a teacher-differentiated curriculum. As compacting leads to an accelerated offer of the standard curriculum, it creates the opportunity to work on enrichment projects in the regular classroom (Bakx et al., 2016; Kuipers, 2017). At this level, the teacher often selects what they consider matches the student's academic needs, and the selected enrichment tasks replace the standard tasks (Kuipers, 2017). The use of enrichment tasks, therefore, are intended not to increase the

8

[286] Appendix B, Table B18, K.4.F.3.
[287] Appendix B, Table B17, K.4.A.2, V.4.A.2.

student's workload, but to replace less meaningful parts of the curriculum with a more meaningful and challenging curriculum (Kieboom, 2015; Kuipers, 2017; Robertson & Pfeiffer, 2016; Tomlinson et al., 2009; Winstanley, 2010).[288] This is where the teacher and the student have to collaborate, as there is joint responsibility between teachers and students regarding a student's time management.[289]

"Meaningful and challenging" implies that the tasks offered contribute positively to the student's development (Tomlinson et al., 2009; Winstanley, 2010); therefore, it should be ensured that work on these tasks is part of the student's regular assignment schedule (Ahlers, 2017; Bakx et al., 2016; Kuipers, 2017; Plucker et al., 2017).[290] In The Netherlands and Flanders, the strategy of compacting and enriching, although not always satisfyingly, is currently an accepted approach to primary curricular adjustments (Ahlers, 2017; Gubbels, 2016; Kuipers, 2017). In this approach, teachers often select enrichment materials based on their availability, instead of selecting educational objectives matching the student's zone of proximal development within a specific domain. This may cause limited success. A second factor causing a lack of success may be that those taking this approach often base it on presumed characteristics of gifted learners which can lead to a single-minded perspective of what is "good" for gifted students (Hocket, 2009). Although their personal interests may be addressed, at the described levels 1 and 2 in Dutch and Flemish education, students have relatively few opportunities to set their personal goals to match their individual talents.

However, levels 3 and 4 in the Dutch and Flemish system offer the opportunity for a learner-centred approach, described by Betts (2004) as level 3 and referred to as such by Hu et al. (2016). At this level learners set their personal educational objectives, choose products that offer proof of their developed expertise and act as experts in their own learning process.[291] Effective teacher support at these levels includes instructional strategies to encourage students to approach learning by using a variety of learning styles and strategies that are appropriate for the selected learning goals (De Bruyckere et al., 2016; Gubbels, 2016; Hattie & Yates, 2015; Johnsen et al., 2016; Roberts & Inman, 2015; Vogelaar, 2017).[292]

In the previous chapter, the topic of *ownership* as a requirement for a didactical approach based on social-constructivism was discussed. Researchers from different paradigms in theories of giftedness take different perspectives. Olszewski-Kubilius et al. (2016) describe how ownership can manifest itself in setting personal educational goals. In their perspective, the most important goal for gifted students is to actualise their wish to achieve the highest level of talent development as possible in their domain of potential. Although based on different premises, scholars from the gifted child paradigm also aim for self-

---

[288] Appendix B, Table B18, K.4.A.3.
[289] Appendix B, Table B18, K.4.B.3, V.4.B.3.
[290] Appendix B, Table B18, V.4.A.3.
[291] Appendix B, Table B19, K.4.D.4, V.4.B.4.
[292] Appendix B, Table B16, V.4.B.1.

actualisation of a person's potential (Kreger-Silverman, 2013). Aiming for talent actualisation requires support in learning how to take on this process of actualisation. In different stages of development, students need to be taught how to recognise relevant educational objectives and how to set realistic goals matching these objectives. This implies that students have to learn how they can divide the process to actualise their dreams into achievable developmental objectives (Kieboom & Venderickx, 2017a, 2017b). The development of metacognition, the ability to reflect on one's actions, and the ability to regulate emotions regarding the learning process is conditional (Haimovitz & Dweck, 2017; Hocket, 2009; Oppong et al., 2018).[293] Based on this educational perspective, an approach where teacher and student collaborate to set educational objectives is required (Chandra Handa, 2009; Johnsen et al., 2016; Peters & Gates, 2010).[294] Setting their own goals in their zone of proximal development is not something that comes easily for children. Involving the student in designing a more personalised curriculum affects the power of the curriculum to support challenging and motivated learning (Hocket, 2009; Tomlinson et al., 2009).[295] This matches with an educational concept where students are challenged to take personal responsibility for their development (Mammadov et al., 2018; Oppong et al., 2018). Referring to D. Matthews and Foster (2006), Latz et al. (2008) conclude that: "If students are able, with the support of educators, to construct their own knowledge, thus participating in curriculum development and differentiation, rather than acting as passive consumers of information, they have reached the pinnacle of learning" (p. 28). In order to make this happen, teachers need broad and deep disciplinary knowledge in order to prepare their students for the future course of education (Olszewski-Kubilius et al., 2016; Tomlinson et al., 2009).[296]

In the differentiation paradigm, the definition of giftedness is an educational one, meaning that giftedness refers to specific educational needs that cannot be met by the regular curriculum and/or regular instructional strategies. Setting up meaningful responses requires a system of "Provide-Identify-Provide". Two approaches, where achievements and observed learning behaviour lead to a similar construct for meaningful responses, are RtI and the change and solution-focused approach. At the same time, where a challenging education is provided, the student's responses to this education are used as a tool for identifying their educational needs. When there is a shift in context around the concept of what a challenging education should contain and which instructional strategies are considered the best way to reach objectives, the result is also a shift in which students are considered to be gifted within that new educational perspective. Based on an increasing emphasis on the development of 21st century skills for all students, it has become necessary to shift to a differently presented curriculum and different instructional strategies. Sternberg's (2017) ACCEL model (Active Concerned Citizenship and Ethical Leadership) can be seen as a

[293] Appendix B, Table B16, K.4.E.1, V.4.E.1.
[294] Appendix B, Table B19, V.4.E.4.
[295] Appendix B, Table B19, K.4.E.4, K.4.B.4.
[296] Appendix B, Table B17, K.4.F.2, V.4.F.2.

response to this. Sternberg argues that the necessity of developing critical thinking skills and translating them into the skill to judge if ethically one's actions are for the good of society, increases in current society. In his ACCEL theory, he states that the ability to apply these skills successfully can be used as an indicator for identifying giftedness (Sternberg, 2017).

In the current didactical paradigm, teaching for creativity and the use of higher order thinking skills, along with self-regulation, are essential (Ambrose & Sternberg, 2016; Chandra Handa, 2009; Gubbels, 2016; Tirri, 2017; Tomlinson et al., 2009).[297] For the specialist in gifted education, this requires a deep understanding of the differences between lower and higher order thinking skills, and the ability to set educational objectives based on these thinking skills (Marzano & Kendall, 2007, 2008).[298] In order to meet these criteria for a modern curriculum for gifted education, different didactical strategies can be used. Research by Hocket (2009) shows how three successful strategies for curricular adjustments for gifted education match the current paradigm shift in general educational strategies. By comparing VanTassel-Baska's integrated curriculum model, Renzulli's multi menu model and the parallel curriculum model from Tomlinson et al. with the growing focus on a social-constructivist approach to general educational strategies, Hocket provided opportunities to strive for changes that may be beneficial to most students. She concludes that both general education curricular experts and experts in gifted education curricula point out that "high-quality curriculum is authentic, outcome driven, flexible for individual differences, and challenging" (Hocket, 2009, p. 412).

The notion of an authentic curriculum refers to solving real problems; the individual and social relevance of knowledge and skills; and problem-solving strategies guided by the way experts work and think. Experts in gifted education suggest that this approach is especially motivating for gifted students, whereas general curriculum specialists stress the importance of making the curriculum more relevant for all students. Meaningful outcomes, being a core issue for both types of curricular specialists, refer to the deep understanding of knowledge and skills taught. However, Hocket concludes that where general curricular specialists' focus on teacher differentiated outcomes (as in level 1 and 2 results), a specialist in gifted education focuses more often on student differentiated outcomes (level 3). Aiming for meaningful outcomes refers to both concrete and abstract products and includes reflection on both learning process and achievements.[299]

The flexibility that allows for individual differences, also mentioned by both groups of curricular specialists, focuses on an approach whereby students are no longer seen as a homogeneous group, regardless of whether they are gifted or not. Again, this matches the concepts of RtI and the change and solution-focused approach. Accepting that all students differ and have differing needs means that a curriculum is only sufficient if it offers teachers the flexibility to vary their approach, pace and interest. Pre-assessment is seen as the

[297] Appendix B, Table B17, V.4.B.2.
[298] Appendix B, Table B17, K.4.B.2.
[299] Appendix B, Table B 20, K.4.E.5.

gatekeeper for differentiation. In the concept of RtI, pre-assessment often takes place based on standardised achievement tests (Coleman & Johnsen, 2011; Robertson & Pfeiffer, 2016). In the change and solution-focused approach, pre-assessment is seen within a broader perspective (Pameijer & van Beukering, 2007; van Meersbergen & de Vries, 2017). In this systemic perspective, educational needs refer both to curricular needs and to pedagogical needs. Adjusting the curriculum for gifted students solely based on achievement, is not considered sufficient (Ziegler et al., 2017).

According to Hocket (2009), recommendations for creating challenging education in general stay vague, whereas specialists in gifted education are very specific about what conditions they consider to be challenging.

Two current examples of those specific directions for challenging education can be found in Chandra Handa's (2015) theory on teaching creativity and in the Learning to Think (LLT) programme develop by Hu et al. (2016). Chandra Handa (2015) describes how educational challenges should focus on matching both academic level and intellectual level. He is referring specifically to tasks that require critical and creative thinking strategies.[300] Chandra Handa distinguishes three aspects to the didactical approach: creative *practice*, as in teacher pedagogy and teaching practice to foster creativity; creative *learning*, as in those aspects of student learning that are essential to initiate and enhance a creative mindset and creative behaviour; and the creative *process* which acts as the bridge between the teacher's creative pedagogy and the student's creative learning.

In a similar way to Chandra Handa, Hu et al. (2016) describe how creating challenging education is influenced by teaching students critical thinking skills. Their LTT programme is based on the assumption that critical thinking is a skill that can be developed. In this programme, Hu et al. take a four-step approach to triggering authentic learning situations. In the first stage, they set conditions around raising the student's interest by creating a so-called cognitive conflict. In stage two, students are encouraged to observe, discuss, experiment and so forth. Stage three focuses on a reflection on the process and content and, in the fourth and final stage, they facilitate the transfer of acquired knowledge and skills to real-life situations.

Although, at first sight, both examples use a different approach, the didactical elements and strategies are quite similar and based on similar assumptions. Both approaches focus on active student participation during the entire learning process. Each student is required to participate in the four stages of learning (instruction; planning and organisation; execution; and evaluation).[301] In both approaches, the teacher purposefully creates situations to ignite curiosity, by choosing situations and problems that can be viewed from different perspectives and solved by using different problem-solving strategies. An interchangeable use of different learning styles is required for the best outcomes (De Bruyckere et al., 2016; Hattie & Yates, 2015).[302] Both strategies are based on offering ample

[300] Appendix B, Table B17, V.4.E.2.
[301] Appendix B, Table B20, V.4.B.5; Table B16, V.4.A.1.
[302] Appendix B, Table B16, K.4.B.1.

opportunities for student collaboration. In the social-constructivist approach, knowledge and skills are not static but develop through interaction between students and their teacher. The teacher creates opportunities for the students to actively learn from each other (Janson, 2017; Johnsen et al., 2016). Students exchange ideas, support each other, evaluate the process and need each other's collaboration for problem solving.[303] Flexibility in learning outcomes and the learning process make it profitable for students to collaborate.[304] The specialist in gifted education knows which factors during collaborative processes contribute to the student reaching their zone of proximal development (Johnsen et al., 2016).[305]

Ability grouping is one of the interventions that stimulates opportunities for collaborative learning. Ability grouping is sometimes possible in the regular classroom, but, in some situations, ability grouping can be better organised by setting up a pull-out programme or even a full-time segregated group (Kitsantas et al., 2017; Oostdam, 2009). Deciding which is the best possible option for organising a provisional structure for meeting the needs of gifted learners cannot be stated based on theory alone. Chapter 2 explained how the pyramid of support is malleable and how this pyramid depends on different factors. What is known, based on research, is that similarly to other students, gifted students need like-minded peers or students of a similar developmental level for a well-balanced development; cognitively, academically, and in social and emotional development (J. Cross, 2012; Hertzog, 2017a, 2017b; Kreger-Silverman, 2013; Neihart et al., 2016; Obergriesser & Stoeger, 2015).[306] Social skills are best developed within a context where other students act and communicate at a similar level (de Bruin-de Boer & van Gerven, 2009). As children reflect on their personal actions by observing the responses of other children, it is important to create opportunities for gifted students to meet and learn with other able students.[307] Teaching students how to communicate effectively with each other helps to create a safe learning environment where collaboration can take place based on equity.[308]

8.4 The Supportive Role of the Teacher

Education based on social-constructivism requires the teacher to play an active role, which is different compared to their role in the past. Social-constructivism requires education to be seen as a two-way street where interplay between teacher and student shows the characteristics of a process of collaboration (Chandra Handa, 2009). Teacher and students are both seen as part of a learning system. The students learn from the teacher as much as the teacher learns from their students (Tomlinson et al., 2009). In order to do this, thinking

---

[303] Appendix B, Table B16, K.4.C.1, V.4.C.1.
[304] Appendix B, Table B17, V.4.C.2.
[305] Appendix B, Table B17, K.4.C.2.
[306] Appendix B, Table B16, K.4.D.1; Table B19, K.4.C.4.
[307] Appendix B, Table B16, V.4.D.1.
[308] Appendix B, Table B18, K.4.C.3, V.4.C.3.

about chances and opportunities is beneficial to both teacher and student (Cauffman & van Dijk, 2009).[309]

Although education tries to develop the student's ability to apply knowledge and skills relatively independently, the teacher is an essential part of the learning process (Johnsen et al., 2016; D. Matthews & Foster, 2009; Tomlinson et al., 2009). Vygotsky (1978) refers explicitly to the teacher's support as being necessary when the student is working on tasks in their zone of proximal development.

Tomlinson et al. (2009) developed the Ascending Intellectual Demand (AID) continuum. Within AID they identified four distinct stages of the learning process, from novice, via apprentice, via practitioner to expert. According to the authors, teachers should aim for each student to have the opportunity and support to become an expert. "With the learner characteristics of each stage of the continuum in mind, the responsive teacher plans curriculum and instruction by aligning models and strategies with the needs of the learner" (Tomlinson et al., 2009, p. 237).[310] This strategy shows similarities to the levels for differentiation developed by Roberts and Inman (2015) for their DAP tool. Roberts and Inman developed this tool for assessing students' products based on a three-tiered structure. For each tier, they developed a rating scale that shows the student's expertise within that specific tier in relation to the specific learning objectives set for the student's task. Both AID and the DAP tool are examples of how teachers can make learning visible and at the same time can provide the appropriate support.[311]

The student's developmental process is supported by providing feed forward (Hattie, 2013; Wiliam, 2011). Feed forward combines assessment of learning and assessment for learning within three steps. First, the teacher provides feedback on the student's achievements by pointing out to what extent learning objectives are mastered (this is already good; this is what you have accomplished). Second, the teacher points out what the student's zone of proximal development is in regard to the set objectives (this is your next goal, this is what you have to learn). Third, the teacher provides feed up. Feed up shows students what they can do to make it better. Teachers can provide feed up at a metacognitive level (take another look, what do you think of, what was the most effective thing you have done) or at the level of concrete suggestions (if you do this, your work will be better, you might want to improve this bit and this is a suggestion of how you can do it).

Rubrics matching the DAP tool, for example, provide both the feedback (this is what you have achieved, and this is why that is already of high quality) and the perspective on the next level of expertise (these are the aspects of your assignment that can be improved). The product itself needs to be reviewed in a way that provides feed up (if you do this, x may be within reach). Vogelaar (2017) shows that gifted students benefit from metacognitive feed up. A metacognitive feed up can be based on different types of questions and tips. It

[309] Appendix B, Table B20, K.4.F.5.
[310] Appendix B, Table B20, K.4.A.5; Table B 18, V.4.F.3.
[311] Appendix B, Table B20, V.4.E.5.

can be divided into five categories: reflect, inquire, suggest, elevate and compliment. Feed up aiming for reflection, may refer to the learning process but also to the learning content. Similarly to an inquisitive feed up, the teacher can stipulate their feed up best through questions. Questions like "what made you do this? what do you think of? where did you find this information? why do you present the information in this context?" have proven to be effective for the learning process. The teacher teaches the student to argue their choices, and by so doing, they stimulate the student to decide whether the choices that were made were also the best choices. Suggest and elevate are two aspects of feed up that are very closely related. Suggestions can be used directly to improve the project on hand, tips that elevate are meant for situations where the student has moved onto a similar project from scratch and wants to use previous feedback to do it better second time around. The last part is to compliment the student, as in pointing out again what is of great value for the process and the product.

In gifted education, some people may find the use of the word "compliment" in the same sentence as the words "achievement" or "product" to be the proverbial red flag. Dweck's theory on mindset advises to be very careful about giving compliments (Dweck, 2006, 2017). In the social-constructivist approach to education, the development of a growth mindset is conditional. A growth mindset influences the emotions that occur during the learning process, motivation and self-regulation, and encourages students to take educational risks. Dweck suggests that compliments should focus on the student's learning process and not on the student's achievements.[312] In the mindset theory, compliments should fuel the student's intrinsic motivation for the act of learning. Giving feed forward helps in the development of a growth mindset.

Modelling of a growth mindset is not enough to help a growth mindset to develop. Direct interventions that encourage behaviour matching the growth mindset are needed (Haimovitz & Dweck, 2017). By providing feed forward, the teacher contributes to the student's ability to reflect on their learning process.[313] The teacher teaches the student how to recognise difficulties during the learning process in such a way that it empowers the student to find their personal solutions to overcoming these difficulties.[314]

Feed forward could be seen as a formative assessment. This type of assessment provides both the teacher and the student with the necessary understanding of what instructional and pedagogical strategies best match the student's developmental process.[315] At the same time, it teaches students where to find clues for a realistic perspective of their own knowledge and skills.[316] As self-efficacy is a core ingredient for motivation, the ability to realistically estimate one's knowledge and skills regarding a specific task, makes the difference to the learning process (Heacox & Cash, 2019; Hoogeveen, 2017b; Siegle & McCoach,

---

[312] Appendix B, Table B16, V.4.F.1.
[313] Appendix B, Table B16, K.4.F.1.
[314] Appendix B, Table B20, V.4.C.5, K.4.C.5
[315] Appendix B, Table B18, K.4.D.3; Table B19, V.4.D.4.
[316] Appendix B, Table B17, V.4.D.2.

2017).[317] Feed forward does not have to wait until a student has handed in their work, it can easily be provided when a teacher offers their support during the learning process while students are still working on their assignments. This type of feed forward is highly effective as it offers students opportunities to redirect their action and keep their eye on the ball (Hattie, 2013; Wiliam, 2011).
Within the context of feed forward being the starting point for regular student and teacher conversations about successes and failures during the learning process, the teacher develops the necessary understanding of what kind of support is most profitable for the student and how best to create meaningful educational responses for individual students.[318] For students, it is helpful to develop an understanding of how their actions during the learning process relate to their achievements (Dweck, 2006, 2017; Hattie, 2013; Wiliam, 2011). In their research regarding the role of emotions, motivation and learning behaviour, Obergriesser and Stoeger (2015) concluded that underachievers in their intervention group discovered that, although they used specific learning strategies, these had little effect if they were not used correctly. Process feedback taught these students that monitoring their own learning behaviour was an effective strategy to becoming more successful learners. Obergriesser and Stoeger's research shows that it is important for specialists in gifted education to understand why a student's results can lag behind what is expected of them, if the student (un)intentionally skips phases of the learning process.[319] This type of *diagnostic* knowledge creates the understanding that is necessary for an educational plan that matches an individual student's needs.
Although matching a student's needs is of great importance, assuring that the teacher can meet those needs under realistic conditions is equally important. Educational plans should work for both teacher and student.[320] The equilateral triangle between student, teacher and other students in the classroom discussed in Chapter 2, should stay intact. If not, teachers will not be able to maintain interventions using a long-term perspective.

8

8.5 Five Competencies in Domain 4

This chapter explored gifted education through the process of responding meaningfully to the educational needs of gifted students. Starting from the viewpoint that gifted students cannot be seen as a homogeneous group, the focus was on an approach that aims for differentiation. As the focus of education should be on the development of students, it is necessary that this is addressed in their zone of proximal development. That zone is affected not only by the level of content, but also by the level of learning skills that are acquired from the assignment. Finding an equilibrium between the levels of self-regulation, motivation and teacher support is a balancing act that may put a strain on classroom management. If the specialist in gifted education has a supportive

[317] Appendix B, Table B17, K.4.D.2.
[318] Appendix B, Table B18, V.4.D.3, V.4.E.3, K.4.E.3.
[319] Appendix B, Table B16, K.4.A.1.
[320] Appendix B, Table B20, V.4.A.5.

Table 8

Domain 4, From Appendix A: Competency Matrix for Specialists in Gifted Education

Generic Teaching Competencies	**4. Responding to educational needs**
	The specialist responds meaningfully to the educational needs of the gifted student from a contextual and developmental perspective. In doing so, (learning) process and (learning) results are equally important.
1. Pedagogical The teacher creates a safe and stimulating learning environment for every student, tuned to their individual educational needs.	**4.1.** The specialist encourages the student's attitude for learning, in a way that stimulates an equal balance between effort, learning and achievement.
2. Didactical In collaboration with their colleagues, the teacher transfers curricular objectives in such a way that they become teachable in the context of school and/or district policy. They must compose the curriculum in a manner that enables their student to master the content.	**4.2.** The specialist adjusts the curriculum in such a way that the student has to reach for their zone of proximal development in all domains.
3. Managerial The teacher provides a structured, well-organised, task-oriented classroom atmosphere that supports the students' developmental process.	**4.3.** The specialist adjusts their classroom management in a way that offers the opportunity to provide instruction/support to the gifted students.
4. Communication The teacher communicates professionally with all educational partners who are involved with the student's education.	**4.4.** The specialist uses the input of the educational partners, designing and implementing meaningful responses to educational needs for individual gifted students.
5. Research and Innovation The teacher adjusts their professional knowledge, perspectives, and actions if and when relevant developments in education, society, and science require it.	**4.5.** The specialist underpins their interventions with current theories on education and giftedness. In doing so they ensure that the interventions are workable in their daily practice.

role in their team, they may be asked to provide a specific structure for classroom management to help teachers to organise the delivery of an enriched curriculum.

A review of the literature showed that meaningful responses affect the curriculum as much as they affect the didactical and pedagogical strategies. At the same time, current changes in education are helping, as there are many similarities between the current educational paradigm and the concept of what is considered to be effective for gifted students. To make the transfer from the previous Dutch and Flemish educational approach to gifted students to the current perspective, the educational partnership is no longer limited to parents, fellow teachers, and school management. The student has become a relevant educational partner as well. Without undertaking this partnership, educational objectives that match 21st century skills lie well beyond reach. Competencies applicable to this domain from the competency matrix are described in Table 8.

8.6 Selection Rates of Indicators for Competency Domain 4

Meaningful responses to the educational needs of gifted learners are highly determined by their educational context and the opportunities that can be created by educators. What meaningful responses are, differs for all learners (Tomlinson et al., 2009; Winstanley, 2010). Educational profiles are unique; meaningful responses take the uniqueness of the student's profile into account (van Meersbergen & de Vries, 2017). Meaningful responses also take the uniqueness of the other actors of the ecological system into account. Teachers, parents, fellow students, they all have individual strengths and weaknesses interplaying with the educational context. Not only do these individual profiles of strengths and weaknesses affect what is considered to be meaningful, but what might be considered a meaningful response is also influenced by the core values and beliefs of the actors in the ecological system. Parents may consider different things important for their children than teachers, while specialists in gifted education may hold different opinions to their colleagues.

The selection of indicators thought to be important and relevant examples of the different competencies in domain 4 are likely to be influenced by what all different stakeholders consider to be meaningful responses. The presentation of data gathered for this part of the research is identical to that of the previous three domains. All tables referred to are included in Appendix H.

8.6.1 Selection Rates per Group of Stakeholders

8.6.1.1 Educational Professionals

Results per Competency

- Competency 4.1 focuses on specialists in gifted education stimulating their students to develop an attitude to learning where effort, learning and achievement are equally balanced. The highest selection rate for this competency amongst educational professionals is for the knowledge

indicator that focuses on understanding how a teacher's attitude contributes to the development of a student's self-concept (24.43%).[321] For the skills indicators, the highest selection rate (22.39%) amongst the educational professionals is for indicator V.4.E.1[322], having the skills to coach students in how to deal with positive and negative emotions towards achievement and teaching them that these emotions are a natural part of the learning process.
Remarkable is their low selection rate (6.87%) for indicator V.4.C.1, encouraging gifted learners to collaborate with other gifted learners. This low selection rate is congruent with the selection rates of other indicators from previous domains which aim at interaction with like-minded peers and peer group education. It becomes even more interesting when the word "gifted" is removed from this indicator, as seen with indicator K.4.C.1 ("knows how collaboration between students can help them optimise their achievement"). This indicator had a selection rate of 17.81%. This suggests that educational professionals acknowledge the value of interpersonal skills for gifted learners, but that they may not prioritise the development of these skills in an educational environment with other gifted learners.

- Competency 4.2 focuses on the specialist's knowledge and skills in adjusting the curriculum for gifted learners so that they have to reach for their zone of proximal development in all domains. The knowledge indicator that showed the highest selection rate (23.16%) is connected to the relevance of the specialist's ability to adjust their didactical strategies when providing educational challenges for the student.[323] On the skills side, the counterpart to this knowledge indicator "encourages creative and critical thinking by developing an investigative attitude in students" has the highest selection rate (26.21%).[324] The educational professionals' second-highest selection rate is for the skills indicator V.4.B.2 (19.85%), focusing on the curriculum being presented in such a way that higher order thinking skills are explicitly required to finish a task.
- Competency 4.3 concerns the specialist adjusting their classroom management in a way that offers the opportunity to provide instruction/support to gifted students. 22.14% of educational professionals found it relevant that specialists of gifted education should understand that enrichment activities are a natural part of a gifted learner's curriculum.[325] The knowledge indicators selected most often all had a strong relation to the instructional needs of gifted learners.[326]

---

[321] K.4.F.1 - Table H1
[322] Table H6
[323] K.4.E.2 - Table H2
[324] V.4.E.2 - Table H7
[325] K.4.A.3 - Table H3
[326] K.4.A.3, K.4.D.3, K.4.F.3 - Table H3

The skills indicator with the highest selection rate (21.63%) is V.4.E.3, "schedules regular teacher-student conferences to monitor a student's perception of their progress on classroom assignments".[327]

- The fourth competency in this domain focuses on collaboration between educational partners in designing and implementing meaningful responses to the educational needs of gifted learners. Within this context, 24.68% of the educational professionals consider it relevant that specialists of gifted education understand how feelings of competence, autonomy, and relationship contribute positively to a student's motivation.[328] 21.63% of them selected the skills indicator that encouraged a student's involvement in the process of setting educational objectives.[329] However, the indicator offering a student choices between the products they could use to show proof of their learning results was only selected by 13.74% of all educational professionals.[330] There is no direct explanation for this contradiction but it is most likely that they thought this issue had been addressed by the selection of indicator V.4.E.4.
- Competency 4.5 focused on the specialist underpinning their interventions with current theories on education and giftedness. 22.14% of educational professionals considered it relevant that specialists understand the importance of discussing both learning process and results with gifted learners.[331] The lowest selection rate (8.65% for knowledge indicator K.4.B.5) covered specialists in gifted education having an awareness of modern didactical strategies; this suggests that most educational professionals consider this to be standard behaviour, as they also have to keep up-to-date in this area. The skills that match this competency, V.4.B.5, focusing on active student participation during the learning process, received the highest selection rate (20.36%) from educational professionals; matching modern didactical strategies applied in a social-constructivist learning environment.

8

Overall Results of Educational Professionals for Domain 4

Combining the selection rates of knowledge and skills indicators, seven out of ten times, the indicators in category E show the highest selection rates amongst educational professionals (see Appendix H, Table H11). However, comparing mean scores per category, it shows that for the knowledge indicators, indicators aiming for the stimulation of the academic development of gifted learners received the highest mean score ($\overline{X} = 3.25$, Appendix H, Table H12). As the difference in what was expected ($\mu = 2.5$) was 1.5 SD, this is considered to be a relevant difference. Therefore, educational professionals consider a strong

[327] Table H8
[328] K.4.E.4 - Table H4
[329] V.4.E.4 - Table H9
[330] V.4.B.4 - Table H9
[331] K.4.E.5 - Table H5

emphasis on academic development for their gifted learners to be important in the competencies of the specialist in gifted education. The indicators they selected most frequently were often directly related to curriculum construction and delivery.
In domain 4, there are six indicators stimulating ownership by gifted learners. Ownership was not defined as a separate category, but can be deduced from indicators spread amongst all categories. In domain 3, educational professionals placed a strong emphasis on indicators that aimed for ownership (through voicing).[332] In domain 4, four of the six indicators focused on ownership did not end up amongst the three most selected indicators per competency. A careful conclusion may be that, while educational professionals consider it relevant to seek the input of gifted learners themselves on the topic of understanding the student's educational needs, they require that input less in designing meaningful responses. Once the understanding of educational needs has been based on inclusion of the student's voice, the opinion of the educational professional appears to be that the design of meaningful responses can best be underpinned with a wider spectrum of knowledge and skills.

8.6.1.2 Specialists in Gifted Education

Results per Competency

- For competency 4.1, the highest and lowest selection rates amongst specialists in gifted education were for the same knowledge indicators as those of educational professionals. Their highest selection rate (22.98%) is for a skills indicator with an emphasis on making a student's efforts for learning visible.[333] Similarly to that selected by educational professionals, the indicator aiming for collaboration between gifted students has the lowest selection rate amongst specialists in gifted education.[334]
- The stratification of selection rates of indicators for both knowledge and skills for competency 4.2 is identical to that of educational professionals, suggesting common ground for this competency.
- For competency 4.3, knowledge indicators with the highest selection rates focus on the provision of an enriched curriculum. The impact of communication between students on the pedagogical classroom climate is the least selected indicator (9.71%).[335] This indicator may be considered as common knowledge and, therefore, was not selected as most relevant for the competencies of a specialist in gifted education. A similarly low selection rate (8.41%) is found for this competency's counterpart amongst the skills indicators, (V.4.C.3, teaching students how to communicate in an open and safe manner); this may also be

---

[332] In domain 3, in six out of ten times, the indicators aiming for ownership ended up amongst the three most selected indicators per competency.
[333] V.4.F.1 - Table H6
[334] V.4.C.1 - 5.83% - Table H6
[335] K.4.C.3 - Table H3

considered to be a commonly developed skill amongst teachers in general.

- Competency 4.4 combines communicative competencies with the ability to respond meaningfully to the needs of gifted learners. In this context, the low selection rate (8.09%) for the indicator stating that gifted learners should be considered as experts of their learning process is interesting.[336] 18.12% of all the specialists selected indicator K.4.C.4, stating that the specialist in gifted education should understand the importance of the gifted learner being involved in regular group activities. 17.80% selected the skills counterpart (V.4.C.4), aimed at organising group activities in their regular classroom groups to prevent the social isolation of gifted learners.
- Looking at the selection rates for competency 4.5, an interesting contradiction between educational professionals and specialists in gifted education appears. Indicator V.4.B.5 had the lowest selection rate amongst specialists in gifted education (11.33%), but the highest selection rate amongst educational professionals (20.36%). Analysing the selection rates for the other skills indicators, this difference could possibly be explained by the difference between their professional roles. 20.06% of the specialists in gifted education selected the indicator V.4.A.5, focusing on the specialist's ability to design educational plans that work for both teacher and student. An identical percentage selected indicator V.4.F.5 aiming for the skill to discuss educational needs from a solution-focused perspective.

Overall Results of Specialists in Gifted Education for Domain 4

For most of the competencies and their indicators, there is a similarity between the stratification of selection rates between specialists in gifted education and educational professionals (Appendix H, Tables H1-H10). However, relevant differences can be observed between both groups of stakeholders for competency 4.1. The skills indicator V.4.F.1 was selected by 22.98% of all the specialists in gifted education. Amongst them this was the highest selection rate for the skills indicator matching this competency. The same indicator, however, did not end up amongst the three most selected indicators for educational professionals. By taking a closer look at the selection rates, the difference between the selection rates for both groups is 3.9%. This suggests that, although it conforms to the set thresholds for relevant differences, in reality, the difference is small.

[336] K.4.D.4 - Table H4

8.6.1.3 Parents

Results per Competency

- For the first competency in this domain, 21.69% of all parents selected indicator K.4.F.1, considering it relevant that specialists in education understand how the teacher affects a student's self-concept. An equal amount selected indicator K.4.D.1. This indicator also focuses on the gifted student's self-concept ("how like-minded peers can contribute to a student's development of a realistic self-concept").[337]
  10.04% of all parents selected indicator K.4.C.1. This indicator has their lowest selection rate for this competency. Compared to the selection rates of educational professionals (17.81%), this reveals a relevant difference. This indicator covers how collaboration between students can contribute to their achievements.
  From the skills indicators, the highest selection rate (24.50%) was for the specialists' skills in supporting their students to use learning styles appropriate for the assignment given.[338] Again, a relevant difference can be observed, this time between parents and specialists in gifted education. The selection rate of this indicator (19.74%) by the specialists in gifted education was not high enough to end up amongst their three most selected indicators.
- For competency 4.2, parents chose the same indicators as educational professionals and specialists in gifted education for their highest and lowest selection rates. For the skills indicators, the stratification of the selection rates was identical to those of educational professionals and specialists in gifted education, suggesting common ground between all three groups of stakeholders.
- The selection rates for the knowledge indicators of competency 4.3 showed common ground with both educational professionals and specialists in gifted education over the indicator with the highest selection rate being 24.10% for K.4.A.3 (understanding that enrichment activities should be a natural part of the gifted learners curriculum). An interesting difference is that 20.08% of all parents selected indicator K.4.B.3. For this indicator, specialists in gifted education had a selection rate of 13.92%, and amongst educational professionals this indicator had their lowest selection rate for this competency (12.21%).[339]
  For the skills indicators, parents had two indicators with an equally high selection rate of 22.09%; V.4.A.3 and V.4.D.3.[340] The aim of the first indicator is that the specialist in gifted education should regularly plan enrichment tasks for their gifted students, while the second indicator aims for the specialist in gifted education to regularly discuss with their

---

[337] Table H1
[338] V.4.B.1 – Table H6
[339] Table H3
[340] Table H8

students what could be considered to be the successes and challenges of their learning activities.

- Similarly to the previous three competencies, in competency 4.4, the highest knowledge indicator selection rate by parents (28.11%) was for an indicator that was also selected most often by educational professionals and specialists in gifted education. This indicator was about understanding that competence, autonomy, and relationship regarding educational objectives contributes positively to a student's motivation. For the skills indicators, common ground between all stakeholders was observed by them sharing almost identically low selection rates for the indicator focused on giving students a choice of the type of product they want to hand in as proof of their achievements.[341]
  There is, however, a relevant difference between parents and specialists in gifted education. 20.08% of parents selected indicator V.4.D.4; the specialist having the skill to ask gifted learners what type of support they consider to be most helpful for their learning process. By comparison, the selection rate of specialists in gifted education themselves was only 14.56%.
- For competency 4.5, 21.69% of parents selected indicator K.4.C.5, considering it necessary that specialists in gifted education have knowledge of how to use solution-focused strategies to help gifted learners to overcome difficulties during the learning process. This indicator was not selected often enough by educational professionals or specialists to end up in the three most selected indicators of either group.
  A second relevant difference can be noticed in the skills indicators. Educational professionals had a selection rate of 20.36% for the indicator that focused on the knowledge of modern didactical strategies. Amongst parents, however, this indicator only got a selection rate of 13.25%.[342] The highest selection rate from parents (22.89%) was for skills indicator V.4.C.5, concerning the specialist teaching the student how to recognise difficulties in their learning process and for the student to make them a topic of conversation with the teacher. This indicator was not selected by enough educational professionals to end up in their three most selected items for this competency.

Overall Results of Parents for Domain 4

The indicator that was most frequently selected by parents was indicator K.4.E.4 at 28.11% (see Appendix H, Table H11). This indicator focuses on the specialist's knowledge of what contributes positively to their student's motivation.
The category of indicators with the highest mean selection rate amongst parents was the D-category, both for knowledge ($\overline{X} = 3.14$) and skills ($\overline{X} = 3.11$) (see

[341] V.4.B.4 – Table H9
[342] V.4.B.5 – Table H10

Appendix H, Tables H37-H38). In the previous competency domains, indicators from category D also showed very high mean selection rates amongst parents. This suggests that parents consider knowledge and skills that stimulate intrapersonal understanding in their children to be important. The category indicators that had the lowest mean selection rate were the C-indicators for knowledge ($\overline{X} = 1.81$) and F-indicators for skills ($\overline{X} = 2.00$). This matches the stratification of their selection rates for all previous domains. This also suggests a trend. Parents may consider a specialist's knowledge of interpersonal functioning less important than direct skills for stimulating interpersonal functioning. At the same time, they seem to find knowledge about the more generic advisory tasks slightly more relevant than direct application of that knowledge, answering the questionnaire from their perspective on what would be the best match for the individual needs of their own children.

8.6.2 Significant Differences Between Groups of Stakeholders

For analyses of the differences between the groups of stakeholders, the ANOVA test and Bonferroni's Multiple Comparisons Test were used. In domain 4, differences between selection rates amongst separate groups of stakeholders are significant for knowledge indicators in the categories A ($p = .010$), D ($p < .001$), and F ($p < .001$) (see Appendix H, Tables H12-H14; see also Figure 32). The most significant differences are noticeable between parents and the other groups of stakeholders. Parents selected less knowledge indicators in category A ($\overline{X} = 2.82$) than educational professionals ($\overline{X} = 3.25$, $p = .013$).
Regarding knowledge indicators in category D (contributing to the intrapersonal understanding of gifted learners), significantly higher selection rates were observed amongst parents ($\overline{X} = 3.14$) compared to the selection rates of educational professionals ($\overline{X} = 2.44$, $p = .001$) and specialists in gifted education ($\overline{X} = 2.00$, $p < .001$). As seen in previous domains, parents emphasised the stimulation of their children's intrapersonal understanding.
In this domain, specialists in gifted education selected indicators from category F more often ($\overline{X} = 2.68$) than educational professionals ($\overline{X} = 2.12$, $p = .002$) and parents ($\overline{X} = 1.84$, $p < .001$). This supports the suggestion that specialists in gifted education find these more generic knowledge and skills indicators relevant as they are a match for the more coordinating role of the specialists in gifted education.
Regarding the skills indicators, significant differences were found for indicators in category D ($p = .001$) and F ($p < .001$) (Appendix H, Tables H15-17; see also Figure 33). Specialists in gifted education selected significantly less D-indicators ($\overline{X} = 2.44$) than educational professionals ($\overline{X} = 3.02$, $p = .001$) or parents ($\overline{X} = 3.11$, $p = .001$). Whilst the specialists in gifted education selected significantly more indicators in category F ($\overline{X} = 2.32$) than parents ($\overline{X} = 2.00$, $p = .001$) and educational professionals ($\overline{X} = 2.08$, $p = .001$). This trend, similar to above, suggests that if specialists are forced to make choices in what they consider to be most relevant, they prefer to include indicators that match their role.

Figure 32

Comparison of Selection Rates of the Knowledge Indicators in Domain 4 Between Groups of Stakeholders - (see also Table H12)

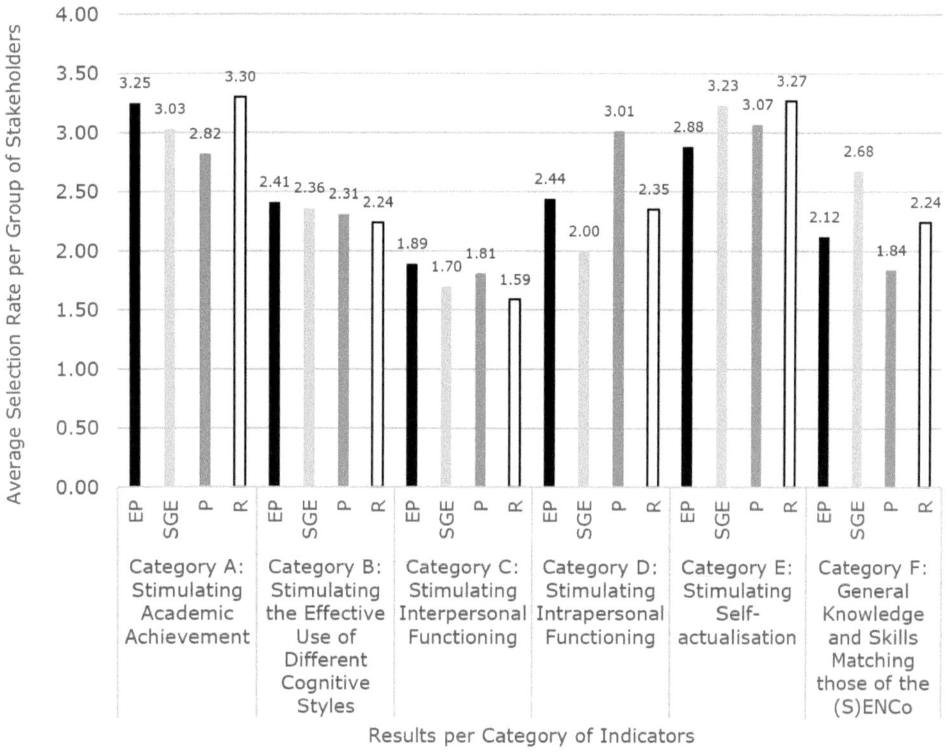

Note. N = 354 divided into four groups of stakeholders: EP = Educational Professionals, SGE = Specialists in Gifted Education, P = Parents, R = Rest. EP: N = 131, SGE: N = 103, P: N = 83, R: N = 37.

Test value = 2.5

Figure 33

Comparison of Selection Rates of the Skills Indicators in Domain 4 Between Groups of Stakeholders - (see also Table H15)

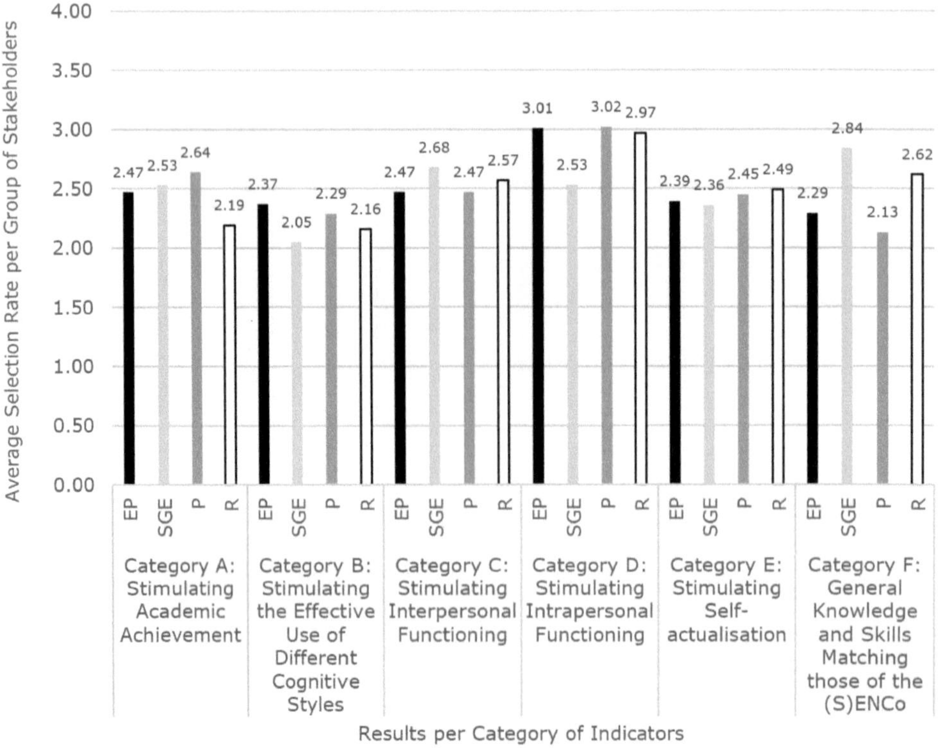

Note. N = 354 divided into four groups of stakeholders: EP = Educational Professionals, SGE = Specialists in Gifted Education, P = Parents, R = Rest. EP: N = 131, SGE: N = 103, P: N = 83, R: N = 37.

Test value = 2.5

8.6.3 Common Ground Between Groups of Stakeholders

Common Ground Amongst All Groups

Initial analysis of the results aimed to identify common ground between all groups of stakeholders. Two frequency analyses were applied. The primary frequency analysis showed that there are 9 knowledge indicators and 8 skills indicators that ended up lying in the three most selected indicators amongst all groups of stakeholders (see Appendix H, Tables H18-H19). Indicators included all 6 categories.

M. Stimulating academic development (5)
N. Stimulating the use of different cognitive styles (1)
O. Stimulating interpersonal functioning (1)
P. Stimulating intrapersonal understanding (1)
Q. Stimulating self-actualisation (6)
R. Generic counselling and advisory skills matching the skills of a (S)ENCo (3)

The secondary frequency analysis aimed to determine if, amongst these 17 indicators, there were three indicators that had the highest selection rates amongst all groups. Two knowledge indicators and one skills indicator met that criterion (see Appendix H, Tables H29-H30).

- Competency 4.2
    - Encourages creative and critical thinking by developing an investigative attitude in students.[343]
- Competency 4.3
    - Understands how enrichment activities become a natural part of a gifted learner's curriculum.[344]
- Competency 4.4
    - Knows how a feeling of competence, autonomy, and relationship regarding educational objectives contributes positively to a student's motivation.[345]

If there is no differentiation between the groups of stakeholders, the selection rates of all knowledge indicators, except for the D-indicators in domain 4, differ significantly from the expected rate (see Appendix H, Tables H20-H21; see also Figure 34). Knowledge indicators in category A have a mean selection rate of $\overline{X} = 3.09$. The difference, 0.590, to the expected rate ($\mu = 2.5$) is significant, $t(353) = 10.983$, $p < .001$, the effect size represents a practical significance, $d = .58$. Knowledge indicators in category E have a selection rate of $\overline{X} = 3.07$, again representing a significant difference, 0.568, $t(353) = 10.107$, $p < .001$, representing a medium sized effect, $d = .54$. Both of these sets of indicators differed positively from the expected mean selection rate, suggesting that all

[343] V.4.E.2 - Table H30
[344] K.4.A.3 - Table H29
[345] K.4.E.4 - Table H29

groups of stakeholders find these indicators important examples of competent professional behaviour for the specialist in gifted education.
Similarly to the previous domains, knowledge indicators in category C showed the lowest selection rates, $\overline{X} = 1.79$. Again, this proved to be a significant difference, -0.715, $t(353) = -12.132$, $p < .001$, $d = .64$, the effect size represents a practical significance.

Figure 34

Mean Selection Rate of the Knowledge Indicators in Domain 4 - (See Table H20)

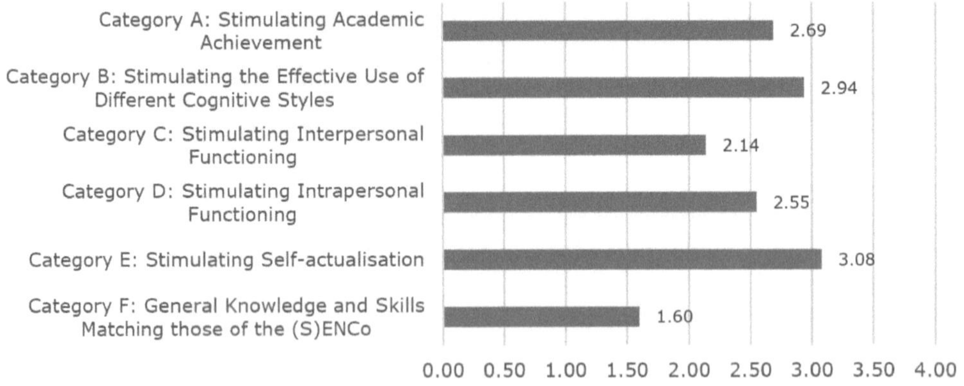

Note. N = 354. Test value = 2.5

Figure 35

Mean Selection Rate of the Skills Indicators in Domain 4 - (See Table H22)

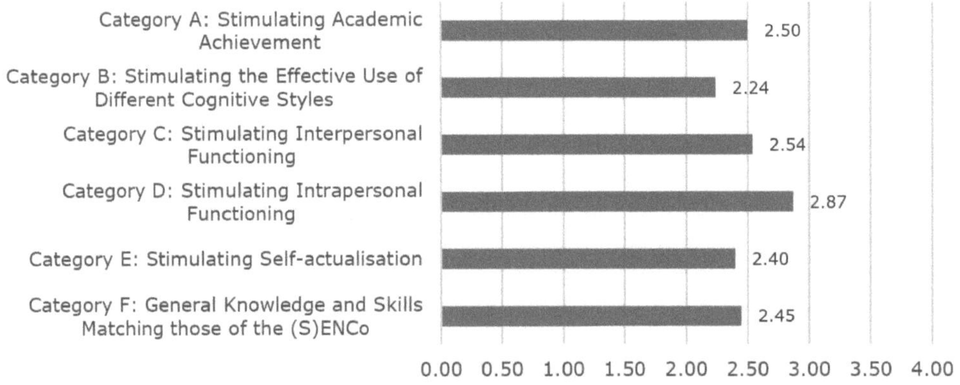

Note. N = 354. Test value = 2.5

For the skills indicators, only those belonging to category B did not differ significantly from the expected value (μ = 2.5) (see Appendix H, Tables H22-H23; see also Figure 35). Of particular interest are the selection rates for the skills indicators in category C. Their selection rates were slightly higher than the expected rate $\bar{X}$ = 2.68 and the difference, 0.181, may be considered significant, $t(353) = 3.093$, $p = .002$, $d = .16$, representing a small effect size. This observed difference between knowledge and skills indicators showed a similar pattern to the three previous domains, suggesting that knowledge about interpersonal functioning was considered less relevant than the specialist's skill to apply meaningful responses regarding interpersonal functioning.
Similar to that seen in domains 1, 2, and 3, the overall results of domain 4 suggest that some categories of indicators were selected more often than others. Again, a clustering in two categories (ADE vs BCF) was possible (see Appendix H, Tables H25-H26). The Paired Samples Statistics Test showed that the selection rates of clustered categories (both for knowledge and skills) differed significantly from the expected selection rate ($p < .001$). The clustered knowledge indicators ADE, $\bar{X}$ = 2.88, were selected significantly more often than the clustered knowledge indicators BCF, $\bar{X}$ = 2.12 ($p < .001$, $d = .75$). The clustered skills indicators ADE, $\bar{X}$ = 2.57 were selected significantly more than the clustered skills indicators BCF, $\bar{X}$ = 2.43 ($p = .012$, $d = .13$) (Appendix H, Tables H27-H28). This suggests that overall, the indicators in category ADE were considered to be more relevant examples for the competencies they were supposed to match.

8.6.4 Comparison Between The Netherlands and Flanders

The final analyses for this domain are around similarities and differences between The Netherlands and Flanders. Using Group statistics and an Independent Samples test, the following results were found (see Appendix H, Tables H33-H36).
Knowledge indicators in category B were selected significantly more often by Dutch stakeholders, $\bar{X}$ = 2.45, compared to Flemish stakeholders, $\bar{X}$ = 1.85. The difference, 0.594, in selection rates between stakeholders from both countries is significant, $p < .001$, the effect size represents a practical significance, $d = .60$. Knowledge indicators in category C had the lowest selection rates for both countries. However, they were more frequently selected by Flemish stakeholders, $\bar{X}$ = 2.16. The difference, 0.448, is also significant, $p = .006$, the effect size represents a practical significance $d = .41$. The mean selection rates for indicators in categories A, B, and F were only slightly different from each other, suggesting that in both countries these indicators were found to be equally important (see Appendix H, Tables H33-H34; see also Figure 36).
Skills indicators were significantly different for indicator D ($p = .001$) and F ($p < .001$) (see Appendix H, tables H35-H36; see also Figure 37). The selection rate of D-indicators was significantly higher amongst Flemish stakeholders ($\bar{X}$ = 3.35) compared to the Dutch stakeholders ($\bar{X}$ = 2.78). The difference, 0.563, is significant, $p < .001$, the effect size meets the threshold for practical significance, $d = .47$. An opposite difference can be observed for indicators in

category F. F-indicators were selected more frequently by Dutch stakeholders ($\bar{X}$ = 2.31) than by Flemish stakeholders ($\bar{X}$ = 1.78). The difference, 0.533, is significant, $p < .001$, the effect size represents a practical significance, $d = .53$. For the previous domains, the differences in selection rates for the F-indicators were explained by the fact that these indicators were in general more frequently selected by specialists in gifted education and that they were underrepresented amongst the Flemish respondents. It is likely that the same explanation is correct for domain 4.

Figure 36

Comparison of Mean Selection Rates in Domain 4 per Category of Knowledge Indicators Between Dutch and Flemish Stakeholders - (See Table H33)

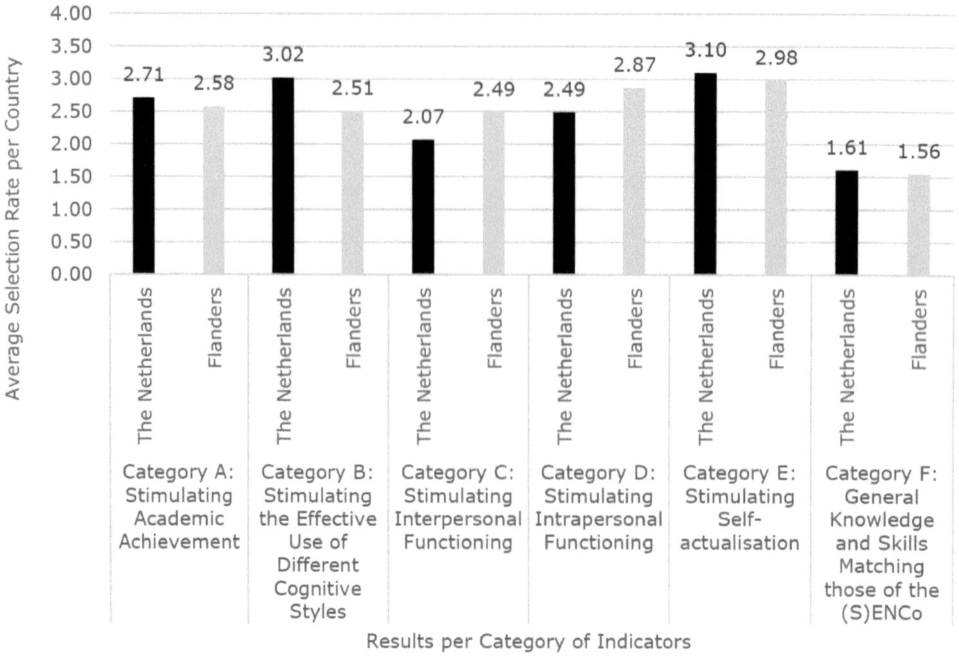

Note. N = 354. The Netherlands N = 299, Flanders N = 55. Test value = 2.5

Figure 37

Comparison of Mean Selection Rates in Domain 4 per Category of Skills Indicators Between Dutch and Flemish Stakeholders - (See Table H35)

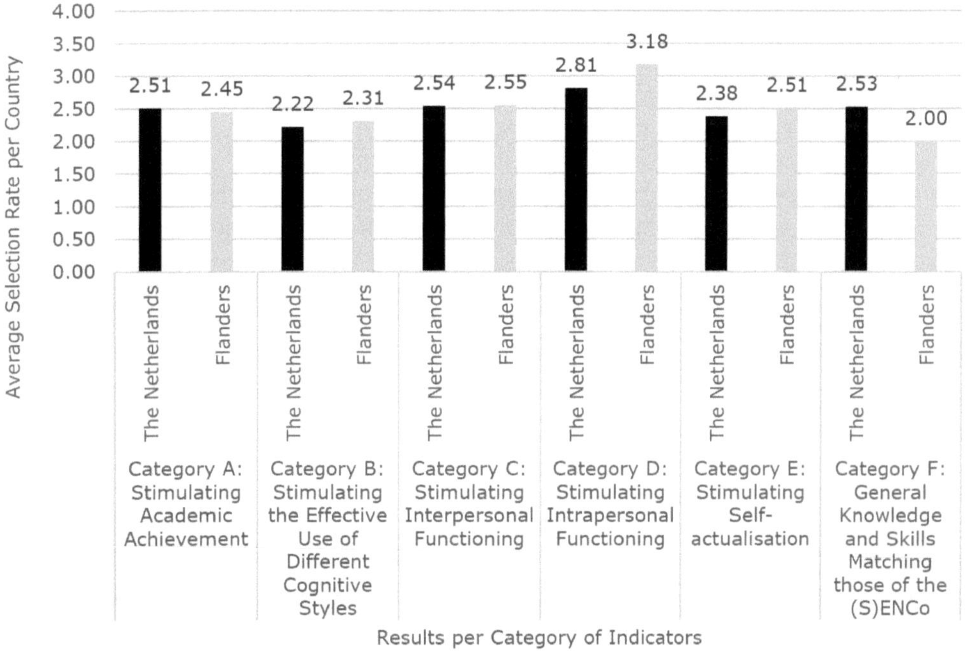

Note. N = 354. The Netherlands N = 299, Flanders N = 55. Test value = 2.5

8.7 Summarising

This chapter focused on domain 4 of the competency matrix, the design and implementation of meaningful responses to the needs of gifted learners. These responses are optimal if they can be tuned to match the individual profiles of gifted students with the individual profiles of the actors in their ecological systems.

- 4.1 The specialist encourages the student's attitude for learning in a way that stimulates an equal balance between effort, learning and achievement.
- 4.2 The specialist adjusts the curriculum in such a way that the student has to reach for their zone of proximal development in all domains.
- 4.3 The specialist adjusts their classroom management in a way that offers the opportunity to provide instruction/support to gifted students.
- 4.4 The specialist uses the input of the educational partners, designing and implementing meaningful responses to educational needs for individual gifted students.

- 4.5 The specialist underpins their interventions with current theories on education and giftedness. In doing so, they ensure that the interventions are workable in their daily practice.

Identically to the previous domains, indicators in category ADE were more frequently selected than indicators in category BCF. There is no difference between knowledge and skills indicators; the higher selection rates for ADE indicators apply for both types.
For the knowledge indicators, the mean selection rates of those indicators belonging to categories A and E were significantly higher than the mean selection rates for other indicators. C-indicators were selected significantly less, showing medium-sized effect sizes. Academic achievements and the stimulation of self-actualisation are considered to be important for gifted learners. Knowledge of the stimulation of interpersonal functioning is considered to be less relevant. However, the skills to use strategies to stimulate interpersonal functioning showed significantly higher selection rates. The conclusion that stakeholders consider the stimulation of interpersonal functioning as irrelevant, may not be drawn.
Significant differences were seen for the skills indicators; however, comparing mean selection rates shows that observed differences are relatively small, with almost all referring to small effect sizes.
In a similar way to that seen in the previous domains, it can be observed that stakeholders select indicators that match their responsibilities in the upbringing and education of gifted learners. Parents select indicators aiming for the wellbeing and development of the individual child more often, while educational professionals show higher selection rates for indicators matching the responsibilities of classroom teachers. Indicators selected most often by specialists in gifted education match their role, with a stronger emphasis on the coordination of education of gifted learners at school level. The results found for domain 4 are consistent with the results found in the previous domains.

9 Domain 5: Assessing the Response to Interventions

9.1 Introduction

In this final chapter about the construction of the competency matrix, the focus is on the fifth domain, assessing the response to interventions: "the specialist monitors the process of responding meaningfully to the gifted student's educational needs and monitors and interprets the gifted student's responses to these interventions contextually" (see Appendix A).
As profiling the educational needs of gifted students leads to intervening in their education and, therefore, leads to meeting their needs in a way that is thought to be beneficial for an individual gifted student's development, the ability to evaluate the success rate of the interventions applied should be a part of the professional repertoire of the specialist in gifted education.
In The Netherlands and Flanders, the change and solution-focused approach aims to change aspects of the learning content, learning process, and learning results. In addition, this approach can be used to improve learning behaviour thought to be beneficial for talent development and to improve the student's wellbeing. This means that the phrase "response to intervention", in this context, exceeds the process of systematically exploring the student's responses to interventions in the curriculum. This chapter discusses what knowledge and skills contribute to the ability of the specialist in gifted education to assess the results of their interventions, as explored in previous chapters. In other words, how can the specialist in gifted education make learning visible when the content being learned is not just academic or may not even be part of a formal curriculum?
A complicating factor in this part of the competency matrix is that, in current research on gifted education, there is almost no specific documentation on how results of non-curricular interventions can be monitored and evaluated. In their chapter on teacher preparation for teachers of the gifted, Johnsen et al. (2017) refer to assessment as a required competency without providing instructions on how specialists in gifted education can apply this in their daily practice. They appear to consider this to be a competency developed during undergraduate teacher training.[346] They do, however, refer to research support in Appendix A in their book for the presented standard. Checking "Standard 4: Assessment" in this specific appendix of their book, it can be recognised that most of their referrals are focused on literature helping teachers to identify gifted learners and are not intended to measure the success of a student's responses to interventions. Referring to the CLEAR curriculum model, Shaunessy-Dedrick (2018) points out the importance of continuous formative assessment: "to determine appropriate student challenge based on learning growth, learning aligned with the NAGC field standards, authentic product development and the incorporation of challenging content and learning experiences beyond basic skill

[346] Appendix B, Table B25, K.5.F.5.

acquisition typical in general education classrooms" (p. 152). However, again, there is no instruction on how teachers can apply this kind of assessment with a broader perspective that is useful for the Dutch and Flemish approach to gifted education. Lo et al. (2019) refer to encouraging educators to use assessment for transactional purposes by adopting growth-oriented assessment frameworks. They state: "In practice, transactional assessment should focus on providing ongoing feedback to learners to encourage further growth and sustaining evidence necessary for justifying referral decisions on more advanced learning opportunities" (Lo et al., 2019, p. 179).
Due to a lack of structured suggestions for a process to assess results from outside of the academic domains, the highlighting of strategies used for the overall evaluation of educational processes based on common practice in The Netherlands and Flanders cannot be avoided in this chapter. The baseline in this process of monitoring and evaluation is communication between all the educational partners involved and the MIA as introduced in Chapter 3.

9.2 Communicating about the Results of Interventions

In Dutch and Flemish schools, responsibility for a student's progress and development can be seen as a shared one between teachers, parents, and the specialist in gifted education (Pameijer & van Beukering, 2007; van Swet, 2009). Support for this approach can be seen in the international literature on gifted education (Hughes, 2017). The ability to share information with colleagues and other educational partners in a way that makes the information understandable and relevant for them is, therefore, conditional (Hughes, 2017).[347] Within that context, communication - considered as an important basic skill for teachers in general - becomes even more relevant (Johnsen & Clarenbach, 2017).
In previous chapters, the importance of involving the educational partners of a student's ecological system in the process of seeing, understanding, and responding to a student's needs, has been discussed. As a result of this involvement, in this final competency domain, the educational partners play an equally important role.
Hughes (2017) describes how collaboration between all different educational partners is beneficial to creating a balance between what is desirable optimally and doable optimally for teachers. She suggests striving for a strong collaboration between all the educational specialists in the school to end the cacophony of voices created when everyone is working towards individual interventions. The collaboration of educational specialists in an inclusive environment can ensure that the interventions that are designed are both workable for teachers and more profitable for their students. Hughes describes the importance of specialists in gifted education attuning their educational purposes to match the purposes of various stakeholders. Interplay between the specialist in gifted education and the classroom teacher when setting educational

[347] Appendix B, Table B22, K.5.F.2.

goals for individual students is, as she describes it, a condition for success.[348] In the context of this final competency domain, this strategy should apply not only when setting educational objectives, but also for evaluating the results of interventions. Education should be seen as an ongoing process, and reached objectives should form a starting point for setting new objectives (Lo et al., 2019). Therefore, at the start of every new term, it would be relevant to share all the information available between the relevant education partners.[349] However, while Hughes pleads for the collaboration of professional educational partners, she does not mention collaboration with parents. In The Netherlands and Flanders, parents are considered to be equally important actors in a student's developmental process as educational professionals (Pameijer, 2012; Pameijer & van Beukering, 2007; van Meersbergen & de Vries, 2017; van Swet, 2009). Hence, they need to have an equally important seat at the table.
A joint strategy between teachers, for documenting responses to interventions and documenting a student's developmental progress could be considered as a starting point for the successful education of gifted students.[350] The assessment of a student's response to intervention, consequently, requires that the specialist in gifted education communicates with the educational partners on how their actions have been of influence on the student's development (Coleman & Johnsen, 2011; Pameijer & van Beukering, 2007; van Meersbergen & de Vries, 2017).[351] Hence, the process of monitoring if and how all partners involved have implemented the agreed interventions at the agreed time and in the agreed manner should be part of the assessment of the response to interventions (Pameijer & van Beukering, 2007; van Meersbergen & de Vries, 2017). This process of monitoring is a two-way stream. Therefore, educational partners should be offered the opportunity to share their observations on a student's development as well as on the process of implementing interventions and the effects of these interventions (Johnsen et al., 2017).[352] This process of sharing information contributes both to the quality of the gathered data and to the partners' commitment to the process of intervening (van Meersbergen & de Vries, 2017). Using the analysed results and the shared information increases the opportunity to tune (curricular) interventions to the needs of the gifted student.[353]

9

9.3 Making Learning Visible

In theory and in practice, the interventions described for gifted students first and foremost aim for curricular changes. A curriculum that matches a gifted student's potential is always mentioned as important. However, in the shift towards a conceptual paradigm where giftedness is no longer seen as a verifiable condition, but rather as a lived process that can be examined

[348] Appendix B, Table B24, V.5.F.4.
[349] Appendix B, Table B23, K.5.A 3.
[350] Appendix B, Table B23, K.5.F.3, V.5.F.3.
[351] Appendix B, Table b24, V.5.A.4, V.5.E.4.
[352] Appendix B, Table B24, V.5.B.4.
[353] Appendix B, Table B24, K.5.A.4, V.5.D.1.

formatively, curricular changes become even more important as giftedness now refers to the mismatch between potential and the education offered (Lo et al., 2019; D. Matthews & Folsom, 2009). Identifying this mismatch and tuning into the student's needs to realise their potential requires this formative assessment to be within an ongoing process.

In The Netherlands and Flanders, schools currently provide curricular changes through the approach of compacting and enrichment (Ahlers, 2017; Bakx et al., 2016; Doolaard & Oudbier, 2010; Janson, 2017; Kuipers, 2017; Vreijs et al., 2017). The way they organise this adjusted curriculum is different for all schools, as is the selection of the educational objectives for which they aim. Setting educational standards outside of the national curriculum is left to the expertise of each school. There is no empirical evidence, above the level of good practice, to show how schools set these objectives or to what extent there is a match between schools in their selection criteria for participation, in this modern perspective on giftedness.

In The Netherlands, as in Flanders, educational adjustments can be divided into curricular changes provided in the regular classroom by the regular classroom teacher, and the opportunity for participation of gifted students in enrichment projects and pull-out programmes (Doolaard & Oudbier, 2010; Janson, 2009; P. Wolters, 2009). Not every school offers the opportunity to participate in special projects. However, most schools offer curricular adjustments within the regular classroom by the regular classroom teacher. In theory, schools have these interventions set up as they are suggested in the literature on gifted education.

Reviewing literature on gifted education and curricular interventions leads to the conclusion that one topic is mentioned by almost every author: gifted students should be challenged, and their education should be meaningful. In the previous chapter, what is seen as challenging and meaningful was discussed. What is experienced as challenge is different for individual learners. Challenging is not only to do with the level of thinking skills that are called on by the assignments; it is also determined by the concept of working in the zone of proximal development (Tomlinson et al., 2009; Winstanley, 2010). Roberts and Inman (2015) describe how the construction of a challenging curriculum should start with setting clear educational objectives. Selecting matching materials and assignments is only the second step. From this perspective, the procedure around creating a curriculum for gifted and able learners is no different to the procedure for setting up the curriculum for other students. However, teachers participating in continuous professional development courses on gifted education report that, in daily practice in primary schools in The Netherlands and Flanders, this first step is often skipped (Slim! Educatief, Postgraduate Teacher Education Course Specialist in Gifted Education, personal communication/class discussions during modules 3 and 4, January-March 2017, January-March 2018; January-March 2019; January-March 2020). This implies that the selection of the student's curriculum is not based upon an objective assessment but is based on opportunity and availability. Under these circumstances, the notion of challenge is not individually defined. It becomes defined by what publishers of learning materials define as challenge, by general assumptions of how gifted learners are assumed to process information and by general assumptions of what gifted students have to learn. According to Winstanley (2010) and Tomlinson et al. (2009), that is exactly the opposite of what challenge defines. Dutch and Flemish teachers mostly select enrichment materials without assessing objectively whether the chosen assignments aim for relevant educational objectives that match a student's zone of proximal development. According to the teachers reporting this practice at the schools in which they work, the selection of materials takes place on the following general principles:

- What materials are available?
- Are these materials fit for gifted students, according to the publisher?
- Does the teacher think that their student will enjoy working with these materials?

As a result, teachers have difficulty in determining how to assess the results of their interventions. The lack of setting concrete learning objectives that can then be objectively assessed also creates difficulties in making learning visible. Teachers currently report what a student has done and to what extent the assignment is executed correctly, but that doesn't necessarily mean that the student has actually learned something. From this perspective, the assessment of the response to intervention is limited. On top of that, there are no standardised tests available in Dutch and Flemish education for most of the enrichment materials and most of the objectives set for gifted students.

9

In international publications on gifted education, the use of standardised test results often underpins the process of response to intervention (Coleman & Johnsen, 2011; Heacox & Cash, 2014). However, as soon as the curriculum exceeds the levels of these tests or aims for domains outside of the range that can be assessed by these tests, the results of the learning process are more difficult to document.

Often, in The Netherlands and Flanders, alternative documentation of a student's development, such as a portfolio of work, replaces the use of standardised tests. Consequently, the importance of developing appropriate knowledge and skills in how to transfer developmental and educational objectives into a quantifiable process increases.[354] Tools like the DAP tool developed and presented by Roberts and Inman (2013, 2015) or suggestions made by Tomlinson et al. (2009) provide the specialists in gifted education with relevant clues of how to make learning visible within the academic domains.[355] These strategies provide ample opportunities for teachers and specialists in gifted education to explore how they can enhance the effectiveness of their instructional strategies and the student's learning.[356] With regard to enrichment coursework, making learning visible then offers opportunities to set new relevant educational objectives and to explore with students the value of set objectives.[357] The specialist in gifted education, in collaboration with the (S)ENCo, coordinates pedagogical-didactical interventions

[354] Appendix B, Table B25, V.5.A.5, V.5.F.5.
[355] Appendix B, Table B22, K.5.A.2, V.5.A.2; Table B23, V.5.B.3.
[356] Appendix B, Table B22, V.5.F.2.
[357] Appendix B, Table B25, K.5.A.5, V.5.E.5.

and monitoring of the student's progress throughout the student's school career, thus taking on the responsibility of setting up an adequate schedule to document the student's learning process (van Gerven & Hoogenberg, 2011).[358]
Nowhere in the international literature on RtI for gifted students, are the effects of specific instructional strategies and pedagogical interventions mentioned as a separate part of a protocol of assessing the response to interventions, nor is it described how these interventions can be assessed (Coleman, 2018; Coleman & Johnsen, 2011). In The Netherlands and Flanders, response to intervention is seen from the perspective of exceeding curricular interventions. Interventions can be focused on behavioural issues, social and emotional issues and so forth (Pameijer & van Beukering, 2007; van Meersbergen & de Vries, 2017). There are no standardised tests to assess the results of these kinds of interventions and assessing the results of these interventions is not addressed in undergraduate teacher education. This implies that specialists in gifted education have to learn how they can monitor their student's response to these types of interventions.
In The Netherlands and Flanders, the regular classroom teacher is the professional who holds primary responsibility for their students' development.[359] However, in a social-constructivist approach, where self-regulated learning is an increasingly implemented educational strategy, the student also has to be taught how to take personal responsibility for this process (Oppong et al., 2018). Considering that setting educational objectives is, therefore, a joint responsibility between teachers and students, the process of making learning visible should be regarded as an intervention on its own merits (Hattie, 2013). For example, discussing a student's progress with them, informs the student both of what their achievements are and about their current zone of proximal development (Wiliam, 2011).[360] Another example is interviewing a student about how emotional regulation influences their achievements in situations of formal testing. By doing so, the specialist develops an understanding of the student's learning behaviour, and at the same time, they can explore effective coping strategies with the student.[361] Providing feed forward, informing a student of their personal zone of proximal development, helps the student to set new educational goals and can therefore be seen as an intervention in the student's learning process (Hattie, 2013; Hattie & Yates, 2015; Lo et al., 2019; Wiliam, 2011).[362] Standardised test results are relevant in this context[363], but are also a one-dimensional way of making learning visible. These results leave the learning process and relevant learning behaviour out of the scope, whereas, in fact, both should be taken into account if someone is setting their own goals (Oppong et al., 2018; Tomlinson et al., 2009; Winstanley, 2010). One of the roles of the specialist of gifted education is to help the student to set new

[358] Appendix B, Table B25, V.5.A.3.
[359] Appendix B, Table B24, K.5.F.4.
[360] Appendix B, Table B22, V.5.E.2.
[361] Appendix B, Table B23, K.5.D.3.
[362] Appendix B, Table B25, K.5.E.5.
[363] Appendix B, Table B23, K.5.E.3.

realistic goals matching their academic abilities, level of effective learning behaviour and the educational context in which the student functions (Heacox & Cash, 2019; Tomlinson et al., 2009; Winstanley, 2010).[364] Creating this type of ownership aimed at self-actualisation requires expertise in how to interpret and understand the relationship between a student's achievements and their ability for self-actualisation (Venderickx & Kieboom, 2017).[365] It also requires the knowledge and skills to show the student how their learning behaviour interacts with their achievements and how to coach the student in striving for achievements commensurate with their abilities (Dweck, 2006; Kieboom & Venderickx, 2017a; Oppong et al., 2018; Stoeger & Obergriesser, 2015; Stoeger et al., 2016; Venderickx & Kieboom, 2017).[366] A joint exploration of the conditions that contribute to the student's feelings of self-efficacy can be helpful at this point. Based on the change and solution-focused approach, this type of ownership not only contributes to achievements within an isolated aspect but also contributes to the student's feelings of autonomy, relationship, and competence. As the ecological system influences the process of development, a student's self-actualisation is therefore always the result of the interplay between the student and the educational partners from this system (Gagné, 2010a; Lo et al., 2019; Sternberg, 2009; Subotnik et al., 2011; Ziegler et al., 2017).[367]

9

9.4 Making Learning Possible

At her inaugural address, Bakx (2019) described how gifted primary school students valued the pedagogical relationship with their teachers above their need for autonomy and the need to feel competent. These findings are similar to those presented by Robinson (2008) in her literature review of what characteristics of teachers define successful teachers of the gifted. To establish this pedagogical relationship, specialists in gifted education have to maintain an intrinsic interest in their students and to show this in the way that they communicate with them (Bakx, 2019; Robinson, 2008).[368] This requires that they are not only capable of reflecting on a student's behaviour, but are also capable of reflecting on their own professional behaviour (Biesta, 2018b; Loughran, 2002).[369] Voicing, discussing with the student how didactical and pedagogical changes in their education may assist the student's development and wellbeing, contributes to this relationship (Bakx, 2019; den Otter, 2015; Mayes et al., 2019; Robinson, 2008).[370] With this teacher-student relationship as a baseline, specialists in gifted education can discuss with students how to use their competencies effectively, how their need for autonomy can be used as a strength without being experienced as rebellious and how social relationships

[364] Appendix B, Table B23, V.5.E.3.
[365] Appendix B, Table B22, K.5.E.2.
[366] Appendix B, Table B24, K.5.D.4.
[367] Appendix B, Table B24, K.5.E.4.
[368] Appendix B, Table B22, V.5.D.2.
[369] Appendix B, Table B22, K.5.D.2.
[370] Appendix B, Table B21, V.5.E.1, V.5.A.1.

with a diverse group of fellow students can add to their learning process as well.[371]
In this relationship, not only does the student but also the specialist in gifted education become a learner. The information students provide during these contacts is useful to monitor the effects of pedagogical-didactical interventions and curricular adjustments. These results help the specialist to adjust their professional behaviour, as well as to guide their colleagues in their professional behaviour towards gifted students. This process matches the process described by the model of intrapersonal adaptation (MIA) in Chapter 3.
In a social-constructivist approach to education, students need to have a realistic perception of their capabilities and abilities (Lo et al., 2019; G. Matthews et al., 2018; Rinn, 2018; Rinn & Majority, 2018).[372] Self-perception, self-understanding, and self-efficacy are relevant for success (Lo et al., 2019; Oppong et al., 2018; Siegle et al., 2017). In the achievement orientation model, Siegle and Mc Coach (2017) described the influence of self-efficacy on the learning process of gifted students. In situations where a gifted student is confronted with course work or projects that call explicitly on their zone of proximal development, it is relevant to teach the student how they can establish self-efficacy around the task that they have to do (Lo et al., 2019).[373] Exploring the student's concept of self-efficacy can be seen both as a pre-assessment and a post-assessment of the chances for successful curricular interventions. This type of assessment can be done by short teacher-student conversations (Siegle et al., 2017). During these conversations, teacher and student explore what knowledge and skills are required to be able to start working on the task. Together they explore when, in previous tasks, the student has also been able to use this knowledge and skills. The next step is to then analyse what knowledge and skills the student will learn by doing the task. This process of assessment can be referred to as reflection before action. The reflective questions the teacher has previously used for reflection *in* action (while doing the task) or reflection *on* action (after the task is finished) can now be used for reflection *before* action.
To be supportive of this aspect of human development, the specialist in gifted education needs to not only be aware of these concepts but should be able to translate their professional knowledge into daily practice where they help students to develop these intrapersonal insights.[374] To do this, it may be helpful to use tools that are also used for identifying educational needs (de Bruin-de Boer & Kuipers, 2011; Houkema, 2009; van Gerven & Drent, 2012). However, the focus should be on identifying how the student's learning behaviour is of influence on the learning process and achievements, and on the exploration and transferral of these insights to the student (Lo et al., 2019).[375] At the same time, focus is also required on how interventions successfully influence a student's

---

[371] Appendix B, Table B25, V.5.C.5.
[372] Appendix B, Table B25, K.5.D.5.
[373] Appendix B, Table B25, V.5.D.5.
[374] Appendix B, Table B21, K.5.D.1, K.5.E.1; Table B24, K.5.D.4.
[375] Appendix B, Table B23, K.5.B.3.

ability to switch strategies and preferences when required by the learning content or the learning context.[376] As learning takes place not only at school, but also in other contexts (Gagné, 2010a; Subotnik et al., 2011), communication with educational partners on their observations of the student's cognitive style can provide relevant information, for instance on the student's ability to adjust their style under different circumstances.[377] When the teacher and the student both develop an understanding of which content-based knowledge and skills, personal learning preferences, and executive skills influence the student's efficacy in a task, they may both be able to adjust their behaviour (Baum et al., 2017; Trail, 2011; Winstanley, 2010). Based on this perspective, coaching in how to adjust learning behaviour and when and how to apply specific skills becomes important (Stoeger & Obergriesser, 2015).[378] As a result, the student develops more effective learning behaviour, and the teacher can determine more adequately if and how learning objectives and learning content meet the individual student's educational needs.[379] As coaching a student in how to become an effective learner is, in fact, a learning process with a set objective, monitoring if and how the coaching takes place is part of this process (Kallenberg et al., 2007; Pameijer & van Beukering, 2007; van Meersbergen & de Vries, 2017). Systemic and systematically reflecting on action during the process of intervening and afterwards on action regarding the effectiveness of the learning process and interventions used, is part of the professional attitude (Biesta, 2012; Korthagen, 2017; Onderwijscoöperatie, 2014a; Vlaamse Regering, 2008). Reflecting on the results and having a clear perspective on both the approach and the method of interventions used is, therefore, relevant.[380]

A student's emotional intelligence, social wellbeing, and ability to initiate and maintain social relations influences learning and achievements (D. Matthews & Foster, 2009; G. Matthews et al., 2018). Therefore, considering the social circumstances that are influencing the student's learning process and achievements is relevant (Peterson & Lorimer, 2011, Prieto & Ferrando, 2009).[381] The statement that all gifted children need additional social support is at least disputable. Research into the vulnerability of gifted students in the interpersonal domain is, at the least, inconclusive (Boor-Klip et al., 2014; G. Matthews et al., 2018; Prieto & Ferrando, 2009). Neither is it realistic to assume that gifted students, due to their intellectual abilities, will not need any support in this developmental domain at all (G. Matthews et al., 2018; Peterson & Lorimer, 2011). Therefore, the assessment of a gifted student's response to pedagogical and didactical interventions should also include a student's social functioning as part of the evaluation process without the assumption that social-emotional problems are characteristic of all gifted children (Prieto & Ferrando, 2009).[382]

[376] Appendix B, Table B21, V.5.B.1, K.5.B.1.
[377] Appendix B, Table B24, K.5.B.4.
[378] Appendix B, Table B22, K.5.B.2.
[379] Appendix B, Table B25, K.5.B.5.
[380] Appendix B, Table B22, V.5.B.2.
[381] Appendix B, Table B23, V.5.C.3.
[382] Appendix B, Table B22, K.5.C.2.

Teacher-student interviews and classroom observations are a way of assessing and monitoring a student's social-emotional development. Documenting this process provides proof of the student's ongoing development.[383]
Research shows that understanding a student's social-emotional functioning increases the quality of their education (G. Matthews et al., 2018).[384] Thus, a student's social functioning and social and emotional wellbeing both provide important information on how successful educational interventions are or can be.[385] Exploring the student's perspective on the curriculum helps the specialist to develop an understanding of the student's social-emotional functioning at school in general and during the learning process in particular (Robinson, 2008).[386]
The exchange of information between educational partners – including the student – in the student's ecological system again becomes relevant.[387] It offers the specialist the opportunity to understand the student's self-perception and their interpersonal functioning.[388] In a change and solution-focused approach, the topic is – as with all other topics – positively addressed from the perspective of opportunities (Cauffman & van Dijk, 2009).[389]

9.5 Five Competences in Domain 5

In this final chapter about the indicators for competent professional behaviour of the specialist in gifted education, what could be considered as relevant for the assessment of the student's responses to applied interventions was analysed. Both formative and summative assessment provides information to create a continuous educational cycle. Making learning visible creates new opportunities to make learning possible. Communication with all educational partners in the student's ecological system (including the student) is conditional for the effectiveness of the assessment process.
The complicating factor for the assessment process at this stage is that, except for the standardised testing used for measuring achievements, there is hardly any form of assessment available to specialists in gifted education outside of those standard to the national curriculum. As interventions are often customised to the specific needs of gifted students, monitoring the process of intervening as well as assessing the results of these interventions are often also customised to a specific situation and educational plan. Identifying educational needs at the start of the cycle is no different from the intervention processes at the end of an intervention cycle. Thus, strategies for this evaluative assessment of gifted students can be based on similar assessment strategies for identifying other educational needs at the start of the cycle (van Meersbergen & de Vries, 2017). Monitoring a student's academic achievements in the regular curriculum can be

[383] Appendix B, Table B22, V.5.C.2; Table B21, V.5.F.1.
[384] Appendix B, Table B21, K.5.C.1.
[385] Appendix B, Table B23, K.5.C.3; Table B25, K.5.C.5.
[386] Appendix B, Table B21, K.5.A.1.
[387] Appendix B, Table B24, K.5.C.4.
[388] Appendix B, Table B21, K.5.F.1.
[389] Appendix B, Table B24, V.5.C.4.

Table 9

Domain 5, From Appendix A: Competency Matrix for Specialists in Gifted Education

Generic Teaching Competencies	5. Assessing the response to intervention
	The specialist monitors the process of responding meaningfully to the gifted student's educational needs and monitors and interprets the gifted student's responses to these interventions contextually.
1. Pedagogical The teacher creates a safe and stimulating learning environment for every student, tuned to their individual educational needs.	**5.1.** The specialist monitors the student's intrapersonal and interpersonal functioning to determine how this development contributes to the student's ability for self-actualisation.
2. Didactical In collaboration with their colleagues, the teacher transfers curricular objectives in such a way that they become teachable in the context of school and/or district policy. They must compose the curriculum in a manner that enables their student to master the content.	**5.2.** The specialist monitors the student's academic achievements and relates these achievements to the level of support given during the learning process.
3. Managerial The teacher provides a structured, well-organised, task-oriented classroom atmosphere that supports the students' developmental process.	**5.3.** The specialist plans how they can monitor and document the student's development periodically based on a set schedule.
4. Communication The teacher communicates professionally with all educational partners who are involved with the student's education.	**5.4.** The specialist discusses the student's overall development with all the relevant actors in the student's ecological system.
5. Research and innovation The teacher adjusts their professional knowledge, perspectives, and actions if and when relevant developments in education, society, and science require it.	**5.5.** Based on current theories of assessment and giftedness, the specialist designs a system of formative and summative assessment for the education of gifted students.

done by using standardised tests. However, to assess the results of enrichment projects and educational objectives outside of the standard curriculum, different strategies are necessary. The same can be said about assessing the results of interventions aiming for effective learning behaviour or aiming for adequate social-emotional functioning. Hence, the specialist's ability to design and plan how they will monitor this development is relevant. Table 9 presents the five competencies identified for this domain.

9.6 Selection Rates of Indicators for Competency Domain 5

In a similar way to the assessment of educational needs, assessing the response to intervention requires the collaboration of all actors in the ecological system (Coleman & Johnsen, 2011; Pameijer & van Beukering, 2007; van Meersbergen & de Vries, 2017). Results can only be understood within the context of the data gathered. This context-bound interpretation calls highly on the specialist's knowledge and skills. In the empirical part of the research, stakeholders were asked to select the indicators that they considered to be the best examples of this knowledge and these skills. Analysis of the data gathered in this research amongst stakeholders is presented in a similar way to previous chapters. All tables referred to are included in Appendix I.

9.6.1 Selection Rates per Group of Stakeholders

9.6.1.1 Educational Professionals

Results per Competency

- Competency 5.1 focuses on the knowledge and skills required by the specialist to monitor the student's intrapersonal and interpersonal functioning. The specialist uses this information to determine how this specific aspect of the student's development contributes to the student's ability for self-actualisation. 22.65% of the educational professionals selected knowledge indicator K.5.D.1, "Understands which factors might have an influence on a gifted student's self-concept".[390] The knowledge of how to develop an understanding of the gifted learner's individual cognitive style (K.5.B.1) had the lowest selection rate (10.18%). For the skills indicators, voicing, including the student's perspective on how didactical and pedagogical changes in their education may contribute their development, was seen as important by 23.92% of all educational professionals.[391] The skill to observe the gifted learner's interaction with other students had the lowest selection rate (11.70%).[392]

[390] Table I1
[391] V.5.E.1 – Table I6
[392] V.5.C.1 – Table I6

- Competency 5.2 aims for the specialist to monitor the student's academic achievements and to relate these achievements to the level of support given during the learning process. The highest selection rate for the knowledge indicators is observed for K.5.D.2, the specialist's understanding how their attitude influences their student's learning process (21.12%)[393]. The knowledge of how to evaluate a student's social functioning as an influencing factor on the student's learning process (K.5.C.2) was selected by the least number of educational professionals (11.45%). For the applied skills, the intrinsic interest of the specialists in their students and the way that they demonstrate this through their communication with their students was selected by 24.43% of all educational professionals.[394] Remarkably, skills indicator V.5.F.2, collaborating with colleagues to explore alternative ways to monitor a student's achievements, was only selected by 11.45% of the educational professionals.
- Competency 5.3 considers the specialist's knowledge and understanding of how to periodically monitor and document the student's development, based on a set schedule. Knowledge indicator K.5.F.3, understanding why a joint protocol for documenting a student's development contributes positively to the quality of education, was selected by 21.37% of all educational professionals. Interestingly the matching skills indicator also showed high selection rates.[395] This indicator (V.5.F.3) included collaboration with colleagues for the development of a joint protocol. Apparently, this indicator was considered a better match for this competency (5.3) than a similar indicator (V.5.F.2) to competency 5.2.
- Competency 5.4 focuses on the educational partnership, discussing the gifted student's development with the relevant actors in the student's ecological system. In this context ecological partners are considered a relevant source of information to help optimise a student's curriculum. The selection rates of this knowledge indicator (26.97%) are similar to indicators in previous domains aiming for this partnership.[396] The matching skills indicator (V.5.A.4) was also selected by the highest percentage of educational professionals.[397]
  The lowest selection rate (9.67%) for this competency was identical for two knowledge indicators, K.5.B.4 (understanding why the observations of the student's cognitive style by educational partners is helpful for planning interventions) and K.5.F.4 (understanding that the classroom teacher has final responsibility for their student's development). Skills indicator V.5.B.4 (offers educational partners the opportunity to share relevant observations on the student's cognitive style) also had a selection rate of 9.67% and was the lowest skills indicator selected in competency 5.4 by the educational professionals.

[393] Table I2
[394] V.5.D.2 – Table I7
[395] V.5.F.3 – Table I8 – 18.58%
[396] K.5.A.4 – Table I4
[397] V.5.A.4 – Table I9 – 23.16%

- The last competency in this domain, 5.5, focuses on the specialist's ability to design a system of formative and summative assessment for the education of gifted students, based on current theories of assessment and giftedness. The understanding of how both formative and summative assessment can contribute to a student's self-actualisation was selected by the highest percentage of the educational professionals (19.08%).[398] The knowledge of how theories on assessment coincide with theories of learning and theories about the teacher's attitude was the least selected knowledge indicator for this competency (K.5.F.5, 11.70%).
  The skills indicator with the highest selection rate was indicator V.5.F.5 (19.08%). Educational professionals consider it relevant that the specialist in gifted education works with classroom teachers to establish assessments beyond commonly used standardised tools to evaluate a student's learning process and achievements.

Overall Results of Educational Professionals for Domain 5

As seen in previous domains, the selection rates for C-indicators for both knowledge and skills were relatively low. This contributes to the observed trend that knowledge and skills regarding interpersonal functioning are not seen as important as other indicators.
Of all the knowledge indicators, the E-indicators had the highest mean selection rate in this domain amongst educational professionals. Their highest mean selection rate for skills indicators is observed for indicators in category D, emphasising the importance of stimulating the student's intrapersonal understanding.
Combining the results for knowledge and skills indicators found that indicators regarding the stimulation of intrapersonal understanding (category D) had selection rates that included them, nine out of ten times, in the three highest selection rates for a competency. Indicators aiming for the stimulation of academic achievement and self-actualisation were also frequently selected (see Appendix I, Tables I1-I10).
In this domain, there were 8 indicators which encouraged ownership in gifted learners. Five out of eight times these indicators had selection rates that were high enough to be included amongst the top three selected indicators per competency. In the previous domains it was concluded that educational professionals selected voicing more in domains that focused on seeing and understanding educational needs. As assessment of the response to interventions is a continuation of the process of seeing and understanding those needs, the current selection rates match earlier findings.

---

398 K.5.E.5 – Table I5

9.6.1.2 Specialists in Gifted Education

Results per Competency

- The first competency in this domain, 5.1, combines pedagogical competencies with competencies that assess whether and to what extent interventions have been beneficial for the gifted learner. The stratification of selection rates amongst specialists in gifted education were identical to those of educational professionals.[399]
- Competency 5.2 combines didactical competencies with competencies that assess the response to interventions. 22.98% of the specialists in gifted education chose indicator K.5.F.2, understanding why it is important for the specialist to share knowledge and skills to monitor a student's development with their colleagues.[400]
  The skills indicator that had the highest selection rate (27.83%) focused on the skill of discussing with students how their achievements relate to the set objectives.[401] The matching knowledge indicator K.5.E.2 also had a relatively high selection rate (18.45%), and belongs to the three most selected knowledge indicators for this competency. Interestingly amongst specialists in gifted education, indicator K.5.A.2 had an identical selection rate. This indicator aims for knowledge of how to assess achievements for educational objectives outside of the regular curriculum.
  The lowest selection rate (9.71%) was for a skills indicator aiming for the specialist's skill to evaluate a student's social functioning as an influencing factor on the student's learning process.[402] This points out the different perspectives held by specialists to educational professionals, as for educational professionals indicator V.5.C.2 was selected in the top three indicators for this competency. The observed difference between the selection rates of both groups of stakeholders is considered as a relevant difference. A second relevant difference between both groups concerns skills indicator V.5.F.2. 14.89% of specialists in gifted education selected this indicator, aiming to share knowledge about assessment strategies with their colleagues. This indicator had the lowest selection rates in the other two groups of stakeholders. Although the percentual difference was small, based on the defined thresholds for relevant differences, this difference should be reported as relevant.
- Competency 5.3 focuses on the combination of competencies in classroom management and organisation, and the assessment of the response to intervention. As for competency 5.1, the stratification of selection rates for the knowledge indicators was identical to those

[399] Table I1
[400] Table I2
[401] V.5.E.2 – Table I7
[402] V.5.C.2 – Table I7

selected by educational professionals. Although percentages of the selection rates differ, for both groups applies that their three highest selection rates were for the identical indicators. The skill to set out a protocol with colleagues on how to document the student's achievements had their highest selection rate (22.01%).[403]

Similarly to the previous competency, a relevant difference can also be observed in the skills indicators. The lowest selection rate (13.27%) was for indicator V.5.C.3 (understanding how social wellbeing affects learning and achievement). A very small percentual difference can be observed between this selection rate and the selection rate of educational professionals, but again the difference meets the threshold to be identified as relevant.

- The fourth competency, 5.4, in this domain focuses on communication with educational partners and the response to interventions. Once more, indicators with the highest and lowest selection rates for both skills and knowledge are identical to those of the educational professionals. This again suggests that there is common ground about the understandings of both educational professional and specialist in gifted education of what is thought to be important.
- The final competency, 5.5, in this domain combines the specialist's attitude to adjusting their knowledge and actions based on professional development, with their knowledge and skills to assess the response to interventions. 23.95% of the specialists in gifted education selected indicator K.5.A.5 - understanding why it is important to set assessable educational objectives.[404] The most relevant skill to demonstrate in competency 5.5 was considered the collaboration with general classroom teachers to establish a way to monitor the gifted learner's learning process and achievements.[405]

  A fourth relevant difference between specialists and educational professionals is found amongst the skills indicators of this competency. 13.27% of the specialists in gifted education selected skills indicator V.5.B.5, discussing with the student how they can improve learning by switching between different cognitive styles, making it the joint lowest selected rate by the specialists for this competency. Amongst educational professionals, indicator V.5.B.5 belonged in the three most selected indicators for this competency. The educational professional's lowest selection rate was for indicator V.5.A.5 (the skill to assess the results of an enriched curriculum)[406] whilst specialists in gifted education selected V.5.A.5 frequently enough (18.12%) to belong in their three most selected indicators for this competency.

---

[403] V.5.F.3 - Table I8
[404] Table I5
[405] V.5.F.5 - Table I10 - 22.01%
[406] Table I10 - 13.74%

Overall Results of Specialists in Gifted Education for Domain 5

Identically to domain 4, there are huge similarities in the selection of knowledge indicators between specialists in gifted education and educational professionals. The top three of the most selected knowledge indicators per competency are highly comparable and differences in the selection rates between both groups were relatively small. For the skills indicators, however, four relevant differences were observed between the specialists in gifted education and educational professionals. These differences cannot be attributed directly to their different roles in the education of gifted learners; therefore there is no straightforward explanation for the differences.
For this competency, indicators belonging to category C (four times) and B (five times) were the least frequently selected indicators by specialists in gifted education. This matches with an overall tendency that shows that BCF indicators are the least frequently selected indicators amongst all groups of stakeholders.

9.6.1.3 Parents

Results per Competency

- For competency 5.1 parents made choices that were different when compared to specialists in gifted education and the educational professionals. Their highest selection rate (22.49%) was for indicator K.5.E.1.[407] This indicator focuses on the specialist's understanding of how a student's self-concept influences the student's self-efficacy. A relevant difference can be observed for indicator K.5.D.1, "understands which factors might have an influence on a gifted student's self-concept". This indicator had the highest selection rate for competency 5.1 amongst the other groups of stakeholders. However, for indicator K.5.D.1, the selection rate amongst parents was not high enough for it to be included in their top three selection of knowledge indicators in competency 5.1. Instead, 18.88% selected indicator K.5.A.1, an indicator focusing on exploring the specialist's understanding of how the student's perspective on the curriculum can be used to develop an understanding of the student's social-emotional functioning.
  For the skills indicators, differences in selections between parents and the other groups of stakeholders were very small. Identical indicators were selected most frequently and, exactly as can be seen with the other groups of stakeholders, an indicator about specialists in gifted education having the skill to observe how gifted students interact with other students in relevant educational situations, had the lowest selection rate.[408]

[407] Table I1
[408] V.5.C.1 - Table I6 – 9.24%

- For competency 5.2, there were hardly any differences between parents and the other two groups of stakeholders with regard to skills indicators. The differences are marginal, with no indication of different views on what should be considered important. Nevertheless, amongst the skills indicators, a small but relevant difference was found. The lowest selection rate (11.65%) for parents was for indicator V.5.F.2 (the specialist collaborates with colleagues, exploring alternative ways to monitor a student's achievements).[409] The most selected skills indicator amongst parents was the indicator where a specialist shows an intrinsic interest in their students through the way that they communicate with them (29.72%, V.5.D.2). This personal relationship is important to parents.
- Competency 5.3 again shows marginal differences between the selection rates of parents and of other groups of stakeholders, this time for both knowledge and skills indicators. Parents consider it relevant for specialists to include them as educational partners when discussing a student's development.[410] They also consider it relevant for the specialist to explore what contributes positively to a student's self-confidence whilst taking a test.[411] They consider the knowledge of how an assessment may contribute to their child's self-actualisation the least relevant knowledge indicator[412], and the skill to plan and organise a formal assessment the least relevant skill.[413]
- Competency 5.4 is the first competency in domain 5 where all groups of stakeholders have the highest selection rate for the same knowledge indicator, K.5.A.4, aiming for the specialist's knowledge of how information from parents and co-workers can help to optimise a student's curriculum. The skills indicator that had the highest selection rate by parents (22.89%) was indicator V.5.D.4, focused on the specialist and student exploring together how they can create a balance between learning, effort, and achievement.
- For competency 5.5, two significant differences with other groups of stakeholders were found. 21.29% of all the parents selected indicator K.5.B.5, focusing on the specialist's knowledge of to what extent their student's cognitive style matches the curriculum offered and how that affects the student's educational needs. This indicator was only selected by 14.25% of all educational professionals and 12.30% of all specialists in gifted education.[414] A similarly large difference can be observed for the matching skills indicator, V.5.B.5. 22.49% of all the parents selected this indicator, compared to 13.27% of the specialists in gifted education. Almost as large a difference (7.28%) is found between specialists in

[409] Table I7
[410] K.5.A.3 - Table I3 – 22.49%
[411] K.5.D.3 - Table I8 – 23.69%
[412] K.5.E.3 - Table I3 – 9.24%
[413] V.5.A.3 - Table I8 – 10.44%
[414] Table I5

gifted education and parents concerning indicator V.5.A.5. However, this time the low selection rate was by parents (10.84%), while specialists selected it 18.12%, implying that parents do not value the specialist's skill in assessing the results of an enriched curriculum as highly as specialists in gifted education. 22.49% of parents preferred the specialists to discuss with their child how switching between different cognitive styles can positively affect their achievements.[415]

Overall Results of Parents for Domain 5

For both the knowledge and skills indicators, the category with the highest mean selection rate amongst parents is category D, aiming for the specialist to stimulate their student's intrapersonal understanding (Appendix 10, Table I11). Analysis per competency showed that, for three competencies, the knowledge indicators concerning the academic achievements of the gifted child had the highest selection rate. For the skills indicators, three indicators in category D had the highest selection rate. Parents consider academic achievements and intrapersonal understanding to be important examples of the competencies of the specialist in gifted education. Although indicators in category C were not selected frequently enough to receive the highest selection rate for one of the competencies, C-indicators were selected frequently enough, five out of ten times, to be included in the three most selected indicators (Appendix I, Tables I1-I10). This suggests that, for parents, the specialist in gifted education can also show competent professional behaviour through stimulating a student's interpersonal functioning.

9.6.2 Significant Differences Between Groups of Stakeholders

For the analyses of the differences between the groups of stakeholders, the ANOVA test and Bonferroni's Multiple Comparisons Test were used. In domain 5, differences between the selection rates of knowledge indicators are significant for the knowledge indicators in categories C ($p = .022$), D ($p < .001$), E ($p = .045$) and F ($p < .001$) (see Appendix I, Tables I12-I14; see also Figure 38). Compared to educational professionals ($\overline{X} = 2.53$, $p = .004$) and specialists in gifted education ($\overline{X} = 2.08$, $p < .001$), parents ($\overline{X} = 3.12$) selected indicators in category D significantly more frequently. Not only is the difference in selection rates between parents and both groups of stakeholders significant, but there is a second significant difference (-0.457, $p = .030$) noticeable between specialists in gifted education and educational professionals (Table I14).
For a second category of indicators, compared to educational professionals, the specialists in gifted education have different selection rates as well (Appendix I, Table I12). The specialists ($\overline{X} = 3.26$) select indicators in category E more frequently than educational professionals ($\overline{X} = 2.89$). Again, this is a significant difference ($p = .045$) representing a small effect size ($d = .02$).

[415] V.5.B.5 – Table I10

Figure 38

Comparison of Selection Rates of the Knowledge Indicators in Domain 5 Between Groups of Stakeholders - (see also Table I12)

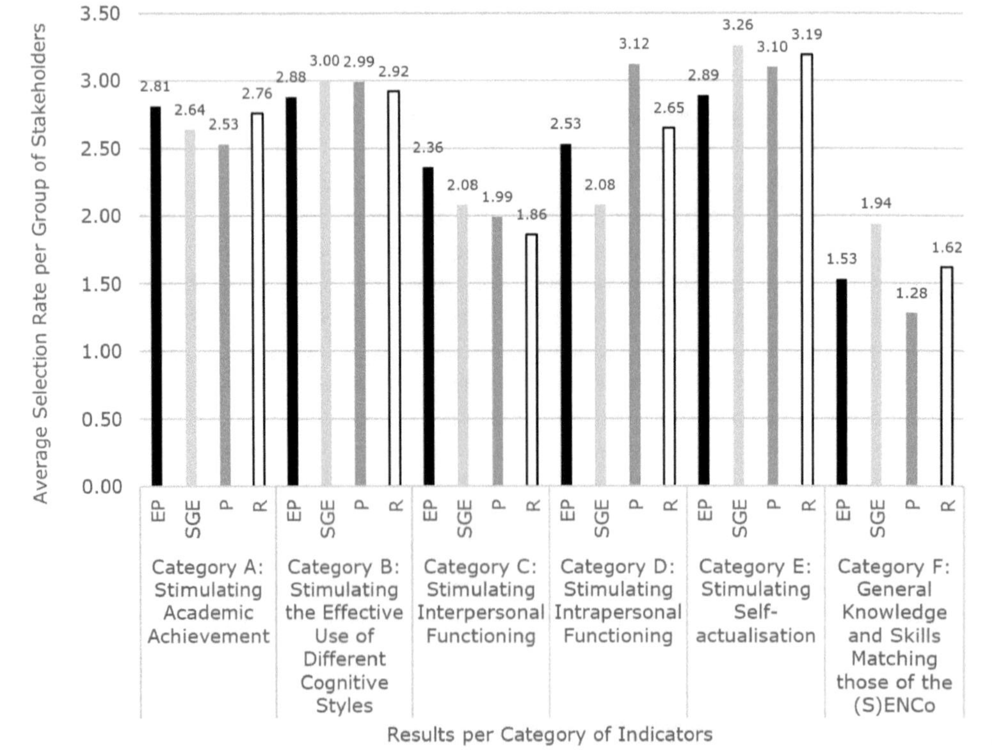

Note. N = 354 divided into four groups of stakeholders: EP = Educational Professionals, SGE = Specialists in Gifted Education, P = Parents, R = Rest. EP: N = 131, SGE: N = 103, P: N = 83, R: N = 37.

Test value = 2.5

The last difference can be observed between specialists in gifted education and both other groups of stakeholders. The specialists had significantly higher selection rates ($\overline{X}$ = 1.94) for indicators in category F compared to educational professionals ($\overline{X}$ = 1.53, p = .015) and parents ($\overline{X}$ = 1.28, p < .001) (Appendix I, Table I14).

Regarding the skills indicators, differences were limited to two categories, D (p = .009) and F (p < .001) (Appendix I, Table I15; see also Figure 39). Specialists in gifted education selected indicators in category D ($\overline{X}$ = 2.53) less frequently than both educational professionals ($\overline{X}$ = 3.01, p = .015) and parents ($\overline{X}$ = 3.22, p = .030) (Appendix I, Table I16). While conversely, as seen for the skills indicators, the selection rate for indicators in category F were significantly higher amongst specialists in gifted education ($\overline{X}$ = 2.84) than amongst educational professionals ($\overline{X}$ = 2.29) and parents ($\overline{X}$ = 2.13).

Figure 39

Comparison of Selection Rates of the Skills Indicators in Domain 5 Between Groups of Stakeholders - (see also Table I15)

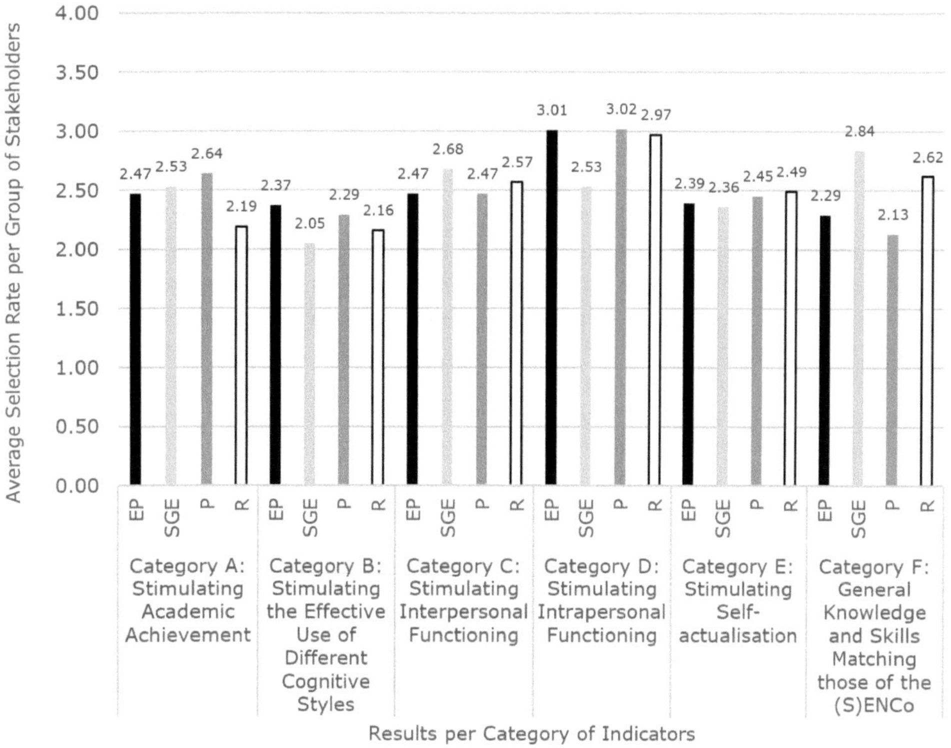

Note. N = 354 divided into four groups of stakeholders: EP = Educational Professionals, SGE = Specialists in Gifted Education, P = Parents, R = Rest. EP: N = 131, SGE: N = 103, P: N = 83, R: N = 37.

Test value = 2.5

9.6.3 Common Ground Between Groups of Stakeholders

Initial analysis of the results aimed to identify common ground between all groups of stakeholders. Two frequency analyses were applied. The primary frequency analysis showed that there are 10 knowledge indicators and 6 skills indicators that ended up in the three most selected indicators amongst all groups of stakeholders (see Appendix I, Tables I18-I19). Indicators included all 5 categories.

S. Stimulating academic development (5)
C. Stimulating interpersonal functioning (2)
D. Stimulating intrapersonal understanding (4)
E. Stimulating self-actualisation (2)
F. Generic counselling and advisory skills matching the skills of a (S)ENCo (3)

The secondary frequency analysis aimed to determine if, amongst these 16 indicators, there were three indicators that had the highest selection rates amongst all groups. One knowledge indicator and one skills indicator were found that met that criterion (see Appendix I, Tables I29-I30).

- Competency 5.1
  - Discusses with the student how didactical and pedagogical changes in their education might contribute to the student's development.[416]
- Competency 5.4
  - Knows how the information from parents and co-workers can help to optimise a student's curriculum.[417]

With no differentiation between groups of stakeholders, the selection rates of all knowledge indicators, except for indicators in category D, differ significantly from the expected rate (see Appendix I, Tables I20-I21; see also Figure 40). Knowledge indicators in category A have a mean selection rate of $\overline{X} = 2.69$. The difference, 0.189, to the expected rate ($\mu = 2.5$) is significant, $t(353) = 3.47$, $p = .001$, $d = .19$, representing a small effect size. Indicators in category B have a mean selection rate of 2.94. Again, a significant difference can be observed, 0.444, $t(353) = 8.10$, $p < .000$, $d = .43$, the effect size represents a practical significance. Similarly to previous domains, the mean selection rate of indicators in category C is relatively low, $\overline{X} = 2.14$, and differs significantly from the expected selection rate ($\mu = 2.5$), $MD = -0.362$, $t(353) = -6.24$, $p < .001$, $d = .33$, the effect size meets the threshold for practical significance. Indicators in category E ($\overline{X} = 3.08$) are selected significantly more than was expected, ($\mu = 2.5$), $MD = 0.579$, $t(353) = 10.62$, $p < .001$, $d = .57$, this represents an effect size for practical significance. Finally, indicators in category F have a very low mean selection rate ($\overline{X} = 1.60$), the difference, -0.901, is significant, $t(353) = -16.03$, $p < .001$, the effect size represents a practical significance, $d = .85$.
Regarding the skills indicators, significant differences from the expected selection rate occur for indicators in categories B ($p < .001$) and D ($p < .001$) (Appendix I, Tables I23-I24; see also Figure 41). Indicators in category B had a slightly lower selection rate ($\overline{X} = 2.24$) than expected ($\mu = 2.5$), $MD = -0.263$, $t(353) = -4.52$, $p < .001$, $d = .24$ representing a small effect size. For skills indicators in category D, the selection rate ($\overline{X} = 2.87$) was higher than expected, ($\mu = 2.5$), $MD = 0.370$, $t(353) = 5.84$, $p < .001$, $d = .31$, the effect size represents a practical significance.
The mean selection rates for domain 5 show a similar pattern to the previous domains. There is a strong emphasis on indicators aiming for the stimulation of the academic achievements of gifted learners. Although there were only two individual indicators in category E that had the highest selection rate amongst all groups of stakeholders, the mean selection rate of indicators in this category was high. This suggests that all stakeholders consider the stimulation of self-

---

[416] V.5.E.1.
[417] K.5.A.4.

actualisation of gifted learners as a relevant example of competent professional behaviour for specialists in gifted education. The low mean selection rate for indicators in category F suggests that these indicators were considered to be relevant by only a small group of stakeholders, and, based on the previous analyses, these indicators were mostly considered relevant by specialists in gifted education.

Figure 40

Mean Selection Rate of the Knowledge Indicators in Domain 5 - (See Table I20)

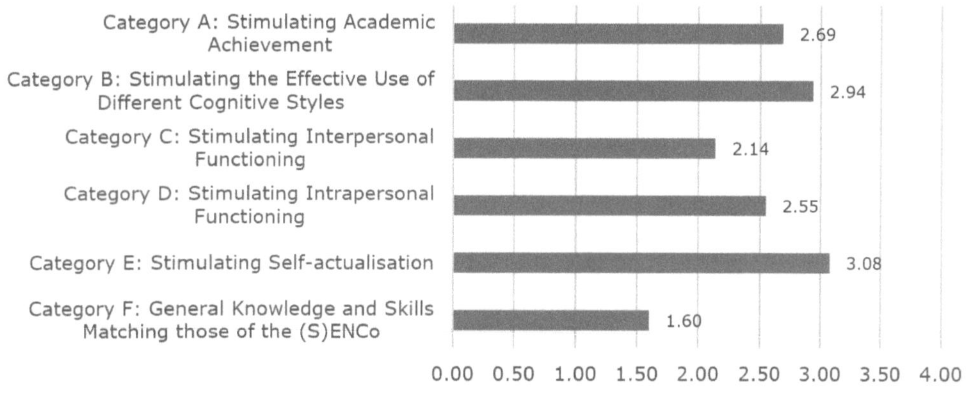

Note. N = 354. Test value = 2.5

Figure 41

Mean Selection Rate of the Skills Indicators in Domain 5 - (See Table I22)

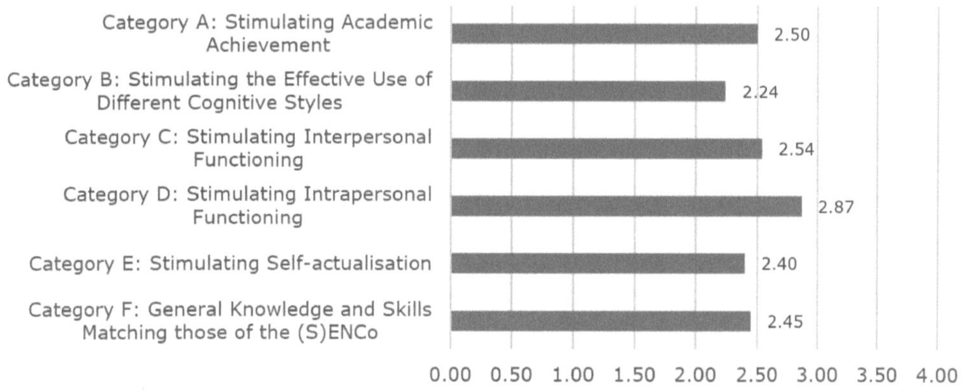

Note. N = 354. Test value = 2.5

In a similar way to all the previous domains, the overall results of domain 5 suggest that some categories of indicators were selected more often than others. Again, a clustering in two categories (ADE vs BCF) was possible. The Paired Samples T Test (Appendix I, Tables I26 and I28) showed that the selection rates of clustered categories (both for knowledge and skills) differed significantly from the expected selection rate ($p < .001$). The clustered knowledge indicators ADE, $\overline{X} = 2.77$ (Appendix I, Table I25), were selected significantly more often than the clustered knowledge indicators BCF, $\overline{X} = 2.23$ ($p < .001$, $d = .58$). The clustered skills indicators ADE, $\overline{X} = 2.59$ (Appendix I, Table I27), were selected significantly more than the clustered skills indicators BCF, $\overline{X} = 2.41$ ($p < .001$, $d = .18$). This suggests that, overall, the indicators in category ADE were considered to be more relevant examples for the competencies they were supposed to match.

9.6.4 Comparison Between The Netherlands and Flanders

The final analyses for this domain regard similarities and differences between The Netherlands and Flanders. Using Group statistics and an Independent Samples test the following results were found (see Appendix I, Tables I31-I35).

Figure 42

Comparison of Mean Selection Rates in Domain 5 per Category of Knowledge Indicators Between Dutch and Flemish Stakeholders - (See Table I32)

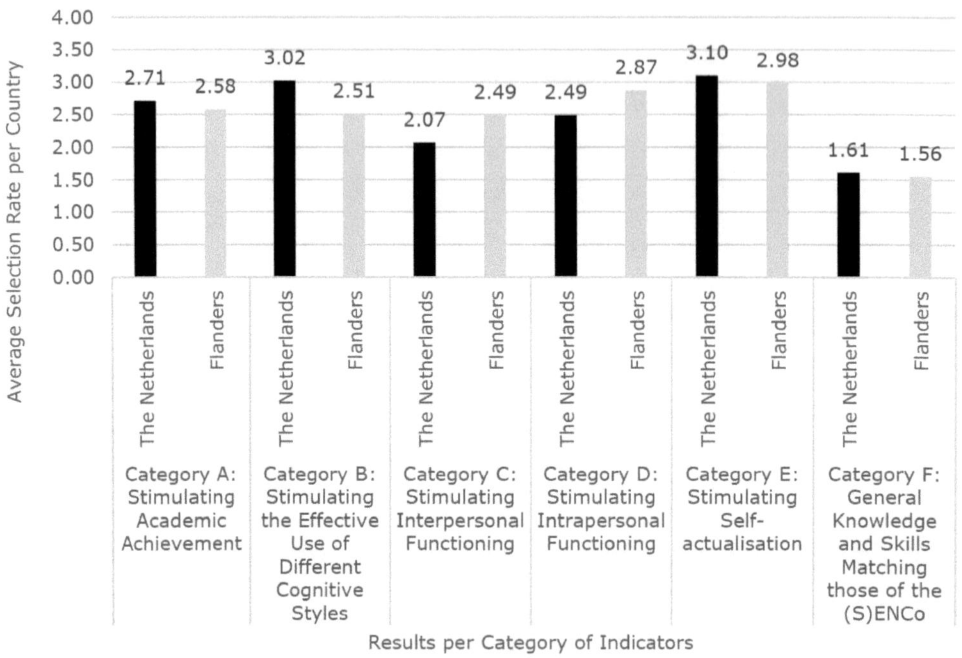

Note. N = 354. The Netherlands N = 299, Flanders N = 55. Test value = 2.5

As seen in the previous domains, there are slight but significant differences for knowledge indicators in categories B (< .001), C (p = .009), and D (p = .42) (Appendix I, Tables I32-I33; see also Figure 42). Knowledge indicators in category B were selected significantly more by Dutch stakeholders ($\overline{X}$ = 3.02) compared to Flemish stakeholders ($\overline{X}$ = 2.51). The difference, 0.514, in selection rates between stakeholders in both countries represents a practical significance, d = .51. Identically to the other domains, indicators in category C had a relatively low selection rate. However, the previously observed difference was again noticeable. C-indicators were selected more frequently by Flemish stakeholders ($\overline{X}$ = 2.49) than by Dutch stakeholders ($\overline{X}$ = 2.07). The difference, -0.417, is significant, and meets the threshold for practical significance (d = .39). D-indicators were also selected more frequently by Flemish stakeholders ($\overline{X}$ = 2.87) than by Dutch stakeholders ($\overline{X}$ = 2.49), the difference, -0.381, is significant. The effect size meets the threshold for practical significance (d = .30).

Figure 43

Comparison of Mean Selection Rates in Domain 5 per Category of Skills Indicators Between Dutch and Flemish Stakeholders - (See Table I34)

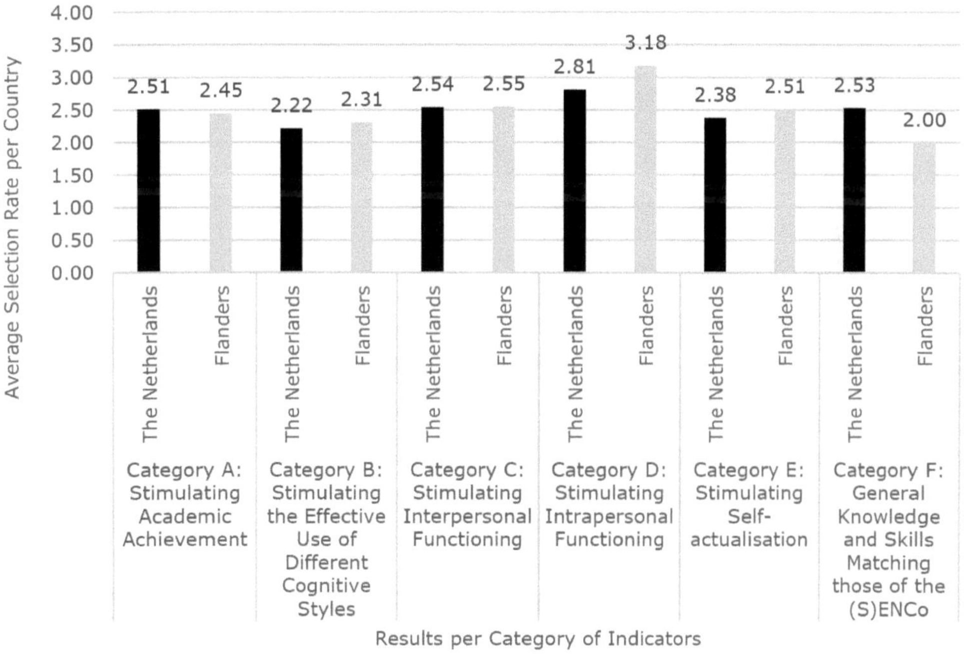

Note. N = 354. The Netherlands N = 299, Flanders N = 55. Test value = 2.5

For the skills indicators, significant differences in selection rates between stakeholders from both countries can be found in categories D (p = .035) and F (p < .001) (Appendix I, Tables I34-I35; see also Figure 43). Flemish stakeholders selected more indicators in category D (MD = 0.369, d = .31), Dutch stakeholders selected more indicators in category F (MD = 0.532, d = .48).

9.7 Summarising

This chapter focused on domain 5 of the competency matrix, assessing the response to interventions applied to gifted students. This process creates the opportunity for the specialist in gifted education to develop educational designs that meet the educational needs of individual gifted learners in the best possible way, while regarding the ecological system in which they are applied.

- 5.1 The specialist monitors the student's intrapersonal and interpersonal functioning, to determine how this development contributes to the student's ability for self-actualisation.
- 5.2 The specialist monitors the student's academic achievements and relates these achievements to the level of support given during the learning process.
- 5.3 The specialist plans how they can monitor and document the student's development periodically based on a set schedule.
- 5.4 The specialist discusses the student's overall development with all the relevant actors in the student's ecological system.
- 5.5 Based on current theories on assessment and giftedness, the specialist designs a system of formative and summative assessment for the education of gifted students.

The highest selection rates were for indicators in the combined cluster ADE, suggesting that indicators for stimulating academic achievements, intrapersonal understanding and self-actualisation are considered to be the most relevant examples of competent behaviour for the specialist in gifted education. Indicators in the clustered category BCF were not selected as frequently and show that stakeholders from Flanders lay a stronger emphasis on stimulation of interpersonal functioning and intrapersonal understanding than Dutch stakeholders.
Although different groups of stakeholders may choose different individual indicators matching a specific competency, at domain level, the significant differences that are observed have mostly small to medium effect sizes. Differences in individual indicators are often consistent with the role different groups of stakeholders have in the upbringing and education of gifted learners. All findings for domain 5 were consistent with findings in the previous domains.

10 A Helicopter View of Stakeholders' Understandings

10.1 Introduction

In this chapter, the overall findings of research into the understandings of stakeholders regarding the competencies required of specialists in gifted education to support the education of gifted students are presented. As previously stated, undertaking this research project was considered important, as, although examination of international literature produced no information on how knowledge and skills indicators for competent behaviour should be prioritised and presented in different categories, there may be relevancy for Dutch and Flemish stakeholders in prioritising those knowledge and skills indicators for competent behaviour to ensure that topics are addressed if they are important for education in general in their countries. The questionnaire used was constructed in such a way that it offered the possibility of analysing whether, when respondents were forced to make a choice between the knowledge and skills indicators, the selection rates of different indicators would still be equally distributed. If not, and if the differences were significant, it would answer the overarching question *"What knowledge and skills indicators are considered by Dutch and Flemish stakeholders as the most relevant and important examples of the competent professional behaviour of the specialist in gifted education?"* The construction of the questionnaire also offered the opportunity to explore whether stakeholders would differ in the emphasis that they placed on specific indicators and if these differences would become visible at group level, suggesting different understandings between different groups of stakeholders. Last, but not least, the construction of the questionnaire offered the opportunity to answer the question of whether there would be noticeable differences between The Netherlands and Flanders regarding their emphasis on specific knowledge and skills indicators.

In previous chapters, the results of the research into the understandings of stakeholders were analysed and presented per competency domain and at a detailed level: per competency, per group of stakeholders (level 1, data pyramid, Figure 13). The results of the different groups of stakeholders were compared (level 2, data pyramid, Figure 13). Analyses showed not only different accents in their selection rates, mostly matching their role in the upbringing and education of gifted children, but also suggested that there is solid common ground between all groups. That detailed approach did justice to the principle of presenting and interpreting findings in the context in which they were observed (Bryman, 2012; Howitt & Cramer, 2009; Kara, 2015; Mitchel et al., 1997). These results provided an understanding of how different groups of stakeholders emphasise different indicators for each competency. Analysing the data at domain level (level 1) it became clear that parents selected indicators for competent behaviour with a different lens. Throughout all the competency domains, the selection rates of parents suggest that, from their perspective, indicators about the stimulation of intrapersonal understanding had the highest relevancy. It was therefore decided to test whether significant differences in selection rates between groups of stakeholders could be noticed at this general level (level 3, see Figure 13) as well. Finding confirmation of these observed differences would support the idea that it is relevant to keep these differences in mind when constructing the curriculum for teacher education programmes for specialists in gifted education. Transferring any observed differences might create opportunities in these programmes to aim for mutual understanding between teachers, parents and specialists in gifted education. The specialist's ability to see things through the lens of the other stakeholders involved, can support educational partnership and contribute to the quality of the education of gifted learners (Broersen & Spreij, 2009; de Wit, 2005; Pameijer, 2012; Pameijer & van Beukering, 2007; van Gerven & Hoogenberg, 2011; van Meersbergen & de Vries, 2017).

Analysis at level 2 of the data pyramid (domain level) suggested that, if there was no differentiation between the separate groups of stakeholders, then the three indicator categories ADE combined showed a consistently higher selection rate than the combined indictors BCF. Together, the indicators with the highest selection rates could provide a general framework for competencies of the specialist in gifted education. It therefore became relevant to test whether these differences were also significant at the most general level of data analyses (level 3, the top of the data pyramid).

At level 2, interesting differences between the selection rates of stakeholders in The Netherlands and Flanders were observed. It was decided to also analyse whether or not these differences would be noticeable and significant at the general level 3. These differences might lead to slightly different emphases in both countries in their teacher education programmes for specialists in gifted education.

Findings at levels 1 and 2 suggested that it would be relevant to analyse the selection rates for three additional issues: stakeholders' selection rates of students' ownership of their education, educational partnership and peer group education. When compared to other indicators, the selection rates for individual indicators in each of these three topics seemed to have deviance. These selection rates were either higher than other indicators, suggesting a practical significance, or lower than other indicators, suggesting that they might not be considered as good examples or at least might be less important compared to other indicators. The results are presented in this chapter.

10.2 Ranking 300 Indicators

Stakeholders had to select three indicators for each competency. This meant that each stakeholder selected 150 indicators. Thus, the 300 indicators (see Appendix B) can be divided into selected/not selected. As a result, all 300 indicators can be ranked from least selected to most selected. This ranking provides a general understanding of what type of professional behaviour is considered to be an important example for the competency of the specialist in gifted education. The ranking is done without differentiating the separate groups of stakeholders. This ranking, presented in Appendix J, Table J1, shows that there is no one indicator that has been selected by *all* individual stakeholders, nor is there an indicator that has not been selected by *any* of the individual stakeholders. 18 indicators had a selection rate that was significantly lower than the mean selection rate ($\overline{X}$ = 177, see Appendix J, Table J2).[418] These selection rates suggest that stakeholders considered these indicators to be the least relevant examples of professional competencies for the specialist in gifted education. 22 indicators had a selection rate that was significantly higher than expected ($\overline{X}$ = 177, see Appendix J, Table J3). These indicators can be considered as the most relevant examples of the professional competency of specialists in gifted education.

Out of 300 indicators, the odd one out is skills indicator V.1.F.4 ("Creates a joint perspective on giftedness between the educational partners"). It is considered that when a joint perspective on giftedness is held by educational partners, that this contributes to the quality of gifted education (Morris Miller, 2008). However, only 1.13% of all the respondents (4/354) selected this indicator. Low selection rates should, however, be interpreted contextually. This means that stakeholders may have found other skills indicators for competency 1.4 more relevant. On the other hand, this extremely low selection rate remains remarkable when compared to the second-lowest selection rate of 12.15% (43/354, item V.2.C.2, "Explores the influences of classmates on the gifted student's achievements"). Although there is no evidence that educational partners are not aware that they may perceive the notion of giftedness differently, it might be a reasonable explanation for the low selection rates. Dai and Chen (2014) suggest that different paradigms in theories of giftedness result in different foci and approaches to education. A lack of awareness of the impact of the theoretical paradigm on the practice of education could explain the extremely low selection rates of indicator V.1.F.4.

The second-highest percentage was 81.64% for indicator K.3.D.2 (289/354, "Knows how mindset influences a student's learning process"). In Chapter 2, the increasing emphasis on the social-constructivist approach to learning is pointed out. Part of this educational approach is the necessity of student's having well-developed self-regulation skills (Obergriesser & Stoeger, 2015; Oppong et al., 2018; Stoeger & Ziegler, 2011). Mindset also affects the effective use of regulation skills (Dweck, 2006, 2017). The high selection rate of indicator K.3.D.2 suggests that there has been a transfer from theory to practice by stakeholders in how to value the notion of mindset as an influencer for educational success.

Competency 1.2 focuses on the integration and coherency between notions of giftedness, educational strategies and educational programming. In this context indicator K.1.D.2 ("Knows different didactical strategies to challenge gifted students in their zone of proximal development") was presented to the stakeholders. This indicator was selected by 86.44% of all respondents (306/354) and had, therefore, the highest selection rate of all 300 indicators. Vygotsky's (1978) concept of the zone of proximal development refers to a general learning theory. This theory was not designed with the education of

[418] The threshold for selection rates that differ significantly from the mean selection rate ($\overline{X}$ = 177) is set at 1.5 SD ≥≤ $\overline{X}$. A difference of 1.5 SD is considered to be of practical significance, similar to the notion of practical significance as defined by Coe (2002).

10

gifted learners in mind. However, Winstanley's (2010) analysis of the ingredients of challenge and the parallel curriculum approach by Tomlinson et al. (2009) point out how a student's opportunity to experience a curriculum calling on their zone of proximal development is essential for high quality in gifted education. The diversification of educational strategies, creating different learning opportunities for different learners in different educational contexts, has become increasingly relevant due to a fast-growing emphasis on education based on an inclusive approach (Borland, 2005; Florian & Black-Hawkins, 2011; Janson, 2017). The interplay between a student's profile of educational needs and educational context creates unique profiles for meaningful educational responses. Having a wide variety of didactical strategies at their disposal and the ability to apply a strategy that seems to fit the individual student's needs is as much a prerequisite for teachers in general as for the specialist in gifted education (Chandra Handa, 2009; Janson, 2017; Renzulli, 2012; Riley, 2011; Roberts & Inman, 2015). Analysis of the selections presented in Appendix J, Tables J4 and J5, reflects the distribution of selection rates presented in Chapters 5-9. As categories, the C-indicators and F-indicators showed relatively low selection rates, suggesting that these indicators were considered to be the least relevant examples for their matching competencies.
If clustered, the observed trend is that indicators in the combined categories of ADE are selected significantly more than indicators in the combined cluster of BCF; this is noticeable even at the highest level of data analyses ($p < .001$).[419]

10.3 Ranking Categories of Indicators

Having found relevant differences at the detailed level of selection rates per competency and significant differences at level 2 (see Figure 13), the data at level 3 was analysed. The literature review, also presented in Chapters 5-9, suggested no prioritisation of indicators for the different categories, nor did it suggest that some indicators should be seen as more relevant than others. However, the previously presented results suggest otherwise, that in practice there is a trend that can be observed, where Dutch and Flemish stakeholders emphasize certain categories of indicators. This would reveal a gap between what is considered to be important from a theoretical perspective and what is considered to be important from a more practical perspective, where individual interests determine the selection. The assumption at the start of this research had been that the selection rates of all the categories would not be substantially different from the expected value ($\mu = 12.5$) and that occurring differences were not likely to be significant.
Data analyses[420] showed that the selection rates of all categories of knowledge indicators differed from the expected value ($\mu = 12.5$), except for indicators in category D.[421] The selection rates of category D ($\overline{X} = 13.13$, $p = .042$) stayed close to the expected value, the effect size is, however, minimal ($d = .10$). The

---

[419] See Appendix J, Tables J26 & J27
[420] All findings are based on $\alpha = .01$ (2-tailed)
[421] Appendix J, Tables J4 & J5

identified differences for the other categories of indicators are significant ($p < .001$). Indicators focused on the specialist in gifted education stimulating interpersonal functioning (category C) were selected significantly less compared to the expected value ($\mu = 12.5$, $\bar{X} = 8.65$, $MD = -3.850$, $t(353) = -15.36$, $p < .001$, $d = .82$, representing a practical significance. These results were completely unexpected. In both The Netherlands and Flanders, there is a relatively strong emphasis on theories describing the different social-emotional needs of gifted learners. The variety in Dutch and Flemish publications about this topic is wide (Buis, 2014; Daeter, 2012; Gevaert & Desmet, 2016; Houkema, 2017; Kieboom, 2015; Kieboom & Venderickx, 2017b; Kieboom & Verheye, 2011; Koenderink, 2012; Kooijman-van Thiel, 2008; Plat, 2016; van der Poel, 2017; C. Wolters, 2020). Combined with the fact that the social-constructivist approach to education in general calls highly on interpersonal skills (Chandra Handa, 2009; Marzano & Heflebower, 2012; Tomlinson et al., 2009; Voogt & Pareja Roblin, 2010; Winstanley, 2010), it had been expected that indicators focused on the specialist's ability to stimulate the interpersonal functioning of gifted students would have a selection rate that was close to the formally expected value. One explanation may be found in the theory that the social classroom climate affects the attitude of all students towards learning and that gifted learners may form no exception to this principle (T. Cross, 2018; D. Matthews & Foster, 2009; Moon, 2009a).

The significantly lower selection rates for indicators aiming for more generic competencies matching the specialist's role as a (S)ENCo (category F) were not entirely unexpected ($\mu = 12.5$, $\bar{X} = 10.19$, MD -2.308, $t(353) = -8.85$, $p < .001$, $d = .47$). The observed differences can be considered of practical significance. The intention of these indicators affects all educational partners. However, it is inherent in the role of the specialists of gifted education to transfer the intention of these indicators into daily practice (Coleman & Johnsen, 2011; Pameijer & van Beukering, 2007; van Gerven & Hoogenberg, 2011; van Meersbergen & de Vries, 2017). The deviance in selection rates can, therefore, be explained by this specific role (Freeman, 1987; Mitchel et al., 1997).

A similar data analysis was completed for the skills indicators.[422] Data showed that the skills indicators in category C stayed close to the expected value ($\mu = 12.5$, $\bar{X} = 12.90$, $MD = .398$). The observed difference was not significant ($p = .109$, $d = .08$). The selection rates of the other indicator categories differed significantly from the expected value. However, based on the statistical threshold, most effect sizes are considered to be small.

For the selection rates of indicators in categories D and F, the differences observed can be considered of practical significance. Skills indicators about the aim of the specialist being to stimulate intrapersonal understanding had the highest mean selection rate ($\bar{X} = 14.49$, $MD = 1.994$, $t(353) = 7.24$, $P < .001$, $d = .38$). Indicators in category F were selected least, ($\bar{X} = 10.53$, $MD = 1.966$, $t(353) = -8.12$, $P < .001$, $d = .43$). Both of the observed differences are practically significant.

[422] Appendix J, Tables J6 & J7

At level 3, results for knowledge indicators differ more from the expected mean rates than skills indicators. At first glance, this may suggest that, if both groups of indicators are combined, the differences observed between categories of indicators may even out. If this were so, most selection rates would stay close to what was expected and the observed differences between theory and practice might no longer be relevant. A final analysis at level 3 was undertaken to explore whether observed trends in selection rates between categories of indicators would still stand if there was no differentiation between knowledge and skills indicators and if there was no differentiation between the separate groups of stakeholders.
All indicators except indicators in category B, differed from the expected mean ($\mu = 25$).[423] Overall, indicators in category E have the highest selection rate ($\bar{X} = 28.17$, MD 3.167, $t(353) = 8.520$, $p < .001$, $d = .45$). Indicators in category F have the overall lowest selection rate ($\bar{X} = 20.73$, MD -4.274, $t(353) = -10.301$, $p < .001$, $d = .55$). Indicators in category C also showed low selection rates ($\bar{X} = 21.55$, MD -3.452, $t(353) = -8.525$, $p < .001$, $d = .45$) (see Figure 44).

Figure 44

Selection Rate of Indicators for the Entire Competency Matrix - (see Table J24)

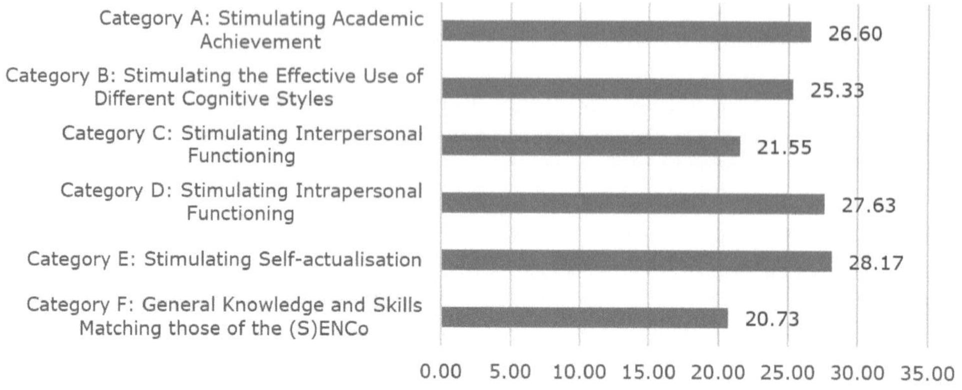

Note. N = 354. Test value = 25

Data analyses at levels 1 and 2 suggested that different groups of stakeholders would make different choices when selecting indicators. This would concur with what is known in theory about differences in the understandings of stakeholders and how these differences affect their choices (Freeman, 1987; Mitchel et al., 1997). These differences would also concur with what teachers in teacher education programmes for gifted education report regarding the differences that they experience between their understandings and the understandings of

[423] Appendix J, Tables J24-J25

Figure 45

Comparison of the Selection Rates of the Indicators for the Entire Competency Matrix Between the Groups of Stakeholders (see Table J30)

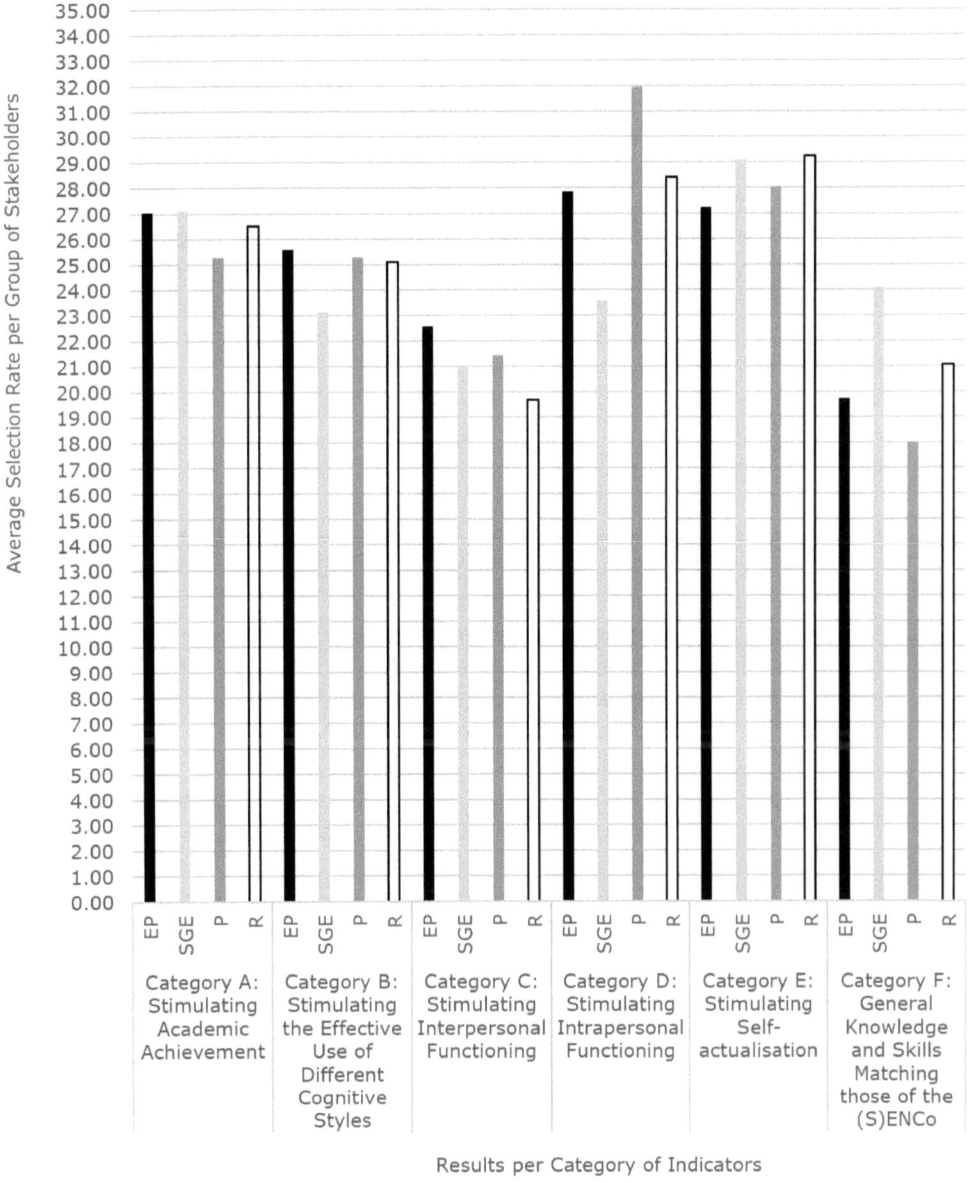

Note. N = 354 divided into four groups of stakeholders: EP = Educational Professionals, SGE = Specialists in Gifted Education, P = Parents, R = Rest. EP: N = 131, SGE: N = 103, P: N = 83, R: N = 37.

Test value = 25

parents (Slim! Educatief, Postgraduate Teacher Education Course Specialist in Gifted Education, personal communication/class discussions during modules 3 and 4, January-March 2017, January-March 2018; January-March 2019; January-March 2020). Tables J18-J23 (see Appendix J) show that findings at level 3 were similar to findings at levels 1 and 2, and that the earlier observed trends were confirmed.

- Educational professionals and specialists in gifted education select knowledge indicators (category B) more often than parents. These knowledge indicators focus on the specialist stimulating their gifted student's ability to recognise and use different cognitive styles depending on which best match the presented task within its context. Parents selected skills indicator B more often than specialists in gifted education. The difference between their selection rates for knowledge and skills indicators suggest that it is likely that these rates are influenced by the context in which they were presented. However, if no differentiation is made between knowledge and skills indicators, then no significant differences can be observed between the separate groups of stakeholders (see Appendix J, Tables J30-J31).
- Educational professionals select knowledge indicators (category D) that aim to stimulate the student's intrapersonal understanding more often than specialists in gifted education. However, parents have the highest mean selection rate for this category. Their selection rate exceeds the selection rate of either educational professionals or specialists in gifted education. Specialists in gifted education select the knowledge and skills indicators from category D less often than the other groups of stakeholders. This suggests that, even though it cannot be said for specialists in gifted education that the intrapersonal understanding of their pupils is not relevant, it can be concluded that, if forced to make a choice, they will prioritise the indicators differently to parents and educational professionals.
  If knowledge and skills indicators are combined, this observed trend is confirmed at the most general level of analysis as well ($p < .001$, $\eta^2 = .10$, see Appendix J, Tables J30-J32).
- Specialists in gifted education select the knowledge and skills indicators in category F more often than educational professionals and parents. F-indicators focus on more general advisory skills, matching the role of the (S)ENCo. At the most general level of data analyses and without differentiating between knowledge and skills indicators, this trend is still observed for category F-indicators ($p < .001$, $\eta^2 = .09$, see Appendix J, Tables J30-J32; see also Figure 45).

10.4 Comparing The Netherlands and Flanders

The differences between the selection rates of stakeholders from both countries when observed at domain level (level 2) are interesting. As a comparison between both countries has never been done before, there is no empirical evidence that explains the observed differences. Based on the previous findings, it now becomes relevant to explore whether the differences between the selection rates of stakeholders from both countries at this general level are also noticeable. The original assumption had been that stakeholders' choices from both countries would be similar. The theoretical study presented in Chapters 2 and 3 shows that the emphasis in both countries on the leading educational paradigms and the general teacher competences conditional for a teaching degree are highly similar. Although the overall conclusion is that the selection rates of stakeholders in both countries match the observed trend described in paragraph 9.3, stakeholders from each country place a slightly different emphasis on the different categories of indicators (see Appendix J, Tables J9-J12). The observed differences are significant.

Among the Dutch stakeholders, knowledge indicators in category B ($\overline{X} = 13.93$) have a higher selection rate than amongst Flemish stakeholders ($\overline{X} = 10.89$). The difference, 3.036, is significant, $t(352) = 5.09$, $p < .001$, the effect size represents a practical significance, $d = .75$.[424] Indicator category B aims for the development of a flexible and effective cognitive style with those skills contributing to academic achievements. Without differentiation between knowledge and skills indicators, these differences can still be observed (see Appendix J, J28-J29; see also Figure 46).

In The Netherlands, the effective use of cognitive skills is currently a popular item in education. Over the past two years, this topic has often been addressed at conferences, in teacher education courses for continuous professional development, educational journals, and book publications (e.g. Dawson & Guare, 2009; Houkema et al., 2019; Kroesbergen et al., 2017; van der Donk et al., 2015; Verheijen, 2019). Until recently, this topic has been addressed less in Flanders (Kieboom & Venderickx, personal communication, November 2019).

The SLO (Dutch foundation for curriculum development) published an "Objectives and Skills List" (2010) to be used for gifted education. The publication of this list stimulated Dutch schools to teach effective cognitive skills to gifted students. During CPD courses on gifted education, participating teachers, working in (part-time) segregated settings for gifted education, explain how they use this list to underpin their curriculum design for educating gifted students (Slim! Educatief, Postgraduate Teacher Education Course Specialist in Gifted Education, personal communication/class discussions during modules 3 and 4, January-March 2017, January-March 2018; January-March 2019; January-March 2020). In these pull-out programmes or (part-time) segregated classes the implementation of this *Objectives and Skills List* (2010) is often addressed by setting up a diverse range of training sessions for gifted students. Teachers attending the previously mentioned CPD courses explain that, at their schools, they want to challenge gifted learners as fully as possible in the regular classroom. This challenging education can be more successful if gifted learners are able to use effective cognitive styles during the learning process. Students who are able to effectively use different cognitive styles and sufficient regulation skills are often considered to be more able to work

[424] Appendix J, Tables J9 & J10

Figure 46

Comparison of Mean Selection Rates on Indicators for the Entire Competency Matrix between Dutch and Flemish Stakeholders - (see Table J28)

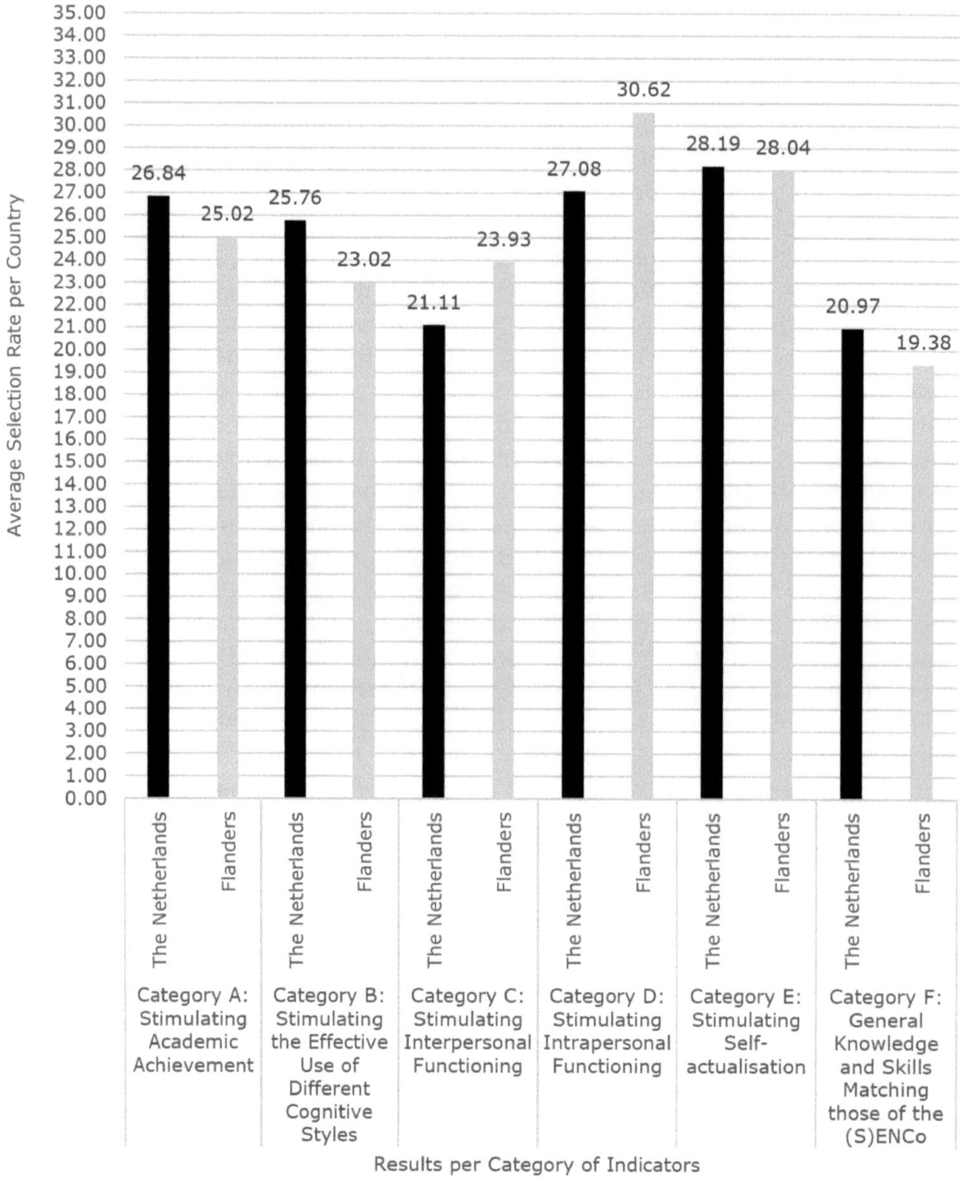

Note. N = 354. The Netherlands N = 299, Flanders N = 55. Test value = 25

independently than those who have not developed an effective use of these styles and skills. Students who operate relatively teacher-independent are, therefore, considered to be easier to challenge in the regular classroom than students who are highly dependent on teacher support. As a result, a focus on the effective use of these skills becomes more important for teachers in the regular classroom. Within this context, pull-out programmes are considered to be facilitators for the training of these effective cognitive styles and regulation skills.

There is no Flemish equivalent to the SLO *Objectives and Skills List* (2010) in use in Flemish pull-out programmes or special classes, to provide systematic underpinning to the curriculum offered in those pull-out programmes. There is also no information available as to whether Flemish schools use the Dutch version of this Objectives and Skills List (Kieboom & Venderickx, personal communication, November 2019). During training courses on gifted education, Dutch teachers who have implemented the use of this list report how goals *of* education (such as the effective use of cognitive skills) often trump goals *in* education, when assessing the results of a student's development through a pull-out programme (Slim! Educatief, Postgraduate Teacher Education Course Specialist in Gifted Education, personal communication/class discussions during modules 3 and 4, January-March 2017, January-March 2018; January-March 2019; January-March 2020).

Flemish stakeholders have a higher selection rate for the knowledge indicators in category C ($\bar{X}$ = 10.49) than Dutch stakeholders ($\bar{X}$ = 8.31). The difference, -3.190, is significant, $t(352) = -3.19$, $p = .002$, the effect size represents a practical significance, $d = .47$.[425] A similar difference is found for indicators in category D, where Flemish stakeholders have a higher selection rate ($\bar{X}$ = 14.65) compared to Dutch stakeholders ($\bar{X}$ = 12.85). The difference, -1.802, is significant, $t(352) = -2.12$, $p = .035$, this represents an effect size of practical significance, $d = .31$.[426] Regarding the skills indicators, Flemish stakeholders have selected D-indicators significantly more than Dutch stakeholders ($\bar{X}$ = 15.96, MD -1.740, $t(352) = 2.30$, $p = .022$, $d = .34$).[427] In The Netherlands, skills indicators in category F ($\bar{X}$ = 10.88) have a higher selection rate than they have in Flanders ($\bar{X}$ = 8.65, MD -2.225, $t(352) = 3.38$, $p = .001$, $d = .50$).[428] Observed significant differences are still noticeable when there is no differentiation between knowledge and skills indicators.[429]

For stakeholders in Flanders, the competency of specialists in gifted education to stimulate interpersonal functioning and intrapersonal understanding, trumps their competency in encouraging the student's academic development. In Flanders, Kieboom and Venderickx's approach to giftedness (Kieboom & Venderickx, 2017a; Venderickx & Kieboom, 2017) seems to be leading currently. As explained in Chapter 2, their perspective on giftedness stresses the

425 Appendix J, Tables J9 & J10
426 Appendix J, Tables J9 & J10
427 Appendix J, Tables J11 & J12
428 Appendix J, Tables J11 & J12
429 Appendix J, Tables J28 & J29

combination of specific personal traits combined with a high developmental potential. They stress the necessity for gifted students to learn how to overcome certain thresholds that they may experience in education and talent development (Kieboom & Venderickx, 2017a). In this approach, Kieboom and Venderickx emphasise the development of interpersonal and intrapersonal skills. Although Kieboom and Venderickx (2017a) support the concept of part-time segregated interventions in addition to interventions in the regular classroom, their approach regarding interpersonal and intrapersonal skills is likely to be more influential. As some of the Flemish respondents were recruited from a call for respondents sent out with the cooperation of Kieboom and Venderickx, it may well be that Flemish respondents were already familiar with this particular view and that this has been of influence on the selection rates for indicators in the domains of interpersonal and intrapersonal development.

10.5 The Additional Analyses: Ownership, Educational Partnership and Peer Group Education

Based on the results of data analyses at levels 1 and 2 presented in Chapters 5-9, analysis was undertaken on three additional topics: (a) voicing; (b) the importance of organising peer-group education for gifted students; and (c) educational partnership. The results of the selection rates were compared with the selection rates of indicators that could not be identified as one of these three categories. The results of the analyses should only be interpreted while bearing in mind that selecting one indicator excludes the opportunity to select another indicator. Each stakeholder could only select 150 indicators out of the 300 indicators presented. Thus, if a stakeholder selected a lot of indicators aiming for ownership, they could select fewer indicators focused on other topics.
Spread throughout categories A to F, indicators addressing a student's ownership of their development, educational partnership, and provision of peer group education can be identified.
Chapter 2 discusses how current educational paradigms in general affect gifted education in The Netherlands and Flanders. In both countries aiming for self-actualisation has become important as this is a prerequisite to a social-constructivist approach to learning (Houkema, 2017; Kieboom & Venderickx, 2017b; Marzano & Heflebower, 2012; Voogt & Pareja Roblin, 2010). An important strategy for self-actualisation is making the student aware of personal responsibilities and therefore also of personal choices in education (Dweck, 2006, 2017; Houkema, 2017). The strategy that is used to involve the student and create ownership is called voicing (den Otter, 2015). Based on the increasing emphasis on ownership, it had been expected that indicators aiming to stimulate ownership and voicing as an educational strategy would get high selection rates. 39 of the 300 indicators can be identified as aiming explicitly for ownership and voicing. 28 of these 39 indicators belong to the categories D and E. As the mean selection rates of both categories of indicators were relatively high, the expectation had been that indicators aiming for ownership and/or using the strategy of voicing would be selected significantly more often than other indicators. Analyses of the data showed that, indeed, these indicators have

a significantly higher selection rate than the other indicators ($\bar{X} = 54.64$, $p < .001$, 95% CI (3.58 – 6.33), $d = .37$, representing a practical significance).[430] The high selection rates for these indicators suggest that, considered as a whole group, the stakeholders consider the knowledge and skills of the specialists, when applied to developing ownership in their gifted learners towards their education, to be important. Analysis per group of stakeholders showed that parents ($\bar{X} = 57.40$) selected indicators regarding ownership and voicing more often than specialists in gifted education ($\bar{X} = 51.83$). The difference, 5.57, is significant ($p = .13$, CI 95% (-10.3503, -0.7879). This suggests that parents consider the inclusion of their child's perspective when developing a good understanding of their educational needs and in designing the best educational interventions possible, to be important. It also suggests that, for specialists in gifted education, other indicators may be equally important and that, depending on the specific competencies the indicators were matched with, they decided to select the best example for that specific indicator.

The second topic to receive a separate analysis is the issue of educational partnership.[431] In a systemic approach to education, educational partnership can be seen as the founding element for interventions (Broersen & Spreij, 2009; Pameijer & van Beukering, 2007; van Meersbergen & de Vries, 2017; van Meersbergen & Jeninga, 2012). If this approach has a strong influence on the professional behaviour of the specialist in gifted education, and if educational partnership is considered to be relevant by the different groups of stakeholders, this should be reflected in the selection rates. However, the first results were inconclusive. There are 36 indicators aiming explicitly for educational partnership. 24 of these indicators belonged to categories B (4), C (4) and F (16). 12 belonged to categories A (9), D (2) and E (1). Based on the previously presented data (with high selection rates for categories A, D and E and lower selection rates for categories B, C and F), it was expected that educational partnership would not have been selected significantly more often than other indicators. Indeed, with no differentiation between groups of stakeholders, indicators concerning educational partnership were not selected more frequently than other indicators. However, a comparison between the groups of stakeholders shows that there are significant differences ($p < .001$, $\eta^2 = .062$). It had been expected that parents would have selected indicators about educational partnership more frequently than the other groups of stakeholders, as the notion of educational partnership can secure the relevancy of their role in the education of their child. However, the highest selection rate for these indicators is found amongst specialists in gifted education ($\bar{X} = 55.56$). The difference between the selection rate of specialists and the selection rate of parents ($\bar{X} = 47.12$, $MD = 8.43$) is significant, $p < .001$, 95% CI (3.0656 – 13.8019). A similar difference is found when comparing the selection rate for educational partnership indicators of specialists of gifted education with

[430] Appendix J, Tables J13-J15
[431] Appendix J, Tables J13-J15

educational professionals ($\overline{X}$ = 47.94, MD = 7.61, $p < .001$, 95% CI (2.8197, 12.4050). The high selection rates of specialists in gifted education for these indicators match directly to their more generic role in the school. Their direct involvement with gifted learners is a more distant one. As they often function as a spider in the web (see also Chapters 2 and 3), their competence in aiming for educational partnership is of high relevance. This confirms the suggestion of Freeman (1987) and Mitchel et al. (1997) that stakeholders make their decisions based on their role in the organisation or in the partnership.
The third topic for which there was separate analysis focused on indicators concerning peer group education.[432] It was noticeable in the literature review study that peer-group education was considered to provide an important opportunity to stimulate both the academic and social-emotional development of gifted students (J. Cross, 2012; T. Cross, 2018; Heacox & Cash, 2019; Janson, 2009; Neihart & Yeo, 2018; Oppong et al., 2018; Rinn & Majority, 2018). It had been expected to see an identical emphasis on peer group education amongst stakeholders. However, in the empirical part of the research a different trend is observed at level 1. The 10 indicators focused explicitly on the organisation of peer group education seemed to have relatively low selection rates compared to other items that were also thought to be important for the development of gifted students. 8 out of 10 belonged to category C, the category that showed the lowest selection rates amongst all indicators. As a separate category, indicators focusing on peer group education had a selection rate ($\overline{X}$ = 37.85) that differs from the selection rate of the other indicators ($\overline{X}$ = 49.69). The difference, -11.834, is significant ($p < .001$, $d = .53$, representing a practical significance). Differences between groups of stakeholders cannot be considered as significant ($p = .342$).[433]
The results of this final analysis cannot be explained by looking through the lens of the different groups of stakeholders, as it has been possible to do for differences in the emphasis on ownership and educational partnership. However, comparing selection rates between stakeholders from the two different countries offered an interesting perspective. Stakeholders from Flanders ($\overline{X}$ = 44.91) selected indicators aiming for peer group education more often than Dutch stakeholders ($\overline{X}$ = 36.56). The difference, -8.35, was significant, $p = .009$, $d = .39$).[434] This suggests that Flemish stakeholders consider the competence of specialists to organise peer group education as more important than Dutch stakeholders do. However, this leaves one important aspect out of the scope. In Chapter 1, gifted education is described as being implemented at a far more advanced level in Dutch schools than in Flemish schools. The Dutch government supports gifted education financially by providing additional funding for schools and regional collaborative projects between local school councils (see Chapters 1 and 2). Due to this additional funding, over the past 15 years Dutch schools have been able to invest more into gifted education than schools in Flanders. Participation in Dutch pull-out programmes often takes place under the condition

[432] Appendix J, Tables J13-J15
[433] Appendix J, Tables J13-J15
[434] Appendix J, Table J16

that, preceding the selection of a student for a pull-out programme, this student will have been challenged in the regular classroom (Houkema et al., 2018). This offers a new perspective on these selection rates. A possible explanation might be that Dutch stakeholders consider peer group education as a service for gifted learners that is already provided for or at least provided for at a sufficient level. Peer group education is an intervention strategy that has become part of the basic repertoire of most schools or school councils in The Netherlands (Houkema et al., 2018). Thus, Dutch stakeholders may consider that this has already been taken care of effectively, suggesting that their need to emphasise these indicators will not be as important as for Flemish stakeholders.

10.6 Summarising

Based on findings at levels 2 and 3, the combined grouping of indicators ADE showed a significantly higher selection rate than the combined grouping of indicators BCF ($p < .001$).
Differences between stakeholders from The Netherlands and Flanders had not originally been expected when this research was started, but, influenced by findings at levels 1 and 2, these expectations were adjusted. Findings at level 3 confirmed the observed differences between both countries. Flemish stakeholders showed a preference for indicators in categories C and D whereas Dutch stakeholders showed a preference for indicators in categories B and F. Without differentiation between knowledge and skills indicators, the significant differences between both countries for the selection rates of indicators in category F can no longer be observed.
Although not as explicitly noticeable as that seen in levels 2 and 3, the different understandings of stakeholders of what they think to be the most important and relevant examples of professional behaviour for specialists in gifted education were ascertained at this general level of data analyses. These findings were in line with the results from a more in-depth level. Parents showed a high selection rate for indicators in category D and specialists in gifted education showed their highest selection rates for indicators in category F. Overall, indicators in category E have the highest selection rates. Analysis of the three additional topics suggests that ownership is considered important, as these indicators had a significantly higher selection rate than other indicators. At the general level, indicators concerning educational partnership were selected no more often than other indicators. This means that, even though individual indicators sometimes have significantly higher selection rates than other indicators, as a separate category the selection rates of educational partnership indicators are not of high significance compared to other indicators. Finally, the small group of separate items focused on the stimulation and organisation of peer group education showed relatively low selection rates amongst all groups of stakeholders. Even though they are selected more often by Flemish stakeholders than by Dutch stakeholders, these higher selection rates do not break with the observed trend.

Part 3

Integrating Theory and Practice

11 Conclusions, Discussion and Recommendations

With this research project, I intended to construct a competency matrix providing a framework for the curriculum development of postgraduate teacher education programmes for specialists in gifted (primary) education. This required that both current educational paradigms as implemented in Dutch and Flemish schools and paradigms in theories of giftedness would underpin the matrix. Use of the matrix would require a clear connection to current teacher education strategies. Combination of these three different conditions led to the central research question:

> *Which competencies are necessary to provide gifted education that meets the needs of gifted students in an educational environment in The Netherlands and Flanders, which knowledge and skills are representative of these competencies, and what is needed to develop these competencies?*

To answer this question, four further questions were required. In this final chapter, I will address all four of them.

This study is considered to be beneficial for different groups of stakeholders of gifted education. Each group may use the results differently but matching their separate positions as stakeholder (see Table 10).

11.1 The Position of Gifted Education in The Netherlands and Flanders

The first question considers the position of gifted education in Dutch and Flemish education. It examines how developments and shifts in educational theories in The Netherlands and Flanders influence the professional context of a specialist in gifted education. To answer this question three sub-questions are addressed: (a) Which developments and shifts in educational theories in The Netherlands and Flanders influence the professional context of primary school teachers in general? (b) How can we currently position gifted education in The Netherlands and Flanders? and (c) Which specific topics are, therefore, recognisably of influence on the professional actions and attitude of the specialist in gifted education?

In The Netherlands and Flanders, five major paradigms affect gifted education: (a) an inclusive approach to education, (b) responding to educational needs, (c) theories on giftedness, (d) social-constructivism, and (e) evidence-informed teaching. In each paradigm, signs of a shift towards renewed perspectives and practices were observed.

The first paradigm affecting gifted education is the notion of inclusion. Based on the Salamanca agreement (Unesco, 1994), both The Netherlands and Flanders strive for an educational concept that no longer excludes children from their right to education (Decreet betreffende maatregelen voor leerlingen met specifieke onderwijsbehoeften, 2014; Wet passend onderwijs, 2012). Taking that as a starting point, over the last 26 years, the Dutch and Flemish government, local authorities, school councils, school management, and teachers have striven

Table 10

Who are Beneficiaries of this Study of the Competencies of the Specialist in Gifted Education

Stakeholders	Rationale for Using the Matrix	Examples of How to Use the Matrix
Teachers	Gifted education is approached as distinct from and different to general education. But teachers may feel encouraged to use existing knowledge and skills as the matrix shows that their general teaching competence forms a strong basis for gifted education.	By using the matrix and the indicators as a directive tool for reflecting on their approach of gifted learners in their classroom, teachers can discover where there is a possibility to gain from small changes in their educational strategies and how this might be beneficial for non-gifted learners as well.
Specialists in gifted education	The matrix consists of a set of 25 coherent competencies offering specialists in gifted education a framework to underpin their professional strategies in their daily practice. This framework may help set professional boundaries to what may or may not be expected from a specialist in gifted education.	Specialists might use the STARR strategy (Situation, Task, Action, Results, Reflection) to describe their professional attitude and strategies. They can use the competency matrix as a directive to assess if their professional attitude and strategies reflect what might be expected of a specialist in gifted education. They can use the matrix to explore in their individual professional context how gifted learners, parents of gifted learners, teachers, and the organisation may benefit from specific knowledge and skills that come with the qualification of being a specialist in gifted education. Similar to competency profiles that are being used by SENCO's and other educational specialists, the matrix provides a wide range of recommendations for professional behaviour for a relatively new group of educational specialists.
School management, Policymakers (at local and national level)	As in this study, gifted education is no longer approached as distinct from and different to general education, it offers suggestions for school management to organise gifted education as part of general education. This would match a more inclusive approach to education in general and may enrich education for all learners.	This study is beneficial for school management to underpin to what extend they consider (part-time) segregation as a necessary way of organising gifted education. It is clarified that the stability of the pyramid of support depends on the competency of educational professionals. The matrix can be used by school management to analyse if and to what extend the professional development of their teachers and educational specialists may be beneficial for the quality of gifted education.

Stakeholders	Rationale for Using the Matrix	Examples of How to Use the Matrix
Teacher educators	Although in the Netherlands and Flanders a wide range of continuous professional development courses focusing on gifted education is available, only few are constructed in a way that explicitly combines gifted education with education in general. By embedding gifted education in education in general, teacher educators may find opportunities to teach teachers how existing knowledge and skills can be effective when applied to gifted education.	This study offers a broad perspective on gifted education. It includes a wide range of knowledge and skills that may increase the quality of gifted education. The matrix offers a set of coherent domains that can be used to set goals for curriculum development on gifted education both in initial teacher education programmes as in continuous professional development courses. Described competencies and knowledge and skills indicators leave ample opportunity for teacher educators to select literature and resources matching the context of the teachers they are educating.
Parents	It is most likely that parents will benefit from this study in an indirect way as the study aims to increase the quality of gifted education by raising the bar for the education of specialists in gifted education. By including parents as stakeholders and distinguish their selection of indicators from the selection of other groups of stakeholders, it has become visible that parents do not per se have an understanding of gifted education that stands diametrically across the understanding that educational professionals and specialists in gifted education have. Therefore, it creates opportunities for educational partnership between all parties involved.	The results of this study provide an opportunity for professionals in education and specialists in gifted education to understand what parents may find important for their gifted child's education. This understanding can be the starting point of a joint exploration between educational partners to what may be the best approach for their child and student under these circumstances with this teacher, these peers, at this school, and with these parents.
Gifted students	Although gifted students were not involved in this study as stakeholders, they can be considered to be stakeholders regarding all the roles and tasks a specialist in gifted education may take on if these roles and actions are directly aiming for the education of the individual student. Among the knowledge and skills indicators, there are several indicators included aiming for direct interaction with the student.	School management, teachers, specialists in gifted education may want to involve gifted learners when it comes to creating a job description for a specialist in gifted education for their organisation. To prevent any possible inequality between gifted learners and non-gifted learners, this involvement should be identical to the manner and extent they would involve other students to create a job description for other educational specialists. If the competency matrix has been used for an initial job description and from that point on students become involved to fine-tune the criteria, it may create opportunities for a stronger relationship between the gifted learners and their specialist in gifted education.

to develop an educational environment open to all learners, meeting their individual needs as fully as possible (Adema et al., 2009; Amsing et al., 2009; Bronneman-Helmers, 2008; Clijsen et al., 2007; Dijksma, 2007; Inspectie van het Onderwijs, 2009; Meijer, 2004; Schuman, 2010; van Hove et al., 2005; Waslander, 2011). The intent of the Dutch educational law on appropriate education (2012) and the Flemish so-called "Begeleidingsdecreet" (Vlaamse Regering, 2019) is to ensure this right for all learners. As a result, a transfer is taking place. A system where students with additional (non-mainstream) needs were segregated is moving to a system where striving for optimal differentiation makes it possible to meet the needs of these learners in a regular educational context (Dekker, 2014a; Doornbos & Bergman, 1991; Inspectie van het Onderwijs, 2009; Vlaamse Regering 2019). This should position gifted education in the regular classroom, giving all learners access to educational challenges instead of having to overcome the barrier of first being identified as gifted, before being given access to what challenges them in education (Florian & Black-Hawkins, 2011; Janson, 2017).

However, in Dutch and Flemish education, a counter movement is also taking place. Under the influence of public opinion, schools have started to consider whether an inclusive approach is the most effective way to meet the needs of gifted learners (Daeter, 2012, 2014; Keeman, 2010). Some consider their needs too complex to be met in the regular classroom and others feel that gifted learners are being held back in their development if being taught in the regular classroom (Daeter, 2012; Keeman, 2010; Mijland, 2010). As a result, in The Netherlands, an increasing number of schools offer full-time segregated education for gifted students, while an even bigger number of schools offer part-time segregated education for gifted students. In Flanders, part-time segregated classes for gifted education have become increasingly popular as well. A small number of schools have started to offer full-time segregated education for gifted learners in addition to their regular grouping. In both countries, access to these special programmes is granted if the student is formally identified as gifted. Schools assess their students for their eligibility to participate in such a programme by either having them take an intelligence test or by using a toolkit which enables schools to identify gifted students, based on the combination of exhibiting specific traits and meeting certain thresholds defined for exceptional results. This development seems almost contradictory to the paradigm shift towards inclusion.

This development has been unintentionally fuelled by a second paradigm shift: the understanding that all children have educational needs and that these needs are different for all students (Florian & Black-Hawkins, 2011; Oostdam, 2009; van Meersbergen & de Vries, 2017). In the Dutch and Flemish context, educational needs are interpreted more widely than for mere academic needs. Educational needs refer to all that is needed, pedagogically and didactically, to stimulate the student's developmental process in the broadest meaning of the word: mentally, social-emotionally, and physically (Pameijer & van Beukering, 2007; van Meersbergen & de Vries, 2017). Biesta's (2014) concept of the three core tasks of education can be recognised in this interpretation: qualification, socialisation, and subjectification. In The Netherlands and Flanders this means

that seeing and understanding educational needs should underpin meaningful responses in education (Pameijer & van Beukering, 2007; van Gerven & Hoogenberg, 2011; van Meersbergen & de Vries, 2017). Different needs require different responses, which should lead to a differentiated curriculum based on differentiated didactical and pedagogical strategies (Bakx et al., 2016; Florian & Black-Hawkins, 2011; Janson, 2017; Mol-Lous, 2011; Oostdam, 2009; Tomlinson et al., 2009; Winstanley, 2010). This underscores the suggestion that inclusion requires a personalised educational environment focusing on opportunities for optimal development for all learners. From the understanding that all children have educational needs, that these needs differ, and that they are systemically influenced, comes the logical thought that a Procrustean-approach to education is no longer sufficient for what our society demands of education (Mol-Lous, 2011; Oostdam, 2009; Stevens, 1997). This implies that differentiation should form the underpinning of all educational responses (Florian & Black-Hawkins, 2011; Tomlinson et al., 2009; Winstanley, 2010). The teacher's pedagogical virtuosity (Biesta, 2014) and pedagogical sensitivity (Bakx, 2015; Bakx et al., 2016) are conditional for optimal differentiation. These skills and attitude call highly on the teacher's competence to accomplish them.
In The Netherlands and Flanders, the educational support system ranks educational needs and responses. The complexity of the needs determines the support level for the teacher's responses (Armstrong et al., 2010; Prodia, 2012). However, this system is highly subjective. What is complex is not only determined by the identified needs but is also determined by what the teacher is able to do under the circumstances where they experience a gap between what is optimally desirable and optimally workable. The interplay between the student's ecological system and the teacher's ecological system outlines what can be done and what becomes difficult to realise (van Gerven & Weterings-Helmons, 2014). Therefore, the level of support that is needed for a student's optimal development can differ per teacher, per year and sometimes even per domain of human development. This realisation is exactly what fuels advocates for full-time segregated education for gifted learners. They argue, quite successful to the general public, that the educational needs of gifted students exceed what should be realistically expected of their teachers, who are responsible for a wide range of educational needs that all need to be addressed in the regular classroom (Daeter, 2012; Smits, 2019). However, as for the long-term effects of segregated educational programmes in The Netherlands and Flanders, the jury is still out. There are hardly any results suggesting that full-time segregation is the most meaningful response for a group of learners who can hardly be called homogeneous. Results seem to indicate that part-time segregated programmes may have a better and longer-lasting effect (Hornstra et al., 2017).
The third paradigm shift affecting gifted education is observed in the discussions of theories about giftedness. Movement from the so-called gifted child paradigm towards a talent development approach and eventually an approach based on the differentiation paradigm has been a gradual one. It represents a shift from the so-called mystery approach to a mastery approach (Dai & Chen, 2014; D. Matthews & Folsom, 2009; D. Matthews & Foster, 2009). Each of these

paradigms in theories of giftedness has different underpinning theories and different consequences for what is considered to be the best educational match (Cohen, 2012; Dai & Chen, 2014). In The Netherlands and Flanders, schools are at liberty to define what they consider to be the best definition of giftedness and gifted education. Criteria for who is identified as gifted differ per school (de Boer et al., 2013; Doolaard & Oudbier, 2010; Haenen & Mol-Lous, 2014). As the paradigm shift from the gifted child paradigm towards a talent development approach or even the differentiation paradigm has not yet been completed, it can be observed that schools try to combine different elements from all paradigms. On the one hand, schools want to differentiate in their responses to the curriculum offered and the educational strategies applied to gifted students. This suggests that they understand that gifted students are a heterogeneous group. On the other hand, this differentiation is mostly based on general assumptions of what schools regard as being "good for gifted learners". This creates a situation where gifted students are considered to be different from other students. However, in their *being different* they are treated as a homogeneous group with only limited options to differentiate based on individual needs (Kuipers, 2017). Special programmes are open for those students who *fit* the programme instead of programmes being tuned to a student's individual needs (Daeter, 2012; Verouden, 2017). In The Netherlands and Flanders, there has been little response to D. Matthews and Foster's (2009) suggested definition of giftedness, that gifted education should be for those students who have educational needs that are hard to meet at that specific moment, under those circumstances, in that specific classroom with that specific teacher. Therefore, although Dutch and Flemish schools are aware that different students have different needs, this does not mean that a shift towards a differentiation paradigm for gifted education, grounded on inclusion and meaningful responses to the needs of individual learners, has been completed yet.

The fourth paradigm shift affecting gifted education is the shift towards social-constructivism and 21st century skills, where education is modified to a system where didactics shift from a knowledge-transfer based approach to a knowledge-construction based approach (Onderwijsraad, 2014; Unesco, 2015; Vlaams Ministerie van Onderwijs en Vorming, n.d.). This approach rests on three domains of skills: thinking skills, social skills, and metacognitive skills (Onderwijsraad, 2014). Consequently, a strong emphasis is placed on the development of these skills (Bambs et al., 2016). This educational paradigm is not new, but what is new, is that nowadays this approach is considered to be good for all students (Mol-Lous, 2011; Thijs et al., 2014; Unesco, 2015). This approach is gradually being introduced into primary education.

Consequently, this affects the profession of teaching. Teachers need to adapt their educational strategies in a way that enables them to model the attitude they want students to develop (Lunenberg et al., 2014; Sternberg, 2009). The characteristics of the social-constructivist approach appear to be a good match for a more mastery-oriented perspective on giftedness (D. Matthews & Foster, 2009; Subotnik et al., 2011). Having a special talent or having specific abilities is no longer considered to be enough to excel in that talent. In this approach, not only the teachers but also the students themselves are held accountable for their individual development. This implies that ownership towards educational objectives becomes more important for students (Mayes et al., 2019; Pameijer & van Beukering, 2007; van Meersbergen & de Vries, 2017). Taking all this into account, this fourth shift in education could be beneficial for the position of gifted education in The Netherlands and Flanders. If these pedagogical-didactical strategies, that create opportunities for diversification, are introduced for all learners, it should be easier for teachers to respond in a way that means that the needs of their gifted learners can be addressed in the regular classroom (Janson, 2017).

The fifth and final paradigm shift affecting the position of gifted education is the increasing value laid upon evidence-informed education (Diery et al, 2020; Hattie, 2013; Marzano, 2007, 2011; Onderwijsraad, 2006; Robinson, 2015; Simpson, 2020). As a result of this development, teachers are required more and more to underpin their actions with solid theoretical knowledge. Dominated by theories of Hattie (2013) and Marzano (2007, 2011), education in The Netherlands and Flanders has turned away from intervention strategies based on evidence of single case studies. This could have a major impact on the strategies used in gifted education in both countries. Currently, most of the strategies used are based on two levels of proof (Robinson, 2015): (a) proof at the level of *soft evidence* or *wisdom*, where personal experiences are generalised to a broader population without checking whether this is a rightful generalisation; (b) proof at the second level of evidence, where the level of good practices is underpinned with evidence mostly based on international research but without contextual interpretation of where this evidence was gathered. As a result, schools experience long periods of trial and error, setting up gifted education. Less successful strategies undermine the opinion of teachers, parents, the government, and in the end the opinion of gifted learners themselves, about the quality of gifted education. Single cases where an outcome has not been as successful as all involved hoped for, reach the media and affect general opinion on the quality of gifted education in general (KRO-NCRV, 2018). Striving for an evidence-informed and contextually implemented approach to gifted education requires high-quality teacher education to develop true specialists in gifted education (Johnsen et al., 2016). Topics of influence on the professional actions and attitude of the specialist in gifted education are, therefore, not only theories of giftedness but also the seeing, the understanding, and the meaningful response to the educational needs of gifted students (Bakx et al., 2016; Haenen & Mol-Lous, 2014; Oostdam, 2009; van Gerven & Hoogenberg, 2011). The assessment of the response to intervention underpins applied educational strategies and strengthens the position of gifted education in the Dutch and Flemish educational system, matching the leading educational paradigms in both countries.

Data-analyses of the results of the questionnaire showed that among the 22 items with the highest selection rates most indicators were focused on understanding and responding to educational needs (see Table J3). There was only one indicator among this top 22 that aimed for deep-level knowledge on giftedness in general. Apparently, although among international professionals in gifted education, theory on giftedness is seen as important in order to be able to provide gifted education, if Dutch and Flemish stakeholders have to choose, they prioritise differently. However, one cannot conclude that they consider knowledge on theories of giftedness as unimportant. Stakeholders may consider this knowledge as conditional for being able to see, understand, and respond meaningfully to educational needs and therefore not see the necessity of pointing this out by selecting these specific indicators. The data do not provide an answer to that hypothesis.

What is interesting is the strong emphasis there is on indicators in this top 22 aiming for knowledge and skills to stimulate ownership and using voicing as a strategy to do so. That matches the shift among the more general educational paradigms where the change- and solution-focused approach and social-constructive learning have become an important strategy to provide appropriate education to all learners. Ownership and voicing are important strategies being used in both approaches.

The system of response to educational needs is based on educational partnership between all the actors in the student's ecological system. This paradigm is, as described in the previous chapters, leading in general education in the Netherlands and Flanders. Among the 22 indicators with the highest selection rates, five indicators reflect educational partnership. However, the average selection rate of the entire cluster of indicators aiming for educational partnership is not significantly higher than the selection rate of other indicators. The fact that these specific indicators ended up among the top 22 of indicators with the highest selection rate suggests that the professional attitude described with these indicators is at least considered to be important for seeing, understanding, and responding to educational needs in gifted education.

The questionnaire does not include indicators suggesting specific knowledge and skills about how to organise (full-time) segregated education. Hence, the data do not provide any information if this observed tendency in the educational practice leads to a situation where stakeholders consider these knowledge and skills as relevant examples of the competency of specialists in gifted education. The selection rates of the few items addressing the organisation of peer group education do not differ significantly from other indicators. What can be concluded is that the specialists' ability to differentiate between students, based on their individual educational needs, is seen as important (see Tables J1 & J3). As the structure of the Dutch and Flemish pyramid of support is based on the interplay of the student's educational needs and what a teacher is able to provide as a meaningful response to those needs, policymakers and school management respond optimally to the results of this study if they create as many options as possible to provide this differentiated response. By leaving the decision how to use these options to the educational partners, they contribute to the ownership of selected strategies for education. This creates an opportunity for educational partners to find a differentiated approach matching the individual student's needs instead of creating opportunities for selecting students who may be fit to join a segregated gifted programme.

11.2 Teacher Education and Gifted Education

The second research question focuses on developments in Dutch and Flemish teacher education and the way that these developments influence the professional actions and attitude of the specialist in gifted education. To answer this question, four sub-questions are addressed: (a) What should reflect the educational level of the specialist in gifted education and how does this match the international demands for teacher education to which both The Netherlands and Flanders have been committed? (b) Which professional standards of teacher education in general interact with the professional standards set for specialists in gifted education? (c) Which factors in current perspectives on teacher education influence the professional standards that can be set for specialists in gifted education? and (d) What are the criteria to be met when stipulating a competency, and how can these criteria form a framework for assessing the quality of the competencies?

The Dublin Descriptors determine the educational level of Dutch and Flemish teacher education (European Higher Education Area, 1999). These descriptors aim for professional development in five domains: (a) knowledge and understanding, (b) applying knowledge and understanding, (c) making judgements, (d) communication skills, and (e) learning skills (Bologna Working Group on Qualifications Frameworks, 2005). The descriptors are set up in a tiered structure. Each level requires that the behaviours described for the previous level be acquired. Initial teacher education takes place at bachelor degree level (Onderwijscoöperatie, 2014a; Vlaamse Regering, 2008). In their professional behaviour, teachers demonstrate the knowledge, skills, and attitude that have been acquired at this level. If we match knowledge, skills, and attitude with the pyramid of support to bachelor degree level, teachers learn how to function at levels 1 and 2 in the pyramid of support (De Bruïne et al., 2015; van Gerven, 2015). If the student requires a more intense or different approach exceeding what the teacher is used to offering at levels 1 and 2 in basic care, then the teacher may call on the SENCo for additional support. The SENCo is a specialised teacher, educated to master's level. In cases where the SENCo has difficulties setting up an effective way of supporting teacher and student at level 2, they may call on the district SENCo. District SENCo's have a similar level of education (master's level), but are often more experienced and used to advising in educational contexts that require a higher level of expertise. The pyramid of support is identical for all learners and their teachers. Specialists in gifted education are likely to be functioning at the first four levels of this pyramid of support (De Bruïne et al., 2015; van Gerven & Hoogenberg, 2011). They become specialised teachers and, depending on their personal professional context, they function either as (a) regular classroom teachers with one or more gifted learners in their group, (b) classroom teachers in a full-time or part-time segregated group for gifted education, or as GSENCo's (Gifted Students' Educational Needs Coordinator) at school or district level (LBBO, n.d.-b). This implies that the educational level of the specialist in gifted education should be identical to the educational level of the SENCo, that is, master's degree level. In daily practice, however, teacher education courses for specialists in gifted education do not have to result in a master's degree. It should be a requirement of institutes providing these courses to use the Dublin Descriptors (Bologna Working Group on Qualifications Frameworks, 2005) at master's degree level to build their curriculum. It should also be a requirement that they assess their students at this level before handing out certification as a specialist in gifted education.

In practice, Dutch and Flemish general teaching degrees are underpinned both by the required level of professional attitude (the Dublin Descriptors) and by the teacher's knowledge and skills as defined by national teaching standards (Onderwijscoöperatie, 2014a, 2014b; Vlaamse Regering, 2008). For both countries applies that the documents describing these teaching standards do not provide any clarity on how these standards were constructed. It seems realistic to assume that professional communities of teachers and teacher educators have contributed to some extent to the construction of these standards but there is no information provided about the methodology used for the empirical underpinning of the standards. References to underpinning research and a theoretical framework are lacking in the national documents that are being used to inform the professional community about these national standards. The Netherlands and Flanders have different sets of teaching standards. However, a comparison between both standards suggests that these standards are highly similar. They may be slightly differently categorised and presented, but there are no differences observed in required knowledge and skills. Five overarching domains can be observed: (1) pedagogics, (2) didactics, (3) classroom management (4) communication, and (5) research and innovation (Onderwijscoöperatie, 2014a, 2014b; Vlaamse Regering, 2008).

A competence-oriented approach to teacher education is a translation of the standards described for these domains. This approach offers opportunities for a solid assessment at the end of a programme. However, the approach should not be translated into a one-dimensional assessment presented as a series of "yes/no" questions based on a Likert-like scale (Biesta, 2012; Ceulemans et al., 2016; Onderwijscoöperatie, 2014b). The competence-oriented approach is intended to develop a professional attitude in future teachers, aiming for life-long learning. Learning is a process of intrapersonal adaptation (Biesta, 2012; European Commission, 2013; Korthagen, 2004; Struyven & Ceulemans, 2012). This interactive process starts with a teacher's professional dilemma or concern and the intention to respond meaningfully to this situation (Korthagen, 2017; Loughran, 2002). In order to achieve this, the teacher needs knowledge and skills that match their concern and the attitude to reflect in and on actions striving for change (Korthagen, 2017; van Braak et al., 2008). In this process, interaction with the student and the student's ecological system is conditional. This requires ownership for actors and educational partnership with all the parties involved (Pameijer & van Beukering, 2007; van Meersbergen & de Vries, 2017). The more complex the situation is, the more important it becomes to use the process of intrapersonal adaptation as a professional attitude. As specialists in gifted education should be able to work at the four different levels in the pyramid of support, it is relevant for them to be educated in a way that stimulates the process of intrapersonal adaptation to develop that professional attitude (van Gerven & Hoogenberg, 2011). In Chapter 3, the notion of skilledge is introduced as the individual's developmental result of this process of intrapersonal adaptation. Intrapersonal adaptation takes place if knowledge and skills are integrated and empowered by reflection in and on action. Knowledge can be taught. Skills can be developed. Skilledge is the result of personal reflection on contextually applied knowledge and skills; therefore, skilledge is, by definition, different for all educational professionals (van Gerven & Hoogenberg, 2011).

To prevent competencies being used as unidirectional checklists, the criteria for stipulating competencies by Merriënboer et al. (2002) should be used. The authors conclude that competencies are contextually bound. They are a coherent, inseparable cluster of knowledge and skills, attitudes, qualities, and apprehensions, and they are action-oriented. They show an interdependency with each other. Competences are learnable and should be stipulated in a way that makes them sustainable. Hence, competencies can best be described in a matrix, visualising how they are intertwined. The necessity for educational professionals to develop skilledge affects the education of future specialists in gifted education. Their education should aim for the development of skilledge in the field of gifted education and in their individual professional context. This requires that teacher educators develop a curriculum that aims for exactly that. The implication is that a directive approach, where teachers are taught to think, act and conform to one dogma on giftedness or education is no longer valid (Bologna Working Group on Qualifications Frameworks, 2005; Diery et al., 2020; van Braak et al., 2008). The education of specialists in gifted education should introduce them to a wide spectrum of theories and approaches. They should be educated to become critical thinkers, able to identify the knowledge and skills necessary for their specific professional context and matching their professional normative concepts by using their ability in pedagogical judgement.

The previously presented perspectives on teacher education affect the construction of the competency matrix. They show the need to combine general knowledge and skills in education with knowledge and skills inferred from theories of giftedness and gifted education. The so-stipulated competencies should offer ample opportunities for the specialist to decide what they consider to be *good* (Biesta, 2012). To prepare the competency matrix to be useful in daily practice, both to build the curriculum for the education of the specialist in gifted education, and to provide a framework for specialists to calibrate their professional behaviour, a translation of each competency into examples of knowledge and skills is necessary.

11.3 Competencies, Knowledge and Skills Indicators

Research question three focuses on the knowledge and skills the specialist in gifted education needs that were pointed out as considered essential for the optimal development of gifted students by the international professional community and international research on gifted education. Again, sub-questions are needed to answer this question: (a) How should we stipulate the competencies to meet the Dublin Descriptors for professional behaviour? (b)

What knowledge and skills can be deduced from international research on gifted education, considering the international educational standards for specialised teachers? and (c) How do the competencies fit into the Dutch and Flemish perspective on education in general?
The five domains considered to be relevant for gifted education (theories on giftedness, seeing educational needs, understanding educational needs, meaningful responses to educational needs, and assessing the results of interventions) were combined with the five domains inferred from general teaching standards (pedagogical, didactical, managerial, communicative, and research and innovation). The so-constructed matrix was used to analyse theories on gifted education. Each domain in the matrix is introduced by an overarching competency combining both axes of the matrix. Using the criteria for the stipulation of competencies by Merriënboer et al. (2002) and combining those criteria with the Dublin Descriptors, meant that the wording of the matrix needed to leave ample opportunity for each specialist in gifted education to interpret these competencies as matching their ability for critical thinking and educational judgement (Onderwijscoöperatie, 2014a, 2014b; Vlaamse Regering, 2008). For instance, if there were five theories available on constructing a curriculum for gifted learners, the competencies needed not to describe what theory to use, but that the use of one of these available theories should underpin the specialist's actions. Consequently, the complete matrix cannot be seen as the ultimate standard for gifted education. It should be seen as a coherent set of recommendations complementary to general teaching standards, that gains its value when they are applied in the context of a specific theory and specific practice. Before addressing points for discussion, results of the theoretical review and the transfer to the competency matrix are summarised.
For domain 1, the overarching competency was stipulated as "the specialist has an in-depth knowledge of the theoretical perspectives of giftedness and gifted education and is able to interpret this knowledge contextually in their professional practice". In practice, it requires that the specialist understands and envisions how their perspective on giftedness influences their ability to see, understand, and respond to the pedagogical needs of gifted students (Borland, 2012; Dai & Chen, 2014; Kreger-Silverman, 2013; Leavitt, 2017; D. Matthews & Foster, 2009; Touron & Freeman, 2017). The specialist should create a coherency between the concepts of giftedness, educational objectives, and educational programming, and the intervention strategies to be used. They need an understanding of how their perspective on giftedness influences the organisation of the educational programme provided (Dai & Chen, 2014; Wood, 2018). Matching the gifted student's needs should be based on interaction with the educational partners (Broersen & Spreij, 2009; Pameijer & van Beukering, 2007; van Meersbergen & de Vries, 2017; Ziegler et al., 2017). This all requires that the specialist relates their in-depth knowledge of giftedness to perspectives in current society and more specifically to the current perspectives on education.
In the other domains, current educational paradigms determining the educational practice in The Netherlands and Flanders are more clearly represented in the text of the competencies. This was conditional for establishing a close connection between general education and gifted education.

Without this connection, in both countries the position of gifted education would remain feeble instead of gaining a more stable position.
The overarching competency for domain 2, focusing on seeing educational needs, states that the specialist takes on a change and solution-focused approach to profiling the educational needs of gifted students (Cauffman & van Dijk, 2009; Coleman & Johnsen, 2011; de Shazer et al., 2012; Pameijer & van Beukering, 2007; van Meersbergen & de Vries, 2017). In this approach, all actors in the student's ecological system are seen as equal educational partners (Broersen & Spreij, 2009). Seeing educational needs requires that the specialist be able to explore the student's strengths and weaknesses within the intrapersonal and interpersonal developmental domain (Freeman, 2008; Neihart et al., 2016; Rinn, 2018; Rinn & Majority, 2018). The ability to assess the student's academic development is conditional for the competences stipulated in domain 4: meaningful educational responses (Ahlers, 2017; Coleman & Johnsen, 2011; Hoogeveen, 2017a, 2017b; Johnsen et al., 2016). To profile the student's educational needs, the specialist should use a systematic approach that contributes to the opportunities to gather all the information needed to draw informed and accurate conclusions. Educational partnership provides the opportunities to gather information and to interpret it contextually (Pameijer, 2012; Pameijer & van Beukering, 2007; van Meersbergen & de Vries, 2017). Therefore, educational partners should be treated as equals and their information regarding the student should be valued as such. Seeing educational needs requires that the specialist is able to set up a personal toolkit for profiling educational needs. The specialist should be able to use the instruments in their toolkit flexibly; matching the situation they are used in and depending on the specific issues that need to be assessed (de Bruin-de Boer & Kuipers, 2011; Houkema, 2009; Johnsen et al., 2016; van Gerven & Drent, 2012).
The importance of the student's ecological system becomes even clearer in domain 3, focusing on the specialist's ability to understand the educational needs they observed for their students. Educational needs must be interpreted in their ecological context (Ziegler et al., 2017). The actors in the student's ecological system interact with the student's strengths and weaknesses (Pameijer & van Beukering, 2007; van Meersbergen & de Vries, 2017). They facilitate the student's development and behaviour by their own actions. The ability to identify this specific interplay creates the necessary understanding for meaningful responses. From a more didactical perspective, the specialist should be able to explain how interaction between the curriculum offered and the student's cognitive style influences their academic development and instructional needs (Houkema, 2009; Renzulli, 2012; Roberts & Inman, 2013). The educational needs of gifted learners are of consequence for the specialist's classroom management (Kuipers, 2017). Understanding the impact of the student's needs on the specialist's actions offers the chance to explore how these needs can best be met. Understanding educational needs is not limited to the specialist in gifted education but should stretch out to all partners in the ecological system. Therefore, a specialist, aware of the importance of including the student's ecological system when designing meaningful responses, should be able to explain to the actors how they contribute to (and affect) the student's

development. Intense collaboration with the student's ecological system, therefore, requires well-developed communication skills (Broersen & Spreij, 2009). The final competency in this domain addresses the specialist's ability to use theory of giftedness and the change and solution-focused approach to interpreting the results of the exploration of the student's educational needs in the context of educational design.
The specialist competencies for domain 4, responding meaningfully to the educational needs of gifted learners, are supported by competencies in the previous domains. Meaningful responses can only be designed if the need for the response is interpreted within the context in which they are displayed and observed (Pameijer & van Beukering, 2007; van Meersbergen & de Vries, 2017). This means that, for the development of these responses, learning process and results and all that affects either one or both are equally important. From a pedagogical perspective this requires the specialist's ability to encourage a student's ability to create a balance between effort, learning, and achievement (Dweck, 2006, 2017; Heacox & Cash, 2019; Kieboom & Venderickx, 2017b; Siegle et al., 2017). Therefore, the specialist should be able to adjust the curriculum in such a way that their students have to reach for their zone of proximal development in all developmental domains (Bakx et al, 2016; Janson, 2017; Tomlinson et al., 2009; Winstanley, 2010). Designing education focused on the student's zone of proximal development creates the need for instruction and support. This calls on the specialist's ability for effective classroom management (Kuipers, 2017; Oostdam, 2009). In domain 3, it was explained how understanding of these needs leads to an understanding of what is needed in the specialist's classroom management; in domain 4, this understanding is transferred to the actual adjustments needed in practice. In this domain, the final competency considers the ability of the specialist to underpin their interventions with current theories on education and giftedness.
The fifth and final domain addresses the importance of the assessment of response to interventions (Coleman & Johnsen, 2011; Johnsen et al., 2016; Pameijer & van Beukering, 2007; van Meersbergen & de Vries, 2017). In the change and solution-focused approach, pedagogical judgement is seen as an overarching quality for educational professionals. They should be able to not only respond to educational needs and to strive for change but, above and beyond all, it is important that the professional should be able to assess whether the changes that are achieved, and the solution implemented, contribute positively to the student's development. This is a moral judgment. The core question is as simple as it is complex: is the achieved change a "good" change (Biesta, 2012)? This question cannot be answered without monitoring the process of responding to educational needs and interpreting the results of one's actions. This requires an attitude of reflecting on action (Korthagen, 2004, 2017; Loughran, 2002; Onderwijscoöperatie, 2014a, 2014b; Vlaamse Regering, 2008). In theories of giftedness, self-actualisation is considered either to be a result of education or the motor for education (Kreger-Silverman, 2013; Subotnik et al., 2011). Different theories of giftedness have different perspectives on this notion. However, for the educational practice of the specialist in gifted education it does not matter if self-actualisation is seen as a result or as a motor. In either case,

the student's intrapersonal and interpersonal development affects the student's ability for self-actualisation (Trail, 2011). Hence, the specialist should be able to monitor this development. Theories in gifted education stress the importance of the academic development of gifted learners (Gagné, 2010a; Heacox & Cash, 2019; Hoogeveen, 2017a; Kreger-Silverman, 2013; Renzulli, 2012; Roberts & Inman, 2013; Sternberg, 2009; Subotnik et al., 2011; Tomlinson et al., 2009; Winstanley, 2010). Consequently, the cycle of seeing, understanding, and responding to educational needs, is perpetual. Monitoring the student's academic achievements and relating these achievements to the level of support given during the learning process, offers opportunities to discover how the student's educational needs have developed (Coleman & Johnsen, 2011; Hattie, 2013; Pameijer & van Beukering, 2007; van Meersbergen & de Vries, 2017; Wiliam, 2011). From a managerial perspective, documenting these developments and results should take place periodically to keep the cyclical process of seeing, understanding, and responding to educational needs ongoing. In order to do this, the specialist should be able to communicate results to all the actors involved, in a way that provides new anchors for future actions (Pameijer & van Beukering, 2007; van Meersbergen & de Vries, 2017). The competency of a specialist in designing a system for formative and summative assessment of their student's learning is conditional to providing a mental and procedural framework for seeing, understanding, and responding to educational needs (Coleman & Johnsen, 2011; Wiliam, 2011; Winstanley, 2010).

The combination of the five domains for gifted education with the five domains inferred from the general teaching standards in The Netherlands and Flanders, provided a framework to embed the theory of gifted education within a more generic perspective of education in general. Concepts such as the inclusive approach to education, responding to educational needs, and the necessity for optimal differentiation, pedagogical and didactical strategies based on social-constructivism, and a growing emphasis on evidence-informed teaching were all taken into account when stipulating the competencies. The transfer from competencies into knowledge and skills indicators provided the 300 examples for competent professional behaviour, divided into knowledge and skills.

However, it could be disputed that these competencies and their matching indicators only refer to gifted education and the needs of gifted learners. All students have educational needs; meaningfully addressing those needs is mandatory for high-quality education (Armstrong et al., 2010; Florian & Black-Hawkins, 2011; Oostdam, 2009; Winstanley, 2010). The difference between the educational needs of gifted learners and the educational needs of other learners can only be explained by theories of giftedness if, and when, authors are able to argue why gifted learners have different needs and what those needs are. However, as soon as we match the needs described in theories of giftedness to needs that are described based on the five major educational paradigms (see research question 1), an overlap between gifted education and general education becomes visible (Bakx, 2019; Borland, 2005; Doolaard & Oudbier, 2010; Florian & Black-Hawkins, 2011; Janson, 2017; D. Matthews & Foster, 2009; Tomlinson et al., 2009; Winstanley, 2010). In the five domains of the

competency matrix this means that the strategies used for general education are also applicable in the context specific to gifted education.
The translation of educational needs of gifted learners into educational strategies considered beneficial for gifted learners loses even more *giftedness-specific language* once the Dublin Descriptors are applied. The importance of developing the specialist's attitude to critical thinking limits the opportunities to describe what theories should be used, how they should be used and what should or could be considered as *good*.
In theory, the inclusive approach in Dutch and Flemish education strives for a situation where the educational needs of students are met without a diagnosis being conditional (Amsing et al., 2009; Bronneman-Helmers, 2008; Clijsen et al., 2007; Decreet betreffende maatregelen voor leerlingen met specifieke onderwijsbehoeften, 2014; Inspectie van het Onderwijs, 2009; Pameijer & van Beukering, 2007; Prodia, 2012; Wet passend onderwijs, 2012). This means that *being gifted* or at least *being officially acknowledged as gifted* should no longer form the condition for having access to specific interventions, regardless of whether they are provided for in the regular classroom, in special programmes or pull-out programmes, and so forth. In theory, this should result in a situation where educational needs define what educational responses are needed and how opportunities for optimal differentiation can be provided (Florian & Black-Hawkins, 2011; Sutherland & Stack, 2014; Winstanley, 2010). In reality, Dutch and Flemish schools are not there yet. They find it difficult to realise this way of seeing, understanding and responding to educational needs.
There is ample room for uncertainty in a situation where an individual student's profile interacts with the ecological system and defines both the need and the opportunities to respond meaningfully to those needs. In this desired approach, practitioners miss the false certainty that a label provides. The gap between what is known and what is understood is still difficult to bridge.
The final research question of this thesis attempts to bridge this gap by offering Dutch and Flemish stakeholders in gifted education the opportunity to select what they consider the most relevant and valuable examples of competent professional behaviour.

11.4 Stakeholders' Understanding of the Competencies of the Specialist in Gifted Education

The fourth and final research question aims to get a better understanding of stakeholders' understandings of the competencies of the specialist in gifted education. This empirical research took place amongst 354 participants, all self-declared stakeholders. They described their stakeholder's position in (gifted) education. Based on their descriptions, three main categories were defined: parents, educational professionals, and specialists in gifted education; respondents not belonging to any of these groups were united in an additional group. One could argue that all specialists in gifted education are educational professionals. However, not all educational professionals are also specialists in gifted education. Therefore, these two groups were separated, based on grouping of the stakeholder respondents by what they declared to be the best

match for their position. The central question for this part of the research is, what do the stakeholders consider to be the most relevant and valuable examples of competent professional behaviour?
The 300 indicators used as these examples were equally divided into knowledge and skills indicators. They were categorised into six groups: (a) stimulating academic development, (b) stimulating the effective use of different cognitive styles, (c) stimulating interpersonal functioning, (d) stimulating intrapersonal understanding, (e) stimulating self-actualisation, and (f) general knowledge and skills matching those of the (S)ENCo. The indicators were presented in sets of six, matching one of the 25 competencies. Respondents were asked to select three indicators, per competency, that they considered best illustrated that competency.
Data analyses focused on two sub-questions: (a) Do different groups of stakeholders differ in the emphasis on specific indicators, and if so, what differences can be found? (b) Are there noticeable differences between Dutch and Flemish stakeholders regarding the emphasis on specific knowledge and skills indicators, and, if so, what differences can be found? The results of the empirical part of the research are described in detail in Chapters 5-9 and at a more general level in Chapter 10.
The theoretical review did not suggest that some indicators would be more relevant and important than others. In theory, all indicators should be equally important. It was assumed that, even though respondents were forced to select their personal "top three" for each competency (separately for knowledge and skills indicators), in the end, all indicators would be equally distributed. However, at all three levels of data analyses, differences between the separate groups of stakeholders were found. These differences not only answer the question of what stakeholders consider to be relevant and valued professional behaviour of the specialists in gifted education, but also provide an insight into what stakeholders consider to be important in gifted education.
The questionnaire was constructed in a way that stakeholders had the opportunity to provide additional information about their choices and suggestions as to what items were missed for each domain. The number of respondents that used this opportunity was very limited (17). These respondents took 2-3 times the opportunity to provide additional information and suggestions. There were no differences between the number of respondents using this opportunity between the separate domains. Their additional information was limited to one or two lines per respondent. Each suggestion was followed up to check if an indicator covering the suggestion was already included in the list of indicators. There were no suggestions that had not already been included in the list.
At the most generic level, differences between groups of stakeholders could be identified per category of indicators. Parents selected indicators aiming for intrapersonal understanding significantly more frequently than the other groups of stakeholders. Although academic development also showed high selection rates amongst parents, teaching their children who they are, how they can understand themselves, and how that can be beneficial later in life was by far considered the most important skill for the specialist in gifted education. These

selection rates match the currently observed trend in The Netherlands and Flanders wherein the notion of intra-personal understanding is seen as an important goal of gifted education. Easily accessible literature for parents and teachers focuses strongly on this aspect (Houkema, 2017; Kreger-Silverman, 2013). At this same generic level, teachers showed the highest selection rates for indicators aiming for intrapersonal understanding as well. However, the difference between their rates for this category of indicators compared to their selection rates for other categories was not as extreme as that found amongst parents. Specialists in gifted education had the highest selection rates for the category of indicators stimulating the gifted students to self-actualisation. Again, for teachers, easily accessible literature on gifted education focuses strongly on this aspect as an educational goal. Self-actualisation can furthermore be recognised in the stronger emphasis on social-constructive educational strategies. Self-regulation, active responsibility, becoming reflective learners, and striving for lifelong learning are all aspects contributing to self-actualisation. These aspects form key elements in theories of social-constructive learning and 21st-century skills (Bambs et al., 2016; Marzano & Kendall, 2008; van der Donk et al., 2015; Vermeulen & Vrieling n.d.; Ylvisaker & Feeney, 2014).
Differences between the groups of stakeholders become more visible and more nuanced once the level of data analyses becomes more in-depth, showing the results in more detail. The results suggest that stakeholders selected indicators matching their role and responsibility in the upbringing and education of gifted children. Parents stayed close to their responsibility of advocating in the best interests of their children. Teachers stayed close to their responsibility to the education of gifted students in the context of their classroom. This means that they advocate not only for the interests of their gifted students but also for the interests of all other students in their classroom. Specialists in gifted education, however, differ from both the other groups of stakeholders. For some competency domains they select identical indicators to teachers and, for other domains, their selection rates are closer to the selection rates of parents. This suggests that they advocate at three levels of interest for gifted learners: at an individual level, at the level of the gifted learners in their regular group and at the level of the gifted learners in the entire school.
The stakeholders of the two countries also differed in their selection of the most relevant and best examples. Flemish stakeholders showed a higher preference for indicators stimulating interpersonal and intrapersonal development. Dutch stakeholders showed a higher preference for indicators stimulating the student's ability to use different cognitive styles effectively. They also selected indicators matching the more generic supportive role of the specialist in gifted education.
Although the sub-questions focused mainly on differences between groups of stakeholders, data suggests that there is also common ground between them. The observed common ground offers opportunities to set up a list of indicators that had the highest selection rates amongst the three main groups of stakeholders. Common ground can be found even more clearly at the most detailed level for data analysis. At that level, frequency rates were presented per competency and compared per group of indicators. A total of 93 indicators were selected by all groups of stakeholders in their top 3. Indicators focused on the

competency of specialists to stimulate the gifted learner's interpersonal functioning were the least selected indicators on this list. This result is particularly interesting as both in literature reflecting empiric research as in more practically based publications there is a wider discussion on this subject of the social and emotional needs of gifted learners (Cross, 2012; Cross, 2018; Davis et al., 2014; Freeman, 2008; Freeman, 2019; Kieboom & Venderickx, 2017; Matthews et al., 2018; Neihart & Yeo, 2018; Neihart et al., 2016; Peterson, 2009). Some studies provide evidence that gifted students have different social-emotional needs from other children. However, there are also studies suggesting that these needs are no different at all.

The distribution of the indicators over the separate categories of A to F matches the findings at the most general level. Indicators in categories ADE are more represented than indicators in category BCF.

The overarching answer to the empirical research question is that indicators aiming for the development of the specialist's knowledge and skills to stimulate the academic achievement of gifted learners, their intrapersonal understanding and the ability for self-actualisation were selected more than indicators aiming for the stimulation of the effective use of different cognitive styles, interpersonal functioning and the more generic competencies around teacher support and matching the role of the SENCo.

If these results are interpreted as not only what stakeholders would prefer to be the knowledge and skills of the specialist in gifted education, but also their preferences for gifted education in general, it provides a starting point for discussing these preferences with the educational partners.

11.5 Limitations and Directions for Further Research

This research offers an insight into what could be considered recommendable for the education of future specialists in gifted education; however, there are also some limitations that need to be addressed.

11.5.1 Limitations

First, calling for respondents meant that everybody who considered themselves as stakeholders of gifted education was encouraged to reply. Although the definition of who may be considered to be stakeholders of gifted education is not limited to teachers and parents of the gifted and specialists in gifted education, almost all respondents belonged to that specific group of stakeholders. Even though approached via their professional channels, school management, policymakers, and teachers not being aware that they may have gifted learners in their classroom did not respond. That implies that the conclusions of this research cannot be generalised towards all involved in primary education. The results of this research do apply to the distinguished groups all being directly involved in the education of gifted learners.

Secondly, at the start of this study, several effects regarding the construction of the questionnaire and the research process were considered. Different ways for data collection were considered as well. Dividing the data collection process into several stages, such as sending respondents every week the questions matching one competency domain, was one option. Using this approach was declined as dividing up the process creates a risk that large groups of respondents might decide to stop participating after submitting the first domain questionnaire. As a result, this would mean that data regarding the last domains could be less reliable than data gathered for the first domains. A second reason to decline this approach was that to gather all the necessary data, chasing respondents to submit their questionnaires may put them off once they had become engaged. Another approach that was considered as an option for data collection was dividing the questionnaire into separate parts, with a certain overlap. By dividing the respondents as well into separate groups, each group could be asked to reply to that specific part of the questionnaire. However, to get reliable results, it meant that at least four times as many respondents were needed. As it was already complex to get as many people as possible interested in the research as such, this option seemed not to be realistic. The current approach represents the balance between what was maximally desirable and maximally realistic and reliable.

Thirdly, the issue of socio-economic and cultural influences needs to be mentioned. This research took place in the socio-economic and cultural context of Dutch and Flemish education. The findings cannot be generalised one-to-one for other countries. Gifted education is positioned at the centre of the educational paradigms influencing Dutch and Flemish general primary education. Although these paradigms can be recognised in other western countries as well, different accents on the interaction between these paradigms may be observed in different countries. That does not mean that these paradigms are not relevant in another socio-economic and educational context, but that they may affect gifted education differently under different circumstances. Theories on giftedness have been interpreted in the Dutch and Flemish context of these five general paradigms.

The fourth issue to address is the impact of Dutch and Flemish teacher general education on the way the competences are stipulated (Onderwijscoöperatie, 2014a, 2014b; Vlaamse Regering, 2008). This issue is a two-edged sword. Using the Dublin Descriptors may circumscribe the use of the matrix and the stipulated competences. Some may find the way that they are described in the current matrix too abstract or even too open to wide interpretation (European Higher Education Area, 1999). They may want to hear at a more concrete level what to know, what to think, what to do, and what results they should strive for. However, that can only be done if we lower the level of the education of the specialist of gifted education by using a lower level of the Dublin Descriptors. The lower the level, the more concrete the matrix of desired competences can be. Lowering the level of the Dublin Descriptors used would degrade the educational level of the specialist in gifted education. This would create a professional distance between the specialist in gifted education and the other educational specialists (LBBO, 2019, n.d.-a).

A fifth limitation is caused by the relationship between the indicators and the specific competency in which they were presented. Respondents were asked to select indicators matching specific competencies. That means that there are limitations to the generalisability of the findings per competency. Although an indicator may not have been selected often enough to end up in the three most frequently selected indicators within a group of stakeholders, that does not mean that this example of knowledge and skills is not considered important. It tells us, that, for that specific competency, other indicators under the circumstances in which the respondents selected them were considered to be a better match. Trends observed at the more general levels were analysed and findings can be generalised as long as it is borne in mind that, under different circumstances, stakeholders may make different selections.

11.5.2 Critical reflections

Reflecting on the process of this research project, if I were to do a study into this topic again, I might make different choices. It would be interesting to compare the results of the analysis of empirical studies on what is suggested to be essential for gifted education and more practically based literature being used by the community of professionals. For this current study, this comparison has not been included as most of the selected publications are based on the empirical studies that were also selected for this study (see also Figures 9 and 10). Major differences between recommended strategies in both types of publications were not expected. This aspect is now taken on as a suggestion for future research.

As stated before, the selection of this practically based literature was a deliberate choice, not only motivated by the limited amount of peer-reviewed literature on Dutch and Flemish gifted education but also because the community of practitioners in gifted education uses these publications to determine what they consider to be as good for gifted education. It is not said that these publications represent the absolute 'good' for gifted education, but educational professionals rely on these publications. Having a teaching degree, graduated at least at a bachelor level, one might expect that based on their initial teacher education, educational professionals are able to critically review the results of applying the strategies in their professional context as presented in these publications. It might be interesting to explore this assumption and to see to what extend Dublin Descriptors that underpin this goal of their initial teacher education (being able to critically review information and reflect on its implications) have become their professional attitude (Diery et al., 2020; Marzano & Kendall, 2007; van Braak et al., 2008).

11.5.3 Recommendations

The construction of the matrix using a systematic literature review and the empirical research using a questionnaire aimed to get a better understanding of how stakeholders in gifted education understand the competencies of the specialist in gifted education. There is no equivalent known within the field of gifted education that can be used to compare the results. A recommendation would be to do a wider exploration. Firstly it would be relevant to compare if the analysis of empirical research would lead to a selection of similar knowledge and skills indicators as the analysis of more practically based literature. For the current study, it has been decided to combine both types of sources. The now suggested comparison may be of added value not only for anchoring the selection of knowledge and skills indicators but also for the valuation of the more practically based literature.

Secondly, it would be interesting to explore to what extent qualified teachers critically review information they use for outlining their educational strategies. This attitude is included in the Dublin Descriptors. Research by van Braak et al. (2008) and Diery et al. (2020) suggests that teachers rely more on their personal professional experience than on empirical research. That may result in a professional attitude wherein practice-based publications that match their personal professional experience are less critically reviewed than publications that do not match their professional experiences. That might affect how these professionals review literature on gifted education as well.

Thirdly, the current study raises a new hypothesis (for instance about the influence of the role of the stakeholder on their selection of categories of indicators) that cannot be addressed with the current results. Hence it is relevant to enrich the results of the current study by a qualitative study by offering stakeholders to select knowledge and skills indicators, provide an additional opportunity to nuance their choices, and to place their choices in the wider perspective of the entire matrix and their specific role in gifted education. This would also provide the opportunity to include gifted students as stakeholders. It would be interesting to explore if their perspective on teacher competencies matches the understandings of what adult stakeholders see as adequate for specialists in gifted education.

Fourthly, it would be relevant to see if stakeholders in general education would make similar selections if the matrix and the knowledge and skills list were slightly adapted and neutralised in language, removing the references to the gifted learner. Comparison between stakeholders of gifted education and stakeholders in general education might provide greater insight into the extent to which stakeholders of gifted education differ from other groups. We might even find that they do not differ in what they consider to be good indicators of competent professional behaviour for educational specialists.

Fifthly, exploring whether stakeholders of gifted education in different countries would make similar choices is recommended. If these choices differ, it would be interesting to see what socio-economic and cultural differences contribute to observed differences. For the future of gifted education, this could provide opportunities for a context-oriented approach to international recommendations for gifted education.

Last but not least, it would be interesting to analyse how different teacher education programmes for educating the gifted provided and implemented in different countries already have included the competencies pointed out in the matrix in their curriculum. Results may provide wider support for the competencies included in the current matrix.

11.6 Implications for Gifted Education and Teacher Education Programmes

The ultimate aim of this project was to find a way to increase the quality of gifted (primary) education by developing a competency matrix for the education of specialists in gifted education. This matrix can be considered as a mental framework, finding its roots in general educational paradigms and teacher education but offering opportunities to specify general strategies for a specific focus group. The results of this research offer opportunities for constructing the curriculum of teacher education courses. The matrix is not a professional standard but a coherent set of competencies, translated into knowledge and skills that can be considered as recommendable for this specific topic in teacher education.

The matrix can be used to compose a curriculum for teacher education focusing specifically on gifted education. This would be a similar approach as is taken to use the recommended standards for professional competencies for SENCO's and other educational specialists. To educate teachers to become specialists in gifted education, the described competencies should be included and developed in their educational programme. Institutes that provide teacher education courses for those who want to become specialists in gifted education, should consider whether their course matches the entire matrix at level 7 of the Dublin Descriptors. If their course does not cover the competencies as described in the matrix and/or aim for the professional attitude of level 7 of the Dublin Descriptors, they may want to consider whether their programme matches up to the title of specialist in gifted education. They may conclude that they need to raise the bar. It is up to each institution to take professional responsibility in using the word *specialist* with care.

However, not all teachers who are interested in gifted education want to become specialists. They may want to develop a level of understanding that helps them to be a better teacher for gifted learners in their classroom without the in-depth knowledge and skills needed by specialists. That professional attitude should be respected and praised. If every student had a teacher who wanted to become a better teacher by taking up personal continuous professional development, education would improve step-by-step, because these teachers would become gradually better in responding to the educational needs of all their students.

The competency matrix provides a theoretical framework for setting up small courses focusing on specific competencies that these teachers want to develop. That matches an approach based on personal concerns as described by Korthagen (2017) and advocates ownership for the content to be learned. Teacher educators can select competency domains or even individual competencies to include in their teacher education programmes. The selected competencies can be used as learning objectives. The list with indicators can be used to address specific learning contents, matching the needs of the participating teachers. The results of the empirical research provide a framework for focusing on an enriched approach to education that could be beneficial for gifted students. Educational professionals selected indicators that they consider to be supportive for gifted education. Their choices relate to the five educational paradigms that affect education in general. Matching a curriculum based on the personal concerns of teachers with specific indicators selected by them as relevant for their practice will raise the bar for short teacher education courses. It creates personal awareness of how to become a better teacher based on their personal ability to review what would be beneficial for them as teachers and therefore aims for ownership regarding new knowledge and skills.

Based on their personal level of expertise, these teacher educators can select literature that provides information to match the knowledge and assignments needed to develop the selected skills.

Including the understandings of all stakeholders of what they consider to be important for the specialist in gifted education, offers the opportunity to create a better understanding between different groups of stakeholders (Broersen & Spreij, 2009). The results of the empirical research show that different stakeholders have different understandings of what they consider to be relevant for the role of specialists in gifted education. Similarities and differences in the emphasis of stakeholders on different categories of knowledge and skills indicators are of interest. Making teachers aware of these different understandings can contribute to an attitude that promotes a nuanced approach of gifted education. It offers the opportunity to see things from a different perspective, which is relevant for building educational partnerships. For curriculum construction, it means that teacher educators should explain to their students how to use these different perspectives in a way that is beneficial for gifted education. If the specialist in gifted education is able to see, understand and respond meaningfully to these differences between all the educational partners involved, they can co-create gifted education in a way that allows its goals and process to be valued by all involved.

Specialists in gifted education can describe their personal practice, using the list of indicators to frame their actions in the context of the matrix. Teacher educators can use the list to see how the specialists understand their role in gifted education. The specialists' understanding of their role affects choices made in daily practice for knowledge and skills to be used. What knowledge and skills can be demonstrated by the specialist in gifted education depends highly on the cases with which they are confronted and the educational context in which these cases take place (Biesta, 2012; Ceulemans et al., 2016; Struyven & Ceulemans, 2012). Not being able to recognise a specific aspect of knowledge or a specific skill does not mean that the specialist considers that specific aspect of knowledge and skills as irrelevant in their role as specialist. It may imply that the specialist has critically reviewed the situation and decided to act differently as the chosen actions seemed to be a better fit for the gifted learner's needs at that time and under those circumstances. Neither can it be expected that the specialist is able to show all the indicators for competent behaviour in one or even several cases if the case and situation do not call on that knowledge and skills.

To use the matrix and the knowledge and skills list as a formal assessment tool for higher education requires additional validation. However, the community of practitioners may have a slightly different opinion when it comes to using the matrix for assessment in their professional context. The Dutch association for

SENCo's and educational specialists (LBBO) uses for each of their professional focus groups (SENCO and Educational Specialist Complex Behaviour) an assessment procedure. These procedures are supported by Dutch schools for teacher education that are providing continuous professional development programmes for each specialism. The professional standards described are similar to those in the matrix that has been constructed for this current study on gifted education (LBBO, 2019; LBBO n.d.-a). In their publications about these standards, there is no information mentioned that these standards have been validated, nor has a theoretical framework been published that underpins the assessment procedure. Hence the community of practitioners may decide that aspects of this matrix can contribute to their assessment procedure for the focus group of specialists in primary gifted education.

Becoming a specialist is not only to do with what one knows. It is also to do with a professional attitude matching the level of the specialists in other fields of expertise. The current matrix addresses both the knowledge and skills and the professional attitude needed to underpin this high level of professionalism.

Teacher education is a profession on its own merits (Creemers, 2002; Lunenberg et al., 2014). This implies that the quality of teacher education and the quality of the education of specialists is affected by the quality of the teacher educator. That teacher educator should be able to meet the standards for teacher education as well as having expertise in the domain of gifted education. The quality of teacher education affects the quality of education and the quality of the education of specialists in gifted education affects the quality of gifted education. This matrix can contribute to the quality of the education of specialists in gifted education and, therefore, raises the bar for gifted education.

11.7 The future of gifted education

The educational landscape is changing rapidly. The need for labels as conditional for providing appropriate education diminishes. Responding to educational needs becomes more solution-focused, having a stronger emphasis on the needs of individual learners. This creates opportunities for educational approaches that are more inclusive and based on the diversity of the population of different learners in each classroom. This is not a local change, but this shift can be observed all over the western countries.

These changes affect gifted education. It becomes less and less defensible to create educational opportunities that are only accessible if a student is identified as being gifted. A lot of these opportunities might be beneficial for all learners, depending on their individual capacities. In reality, inequity crops these opportunities. Education involves not only competent teachers and students that are willing to learn. Education involves socio-economic opportunities for learning as well. Without facilities for learning, there will be many children missing out on the opportunities to develop themselves commensurate to their innate capacities. If we do not stay aware of the effects of inequity on education and act upon them, gifted education will miss important chances for change.

In education, approaching students based on a dichotomy of strengths and weaknesses is still an often-used approach. However, strengths and weaknesses can only be determined if we relate an individual student's profile to what is determined as the set standard or set objective for education. This ignores that what can be considered as a strength in one situation may be a weakness in a different context. Thinking in strengths and weaknesses represents a normative mental framework to what should be or what should not be. Moving away from this perspective means there are no 'difficult' students and that there is no 'complex behaviour'. Difficulties and complexity refer to our personal ability to cope with the educational needs of our learners and our personal mental framework of what we consider to be 'good'. The more we miss the competence to respond meaningfully to those needs, the more we experience these needs as difficult and complex. Gifted education could benefit from an approach wherein what is considered to be 'good for gifted learners' can be seen in a wider perspective of how do the suggested strategies contribute to the qualification, socialisation, and subjectivation of this learner.

2020 was the year of a global pandemic. This pandemic changed the world and there is no way back. This forces us to review the way we educate our next generation. We not only have to review our educational goals and what we teach but above and beyond all, we have to review the way how we educate this generation. Facing a reality in which learners no longer have been able to come to school every day in a traditional way, implies that we have to find new ways of teaching. In a single group of 25 teachers participating in a teacher education programme to become specialists in gifted education, 23 teachers reported that since the pandemic disrupted our lives, their programmes for gifted learners have been paused or even stopped entirely. As these teachers come from different schools and work in different regions in The Netherlands, it seems likely that their experience does not stand on its own. If this is a trend that can be observed in other countries, we may need to consider that we have to reinvent gifted education, no longer seeing it as distinct and different from education in general.

It seems likely to expect that in the foreseeable future long-distance learning, electronic learning, has to be flexibly combined with learning in traditional ways and places and in the traditional context of schools. Blending all opportunities for learning may provide us with new chances to develop more personalised and more individualised educational programmes than we offer our students now. In doing so, qualifying as being gifted may no longer be a relevant criterium for selection. As soon as we are to change our understandings of what is good for gifted education into what is good for the development of this learner under the current circumstances in this situation, we might be able to strive for a more equitable approach of education in general and gifted education in particular. And that would raise the bar for gifted education even more.

Samenvatting

In Nederland en Vlaanderen is sinds het eind van de jaren tachtig van de vorige eeuw een snel groeiende groep personen betrokken bij het zogenoemde begaafdheidsonderwijs. Er zijn inmiddels diverse instanties die zich richten op het opleiden van deze betrokkenen. Cursussen en opleidingen worden aangeboden onder titels als talentbegeleider, begaafdheidscoach, specialist hoogbegaafdheid, specialist begaafdheid, begaafdheidsexpert. De kaders van waaruit deze opleidingen werken, zijn breed en voor buitenstaanders niet altijd duidelijk. Ze verschillen voor elk opleidingstraject. De achtergronden van de deelnemende studenten/cursisten aan deze trajecten zijn evenzeer divers. Niet voor alle opleidingstrajecten geldt dat een onderwijsbevoegdheid een vereiste is voor deelname. Tevens zijn er sterke verschillen tussen het opleidingsniveau en de professionele achtergrond van de opleiders die deze trajecten verzorgen. Er zijn bijvoorbeeld maar weinig Velon-geregistreerde lerarenopleiders bij deze trajecten betrokken. Het uitstroomprofiel van deze nascholingstrajecten over begaafdheidsonderwijs varieert daardoor sterk. Er zijn grote verschillen waarneembaar in zowel de breedte als de diepgang en de betrouwbaarheid van de verworven kennis. Deelnemers aan de diverse trajecten laten een zeer ruime variatie zien in het door hen ontwikkelde vermogen om verworven kennis om te zetten in concrete toegepaste vaardigheden. Tot slot verschilt ook per nascholingstraject op welke aspecten van de professionele attitude er accenten ontwikkeld worden en in hoeverre opleiders bewust bezig zijn met de ontwikkeling van die professionele attitude bij de deelnemers. Na afloop van deze nascholingstrajecten scharen alle opgeleide personen zichzelf onder een vergelijkbare noemer: specialist begaafdheidsonderwijs. Alle betrokkenen streven daarbij naar hetzelfde: een hogere kwaliteit van het begaafdheidsonderwijs.

De kwaliteit van het begaafdheidsonderwijs wordt mede bepaald door de professionele competentie van de specialist begaafdheidsonderwijs. Als de competenties van de specialist begaafdheidsonderwijs naar een hoger niveau worden getild, zal dit een gunstig effect hebben op de kwaliteit van het begaafdheidsonderwijs in Nederland en Vlaanderen.

Maar wat is nu een specialist begaafdheidsonderwijs voor het primair onderwijs? Binnen welke professionele kaders begeeft deze specialist zich? Hoe zou het competentieprofiel van die specialist begaafdheidsonderwijs eruit kunnen zien en welke professionele attitude zou gekoppeld moeten zijn aan de term specialist begaafdheidsonderwijs? Wat betekent dat voor professionele eisen die aan de opleiders van specialisten begaafdheidsonderwijs gesteld mogen worden? Dit onderzoek richtte zich op dit vraagstuk.

Het doel van dit onderzoek was het opzetten van een competentiematrix die het kader kan vormen voor het niveau en de inhoud van het aanbod van opleidingen tot specialist begaafdheidsonderwijs. Het onderzoek moest leiden tot de omschrijving van de competenties voor de specialist begaafdheidsonderwijs waarvoor er in de toekomst onder betrokkenen bij het begaafdheidsonderwijs een breed draagvlak zou kunnen gaan ontstaan. Daarom heeft er naast een

breed literatuuronderzoek ook een onderzoek in het begaafdheidsonderwijs plaatsgevonden onder stakeholders (leraren, educatief specialisten, specialisten begaafdheid, ouders en anderszins betrokkenen). Begaafde leerlingen hebben geen deel uitgemaakt van het stakeholdersonderzoek. Dat heeft te maken met het volgende.
De specialist begaafdheidsonderwijs kan verschillende rollen vervullen. Dit varieert van groepsleraar zijn in een reguliere groep of een plusgroep, groepsleraar zijn van voltijds begaafdheidsonderwijs, of - vergelijkbaar met de intern begeleider - vooral gericht zijn op het begeleiden van de collega's al dan niet in een bovenschoolse setting. Er zijn veel taken en verantwoordelijkheden die zich geheel of gedeeltelijk buiten het zicht van de leerling afspelen maar daarmee niet minder belangrijk zijn voor het totaalbeeld van de competenties van de specialist begaafdheidsonderwijs. Omdat de leerling op die taken geen zicht heeft, is het niet realistisch de leerling een uitspraak te laten doen over wat dan competent gedrag zou zijn passend bij die taken.
Begaafde leerlingen zijn, zoals ook blijkt uit onderzoek van Bakx (2019), heel goed in staat om te verwoorden wat ze nodig hebben van hun *leraar*. Maar die leraar hoeft niet altijd een specialist begaafdheidsonderwijs te zijn. Uiteraard is het wel zo dat uitspraken die leerlingen in een onderzoek als dat van Bakx gedaan hebben, in de praktijk wél belangrijk zijn als de specialist begaafdheid de rol van groepsleraar vervult.

Het onderzoek viel in drie delen uiteen: (a) het scheppen van het onderwijskundig kader waarin begaafdheidsonderwijs in Nederland en Vlaanderen plaatsvindt; (b) de constructie van de matrix en de omschrijving van de indicatoren voor kennis en vaardigheden; (c) het onderzoeken van het perspectief van stakeholders op wat zij vanuit de aangeboden structuur de meest relevante voorbeelden vonden van professioneel competent gedrag van de specialist begaafdheidsonderwijs.

In het eerste deel van dit proefschrift wordt aan de hand van algemeen literatuuronderzoek beschreven wat de positie is van het begaafdheidsbasisonderwijs in Nederland en Vlaanderen. Vanuit vijf grote paradigmaverschuivingen die voor het hele onderwijs gelden, is verkend wat die verschuivingen betekenen voor de wijze waarop het begaafdheidsonderwijs vorm wordt gegeven. De vijf paradigma's betreffen (a) het principe van inclusief onderwijs; (b) de ontwikkeling naar veranderingsgericht werken; (c) de verschuivende theoretische visies op begaafdheid als zodanig; (d) de in toenemende mate sociaal-constructivistische aanpak van het onderwijs met het bijbehorende accent op de zogeheten 21e-eeuwse vaardigheden; (e) het groeiende belang dat wordt toegekend aan evidence-informed onderwijzen. Deze paradigma's beïnvloeden niet alleen direct het begaafdheidsonderwijs, maar ook indirect. De wisselwerking tussen de paradigma's onderling leidt ook weer tot gevolgen voor het begaafdheidsonderwijs. Daarmee wordt duidelijk dat begaafdheidsonderwijs niet gezien kan worden als een op zichzelf staande invulling van de educatieve behoeften van begaafde leerlingen.
Begaafdheidsonderwijs maakt integraal deel uit van het algemene onderwijs-

aanbod aan alle leerlingen op een school. De ontwikkelingen in het denken over begaafdheidsonderwijs kunnen derhalve niet los gezien worden van ontwikkelingen in het onderwijs in het algemeen.

Deze verkenning leverde vijf kennis- en vaardigheidsdomeinen op als thema's voor de competentiematrix voor de specialist begaafdheidsonderwijs: (1) generieke kennis over theoretische concepten over begaafdheid; (2) het zien van educatieve behoeften; (3) het begrijpen van educatieve behoeften; (4) het betekenisvol kunnen reageren op educatieve behoeften; en (5) het kunnen beoordelen van het effect van toegepaste interventies op de ontwikkeling van de leerling. Deze onderscheiden kennis- en vaardigheidsdomeinen vormen de horizontale as van de competentiematrix voor de specialist begaafdheid.
In dit deel van het onderzoek is ook verkend hoe het opleidingsniveau van de specialist begaafdheidsonderwijs zich zou moeten verhouden tot de opleidingsniveaus van andere specialisten in het primair onderwijs zoals de masters in Educational Needs (gedragsspecialisten, specialisten intern begeleiden, leren en innoveren, lees- en rekenspecialisten). Deze verkenning maakte duidelijk dat om je specialist begaafdheidsonderwijs te kunnen noemen ten minste een bevoegdheid in het (primair) onderwijs vereist is. Dit is een randvoorwaarde die voor alle educatief specialisten geldt; zij hebben allemaal hun bevoegdheid als leraar behaald. Deze bevoegdheid is ingekaderd in de eerste cyclus van de Dublin Descriptoren voor hoger onderwijs (niveau bachelor). De masters in Educational Needs worden daarna in hun vervolgstudie opgeleid op het niveau van de tweede opleidingscyclus van de Dublin Descriptoren voor hoger onderwijs. Als de specialist begaafdheidsonderwijs op hetzelfde niveau wil kunnen functioneren als deze andere educatief specialisten, is een opleidingsniveau vereist waarbij aantoonbaar op een professionele attitude wordt ingezet overeenkomstig die tweede cyclus. Concreet betekent dit dat de specialist begaafdheidsonderwijs altijd een gespecialiseerde leraar is die na het behalen van zijn bachelordiploma heeft doorgestudeerd op het niveau van een master. Door deze directe koppeling aan de bevoegdheid van leraar voor het primair onderwijs komen de generieke competenties van de leraar in beeld. Een vergelijking tussen Nederland en Vlaanderen maakte grote overeenkomsten in die omschreven competenties/beroepsbekwaamheden duidelijk. Door de bekwaamheidseisen uit beide landen te combineren werd een tweede reeks van vijf domeinen zichtbaar die het begaafdheidsonderwijs beïnvloeden: (1) pedagogische bekwaamheid; (2) didactische bekwaamheid; (3) organisatorische bekwaamheid; (4) communicatieve bekwaamheid; (5) bekwaamheid in onderzoek en reflectie. Deze vijf domeinen vormen de verticale as van de competentiematrix.

In het tweede deel van het onderzoek is vanuit de bestaande internationale literatuur over begaafdheidsonderwijs naar een invulling gezocht van de matrix. Zo zijn per domein steeds vijf competenties beschreven. De verschillende competenties hebben een sterke onderlinge verwevenheid. Ze zijn daarbij zodanig geformuleerd dat voor de toepassing ervan de professionele attitude op het niveau van de tweede cyclus van de Dublin Descriptoren is vereist. Dat betekent dat elke competentie ruimte overlaat aan de specialist om op basis van theorie en eigen inzicht evidence-informed tot een invulling van de competentie te komen. Op basis van ditzelfde literatuuronderzoek zijn 300 kennis- en vaardigheidsindicatoren beschreven. Elke indicator is een voorbeeld van wat competent professioneel gedrag zou kunnen zijn voor de specialist begaafdheidsonderwijs. Elk domein van de horizontale as van de competentiematrix is gedocumenteerd in een eigen hoofdstuk.

Het empirisch onderzoek richtte zich op het perspectief van stakeholders op deze kennis- en vaardigheidsindicatoren. Zij selecteerden per competentie drie kennis- en drie vaardigheidsindicatoren die zij als het beste voorbeeld van professioneel competent gedrag beschouwden. De resultaten van de stakeholders zijn vervolgens ingedeeld in vier subgroepen: educatieve professionals, specialisten begaafdheid, ouders en een restgroep. De restgroep was in verhouding tot de andere groepen stakeholders echter zo klein dat de resultaten weliswaar in alle tabellen in de bijlagen zijn opgenomen, maar dat ze niet in de lopende tekst zijn verwerkt. Door stakeholders in te delen in groepen konden overeenkomsten en verschillen in perspectieven geanalyseerd en getoetst worden. Waar dit relevant was, is tevens een vergelijking gemaakt tussen selecties van Nederlandse en Vlaamse stakeholders.
De resultaten van het empirisch onderzoek (uitgevoerd als derde component van het onderzoek) zijn vervolgens domeingebonden gedocumenteerd. Dat wil zeggen dat direct achter elk deel van het literatuuronderzoek voor het betreffende domein de resultaten van het empirisch onderzoek zijn beschreven. Per groep stakeholders en per competentie wordt een beeld geschetst van de selectie van indicatoren. Omdat de indicatoren in de context van een specifieke competentie aan de deelnemers zijn aangeboden, is op detailniveau deze context ook weer gebruikt om de resultaten te presenteren. Vervolgens worden overeenkomsten en verschillen in keuzes tussen de groepen stakeholders besproken. Tot slot is er een apart hoofdstuk waarbij de resultaten van het empirisch onderzoek op domeinniveau en uiteindelijk ook op het niveau van de gehele matrix worden besproken.

Het onderzoek maakt zichtbaar dat de selectie van indicatoren samenhangt met de rol die verschillende groepen stakeholders hebben in het onderwijs en de opvoeding van begaafde leerlingen. Elke groep stakeholders legt eigen accenten. Op een aantal punten zijn deze accentverschillen relevant en significant voor de onderwijspraktijk. Het inzicht in de onderlinge verschillen en overeenkomsten biedt voldoende aanknopingspunten voor het ontwikkelen van wederzijds begrip tussen de groepen stakeholders. Lerarenopleiders kunnen in hun opleidingsaanbod verschillen gebruiken om een positieve communicatie tussen de ouders en scholen te bevorderen. Dat komt de onderlinge verstandhouding ten goede en draagt daardoor bij aan de verhoging van de kwaliteit van het begaafdheidsonderwijs.

De competentiematrix en het overzicht van indicatoren kunnen door lerarenopleiders gebruikt worden om hun curriculum mee in te richten. Afhankelijk van de diepgang van het onderwijsaanbod dat zij willen verzorgen kunnen zij zich richten op de ontwikkeling van alle 25 competenties (startbekwaamheid voor de specialist begaafdheidsonderwijs) of op de ontwikkeling van competenties in deelgebieden (geschikt voor korte cursussen).

Specialisten begaafdheidsonderwijs kunnen de matrix en de indicatoren gebruiken om hun professioneel handelen te (her)ijken. Aan de hand van casuïstiek uit hun eigen praktijk kunnen zij met behulp van de matrix en indicatoren voor derden inzichtelijk maken wat de professionele onderbouwing is van hun handelen. Daarmee schept de matrix een professioneel kader dat past bij de beroepsgroep van specialisten begaafdheidsonderwijs. Dit kader is niet bedoeld als checklist waarmee eendimensionaal geselecteerd kan worden op "geschikt" of "ongeschikt" voor de functie of als specialist begaafdheidsonderwijs. Een dergelijke checklist laat altijd de contextgebonden beroepsuitoefening van de specialist buiten beschouwing. Daardoor biedt een checklist geen ruimte voor een door de specialist zelf toegepaste normatieve interpretatie van kennis en vaardigheden. Wat "goed" is in onderwijs kent vele gezichten en wordt bepaald door het samenspel van alle educatieve partners onder gegeven omstandigheden.

References

Ackerman, C. (2009). The essential elements of Dabrowski's theory of positive disintegration and how they are connected. *Roeper Review*, *31*(2), 81-95. https://doi.org/10.1080/02783190902737657

Adema, E., Loeffen, E., Miltenburg, E., & Voets, M. (2009). *Passend en competent* [Appropriate and competent]. KPC groep.

Ahlers, L. (2017). Compacten van de leerstof [Curriculum compacting]. In E. van Gerven (Ed.), *De gids: over begaafdheid in het basisonderwijs* [Guide for gifted education in primary schools] (2nd ed., pp. 143-156). Leuker.nu.

Ainscow, M., Booth, T., & Dyson, A. (2006). *Improving schools, developing inclusion*. Routledge.

Ajzen, I. (1991). Theory of planned behavior. *Organizational Behavior and Human Decision Processes*, *50*(2), 179-211. https://doi.org/10.1016/0749-5978(91)90020-T

Ambrose, D., & Sternberg, R. (Eds.). (2016). *Giftedness and talent in the 21st century*. Sense.

Amend, E., Schuler, P., Beaver-Gavin, K., & Beights, R. (2009). A unique challenge: Sorting out the differences between giftedness and Asperger's disorder. *Gifted Child Today*, *32*(4), 57-63. https://doi.org/10.1177/107621750903200414

Amkreutz, R. (2017, December 14). Nieuw in Vlaanderen: haal je master aan de hogeschool [New in Flanders: get your master's degree at a university of applied sciences]. *De Morgen*. https://www.demorgen.be/binnenland/nieuw-in-vlaanderen-haal-je-master-aan-de-hogeschool-b3177ff8/

Amsing, M., Bertu, M., & Van der Maas, M. (2009). *1-zorgroute in het basisonderwijs: bevindingen van onderzoek naar de resultaten en implementatie van 1-zorgroute* [Single support route for special educational needs: Results of research into the results and implementation of the single support route]. KPC Groep.

Anderson, V., Anderson, P., Jacobs, R., & Spencer Smith, M. (2014). Development and assessment of executive functioning: from preschool to adolescence. In V. Anderson, R. Jacobs, & P. J. Anderson (Eds.), *Executive functions and the frontal lobes* (pp. 123-154). Psychology Press.

Armstrong, A. C., Armstrong, D., & Spandagou, I. (2010). *Inclusive education: International policy and practice.* Sage.

Audenaert, M., Lambert, R., Timbremont, B., & De Winter, M. (2015). Zorg realiseren bij (hoog)begaafdheid [Organising support for gifted learners]. *Caleidoscoop*, *27*(3), 15-21.

Bakx, A. (2015). *De pedagogisch sensitieve leraar* [Pedagogical sensitive teachers]. Coutinho.

Bakx, A. (2019). *Begaafde leerling zoekt leerkracht* [Gifted pupils looking for empowerment by their teachers]. Inaugural address RU, 12 April, 2019. Radboud Universiteit. https://repository.ubn.ru.nl/handle/2066/202618

Bakx, A., de Boer, E., van den Brand, M., & van Houtert, T. (2016). *Werken met begaafde leerlingen in de klas* [Working with high-ability pupils in the classroom]. Koninklijke Van Gorcum.

Balchin, T., Hymer, B., & Matthews, D. (Eds.). (2009). *The Routledge International Companion to Gifted Education.* Routledge.

Bambs, H., Steyaert, S., Van Respaille, L., Steyaert, J., Nuytemans, M., & Berghmans, E. (2016). *Van lerensbelang* [The importance of learning]. Retrieved October 1, 2018, from https://onderwijs.vlaanderen.be/sites/default/files/atoms/files/Van-Lerensbelang-eindrapport.pdf

Barkley, R. A. (2012). *Executive functions: What they are, how they work, and why they evolved*. Guilford Press.

Baum, S., Schader, R., & Owen, S. (2017). *To be gifted and learning disabled* (3rd ed.). Prufrock Press.

Bekina. (n.d.). *Over Bekina* [About Bekina]. Retrieved May 16, 2018, from https://www.bekina.org/content/over-bekina

Berlin Communiqué. (2003). *Realising the European Higher Education Area*. https://enqa.eu/wp-content/uploads/2013/03/BerlinCommunique1.pdf

Betts, G. (2004). Fostering autonomous learners through levels of differentiation. *Roeper Review*, *26*(4), 190-191.

Betts, G., & Neihart, M. (2010). *Revised profiles of the gifted and talented.* Retrieved May 21, 2016, from http://talentstimuleren.nl/thema/stimulerend signaleren/publicatie/269-revised-profiles-of-the-gifted-and-talented

Biesta, G. (2007). Why "what works" won't work: evidence-based practice and the democratic deficit in educational research. *Educational Theory*, *57*(1), 1-22. https://doi.org/10.1111/j.1741-5446.2006.00241.x

Biesta, G. (2010). Why "what works" still won't work: from evidence-based education to value-based education. *Studies in Philosophy and Education*, *29*, 491-503. https://doi.org/10.1007/s11217-010-9191-x

Biesta, G. (2012). The future of teacher education: Evidence, competence or wisdom? *RoSE*, *3*(1), 8-21.

Biesta, G. (2014). *The beautiful risk of education*. Paradigm Publishers.

Biesta, G. (2018a). *De terugkeer van het lesgeven* [The return of teaching]. Phronese.

Biesta, G. (2018b). Persoonsvorming in het onderwijs: Over vorming van personen en vorming tot persoon willen zijn [Personal development in education: formation of the person or formation for wanting to be a person]. In G. Geerdink & F. de Beer (Eds.), *Kennisbasis lerarenopleiders: Katern 6. Vorming in de lerarenopleidingen* [Fundaments of professional knowledge for teacher educators: Part 6. Pedagogy at teacher education centres] (pp. 21-34). Velon.

Blok, H., Ledoux, G., & Roeleveld, J. (2013). *Opbrengstgericht werken in het primair onderwijs: Theorie en praktijk* [A results-based approach in primary education: Theory and practice]. Kohnstamm Instituut.

Bologna Working Group on Qualifications Frameworks. (2005). *A Framework for Qualifications of The European Higher Education Area.* Ministry of Science, Technology and Innovation. http://ecahe.eu/w/images/7/76/A_Framework_for_Qualifications_for_the_European_Higher_Education_Area.pdf

Boor-Klip, H., Cillessen, A., & van Hell, J. (2014). Social understanding of children in middle and late childhood. *Gifted Child Quarterly*, *58*(4), 259-271. https://doi.org/10.1177/0016986214547634

Borland, J. (2005). Gifted education without gifted children: The case for no conceptions of giftedness. In R. Sternberg, & J. Davidson (Eds.), *Conceptions of Giftedness* (2nd ed., pp. 1-19). Cambridge University Press.

Borland, J. (2012). You can't teach an old dogmatist new tricks. Dogmatism and Gifted Education. In D. Ambrose, R. Sternberg, & B. Sriraman (Eds.), *Confronting Dogmatism in Gifted Education* (pp. 11-24). Routledge.

Brinkman, J. (2014). *De vragenlijst: een goed meetinstrument voor toepasbaar onderzoek* [Questionnaires: valid and reliable instruments for applied sciences]. Noordhoff Uitgevers.

Broersen, A., & Spreij, L. (2009). Parameters in de dynamiek van educatief partnerschap [Parameters in the dynamics of educational partnership]. *Tijdschrift voor Orthopedagogiek*, *48*, 483-492.

Bronfenbrenner, U. (1979). *The ecology of human development: experiments by nature and design*. Harvard University Press.

Bronneman-Helmers, H. (2008). *Vijftien jaar onderwijsvernieuwingen in Nederland: Deelonderzoek bij het Parlementair onderzoek onderwijsvernieuwingen* [The review of fifteen years of changes in education in The Netherlands: Parlementary inquest into the modernisation of education]. Sociaal en Cultureel Planbureau.

Bryman, A. (2012). *Social research methods* (4th ed.). Oxford University Press.

Buis, S. (2014). *Mijn hoogbegaafde kind en ik* [Me and my gifted child]. Uitgeverij Nanda.

Burger-Veltmeijer, A., & Minnaert, A. (2016). De waarde van de Sterkte en Zwakte Heuristiek in diagnostiek bij het vermoeden van hoogbegaafdheid [The added value of the Strengths and Weaknesses Heuristic for assessments in case giftedness is suspected]. *Orthopedagogiek: Onderzoek en Praktijk*, *55*(7-8), 160-174.

Buyse, E., Verschueren, K., Verachtert, P., & van Damme, J. (2009). Predicting school adjustment in early elementary school: Impact of teacher-child relationship quality and relational classroom climate. *Elementary School Journal* (110), 119-141. https://doi.org/10.1086/605768

Calahan, C., & Hertberg-Davis, H. (Eds.). (2018). *Fundamentals of gifted education: Considering multiple perspectives* (2nd ed.). Routledge.

Campbell, C., Pollock, K., Brisco, P., Carr-Harris, S., & Tuters, S. (2017). Developing a knowledge network for applied education research to mobilise evidence in and for educational practice. *Educational Research*, *59*(2), 209-227. https://doi.org/10.1080/00131881.2017.1310364

Carrier, N. (2015). *What catches on? The role of evidence in the promotion and evaluation of educational innovations*. Ontario Institute for Studies in Education.

Cauffman, L., & van Dijk, D. (2009*). Handboek oplossingsgericht werken in het onderwijs* [Manual for a solution-oriented approach in education]. Coutinho.

Ceulemans, C., Struyf, E., & Simons, M. (2016). Handig instrument of sturend mechanisme? Over de werking van beroepstandaarden voor leraren en lerarenopleiders [Professional standards for teachers and teacher educators: Convenient or directive?]. *Tijdschrift voor Lerarenopleiders*, *37*(4), 53-64.

Chandra Handa, M. (2009). Learner-centred differentiation model: a new framework. *The Australasian Journal of Gifted Education*, 55-66. https://doi.org/10.1177/0261429413489161

Chandra Handa, M. (2015). Imagination first: unleash the power of possibility. *Gifted Education International*, *31*(2), 117-141. https://doi.org/10.1177/0261429413489161

Choochem. (n.d.). *Over Choochem* [About Choochem]. Retrieved May 16, 2018, from https://www.choochem.nl/over-choochem

Clijsen, A., Gijzen, W., de Lange, S., & Spaans, G. (2007). *1-zorgroute: naar handelingsgericht werken* [Single support route for special educational needs: Developments toward an action-oriented approach]. KPC groep.

Clijsen, A., Spaans, G., Hollenberg, J., Pameijer, N., & Struijksma, C. (2013). *Ontwikkelingsperspectief in het basisonderwijs* [Developmental perspectives for students in primary education]. PO-Raad.

Coe, R. (2002, September 12-14). *It's the effect size, stupid. What effect size is and why it is important*. [Paper presentation] Annual Conference of the British Educational Research Association. Exeter, UK. https://www.leeds.ac.uk/educol/documents/00002182.htm

Coffield, F., Mosley, D., Hall, E., & Ecclestone, K. (2004). *Learning styles and pedagogy in post-16 learning: A systematic and critical review.* Learning and Skills Research Centre.

Cohen, L. (2012). Dogma and definitions of giftedness and talent. In D. Ambrose, R. Sternberg, & B. Sriraman (Eds.), *Confronting dogmatism in gifted education* (pp. 25-40). Routledge.

Coleman, M. (2018). Response to Intervention (RtI) approaches to identification in gifted education. In C. Calahan, & H. Hertberg-Davis (Eds.), *Fundamentals of gifted education: Considering multiple perspectives* (2nd ed., pp. 146-152). Routledge.

Coleman, M. R., & Johnsen, S. K. (Eds.). (2011). *RtI for gifted students*. Prufrock Press.

Collangelo, N., & Davis, G. (Eds.). (2003). *Handbook of gifted education* (3rd ed.). Allyn & Bacon.

Cooper, A., Klinger, D., & McAdie, P. (2017). What do teachers need? An exploration of evidence-informed practice for classroom assessment in Ontario. *Educational Research*, *59*(2), 190-208. https://doi.org/10.1080/00131881.20171310392

Creemers, B. (Ed.). (2002). *De professionalisering van de leraar: onderwijskundig lexicon* [Professional development of teachers: An educational lexicon]. Samsom.

Cross, J. (2012). Peer relations. In T. Cross, & J. Cross (Eds.), *Handbook for counselors serving students with gifts & talents: development, relationships, school issues and counseling needs/interventions* (pp. 409-425). Prufrock Press.

Cross, T. (2018). *On the social and emotional lives of gifted children*. Prufrock Press.

Daeter, B. (2012). *Hoogbegaafde kinderen: Leonardo-onderwijs, recht van het kind - uitdaging voor ons* [Gifted children: Leonardo-schools, children's rights and educational opportunities]. ASPEKt.

Daeter, B. (2014). *Onderwijs aan hoogbegaafden: spil voor onderwijsvernieuwing* [Educating gifted learners: The essentials for the modernization of education]. Boekscout.nl.

Dai, D. (2008). Thinking and learning styles. In J. Plucker, & C. Callahan (Eds.), *Critical issues and practices in gifted education* (pp. 695-706). Prufrock Press.

Dai, D., & Chen, F. (2014). *Paradigms of gifted education: A guide to theory-based practice-focused research.* Prufrock Press.

Danielian, J., Fugate, C., & Forgarty, E. (Eds.). (2018). *Teaching gifted children: Success strategies for teaching high ability learners.* Prufrock Press.

Davis, G., Rimm, S., & Siegle, D. (Eds.). (2014). *Education of the gifted and talented*. Pearson Education.

Dawson, P., & Guare, R. (2009). *Slim maar... Help kinderen hun talenten benutten door hun executieve functies te versterken* [Smart but scattered. The revolutionary "Executive Skills" approach to helping kids reach their potential]. Hogrefe.

Day, D. (2010). *The nature and nurture of giftedness: A new framework for understanding gifted education.* Teachers College Press.

de Bie, B. (2016). *Ontwikkeling van een zelfevaluatietool voor de implementatie van 21ste-eeuwvaardigheden in Vlaamse basisscholen* [Developing a self-evaluation tool for the implementation of 21st century skills in Flemish primary schools]. Open Universiteit.

de Boer, G., Minnaert, A., & Kamphof, G. (2013). Gifted education in the Netherlands. *Journal for the Education of the Gifted*, *36*(1), 133-150. https://doi.org/10.1177/0162353212471622

De Bruyckere, P., Kirschner, P., & Hulshof, C. (2016). *Jongens zijn slimmer dan meisjes* [Boys are smarter than girls]. Lannoo.

de Bruin-de Boer, A., & Kuipers, J. (2011). *SiDi-3* [Identifying gifted students]. Cedin.

de Bruin-de Boer, A., & van Gerven, E. (2009). De sociaal-emotionele ontwikkeling van begaafde leerlingen [Social-emotional development of gifted learners]. In E. van Gerven (Ed.), *Handboek hoogbegaafdheid* [Handbook of giftedness] (pp. 188-211). Koninklijke Van Gorcum.

de Bruïne, E., Franssens, J., Jansen, R., & Van Veen, D. (2015). Passend onderwijs? Passend opleiden! [Adequate teaching requires adequate teacher education]. *Tijdschrift voor Lerarenopleiders*, *36*(3), 29-40.

de Shazer, S., Dolan, Y., Korman, H., Trepper, T., McCollum, E., & Insoo Berg, K. (2012). *More than miracles: The state of the art of solution-focused brief therapy*. Routledge.

de Vries, P. (2017). *Handboek ouderbetrokkenheid* 3.0 [Manual on parental involvement in schools 3.0]. CPS.

De Wit, C. (2005). Ouders als educatieve partner: een handreiking voor scholen [Educational partnerships with parents: Suggestions for schools]. *Qprimair*. https://docplayer.nl/2479-Ouders-als-educatieve-partner.html

Deci, E., & Ryan, R. (2000). The 'what' and 'why' of goal pursuits: human needs and the self-determination of behavior. *Psychology Inquiry*, *11*, 227-268. https://doi.org/10.1207/S15327965PLI1104_01

Decreet betreffende gelijke onderwijskansen-1 [Decree regarding equity in educational opportunities]. (2012, September 1). Retrieved December 11, 2017, from https://codex.vlaanderen.be/Zoeken/Document.aspx?DID=1009657

Decreet betreffende maatregelen voor leerlingen met specifieke onderwijs-behoeften [Decree regarding the support structure for students with special educational needs]. (2014, March 21). https://codex.vlaanderen.be/Portals/Codex/documenten/1024474.html

Dekker, S. (2014a, April 23). *Passend onderwijs en Hoogbegaafdheid* [Kamerbrief]. [Letter to the Dutch House of Commons regarding "appropriate education" in relation to giftedness]. Retrieved August 1, 2016, from https://www.rijksoverheid.nl/documenten/kamerstukken/2014/04/23/kamerbrief-passend-onderwijs-en-hoogbegaafdheid

Dekker, S. (2014, March 10 (b)). *Plan van aanpak toptalenten 2014-2018* [Kamerbrief] [Letter to the Dutch House of Commons regarding the education of talented students]. Retrieved August 1, 2016, from https://www.rijksoverheid.nl/documenten/kamerstukken/2014/03/10/plan-van-aanpak-toptalenten-2014-2018

den Otter, M. (2015). *Voicing: geef kinderen een stem* [Voicing. Including children's perspectives and opinions towards their education]. Instondo.

Diery, A., Vogel, F., Knogler, M., & Seidel, T. (2020). Evidence-based practice in higher education: teacher-educators' attitudes, challenges, and uses. *Frontiers in Education*, *5*(62), 1-13. https://doi.org/10.3389/feduc.2020.00062

Dijksma, S. (2007). *Uitwerking passend onderwijs* [Kamerbrief]. [Letter to the Dutch House of Commons regarding the resolutions towards "appropriate education"] PO/ZO/07/26259. https://www.werkhoezithet.nl/doc/kennisbank/uitwerking_passend_onderwijs.pdf

Dijksma, S. (2008, August 25). *Stimuleren van excellentie in het basisonderwijs* [Kamerbrief]. [Stimulating excellence in primary education. Letter to the Dutch House of Commons]. https://www.tweedekamer.nl/kamerstukken/brieven_regering/detail?id=2008Z01249&did=2008D01454

Dijkstra, W., Ongena, Y., & Loosveldt, G. (2014). *Onderzoek doen met vragenlijsten* [Using questionnaires for research]. VU University Press.

Doolaard, S., & Oudbier, M. (2010). *Onderwijsaanbod aan (hoog)begaafde leerlingen in het basisonderwijs* [Educational provision for gifted students in primary education]. GION/RUG.

Doornbos, K., & Bergman, J. (1991). *Samen naar school: aangepast onderwijs in gewone scholen* [Going to school together: Educational adaptations for special needs students in regular schools]. Ambo.

Drent, S. (2009). Compacten [Compacting]. In E. van Gerven (Ed.), *Handboek Hoogbegaafdheid* [Handbook of giftedness] (pp. 76-85). Koninklijke Van Gorcum.

Drent, S., & van Gerven, E. (2002). *Professioneel omgaan met hoogbegaafde leerlingen in het basisonderwijs* [A professional approach to gifted education]. Lemma.

Drent, S., & van Gerven, E. (2012). *Passend onderwijs voor begaafde leerlingen* [Appropriate education for gifted learners]. Koninklijke Van Gorcum.

Dweck, C. (2006). *Mindset*. Ballantine.

Dweck, C. (2017). *Mindset*. Updated edition. Robinson.

Edinger, M. (2017). Online teacher professional development for gifted education: Examining the impact of a new pedagogical model. *Gifted Child Quarterly*, *61*(4), 300-312. https://doi.org/10.1177/0016986217722616

Engeström, Y. (2001). Expansive learning at work: toward an activity theoretical reconceptualization. *Journal of Education and Work*, *14*(1), 133-156. https://doi.org/10.1080/13639080020028747

Enthoven, M., & de Bruijn, E. (2010). Beyond locality: the creation of public practice-based knowledge through practitioner research in professional learning communities and communities of practice: A review of three books on practitioner research and professional communities. *Educational Action Research*, *18*(2), 289-298. https://doi.org/10.1080/09650791003741822

European Commission. (2007). *Improving the quality of teacher education*. https://ec.europa.eu/transparency/regdoc/rep/1/2007/EN/1-2007-392-EN-F1-1.Pdf

European Commission. (2013). *Supporting teacher competence development for Better Learning Outcomes.* https://ec.europa.eu/assets/eac/education/experts-groups/2011-2013/teacher/teachercomp_en.pdf

European Higher Education Area. (1999). Joint declaration of the European Ministers of Education - The Bologna declaration of 19 June 1999. http://www.ehea.info/media.ehea.info/file/Ministerial_conferences/02/8/1999_Bologna_Declaration_English_553028.pdf

Eyre, D. (2001). *Able children in ordinary schools*. David Fulton.

Eyre, D. (2007). Structured tinkering: Improving provision for the gifted in ordinary schools. *Gifted & Talented International*, *4*(1), 31-38. https://doi.org/10.1080/15332276.2007.11673483

Figg, S., Rogers, K., McCormick, J., & Low, R. (2012). Differentiating low performance of the gifted learner: achieving, underachieving, and selective consuming students. *Journal of Advanced Academics*, *23*(1), 53-71. https://doi.org/10.1177/1932202X11430000

Florian, L., & Black-Hawkins, K. (2011). Exploring inclusive pedagogy. *British Educational Research Journal*, 813-828. https://doi.org/10.1080/01411926.2010.50196

Franco-Santos, M., & Bourne, M. (2005). An examination of the literature relating to issues affecting how companies manage through measures. *Production Planning and Control*, *16*(2), 114-124. https://doi.org/10.1080/09537280512331333020

Freeman, J. (2008). The emotional development of the gifted and talented. Conference proceedings. *Gifted and Talented Provision*. Optimus Educational.

Freeman, J. (2010). *Gifted lives*. Routledge.

Freeman, J. (2017). The long-term effects of families and educational provision on gifted children. *Educational and Child Psychology*, 30(2), 7-17.

Freeman, J. (2019). The emotional development of the gifted and talented. In B. Wallace, D. Sisk, & J. Senior (Eds.), *The Sage handbook of gifted and talented education* (pp. 169-183). Sage.

Freeman, J., Raffan, J., & Warwick, I. (2010). *Worldwide provision to develop gifts and talents: An international survey. Research report*. CfBT Education Trust.

Freeman, R. (1987). *Strategic management: A stakeholder approach*. Pitman.

Frumau, M., Derksen, J.J.L., & Peters, W. (2011). Hoogbegaafdheid: het label voorbij [Giftedness, there is no need for labels]. *GZ-Psychologie*, (5), 32-37.

Gagné, F. (2010a). *Building gifts into talents: Brief overview of the DMGT 2.0.* Université du Québec à Montréal.

Gagné, F. (2010b). Motivation within the DMGT 2.0 framework. *High Ability Studies*, *21*(2), 81-99. https://doi.org/10.1080/13598139.2010.525341

Galbraith, J., & Delisle, J. (2015). *When gifted kids don't have all the answers.* Free Spirit Publishers.

Garces-Bacsal, R. (2012). Old, new, borrowed, blue: reconceptualizing the systems framework. *High Ability Studies*, *23*(1), 57-59. https://doi.org/10.1080/13598139.2012.679093

Geerdink, G. (2018). Lerarenopleiders en onderzoek: een historisch perspectief op de relatie [Teacher educators practising research: A historical perspective]. In F. Boei, & M. Willemse (Eds.), *Kennisbasis lerarenopleiders: Katern 5. Onderzoek in de lerarenopleidingen* [Fundaments of professional knowledge for teacher educators: Part 5. Research at teacher education centres] (pp. 21-33). Velon.

Gerritsma, A., de Haan, H., den Hollander, I., Mitzschke, M., & van der Veen, Y. (2013). *Onderzoeksvaardig in onderwijs* [Action research in education]. Koninklijke Van Gorcum.

Gevaert, T., & Desmet, O. (2016). *Slim onderpresteren aanpakken* [Smart approaches: dealing with underachieving students]. Garant.

Glass, G., McGaw, B., & Smith, M. (1981). *Meta-analyses in social research*. Sage.

Godor, B. (2019). Gifted metaphors: exploring the metaphors of teachers in gifted education and their impact on teaching the gifted. *Roeper Review*, *41*(1), 51-60. https://doi.org/10.1080/02783193.2018.1553219

Gore, J., Lloyd, A., Smith, M., Bowe, J., & Ellis, H. (2017). Effects of professional development on the quality of teaching: results from a randomised controlled trial of quality teaching rounds. *Teacher and Teacher Education*, *68*, 99-113. https://doi.org/10.1016/j.tate.2017.08.007

Grantham, T. (2012). Eminence-focused gifted education: concerns about forward movement void of an equity vision. *Gifted Child Quarterly*, 56(4), 215-220. https://doi.org/10.1177/0016986212456074

Gubbels, J. (2016). *The dynamics of giftedness in the upper primary grades.* Radboud Universiteit.

Haenen, J., & Mol-Lous, A. (2014). Succesfactoren voor passend onderwijs aan begaafde leerlingen: adviezen voor (aankomende) leerkrachten en opleidingen [Opportunities for success in "appropriate education" for gifted learners: advice for (undergraduate) teachers and teacher education institutes]. *Tijdschrift voor Hoger Onderwijs*, 66-79.

Haimovitz, K., & Dweck, C. (2017). The origins of children's growth and fixed mindsets: new research and a new proposal. *Child Development*, 1-11. https://doi.org/10.1111/cdev.12955

Hattie, J. (2013). *Leren zichtbaar maken* [Visible learning for teachers]. Bazalt Educatieve Uitgaven.

Hattie, J. (2014). *De impact van leren zichtbaar maken* [Visible learning]. Bazalt Educatieve Uitgaven.

Hattie, J., & Yates, G. (2015). *Leren zichtbaar maken met de kennis over hoe wij leren* [Visible learning and the science of how we learn]. Bazalt Educatieve Uitgaven.

Heacox, D., & Cash, R. (2017). *Differentiation for gifted learners*. Free Spirit Publishing.

Heacox, D., & Cash, R. (2019). *Differentiation for gifted learners: Going beyond the basics*. Free Spirit Publishing.

Heller, K. (2010). The Munich model of giftedness and talent. In *Munich Studies of Giftedness* (pp. 3-13). LIT Verlag.

Heller, K., Mönks, F., Sternberg, R., & Subotnik, R. (Eds.). (2000). *International Handbook of Giftedness and Talent.* Elsevier.

Hendriksen, J. (2005). *Cirkelen rond Kolb* [Kolb's learning theory]. Boom/Nelissen.

Henshon, S. (2018). Toward a visionary future: An interview with Sally Reis. *Roeper Review*, *40*(4), 215-221. https://doi.org/10.1080/02783193.2018.1501645

Hertzog, N. (2017a). Designing the learning context in school for talent development. *Gifted Child Quarterly*, *61*(3), 219-228. https://doi.org/10.1177/0016986217705712

Hertzog, N. (2017b). Provisions for young advanced learners. In J. Riedl-Cross, C. O'Reilly, & T. Cross (Eds.), *Providing for the special needs of students with gifts and talents* (pp. 153-172). CTYI Press.

HINT. (2018). *HINT Nederland*. Retrieved May 16, 2018, from https://www.hintnederland.nl/

Hocket, J. (2009). Curriculum for highly able learners that conforms to general education and gifted education quality indicators. *Journal for the Education of the Gifted*, *32*(3), 394-440.

Hollenberg, J., van Boxtel, H., & Keuning, J. (2014). Werken met leerlingen met hoge vaardigheidsscores in het Cito Volgsysteem primair en speciaal onderwijs [Working with students who get high scores on standardised educational tests]. *Tijdschrift voor Orthopedagogiek*, *53*, 347-356.

Holvoet, E., Elen, J., Feremans, K., & Depaepe, F. (2017). De leraar staat er niet alleen voor: naar een passend ontwerp voor een op de toekomst gerichte lerarenopleiding [Support for teachers. Developing future-focused teacher education]. *Tijdschrift voor Lerarenopleiders*, *38*(4), 89-95.

Hoogeveen, L. (2008). *Social emotional consequences of accelerating gifted students*. [Doctoral dissertation, Radboud Universiteit]. Radboud Repository. https://repository.ubn.ru.nl/bitstream/handle/2066/73315/73315.pdf?sequence=1

Hoogeveen, L. (2017a). Leertijdgerichte interventies [Acceleration]. In E. van Gerven (Ed.), *De gids: over begaafdheid in het basisonderwijs* [Guide for gifted education in primary schools] (2nd ed., pp. 185-193). Leuker.nu.

Hoogeveen, L. (2017b). Motivatie [Motivation]. In E. van Gerven (Ed.), *De gids: over begaafdheid in het basisonderwijs* [Guide for gifted education in primary schools] (2nd ed., pp. 89-100). Leuker.nu.

Hoogeveen, L., Hell, J., Mooij, T., & Verhoeven, L. (2004). *Onderwijsaanpassingen voor hoogbegaafde leerlingen: meta-analyses en overzicht van internationaal onderzoek* [Educational modifications for gifted learners: meta-analyses and review of international research]. Radboud Universiteit, CBO/ITS.

Hornstra, L., van der Veen, I., & Peetsma, T. (2017). Effects of full-time and part-time high-ability programs on developments in students' achievement emotions. *High Ability Studies*, *28*(2), 199-224. https://doi.org/10.1080/13598139.2017.1332575

Houkema, D. (2009). *Hoogbegaafdheid in zicht* [Giftedness identified]. Retrieved November 15, 2017, from http://www.hoogbegaafdheid-in-zicht.nl/index.php/talent-in-ontwikkeling

Houkema, D. (2017). Een psychologisch perspectief op de profielen van begaafde leerlingen [A psychological perspective on the profiles of gifted learners]. In E. van Gerven (Ed.), *De gids: over begaafdheid in het basisonderwijs* [Guide for gifted education in primary schools] (2nd ed., pp. 75-88). Leuker.nu.

Houkema, D., Janssen, Y., & Steenbergen-Penterman, N. (2018). *Passend onderwijs voor begaafde leerlingen binnen samenwerkingsverbanden: rapportage van een inventariserend onderzoek*. ["Appropriate education" for gifted learners provided by school districts: An inventory]. SLO.

Hovius, M., & van Kessel, N. (2010). *Professionalisering van leraren in het buitenland: een inventarisatie van de stand van zaken in twaalf Europese landen* [The professional development of teachers in twelve European countries]. Open Universiteit: Ruud de Moor Centrum.

Howitt, D., & Cramer, D. (2009). *Statistiek in de sociale wetenschappen* [Statistics and social sciences]. Pearson Education.

Hu, W., Jia, X., Plucker, J., & Shan, X. (2016). Effects of a critical thinking skills program on the learning motivation of primary school students. *Roeper Review*, *38*(2), 70-83. https://doi.org/10.1080/02783193.2016.1150374

Hughes, C. (2017). From cacophony to chorus: collaborating in a time of multiple standards. In S. Johnsen, & J. Clarenbach (Eds.), *Using the national gifted education standards for pre-K grade 12 professional development* (pp. 53-68). Prufrock Press Inc.

Inspectie van het Onderwijs. (2009). *De sterke basisschool: definitie en kenmerken* [High-quality primary schools: Definitions and characteristics]. Inspectie van het Onderwijs: De Meern.

Janson, D. (2009). Verrijkingsgroepen [Pull-out programmes]. In E. van Gerven (Ed.), *Handboek hoogbegaafdheid* [Handbook of giftedness] (pp. 114-127). Koninklijke Van Gorcum.

Janson, D. (2017). Rijk onderwijs in de 21e eeuw [Challenging education in the 21st century]. In E. van Gerven (Ed.), *De gids: over begaafdheid in het basisonderwijs* [Guide for gifted education in primary schools] (2nd ed., pp. 157-170). Leuker.nu.

Johnsen, S., & Clarenbach, J. (Eds.). (2017). *Using the national gifted education standards for Pre-K-12-Grade 12 professional development.* Prufrock Press Inc.

Johnsen, S., VanTassel-Baska, J., Robinson, A., Cotabish, A., Dailey, D., Jolley, J., & Adams, C. (Eds.). (2016). *Using the national gifted education standards.* Prufrock Press Inc.

Joint Quality Initiative. (2004, March 23). *Shared 'Dublin' descriptors for the bachelor's, master's and doctoral Awards.* Retrieved March 7, 2018, from https://www.vitae.ac.uk/policy/dublin-descriptors-for-doctorate-mar-2004-vitae.pdf/%40%40download/file/Dublin-descriptors-for-doctorate-Mar-2004-Vitae.pdf

Jolles, J. (2011). *Ellis en het verbreinen* [Ellis; or how children's brains develop]. Neuropsych Publishers.

Jutte, J. (2008). *Ont-moeten: boeiend onderwijs in een lerende school* [Education provided by schools who invest in continuing professional development for their team members]. Natuurlijk leren B.V.

Kallenberg, T., Koster, B., Onstenk, J., & Scheepsma, W. (2007). *Ontwikkeling door onderzoek: een handreiking voor leraren* [Continuous professional development by conducting research. Suggestions for teachers]. Thieme Meulenhoff.

Kaplan, A., Neuber, A., & Garner, J. (2019). An identity systems perspective on high ability in self-regulated learning. *High Ability Studies*, *30*(1-2), 53-78. https://doi.org/10.1080/13598139.2019.1568830

Kappen, A. & Förren, M. (n.d.). *Kwaliteitskaart Opbrengstgericht werken aan technisch lezen met datamuur en groepsplan* [Data driven education for developing technical reading skills by using a data wall and action plans for group clustering]. School aan Zet. Retrieved December 13, 2020.

https://www.onderwijsmaakjesamen.nl/uploads/2018/12/SaZ-H-Opbrengstgericht-werken-Datamuur-Werken-met-groepsplannen.pdf

Kara, H. (2015). *Creative research methods in the social sciences: A practical guide*. Policy Press.

Keeman, N. (2010). Er zullen altijd kinderen blijven die Leonardo-onderwijs nodig hebben [There will always be students who need to be educated at "Leonardo" schools"]. *Talent, 12*(3), pp. 5-7.

Kelchtermans, G. (2012). *De leraar als (on)eigentijdse professional: reflecties over de "moderne" professionaliteit van leerkrachten. Notitie in opdracht van de Nederlandse onderwijsraad* [Teachers as modern professionals. Reflections on the professionalism of teachers]. Centrum voor Onderwijsbeleid, -Vernieuwing en Lerarenopleiding.

Keuchenius, A. (2001). Plato krijgt miljoen subsidie [€ 500,000 state funding for society for the advocacy of giftedness]. *Talent, 3*(5), pp. 27-28.

Kieboom, T. (2015). *Hoogbegaafd: als je kind (g)een Einstein is* [Giftedness: What to do when your child is (not) another Einstein]. Lannoo.

Kieboom, T., & Venderickx, K. (2017a). Begaafdheid als zijnsfactor [Giftedness and personality]. In E. van Gerven (Ed.), *De gids: over begaafdheid in het basisonderwijs* [Guide for gifted education in primary schools] (2nd ed., pp. 37-46). Leuker.nu.

Kieboom, T., & Venderickx, K. (2017b). *Meer dan intelligent alleen* [Giftedness exceeds intelligence]. Lannoo.

Kieboom, T., & Verheye, D. (2011). *De hoogbegaafdheid survivalgids* [A children's survival guide to giftedness]. Abimo.

King, J., & Lau-Smith, J. (2013). Teaching from the inside out: Discovering and developing the self-that-teaches. In F. Korthagen, Y. Kim, & W. Greene (Eds.), *Teaching and learning from within: A core reflection approach to quality and inspiration in education* (pp. 58-73). Routledge.

Kitsantas, A., Bland, L., & Chirinos, D. (2017). Gifted students' perceptions of gifted programs: an inquiry to their academic and social functioning. *Journal for the Education of the Gifted, 40*(3), 266-288. https://doi.org/10.1177/0162353217717033

Koenderink, T. (2012). *De 7 uitdagingen: de 7 uitdagingen in het onderwijs aan hoogbegaafde kinderen* [The seven challenges for gifted education]. Novilo.

Kooijman-van Thiel, M. (2008). *Hoogbegaafd dat zie je zo!* [This is how you recognise giftedness]. OYA Productions.

Korthagen, F. (2004). Zin en onzin van competentiegericht opleiden [Competence-oriented teacher education. Does it make sense or not?]. *Tijdschrift voor Lerarenopleiders, 25*(1), 13-23.

Korthagen, F. (2017). Inconvenient truths about teacher learning: towards professional development 3.0. *Teachers and teaching*, *23*(4), 387-405. https://doi.org/10.1080/13540602.2016.1211523

Korthagen, F., & Nuijten, E. (2018). Bijdragen aan persoonsvorming in de lerarenopleiding [Contributions to the development of a professional attitude and identity]. In G. Geerdink & F. de Beer (Eds.), *Kennisbasis lerarenopleiders*: *Katern 6. Vorming in de lerarenopleidingen* [Fundaments of professional knowledge for teacher educators: Part 6. Pedagogy at teacher education centres] (pp. 89-102). Velon.

Kramer, F., & Wildeboer, T. (2017). *Slim! De vier sleutels voor een effectieve les* [Brilliant! Four keys for effective teaching]. Leuker.nu.

Kreger-Silverman, L. (2013). *Giftedness 101.* Springer Publishing Company.

KRO-NCRV [Broadcaster]. (2018, June 12). Hoogbegaafdheid [Giftedness] [Tv documentary]. De Monitor, KRO-NCRV. https://demonitor.kro-ncrv.nl: https://demonitor.kro-ncrv/uitzendingen/hoogbegaafdheid

Kroesbergen, E. (2017). Begaafde leerlingen met dyslexie [Gifted learners with dyslexia]. In E. van Gerven (Ed.), *De gids: over begaafdheid in het basisonderwijs*. [Guide for gifted education in primary schools] (2nd ed., pp. 243-254). Leuker.nu.

Kroesbergen, E., van Hooijdonk, M., van Viersen, S., & Middel-Lalleman, M. (2015). The psychological well-being of early identified gifted children. *Gifted Child Quarterly*, 1-15. https://doi.org/10.1177/0016986215609113.

Kroesbergen, E., Verkaik, D., & van Gerven, E. (2017). Executieve functies en begaafdheid [Executive function and giftedness]. In E. van Gerven (Ed.), *De gids: over begaafdheid in het basisonderwijs* [Guide for gifted education in primary schools] (2nd ed., pp. 225-243). Leuker.nu.

Kuhn, T. (1962). *The structure of scientific revolutions*. University of Chicago Press.

Kuipers, J. (2017). Passend klassenmanagement voor excellente en (hoog)begaafde leerlingen [Adequate classroom management for excellent and (highly) able learners]. In E. van Gerven, *De gids: over begaafdheid in het basisonderwijs* [Guide for gifted education in primary schools] (2nd ed., pp. 113-123). Leuker.nu.

Kwakman, K. (2002). Continue professionele ontwikkeling tijdens de beroepsuitoefening [Continuous professional development]. In B. Creemers (Ed.), *De professionalisering van de leraar: onderwijskundig lexicon* [Teachers' professional development: Educational lexicon] (pp. 93-114). Samsom.

Landelijke Beroepsgroep Begeleiders in het Onderwijs. (n.d.). *Organogram*. Retrieved March 31, 2020, from https://www.lbbo.nl/over-ons/organogram

Latz, A., Speirs Neumeister, K., Adams, C., & Pierce, R. (2008). Peer coaching to improve classroom differentiation: perspectives from project CLUE. *Roeper Review*, *31*(1), 27-39.

Lavrijssen, J., & Verschueren, K. (2019). *Begaafde jongeren, moeilijke gevallen? Het belang van systematisch onderzoek naar het functioneren van cognitief sterke jongeren* [Gifted adolescents, a complex group? The importance of systematic research]. KU Leuven.

LBBO. (2019). *De beroepsstandaard voor de intern begeleider: werkzaamheden, bekwaamheidseisen en ethische code* [The professional standard for SENCo's: Tasks, competence and ethics]. Pica.

LBBO. (n.d.-a). *Beroepsstandaard voor de gedragsspecialist* [The professional standard for specialists in complex behaviour]. Pica.

LBBO. (n.d.-b). "Over ons" [About us]. (https://www.lbbo.nl/over-ons/onze-beroepsgroepen/afdeling-specialisten-begaafdheid

Leavitt, M. (2017). *Your passport to gifted education*. Springer.

Lee, S., Olszewski-Kubilius, P., & Turner Thompson, D. (2012). Academically gifted students' perceived interpersonal competence and peer relationships. *Gifted Child Quarterly*, *56*(2), 90-104. https://doi.org/10.1177/0016986212442568

Leeman, Y., & Wardekker, W. (2004). *Onderwijs met pedagogische kwaliteit* [Education and pedagogical quality]. Christelijke Hogeschool Windesheim.

Leonardo-onderwijs, voor (hoog)begaafde kinderen [Gifted learners at "Leonardo-schools"]. (n.d.). Retrieved November 12, 2017 from https://educatie-en-school.infonu.nl/diversen/21412-leonardo-onderwijs-voor-hoogbegaafde-kinderen.html

Little, C. (2017). Teaching strategies to support the education of gifted learners. In S. Pfeiffer, E. Shaunessy-Dedrick, & M. Foley-Nicpon (Eds.), *APA handbook of giftedness and talent* (pp. 371-386). American Psychological Association.

Lo, O. C., & Porath, M. (2017). Paradigm shifts in gifted education: An examination vis-à-vis its historical situatedness and pedagogical sensibilities. *Gifted Child Quarterly*, 61(4), 343-360. https://doi.org/10.1177/0016986217722840

Lo, O., Porath, M., Hiu, H., Chen, C., Tsai, K., & Wu, I. (2019). Giftedness in the making: a transactional perspective. *Gifted Child Quarterly*, *63*(3), 173-83. https://doi.org/10.1177/0016986218812474

Loughran, J. (2002). Effective reflective practice in search of meaning in learning about teaching. *Journal of Teacher Education*, *53*(1), 33-43. https://doi.org/10.1177/0022487102053001004

Loughran, J. (2015). The demands, challenges and expectations of teacher education. *Tijdschrift voor Lerarenopleiders*, *36*(3), 5-15.

Lunenberg, M., Dengerink, J., & Korthagen, F. (2014). *The professional teacher educator: roles, behaviour, and professional development of teacher educators.* Sense Publishers.

Mammadov, S., Cross, T., & Ward, T. (2018). The big five personality predictors of academic achievement in gifted students: mediation by self-regulatory efficacy and academic motivation. *High Ability Studies*, *29*(2), 111-134. https://doi.org/10.1080/13598139.2018.1489222

Marzano, R. (2007). *Wat werkt op school* [What works at schools, translating research into action]. Bazalt Educatieve Uitgaven.

Marzano, R. (2011). *De kunst en wetenschap van het lesgeven: een evidence-based denkkader voor goed, opbrengstgericht onderwijs. Tien vragen (én antwoorden) om uw lessen sterker te maken* [The art and science of teaching]. Bazalt Educatieve Uitgaven.

Marzano, R., & Heflebower, T. (2012). *Klaar voor de 21e eeuw* [Teaching and assessing 21st century skills]. Bazalt Educatieve Uitgaven.

Marzano, R., & Kendall, J. (2007). *The new taxonomy of educational objectives*. Corwin Press.

Marzano, R., & Kendall, J. (2008). *Designing and assessing educational objectives: Applying the new taxonomy.* Corwin Press.

Maslow, A. (1954). *Motivation and personality*. Harper & Brothers.

Matthews, D., & Folsom, C. (2009). Making connections: Cognition, emotion and a shifting paradigm. In T. Balchin, B. Hymer, & D. Matthews (Eds.), *The Routledge international companion to gifted education* (pp. 18-25). Routledge.

Matthews, D., & Foster, J. (2006). Mystery to mastery: Shifting paradigms in gifted education. *Roeper Review*, *28*, 64-69. https://doi.org/10.1080/02783190609554340

Matthews, D., & Foster, J. (2009). *Being smart about gifted education*. Great Potential Press.

Matthews, G., Lin, J., Zeidner, M., & Roberts, R. (2018). Emotional intelligence and giftedness. In S. Pfeiffer, E. Shaunessy-Dedrick, & M. Foley-Nicpon (Eds.), *APA handbook of giftedness and talent* (pp. 585-600). American Psychological Association.

Mayes, E., Finneran, R., & Black, R. (2019). The challenges of students' voice in primary schools: students "having a voice"and "speaking for" others. *Australian Journal for Education*, *63*(2), 157-172. https://doi.org/10.1177/0004944119859445

McBee, M. (2010). Examining the probability of identification for gifted programs for students in Georgia elementary schools: A multilevel path analysis study. *Gifted Child Quarterly*, *54*(4), 283-297. https://doi.org/10.1177/0016986210377927

McCallum, R., Bell, S., Caldwell-Miller, K., Hopkins, M., & Hilton-Prillhart, A. (2013). A model for screening twice-exceptional students (gifted with learning disabilities) within a response to intervention paradigm. *Gifted Child Quarterly*, *57*(4), 209-221. https://doi.org/10.1177/0016986213500070

Meijer, W. (2004). *WSNS welbeschouwd* [WSNS "Going to School Together" reviewed]. Garant.

Meltzer, L. (Ed.). (2018). *Executive function in education: From theory to practice* (2nd ed.). Guilford Press.

Merriënboer, J., van der Klink, M., & Hendriks, M. (2002). *Competenties: van complicaties tot compromis. Over schuifjes en begrenzers* [Competences: from complications to compromise. About sliders and limiters]. Onderwijsraad.

Mijland, E. (2010). Deskundigheid én een klik met leerlingen [Being competent ánd having a good relationship with your students]. *Talent*, *12*(3), 8-9.

Ministerie Onderwijs en Wetenschappen (2006) *Bekwaamheidseisen leraren per 1 augustus 2006*. https://instellingsinformatie.duo.nl/Public/CFI-online/Images/CFINL06wk11APOKP20067375_tcm2-38566.pdf

Mitchel, R., Agle, B., & Wood, D. (1997). Toward a theory of stakeholder identification and salience: defining the principle of who and what really counts. *The Academy of Management Review*, *22*(4), 853-886. Retrieved January 10, 2018, from https://www.jstor.org/stable/pdf/259247.pdf?refreqid=excelsior%3A0833b017d7e8d83581167934ebfcb14c

Mol-Lous, A. (2011). *Passend onderwijs: Haute couture of Zeeman? Inaugurele rede bij de installatie van het lectoraat passend onderwijs/inclusive education* [Appropriate education: Haute couture or ready-to-wear? Inaugural address]. Hogeschool Leiden.

Mooij, T., Hoogeveen, L., Driessen, G., van Hell, J., & Verhoeven, L. (2007). *Succescondities voor onderwijs aan hoogbegaafde leerlingen. Eindverslag van drie deelonderzoeken* [Conditions for successfully educating gifted learners. Final results of three coherent research projects]. Radboud Universiteit.

Moon, S. (2009a). Myth 15: High-ability students don't face problems and challenges. *Gifted Child Quarterly*, *53*(4), 274-276. https://doi.org/10.1177/0016986209346943

Moon, S. (2009b). Theories to guide effective curriculum development. In J. VanTassel-Baska, T. Cross, & F. Olenchack (Eds.), *Social-emotional curriculum with gifted and talented students* (pp. 11-40). Prufrock Press.

Morelock, M. (1996). On the Nature of Giftedness and Talent: Imposing Order on Chaos. *Roeper Review, 19*(1), 4-12. https://doi.org/10.1080/02783199609553774

Morris Miller, E. (2008). Conceptions of giftedness. In J. Plucker, & C. Callahan (Eds.), *Critical issues and practices in gifted education* (pp. 107-117). Prufrock Press.

Mulder, M., Weigel, T., & Collins, K. (2007). The concept of competence in the development of vocational education and training in selected EU member states: A critical analysis. *Journal of Vocational Education and Training, 59*(1), 65-85. https://doi.org/10.1080/13636820601145630

Murray, J., & Male, T. (2005). Becoming a teacher educator: Evidence from the field. *Teaching and Teacher Education, 21*(2), 125-142. https://doi.org/10.1016/j.tate.2004.12.006

Neihart, M., & Yeo, L. (2018). Psychological issues unique to the gifted student. In S. Pfeiffer, E. Shaunessy-Dedrick, & M. Foley-Nicpon (Eds.), *APA handbook of giftedness and talent* (pp. 497-510). American Psychological Association.

Neihart, M., Pfeiffer, S., & Cross, T. (Eds.). (2016). *The social and emotional development of gifted children. What do we know?* (2nd ed.). Prufrock Press.

Nelson, J., & Campbell, C. (2017). Evidence-informed practice in education: meanings and applications. *Educational Research, 59*(2), 127-135. https://doi.org/10.1080/00131881.2017.1314115

Nieuwboer, A. (2015). *Veilige basis in de klas* [A (pedagogically) safe classroom environment]. [Master's thesis, Faculty of Social and Behavioural Sciences, Universiteit Utrecht]. Utrecht University Repository. https://dspace.library.uu.nl/handle/1874/311796.

Obergriesser, S., & Stoeger, H. (2015). The role of emotions, motivation, and learning behavior in underachievement and results of an intervention. *High Ability Studies, 26*(1), 176-190. https://doi.org/10.1080/13598139.2015.1043003

Oliver, M. (1996). *Understanding disability: From theory to practice*. Macmillan Press.

Onderwijscoöperatie. (2014a). *Bekwaamheidseisen* [Professional standards for teachers]. Ingangsdatum August 1, 2017. De onderwijscoöperatie.

Onderwijscoöperatie. (2014b). *Herijking bekwaamheidseisen. Verantwoording* [Reevaluating professional standards for teacher education]. De onderwijs-coöperatie.

Onderwijsraad. (2006). *Naar meer evidence based onderwijs* [Towards evidence-based education]. Retrieved February 22, 2016, from https://www.onderwijsraad.nl/upload/documents/

Onderwijsraad. (2014). *Een eigentijds curriculum* [Modern curriculum]. Retrieved November 15, 2017, from https://www.onderwijsraad.nl/publicaties/2014/een-eigentijds-curriculum/volledig/item7128

Oostdam, R. (2009). *Tijd voor dikke leerkrachten: over maatwerk als kern van goed onderwijs* [It is time for excellent teachers: Customisation as core concept for high quality education]. Hogeschool van Amsterdam.

Oostdam, R., Peetsma, T., & Blok, H. (2007). *Het nieuwe leren in basisonderwijs en voortgezet onderwijs: een verkenningsnotitie voor het Ministerie van Onderwijs, Cultuur en Wetenschap*. [Current educational strategies in primary and secondary education: An exploration for the Dutch ministry of education]. SCO-Kohnstamm Instituut.

Oppong, E., Shore, B., & Muis, K. (2018). Clarifying the connections among giftedness, metacognition, self-regulation, and self-regulated learning: Implications for theory and practice. *Gifted Child Quarterly*, *62*(1), 1-18. https://doi.org/10.1177/0016986218814008

Pameijer, N. (2012). *Samen sterk* [Benefits of educational partnership]. Acco.

Pameijer, N. (2017). *Handelingsgericht werken: samen werken aan schoolsucces* [The action-oriented approach: Collaboration and successful education]. Acco.

Pameijer, N., & Denys, A. (2018). *Handelingsgericht werken: samen werken aan schoolsucces (Vlaamse editie)* [The action-oriented approach: Collaboration and successful education (Flemish edition)]. Acco.

Pameijer, N., & van Beukering, T. (2007). *Handelingsgericht werken: een handreiking voor de intern begeleider* [The action-oriented approach: suggestions for SENCo's]. Acco.

Pashler, H., McDaniel, M., Rohrer, D., & Björk, R. (2008). Learning styles: Concepts and evidence. *Psychological science in the public interest*, 9(3), https://doi.org/10.1111/j.1539-6053.2009.01038.x

Peters, S., & Gates, J. (2010). Teacher observation form: revisions and updates. *Gifted Child Quarterly*, 54(3), 179-188. https://doi.org/10.1177/0016986210369258

Peters, S., Erstad, H., & Matthews, M. (2017). Advanced Academics: A response to intervention perspective on gifted education. In J. Plucker, A. Rinn, & M. Makel (Eds.), *From giftedness to gifted education* (pp. 267-282). Prufrock Press.

Peters, S., Matthews, M., McBee, M., & McCoach, D. (2014). *Beyond gifted education: Designing and implementing advanced academic programs.* Prufrock Press.

Peters, S., & Oveross, M. (2020). Response to intervention. In J. Plucker, & C. Callahan (Eds.), *Critical issues and practices in gifted education* (pp. 377-386). Prufrock Academic Press.

Peterson, J. (2008). *The essential guide to talking with gifted teens.* Free Spirit.

Peterson, J. (2009). Myth 17: Gifted and talented do not have unique social and emotional needs. *Gifted Child Quarterly*, *53*, 280-282. https://doi.org/10.1177/0016986209346946

Peterson, J., & Lorimer, M. (2011). Student response to a small-group affective curriculum in a school for gifted children. *Gifted Child Quarterly*, *55*(3), 167-180. https://doi.org/10.1177/0016986211412770

Pfeiffer, S. (Ed.). (2008). *Handbook of giftedness in children.* Springer Science and Business Media.

Pfeiffer, S., & Muniz Prado, R. (2018). Counselling the gifted: current status and future prospects. In S. Pfeiffer (Ed.), *Handbook of giftedness in children. Psychoeducational theory, research and best practices* (2nd ed., pp. 299-313). Springer Science and Business Media.

Piechowski, M. (2014). *'Mellow out', they say. If only I could. Intensities and sensitivities of the young and bright* (2nd ed.). Royal Fireworks Press.

Pillen, M., Willemse, M., & Boei, F. (2018). Professionalisering van leraren-opleiders bij het doen en begeleiden van onderzoek [The professional development of teacher educators in conducting and coaching research processes]. In F. Boei, & M. Willemse (Eds.), *Kennisbasis lerarenopleiders: Katern 5. Onderzoek in de lerarenopleidingen* [Fundaments of professional knowledge for teacher educators: Part 5. Research at teacher education centres] (pp. 165-179). Velon.

Plat, M. (2016). *Eerste hulp bij hoogbegaafdheid.* SlimVaardig.

Plucker, J., Rinn, A., & Mackel, M. (2017). *From giftedness to gifted education: Reflecting theory in practice.* Prufrock Press.

Pols, W. (2017). De generieke kennisbasis van de lerarenopleider op de pabo [The foundation of generic knowledge of teacher educators in under-graduate teacher educating programmes]. In G. Geerdink, & I. Pauw (Eds.), *Kennisbasis lerarenopleiders: Katern 3. Inhoud en vakdidactiek op de lerarenopleidingen* [Fundaments of professional knowledge for teacher educators: Part 3. Subject-specific knowledge and didactics in teacher education] (pp. 29-40). Velon.

Portenga, S. (2017). Athletic talent development: from theory to practice. In J. Plucker, A. Rinn, & M. Mackel, *From giftedness to gifted education: Reflecting theory in practice* (pp. 115-148). Prufrock Press.

Prieto, M., & Ferrando, M. (2009). Emotional intelligence: re-examining some preconceptions. In T. Balchin, B. Hymer, & D. Matthews (Eds.), *The Routledge international companion to gifted education* (pp. 149-154). Routledge.

Prodia. (2012). *Protocollering van diagnostiek: het protocol bij vermoeden van hoogbegaafdheid* [Protocol for the assessment of students considered to be gifted]. http://www.prodiagnostiek.be/

Randolph, J. (2009). A guide to writing the dissertation literature review. *Practical Assessment, Research and Evaluation*, *14*(13).

Registerleraar.nl. (2017). *Bepalingen en procedures voor aanbieders van professionaliseringsactiviteiten (4.0). Bijlage bij het reglement van registratie en herregistratie 2015, versie 3.0.* [Rules and regulations for providing continuous professional development for teachers]. Registerleraar.nl.

Reis, S., Baum, S., & Burke, E. (2014). An operational definition of twice-exceptional learners: Implications and applications. *Gifted Child Quarterly*, *58*(3), 217-230. https://doi.org/10.1177/0016986214534976

Renzulli, J. (1985). The three-ring conception of giftedness: A developmental model for creative productivity. In R. Sternberg, & J. Davidson (Eds.), *Conceptions of giftedness* (pp. 53-92). Cambridge University Press.

Renzulli, J. (2012). Reexamining the role of gifted education and talent development for the 21st century: a four-part theoretical approach. *Gifted Child Quarterly*, *56*, 150-159. https://doi.org/10.1177/0016986212444901

Resing, W., & Vogelaar, B. (2017). Dynamisch testen: ook begaafde leerlingen verschillen in instructiebehoefte [Dynamic testing of gifted students: how gifted learners differ in their needs for instruction]. In E. van Gerven (Ed.), *De gids: over begaafdheid in het basisonderwijs* [Guide for gifted education in primary schools] (2nd ed., pp. 47-58). Leuker.nu.

Riedl-Cross, J., & Cross, T. (2017). Providing for the positive psychological development of students with gifts and talents. In J. Riedl-Cross, C. O'Reilly & T. Cross (Eds.), *Providing for the special needs of students with gifts and talents* (pp. 199-234). CTYI Press.

Rijksoverheid [Dutch government]. (n.d.). *Passend onderwijs* [Appropriate education]. Retrieved July 15, 2019, from https://www.passendonderwijs.nl/in-en-om-de-school/leraren-en-passend-onderwijs/hoogbegaafdheid/.

Riley, T. (2011). *Teaching gifted students in the inclusive classroom.* Prufrock Press Inc.

Rinn, A. (2018). Social and emotional considerations for gifted students. In S. Pfeiffer, E. Shaunessy-Dedrick, & M. Foley-Nicpon (Eds.), *APA handbook of giftedness and talent* (pp. 453-464). American Psychological Association.

Rinn, A., & Majority, K. (2018). The social and emotional world of the gifted. In S. Pfeiffer (Ed.), *Handbook of giftedness in children: Psychoeducational theory, research and best practices* (2nd ed., pp. 49-63). Springer International Publishing.

Roberts, J. L., & Inman, T. F. (2013). *Teacher's survival guide: Differentiating instruction in the elementary classroom.* Prufrock Press.

Roberts, J., & Inman, T. (2015). *Strategies for differentiating instruction: Best practices for the classroom.* Prufrock Press.

Robertson, S., & Pfeiffer, S. (2016). Development of a procedural guide to implement response to intervention (RtI) with high-ability learners. *Roeper Review, 38*(1), 9-23. https://doi.org/10.1080/02783193.2015.1112863

Robinson, A. (2008). Teacher characteristics. In J. Plucker, & C. Callahan (Eds.), *Critical issues and practices in gifted education* (pp. 669-681). Prufrock Press.

Robinson, A. (2015, August 10-14). *What makes a practice best? Evidence-based recommendations in gifted education.* [Conference Keynote Speech]. World Council Gifted & Talented Children, 2015: Odense, Denmark. [Video]. YouTube. https://www.youtube.com/watch?v=dNs0L7jJ3pg

Rodriguez, J. (2019). Moving towards inclusive education for all: examining stakeholder perceptions of educating students with special needs in schools operated by the Jordan field of UNRWA. *International Journal of Disability, Development and Education*. https://doi.org/10.1080/1034912X.2019.1671580

Roelofs, T. (2017). Samen beoordelen van beroepsvaardigheden: de praktijk-toets [Collaborative assessment of professional competence: assessing practice]. In M. Timmermans, & C. van Velzen (Eds.), *Kennisbasis lerarenopleiders: Katern 4. Samen in de school opleiden* [Fundaments of professional knowledge for teacher educators: Part 4. Educating aspiring teachers during apprenticeships] (pp.141-142). Velon.

Rogers, K. (2007). Lessons learned about educating the gifted and talented: A synthesis of the research of the educational practice. *Gifted Education Quarterly, 51*, 382-396. https://doi.org/10.1177/016253215592498

Schleiger, A. (Ed.). (2012). *Preparing teachers and developing school leaders for the 21st century: Lessons from around the world*. OECD Publishing. http://www.oecd-ilibrary.org/docserver/download/9812021e.pdf

Schrover, E. (2015). *Uitdagend onderwijs aan begaafde leerlingen* [Challenging education for gifted learners]. Koninklijke Van Gorcum.

Schuman, H. (2010). *Inclusief onderwijs: Dilemma's en uitdagingen* [Inclusive education: Dilemmas and challenges]. Garant.

Schuman, H., & de Vries, P. (2020). *Passend Onderwijs in de praktijk: kern-thema's & casuïstiek* [Appropriate education in practice. Core themes and case-studies]. Perspectief Uitgevers.

Scott, T., Callahan, C., & Urquhart, J. (2008). Paint-by-number teachers and cookie-cutter students: The unintended effects of high-stakes testing on the education of gifted students. *Roeper Review*, *31*(1), 40-52.

Segers, E., & Hoogeveen, L. (2012). *Programmeringstudie inzake excellentie-onderzoek primair, voortgezet en hoger onderwijs* [Suggestions for research into excellence at primary, secondary and higher education]. Radboud Universiteit.

Sharples, J. (2013). *Evidence for the frontline: Alliance for Useful Evidence*. Retrieved August 21, 2020 from https://www.alliance4usefulevidence.org/publication/evidence-for-the-frontline/

Shaunessy-Dedrick, E. (2018). Looking back and looking forward: curriculum for gifted and talented students. In S. Pfeiffer (Ed.), *Handbook of giftedness in children* (2nd ed., pp. 149-162). Springer.

Shavinina, L. (Ed.). (2009). *International handbook of giftedness.* Springer.

Siegle, D., & McCoach, D. (2017). Underachievement and the gifted child. In S. Pfeiffer, E. Shaunessy-Dedrick, & M. Foley-Nicpon (Eds.), *APA handbook of giftedness and talent* (pp. 559-573). American Psychological Association.

Siegle, D., & McCoach, D. (2020). Underachievers. In J. Plucker, & C. Callahan (Eds.), *Critical issues and practices in gifted education* (pp. 521-534). Prufrock Academic Press.

Siegle, D., McCoach, D., & Bloomfield, E. (2017). Achievement orientation model: understanding how what we believe determines whether we achieve. In J. Plucker, A. Rinn, & M. Makel (Eds.), *From giftedness to gifted education* (pp. 301-318). Prufrock Press.

Simpson, A. (Ed.). (2020). *The evidential basis of "evidence-based education".* Routledge.

Sisk, D. (2009). Myth 13: The Regular Classroom. *Gifted Child Quarterly*, *53*(4), 269-271. https://doi.org/10.1177/0016986209346939

SLO (n.d.). *21e-eeuwse vaardigheden* [21st century skills]. https://slo.nl/thema/meer/21e-eeuwsevaardigheden/

SLO (2010). Doelen- en vaardighedenlijst [Overview of educational objectives and skills]. https://talentstimuleren.nl/thema/stimulerend-signaleren/hulpmiddel/259-doelen-en-vaardighedenlijst

Slob, A. (2018a, March 14). *Passend onderwijs en hoogbegaafdheid* [Kamerbrief] ["Appropriate Education" and giftedness. Letter to the Dutch House of Commons]. https://www.rijksoverheid.nl/documenten/kamerstukken/2018/03/14/kamerbrief-over-uitvoering-regeerakkoord-middelen-hoogbegaafdheid

Slob, A. (2018b, December 7). *Regeling subsidie begaafde leerlingen po en vo* [Governmental funding for gifted education in primary and secondary education]. https://wetten.overheid.nl/BWBR0041640/2019-01-01

Sluijsmans, D., & van der Klink, M. (2017). Het beoordelen van de beroepsbekwaamheid van startende leraren: naar een evenwichtig samenspel tussen opleidingsinstituut en school [The assessment of professional competence of new teachers: the search for a balanced collaboration between teacher-education and schools]. In M. Timmermans, & C. van Velzen (Eds.), *Kennisbasis lerarenopleiders: Katern 4. Samen in de school opleiden* [Fundaments of professional knowledge for teacher educators: Part 4. Educating aspiring teachers during apprenticeships] (pp. 123-140). Velon.

Smit, F., Sluiter, R., & Driessen, G. (2006). *Literatuurstudie ouderbetrokkenheid in internationaal perspectief* [Educational partnership between parents and schools. International literature review]. ITS – Radboud Universiteit.

Smits, J. (2019). "Speciaal onderwijs" voor slimme leerlingen ["Special education" for bright students]. https://www.onderwijsbrabant.nl/wp-content/uploads/2019/03/OB2019-Rapport-Smits-hoogbegaafdheid.pdf

Snyder, K., Nietfeld, J., & Linnenbrink-Garcia, L. (2011). Giftedness and metacognition: A short-term longitudinal investigation of metacognitive monitoring in the classroom. *Gifted Child Quarterly*, *5*(3), 181-193. https://doi.org/10.1177/0016986211412769

Stack, N., & Sutherland, M. (2017). Inclusive school and classroom provision for gifted and talented students. In J. Riedl-Cross, C. O'Reilly & T. Cross (Eds.), *Providing for the needs of students with gifts and talents* (pp. 57-86). CTYI Press.

Sternberg, R. (2002). *Succesvolle intelligentie* [Successful intelligence]. Swets & Zeitlinger.

Sternberg, R. (2009). Teaching for successful intelligence: Principles, practices and outcomes. In J. C. Kaufman, & E. L. Grigorenko (Eds.), *The essential Sternberg* (pp. 183-194). Springer Publishing Company.

Sternberg, R. (2016) *Giftedness in the 21st century. How to make this world a better place?* 15th Echa conference, 2-5 March, Vienna.

Sternberg, R. (2017). ACCEL: A new model for identifying the gifted. *Roeper Review*, *39*(3), 152-169.

Stevens, L. (1997). *Over denken en doen: Een pedagogische bijdrage aan adaptief onderwijs* [About thinking and acting: A pedagogical contribution to appropriate education]. Procesmanagement Primair Onderwijs.

Stoeger, H., & Ziegler, A. (2011). Self-regulatory training through elementary-school students' homework completion. In B. Zimmerman, & D. Schunk (Eds.), *Handbook of self-regulation of learning and performance* (pp. 87-101). Routledge.

Stoeger, H., & Obergriesser, S. (2015). The role of emotions, motivation, and learning behavior in underachievement and results of an intervention. *High Ability Studies*, *26*(1), 167-196. https://doi.org/10.1080/13598139.2015.1043003

Stoeger, H., Fleischman, S., & Obergriesser, S. (2016). Self-regulated learning (SRL) and the gifted learner in primary school: the theoretical basis and empirical findings on a research program dedicated to ensuring that all students learn to regulate their own learning. *Asia Pacific Educational Review*, *16*, 257-267. https://doi.org/10.1007/s12564-015-9376-7

Struyven, K., & Ceulemans, C. (2012). Zorg om de goede leerkracht [Concerns about good teachers]. *Tijdschrift voor Lerarenopleiders*, *34*(2), 17-28.

Subotnik, R., Olszewski-Kubilus, P., & Worrel, F. (2011). Rethinking giftedness and gifted education: A proposed direction forward based on psychological science. *Psychological Science in the Public Interest*, *12*(1), 3-54. https://doi.org/10.1177/1529100611418056.

Subotnik, R., Stoeger, H., & Olszweski-Kubilius, P. (2017). Talent development research, policy, and practice in Europe and the United States: Outcomes from a summit of international researchers. *Gifted Child Quarterly*, *61*(3), 262-269. https://doi.org/10.1177/0016986217701839

Sutherland, M., & Stack, N. (2014). Ability as an additional support need: Scotland's inclusive approach to gifted education. *CEPS Journal*, *4*(3), 73-87. https://doi.org/URN:urn:nbn:de:0111-opus-96180

Thijs, A., Fisser, P., & van der Hoeven, M. (2014). *21e-eeuwse vaardigheden in het curriculum van het funderend onderwijs: Een conceptueel kader* [21st century skills as part of the curriculum for primary education: A conceptual framework]. SLO.

Tirri, K. (2017). Teacher education is the key to changing the identification and teaching of the gifted. *Roeper Review*, *39*(3), 210-2012. https://doi.org/10.1080/02783193.2017.1318996

Tomlinson, C., Kaplan, S., Renzulli, J., Purcell, J., Leppien, J., Burns, D., Strickland, C., & Imbeau, M. (2009). *The parallel curriculum: a design to develop learner potential and challenge advanced learners.* Corwin Press.

Touron, J., & Freeman, J. (2017). Gifted education in Europe: implication for policymakers and educators. In S. Pfeiffer, E. Shaunessy-Dedrick, & M. Foley-Nicpon (Eds.), *APA handbook of giftedness and talent* (pp. 55-70). American Psychological Association.

Trail, B. (2011). *Twice exceptional gifted children: Understanding, teaching and counseling gifted students.* Prufrock Press.

Troxclair, D. (2010). Differentiating instruction for gifted students in regular education social studies classes. *Roeper Review*, 22(3), 195-198. https://doi.org/10.1080/02783190009554033

Troxclair, D. (2013). Pre-service teacher attitudes towards giftedness. *Roeper Review*, 35, 58-64. https://doi.org/10.1080/02783193.2013.740603

Unesco. (1994). *The Salamanca statement and framework for action on special needs education.* United Nations Educational, Scientific and Cultural Organization/Ministry of Education and Science.

Unesco. (2015). *Rethinking education: Towards a global common good?* Unesco.

Van Braak, J., VanderLinden, R., & Aelterman, A. (2008). De wisselwerking tussen onderwijsonderzoek en onderwijspraktijk: de rol van de lerarenopleiding [The interplay between educational research and educational practice]. *Tijdschrift voor Lerarenopleiders*, *29*(4), 5-12.

van den Berg, E., & Suasso, E. (n.d.). De beroepspraktijk en het leren en de continue professionalisering van leraren [Professional practice, learning and the continuous professional development of teachers]. In M. Timmermans, & C. van Velzen (Eds.), *Kennisbasis lerarenopleiders: Katern 4. Samen in de school opleiden* [Fundaments of professional knowledge for teacher educators: Part 4. Educating aspiring teachers during apprenticeships] (pp. 105-122). Velon.

van den Heuvel, D. (2007). *Oudervereniging Pharos 20 jaar*. Retrieved May 16, 2018, from http://home.wxs.nl/~heuve533/Twintig%20jaar%20Pharos.htm

van der Donk, M., Tjeenk-Kalff, A., & Hiemstra-Beernink, A. (2015). *Beter bij de les: training executieve functies* [Attention in education: Training your students' executive functions]. Lannoo.

van der Meer, N. (2016). *Minder onderwijzen, meer leren: Handelingsgericht werken 2.0* [Less teaching, more learning*:* The action-oriented approach 2.0]. Bazalt Educatieve Uitgeverijen.

van der Poel, L. (2017). *Bijdehante baby's en pittige peuters* [Bright babies and clever toddlers]. Leuker.nu.

van Gerven, E. (2008). *Slim beleid: keuzes en consequenties bij beleid voor hoogbegaafde leerlingen in het basisonderwijs* [Smart policies on gifted education: Choices and consequences, developing a school policy for primary schools]. Koninklijke Van Gorcum.

van Gerven, E. (2014). *Knapzak praktijkgidsen: uitdagend onderwijs* [Back to basics in education: Challenging education]. Leuker.nu.

van Gerven, E. (2015). *Knapzak praktijkgidsen: de cirkel van zorg voor intern begeleiders* [Back to basics in education: the change and solution-focused approach]. Leuker.nu.

van Gerven, E. (2016). *Module 2: Profileren van educatieve behoeften. Opleiding specialist begaafdheid* [Profiling educational needs. Postgraduate teacher education for specialists in gifted education]. Slim! Educatief.

van Gerven, E. (2017). Inclusie en onderwijs aan begaafde leerlingen: Twee geloven op één kussen? [Inclusion and educating gifted learners: Sleeping with the devil?] In E. van Gerven (Ed.), *De gids: over begaafdheid in het basisonderwijs* [Guide for gifted education in primary schools] (pp. 19-36). Leuker.nu.

van Gerven, E. (2020). Executive functions, executive skills and gifted learners. In C. Fugate, C. Boswell, & W. Behrens (Eds.), *Understanding the needs of twice-exceptional learners: Building a bridge between research and practice* (pp. 33-70). Prufrock Press.

van Gerven, E., & Drent, S. (2012). *Digitaal handelingsprotocol hoogbegaafdheid* [Digital toolkit for identifying and educating gifted learners]. Koninklijke Van Gorcum.

van Gerven, E., & Hoogenberg, I. (2011). *Begaafd begeleiden: competentie-matrix voor de specialist begaafdheid* [Gifted education: Competency matrix for specialists in gifted education]. Koninklijke Van Gorcum.

van Gerven, E., & Weterings-Helmons, A. (2014). *Specialist dubbel-bijzondere leerlingen* [Specialist twice-exceptional learners]. Slim! Educatief. https://www.slimeducatief.nl/cursus/specialist-dubbel-bijzonder/

van Hove, G., Mortier, K., & De Schrauwer, E. (2005). *Onderzoek inclusief onderwijs* [Research on inclusive education]. Universiteit vakgroep Orthopedagogiek: Ministerie van de Vlaamse Gemeenschap.

van Meersbergen, E., & de Vries, P. (2017). *Handelingsgericht werken in passend onderwijs: achtergronden, aanpakken en hulpmiddelen* [The action-oriented approach for "Adequate education"]. Perspectief uitgevers.

van Meersbergen, E., & Jeninga, J. (2012). De ecologie van de leerling: Een systeemgericht model voor het onderwijs [The student's ecological system: A systemic approach to education]. *Tijdschrift voor Orthopedagogiek, 51*(4), 175-185.

van Swet, J. (2009). Diagnostiek vanuit een oplossingsgericht perspectief. In E. van Gerven (Ed.), *Handboek hoogbegaafdheid* [Handbook of giftedness] (pp. 38-57). Koninklijke Van Gorcum.

van Veen, K., Zwart, R., Meirink, J., & Verloop, N. (2010). *Professionele ontwikkeling van leraren: Een reviewstudie naar effectieve kenmerken van professionaliseringsinterventies van leraren* [The professional development of teachers: Review study into characteristics of effective interventions for professionalisation of teachers]. ICLON/Expertisecentrum Leren van Docenten. https://www.rug.nl/research/portal/files/15093764/Review_professionele_ontw_leraren.pdf

Vandenberghe, R. (2005). Samenwerking onderzoek en praktijk: mogelijkheden en grenzen [Collaboration for research and practice: opportunities and limitations]. *Pedagogische Studiën*, *82*, 262-274.

Venderickx, K., & Kieboom, T. (2017). Sociaal-emotionele ontwikkeling. In E. van Gerven (Ed.), *De gids: over begaafdheid in het basisonderwijs* [Guide for gifted education in primary schools] (2nd ed., pp. 101-112). Leuker.nu.

Vermeulen, M. (2016). *Leren te organiseren* [Learning how to organise]. [Inaugural address]. Oratie Open Universiteit.

Vermeulen, M., & Vrieling, E. (n.d.). *21e-eeuwse vaardigheden: achtergronden en onderwijsimplicaties in zeventien vragen en antwoorden* [21st century skills: background and implications for education; Seventeen questions and answers]. Welten-instituut, Open Universiteit.

Verouden, J. (2017). Peergroeponderwijs. In E. van Gerven (Ed.), *De gids: over begaafdheid in het basisonderwijs* [Guide for gifted education in primary schools] (2nd ed., pp. 171-185). Leuker.nu.

Vlaams Ministerie van Onderwijs en Vorming. (n.d.). *Pre-ambule eindtermen voor de 21ste eeuw: "Een baken met een vliegwiel"* [National standards for 21st century skills]. http://docs.vlaamsparlement.be/pfile?id=1140369

Vlaamse Onderwijsraad. (2017). *Advies over de relatie onderwijscurricula en beroepskwalificaties* [Advice on the interplay between curriculum and professional standards]. https://assets.vlor.be/www.vlor.be/import/ar-ar-adv-1718-004.pdf

Vlaams parlement [Flemish parliament]. (2019, December 18). Plenaire vergadering. Actuele vraag over de begeleiding van hoogbegaafde kinderen op school van Kathleen Krekels aan minister Ben Weyts [Plenary assembly. Question asked by Kathleen Krekels to minister Ben Weyts of the Flemish parliament about the support received by gifted children at school] [Filmfragment]. Retrieved January 3, 2020 from https://www.vlaamsparlement.be/plenaire-vergaderingen/1347160/verslag/1353006

Vlaamse Regering. (2008, January 17). *Bijlage bij het besluit van de Vlaamse Regering van 5 oktober 2007 betreffende de basiscompetenties van de leraren*. [Appendix to the decision of the Flemish Government regarding basic teacher competences]. Belgisch Staatsblad (pp. 1594-1631). http://eindtermen.vlaanderen.be/lerarenopleiding/documenten/Basiscompetenties_2007.pdf

Vlaamse Regering. (2019). *Regeerakkoord van de Vlaamse Regering* 2019-2024 [Flemish Coalition Agreement 2019-2024]. https://www.vlaanderen.be/publicaties/regeerakkoord-van-de-vlaamse-regering-2019-2024

Vlaams Ministerie van Onderwijs en Vorming. (n.d.). Hoogbegaafde leerlingen [Gifted students]. Vlaamse Overheid, Vlaams Ministerie van Onderwijs en Vorming. Retrieved December 11, 2017, from http://onderwijs.vlaanderen.be/nl/hoogbegaafde-leerlingen

Vlaanderen in Onderwijs en Vorming. (n.d.). Onderwijsdoelen [Educational objectives]. Retrieved April 29, 2020, from https://www.kwalificatiesencurriculum.be:onderwijsdoelen

Vogelaar, B. (2017). *Dynamic testing and excellence: Unfolding potential*. Universiteit Leiden.

Voogt, J., & Pareja Roblin, N. (2010). *21st century skills: Discussienota* [21st century skills: Discussion paper]. Kennisnet.

Vreijs, C., Ndanjo Ndungbogun, G., Kieboom, T., & Venderickx, K. (2017). Training effects on Belgian preschool and primary school teachers' attitudes towards the best practices for gifted children. *High Ability Studies,* (1), 1-20. https://doi.org/10.1080/13598139.2017.1312295

Vuyk, M., Krieshok, T., & Kerr, B. (2016). Openness to experience rather than overexcitabilities: Call it like it is. *Gifted Child Quarterly*, 60(3), 191-211. https://doi.org/10.1177/0016986216645407

Vygotsky, L. (1978). *Mind in society: The development of higher psychological processes*. Harvard University Press.

Wallace, B., & Adams, H. (2019). TASC: thinking actively in a social context: a universal framework for developing thinking skills and problem solving across the curriculum. In B. Wallace, D. Sisk, & J. Senior (Eds.), *The SAGE handbook of gifted and talented education* (pp. 246-261). Sage Publishers.

Wallace, B., Leyden, S., Montgomery, D., Winstanley, C., Pomerantz, M., & Fitton, S. (2010). *Raising the achievement of all pupils within an inclusive setting: Practical strategies for developing best practice.* Routledge.

Wallace, B., Sisk, D., & Senior, J. (Eds.). (2019). *The SAGE handbook of gifted and talented education.* Sage.

Waslander, S. (2011). *Passend onderwijs, passend beleid?* [Appropriate education, appropriate policies?]. ECPO.

Wassink, H., & Bakker, C. (2014). Het goede leren: over normatieve professionalisering van leraren. In R. Kneyber, & J. Evers (Eds.), *Het alternatief: weg met de afrekencultuur in het onderwijs!* (pp. 223-236). Boom.

Wassink, H., Ruijgrok, C., & Hermsen, B. (2018). Het fysieke aspect van reflectie: Een speelse ingang tot betekenisvol leren van leraren. [The physical aspect of learning: a different approach of meaningful learning by teachers]. *Tijdschrift voor Lerarenopleiders*, *39*(1), 5-14.

Webb, J., Amend, E., Beljan, P., Webb, N., Kuzujanakis, M., Olenchak, R., & Goerss, J. (2016). *Misdiagnosis and dual diagnosis of gifted children and adults: ADHD, bipolar, OCD, Asperger's disorder, depression and other disorders*. Great Potential Press.

Wet op het primair onderwijs (WPO) [Law on primary education]. (2005). https://wetten.overheid.nl/BWBR0003420/2005-01-01/0/informatie

Wet passend onderwijs [Law on appropriate education]. (2012, October 11). https://wetten.overheid.nl/BWBR0032176/2018-01-01/0/

Wet primair onderwijs BES [Law on primary education]. (2016, January 18). http://wetten.overheid.nl/BWBR0030280/2016-01-18

Weterings, A. (2017). Begaafde leerlingen met ernstige reken- en wiskundeproblemen en dyscalculie [Gifted learners with severe mathematical problems and dyscalculia]. In E. van Gerven (Ed.), *De gids: over begaafdheid in het basisonderwijs* [Guide for gifted education in primary schools] (2nd ed., pp. 255-278). Leuker.nu.

Wilder, E., Greene, W., & Kim, Y. (2013). Core reflection as a catalyst for change in teacher education. In F. Korthagen, Y. Kim, & W. Greene (Eds.), *Teaching and learning from within: A core reflection approach to quality and inspiration in education* (pp. 186-198). Routledge.

Wiliam, D. (2011). *Embedded formative assessment*. Solution Tree Press.

Willingham, D. (2009). *Why don't students like school: A cognitive scientist answers questions about how the mind works and what it means for your classroom*. Jossey-Bass.

Winstanley, C. (2010). *The ingredients of challenge*. Trentham Books.

Wolters, C. (2020). *Meer dan slim: Talentgerichte benadering van hoogbegaafde kinderen en jongeren* [They are not only bright: A talent-focused approach for gifted children and adolescents]. Pica.

Wolters, P. (2009). Werken met protocollen [The use of a protocol]. In E. van Gerven, *Handboek hoogbegaafdheid* [Handbook of giftedness] (pp. 22-36). Koninklijke Van Gorcum.

Wood, S. (2018). Professional school counselors and gifted educators: Working towards solutions. In J. Danielian, C. Fugate, & E. Fogarty (Eds.), *Teaching gifted children: Success strategies for teaching high-ability learners* (pp. 379-391). Prufrock Press.

Ylvisaker, M., & Feeney, T. (2014). Helping children without making them helpless: Facilitating development of executive self-regulation in children and adolescents. In V. Anderson, R. Jacobs, & P. J. Anderson (Eds.), *Executive functions and the frontal lobes: A lifespan perspective* (pp. 409-438). Psychology Press.

Zelazo, P., & Carlson, S. (2012). Hot and cool executive function in childhood and adolescence: Development and plasticy. *Child Development Perspectives*, *6*, 354-360. https://doi.org/10.1111/j.1750-8606.2012 .00246.x

Zhang, J., Liu, G., & Lin, C. (2012). An action-oriented approach to gifted education: Evidence from the field of scientific creativity. *High Ability Studies*, *23*(1), 123-125.

Ziegler, A., Stoeger, H., & Balestrini, D. (2017). Systemic Gifted Education. In J. Riedl-Cross, C. O'Reilly, & T. Cross (Eds.), *Providing for the special needs of students with gifts & talents* (pp. 15-56). CTYI Press.

Appendix A

Competency Matrix for Specialists in Gifted Education

	1. Concepts of giftedness	2. Seeing educational needs	3. Understanding educational needs	4. Responding to educational needs	5. Assessing the response to intervention
	The specialist has an in-depth knowledge of the theoretical perspectives of giftedness and gifted education and is able to interpret this knowledge contextually in their professional practice.	The specialist takes a change and solution-focused approach to profiling the educational needs of gifted students. In this approach, all actors in the student's ecological system are seen as equal educational partners.	The specialist interprets and understands the educational needs of gifted students in their ecological context i.e., an interactive system of the student, the teacher, the curriculum, physical learning environment, the family, and peers.	The specialist responds meaningfully to the educational needs of the gifted student from a contextual and developmental perspective. In doing so, (learning) process and (learning) results are equally important.	The specialist monitors the process of responding meaningfully to the gifted student's educational needs and monitors and interprets the gifted student's responses to these interventions contextually.
1. Pedagogical The teacher creates a safe and stimulating learning environment for every student, tuned to their individual educational needs.	***1.1.*** The specialist in gifted education understands and envisions how their perspective on giftedness influences their ability to see, understand and respond to the pedagogical needs of gifted children.	***2.1.*** The specialist explores the strengths and weaknesses of the gifted student within the intrapersonal and interpersonal developmental domain.	***3.1.*** The specialist develops an understanding of how the actors in the student's ecology influence the student's strengths and weaknesses.	***4.1.*** The specialist encourages the student's attitude for learning in a way that an equal balance between effort, learning and achievement is stimulated.	***5.1.*** The specialist monitors the student's intrapersonal and interpersonal functioning to determine how this development contributes to the student's ability for self-actualisation.
2. Didactical In collaboration with their colleagues, the teacher transfers curricular objectives in such a way that they become teachable in the context of a primary school, matching school policy. They must compose the curriculum in a manner that enables their student to master the content.	***1.2.*** The specialist creates a coherency between the concepts of giftedness, educational objectives and educational programming and the intervention strategies to be used.	***2.2.*** The specialist assesses the student's developmental levels in all academic domains offered, considering the student's cognitive style.	***3.2.*** The specialist explains how the interplay between the curriculum and the student's cognitive style influences their development and instructional needs.	***4.2.*** The specialist adjusts the curriculum in such a way that the student has to reach for their zone of proximal development in all domains.	***5.2.*** The specialist monitors the student's academic achievements and relates these achievements to the level of support given during the learning process.

	1. Concepts of giftedness	2. Seeing educational needs	3. Understanding educational needs	4. Responding to educational needs	5. Assessing the response to intervention
3. Managerial The teacher provides a structured, well-organised, task-oriented classroom atmosphere that supports the students' developmental process.	***1.3.*** The specialist in gifted education understands how their perspective of giftedness influences the organisation of the educational programme provided.	***2.3.*** The specialist takes on a systematic approach to the process of profiling educational needs and manages their time and actions in a way that contributes to the optimal results of this process.	***3.3.*** The specialist understands how the individual student's needs are of consequence for their classroom management.	***4.3.*** The specialist adjusts their classroom management in a way that offers the opportunity to provide instruction/support to the gifted students.	***5.3.*** The specialist plans how they can monitor and document the student's development periodically based on a set schedule.
4. Communication The teacher communicates professionally with all educational partners who are involved with the student's education.	***1.4.*** Based on interaction between the educational partners, the specialist in gifted education develops a perspective of giftedness, and an approach to gifted students that matches their individual needs.	***2.4.*** The specialist approaches all actors in the student's ecological system as equal partners and values their information about the student's educational needs as such.	***3.4.*** Based on the results of their exploration of the student's educational needs, the specialist explains to the actors in the student's ecological system how their behaviour influences the student's development.	***4.4.*** The specialist uses the input of the educational partners; designing and implementing meaningful responses to educational needs for individual gifted students.	***5.4.*** The specialist discusses the student's overall development with all the relevant actors in the student's ecological system.
5. Research and innovation The teacher adjusts their professional knowledge, perspectives and actions if and when relevant developments in education, society and science require it.	***1.5.*** The specialist in gifted education relates their in-depth knowledge of giftedness to perspectives of current society in general and more specifically to the current perspectives on education.	***2.5.*** The specialist is able to develop a personal toolkit for profiling the educational needs of gifted students flexibly and appropriately, using different instruments in different contexts.	***3.5.*** The specialist uses theory on giftedness and RtI to interpret the results of the exploration of the student's educational needs in the context of educational development.	***4.5.*** The specialist underpins their interventions with current theories on education and giftedness. In doing so they ensure that the interventions are workable in their daily practice.	***5.5.*** Based on current theories of assessment and giftedness, the specialist designs a system of formative and summative assessment for the education of gifted students.

Table B1

Domain 1, Competency 1.1

Pedagogical	Concepts of Giftedness			
	The specialist in gifted education understands and envisions how their perspective on giftedness influences their ability to see, understand and respond to the pedagogical needs of gifted children. The specialist...			
The teacher creates a safe and stimulating learning environment for every student, tuned to their individual educational needs.	K.1.A.1	Knows that the interaction between aptitude and development influences a student's achievements.	V.1.A.1	Stimulates the student's understanding of their personal ability.
	K.1.B.1	Knows how personal traits during developmental processes influence a student's achievements.	V.1.B.1	Discusses with the student how the student is of influence on their personal development.
	K.1.C.1	Knows how peers influence a student's developmental process.	V.1.C.1	Stimulates gifted students to interact with gifted peers.
	K.1.D.1	Knows which risk factors might influence a student's wellbeing.	V.1.D.1	Explores how they can improve the student's well-being in education.
	K.1.E.1	Knows why autonomy, competency and relationship are basic pedagogical needs for gifted students.	V.1.E.1	Makes gifted students responsible for their development, matching their individual ability to take up this responsibility.
	K.1.F.1	Knows different current theories on giftedness.	V.1.F.1	Explains to the educational partners how their perspective of giftedness influences their choices for the education of a gifted child.

Table B2

Domain 1, Competency 1.2

Didactical	Concepts of Giftedness			
	The specialist creates a coherency between the concepts of giftedness, educational objectives and educational programming and the intervention strategies to be used. The specialist...			
In collaboration with their colleagues, the teacher transfers curricular objectives in such a way that they become teachable in the context of a primary school, matching school policy. They must compose the curriculum in a manner that enables their students to master the content.	K.1.A.2	Knows why continuity is important for the curriculum of gifted students.	V.1.A.2	Sets educational objectives that match the individual needs of gifted students.
	K.1.B.2	Knows how curricular adjustments can be of consequence for the way that they provide instruction and support for gifted students.	V.1.B.2	Ensures the match between instruction and the individual needs of gifted students.
	K.1.C.2	Knows how they can improve their didactical interventions by focusing on cooperative and collaborative learning.	V.1.C.2	Creates situations wherein cooperative and collaborative learning is encouraged.
	K.1.D.2	Knows different didactical strategies to challenge gifted students in their zone of proximal development.	V.1.D.2	Discusses with the individual student what emotional responses can occur during the learning process.
	K.1.E.2	Knows that each theoretical perspective has a different focus on useful didactical interventions.	V.1.E.2	Selects interventions that are a match with the student's needs.
	K.1.F.2	Knows that a different perspective on giftedness also means a different focus on the ultimate educational objectives.	V.1.F.2	Sets educational objectives that are fit for the school's perspective of giftedness.

Table B3

Domain 1, Competency 1.3

Managerial	Concepts of Giftedness			
	The specialist in gifted education understands how their perspective on giftedness influences the organisation of the educational programme provided. The specialist...			
The teacher provides a structured, well-organised, task-oriented classroom atmosphere that supports the students' developmental process.	K.1.A.3	Knows how their perspective on giftedness influences the selection of students who are offered an adjusted curriculum.	V.1.A.3	Ensures that a challenging curriculum is available for all their students.
	K.1.B.3	Knows why it is important to pay attention to a student's executive skills.	V.1.B.3	Explores whether their expectations of the student's self-regulation skills equal the skills the student has in reality.
	K.1.C.3	Knows that there are different strategies for cooperative and collaborative learning for peer group education.	V.1.C.3	Deliberately organises interaction between gifted students of an equal level to stimulate their learning process.
	K.1.D.3	Knows which conversation techniques are helpful for the coaching of gifted students.	V.1.D.3	Discusses with the student what they think might contribute positively to the effectivity of their learning process.
	K.1.E.3	Knows why they have to assess the student's learning process and achievements.	V.1.E.3	Provides feed forward on a daily basis on the gifted student's learning process and achievements, as they do for their other students as well.
	K.1.F.3	Knows which classroom management strategies may be helpful to help them organise instruction for gifted students.	V.1.F.3	Organises pedagogical and didactical interventions within realistic limits in their daily practice.

Table B4

Domain 1, Competency 1.4

Communication	Concepts of Giftedness			
	Based on interaction between the educational partners, the specialist in gifted education develops a perspective of giftedness, and an approach to gifted students that matches their individual needs. The specialist...			
The teacher communicates professionally with all educational partners who are involved with the student's education.	K.1.A.4	Knows why it is important to organise gifted education from the joint perspective of the educational partners.	V.1.A.4	Discusses with the educational partners which educational objectives should be set for the individual gifted student.
	K.1.B.4	Knows what the theoretical perspectives are on the learning process of gifted students.	V.1.B.4	Discusses with the educational partners how they can contribute positively to the gifted student's developmental process.
	K.1.C.4	Knows how communication between educational partners contributes to the social functioning of gifted students.	V.1.C.4	Encourages open communication between all educational partners involved.
	K.1.D.4	Knows how they can tune their way of communicating to the personal traits of individual gifted students.	V.1.D.4	Sets up an open dialogue with the gifted student in order to stimulate the student's construct of a realistic self-concept.
	K.1.E.4	Knows how the interaction between them and the gifted student contributes to a well-balanced developmental process for this student.	V.1.E.4	Discusses chances and thresholds the gifted student encounters in their developmental process.
	K.1.F.4	Knows why collaboration between the educational partners is important to ensure high quality in gifted education.	V.1.F.4	Creates a joint perspective on giftedness between the educational partners.

Table B5

Domain 1, Competency 1.5

Research and innovation	Concepts of Giftedness			
	The specialist in gifted education relates their in-depth knowledge of giftedness to the perspectives on current society in general and more specifically to the current perspectives on education. The specialist...			
The teacher adjusts their professional knowledge, perspectives and actions if and when relevant developments in education, society and science require it.	K.1.A.5	Is aware of the different perspectives on the risks and effects of acceleration for gifted students.	V.1.A.5	Makes a risk analysis considering the pros and cons of acceleration of an individual student.
	K.1.B.5	Invests in personal continuous professional development regarding theories of giftedness.	V.1.B.5	Monitors in daily practice how new didactical and pedagogical approaches influence a student's learning process.
	K.1.C.5	Is aware of different theoretical perspectives on the social functioning of gifted students.	V.1.C.5	Provides a safe social classroom climate to stimulate the learning process of gifted students.
	K.1.D.5	Understands how they can monitor the well-being of gifted students.	V.1.D.5	Chooses interventions that can contribute to the well-being of gifted students.
	K.1.E.5	Understands how current perspectives in education match current perspectives of theories on giftedness.	V.1.E.5	Uses current educational strategies to teach students how to take responsibility for their learning.
	K.1.F.5	Is aware of the levels and domains of expertise of their colleagues and knows how they can put this expertise to use.	V.1.F.5	Offers feedback to their colleagues to improve the quality of gifted education.

Table B6

Domain 2, Competency 2.1

Pedagogical	Seeing Educational Needs			
	The specialist explores the strengths and weaknesses of the gifted student within the intrapersonal and interpersonal developmental domain. The specialist...			
The teacher creates a safe and stimulating learning environment for every student, tuned to their individual educational needs.	K.2.A.1	Knows how the curriculum offered affects a student's well-being.	V.2.A.1	Has teacher-student conversations about how the gifted student experiences and values the curriculum offered.
	K.2.B.1	Knows how the gifted student's cognitive style affects their interaction with classmates.	V.2.B.1	Stimulates students to work in groups where they complement each other and support each other's strengths.
	K.2.C.1	Knows how the gifted student's strengths can influence their interpersonal relationships with classmates.	V.2.C.1	Observes how gifted students use their strengths when interacting with classmates.
	K.2.D.1	Knows what types of behaviour by the gifted student may be a sign for the development of negative self-concept.	V.2.D.1	Selects appropriate techniques to explore and describe the student's self-concept.
	K.2.E.1	Knows how a peer group can influence a gifted student's well-being.	V.2.E.1	Encourages the interaction between a gifted student and their classmates in a way that teaches them how to achieve to the best of their joint ability.
	K.2.F.1	Knows how instruments for social observation can provide understanding in a gifted student's interpersonal functioning.	V.2.F.1	Selects the appropriate instruments to explore a gifted student's social functioning.

Table B7

Domain 2, Competency 2.2

Didactical	Seeing Educational Needs			
	The specialist assesses the student's developmental levels in all academic domains offered, considering the student's cognitive style. The specialist...			
In collaboration with their colleagues, the teacher transfers curricular objectives in such a way that they become teachable in the context of a primary school, matching school policy. They must compose the curriculum in a manner that enables their students to master the content.	K.2.A.2	Knows how to determine a student's zone of proximal development.	V.2.A.2	Uses valid instruments to determine the student's zone of proximal development.
	K.2.B.2	Has sufficient knowledge of current theories on learning styles.	V.2.B.2	Observes how differentiation in the addressed learning style of the offered curriculum influences the student's learning behaviour.
	K.2.C.2	Knows how actors in the student's ecological system affect a student's achievements.	V.2.C.2	Explores the influences of classmates on the gifted student's achievements.
	K.2.D.2	Knows how underachievement affects a student's self-concept.	V.2.D.2	Discusses with the gifted student the emotions that occur during their learning process.
	K.2.E.2	Knows how executive skills affect the developmental process of gifted students.	V.2.E.2	Profiles a gifted student's strengths and weaknesses regarding their executive skills.
	K.2.F.2	Knows the limitations of school achievements as a reliable indicator for giftedness.	V.2.F.2	Takes other indicators besides school achievements into consideration when estimating if the student may be gifted.

Table B8

Domain 2, Competency 2.3

Managerial	Seeing Educational Needs			
	The specialist takes on a systematical approach to the process of profiling educational needs and manages their time and actions in a way that contributes to the optimal results of this process. The specialist...			
The teacher provides a structured, well-organised, task-oriented classroom atmosphere that supports the students' developmental process.	K.2.A.3	Knows why it is important to follow up on a gifted student's school achievements.	V.2.A.3	Systematically acquires information on a student's academic achievements.
	K.2.B.3	Knows why it is relevant to offer the student different opportunities through a choice of products to show the results of the learning process.	V.2.B.3	Encourages the student to choose different products to show the results of the learning process.
	K.2.C.3	Knows how educational needs are affected by social group processes.	V.2.C.3	Explores how, in the case of unexpected results, the student's social context influences the student's learning process.
	K.2.D.3	Knows how they can profile a student's self-concept.	V.2.D.3	Explores the student's self-concept in such a way that it fits their assessment question and matches the student's abilities.
	K.2.E.3	Knows how they can use observation techniques as a way to explore the student's ability for self-regulation.	V.2.E.3	Observes the student's self-regulating skills in a situation where the student is required to use these skills to the best of their ability.
	K.2.F.3	Knows how (f)actors in the student's ecological system affect a student's academic achievements.	V.2.F.3	Explores how (f)actors in the student's ecological system influence the student's academic achievements.

Table B9

Domain 2, Competency 2.4

Communication	Seeing Educational Needs			
	The specialist approaches all actors in the student's ecological system as equal partners and values their information about the student's educational needs as such. The specialist...			
The teacher communicates professionally with all educational partners who are involved with the student's education.	K.2.A.4	Knows which information the educational partners need in order to provide a matching curriculum for the student.	V.2.A.4	Writes a research report with a concrete suggestion of how to adjust the student's curriculum.
	K.2.B.4	Knows that students may show different task-oriented strategies in different circumstances.	V.2.B.4	Asks educational partners to describe how the student executes an assignment.
	K.2.C.4	Knows how information about the student's social functioning can contribute to a student's well-being.	V.2.C.4	Explores similarities and differences in the student's social behaviour under different circumstances.
	K.2.D.4	Knows how solution-focused conversation techniques can contribute to the development of a student's self-concept.	V.2.D.4	Is focused on solutions and learnable skills instead of being focused on problems when talking with a student.
	K.2.E.4	Understands the student's ability to assess the consequences of their choices.	V.2.E.4	Consults the student when prioritising intervention strategies.
	K.2.F.4	Understands why it is important to explain the objectives of the assessment of educational needs to the educational partners.	V.2.F.4	Confers with educational partners on how assessment can contribute to the improvement of the student's education.

Table B10

Domain 2, Competency 2.5

Research and innovation	Seeing Educational Needs			
	The specialist is able to develop a personal toolkit for profiling educational needs of gifted students flexibly and appropriately using different instruments in different contexts. The specialist...			
The teacher adjusts their professional knowledge, perspectives and actions if and when relevant developments in education, society and science require it.	K.2.A.5	Knows how they can determine the academic level of an underachieving student.	V.2.A.5	Sets up the didactical assessment to challenge the student intrinsically to strive for optimal results.
	K.2.B.5	Knows different theories on higher and lower order thinking skills.	V.2.B.5	Profiles strengths and weaknesses in the student's use of thinking skills.
	K.2.C.5	Knows why it is important to take on different perspectives when exploring a gifted student's problematic social-emotional behaviour.	V.2.C.5	Explores the social-emotional problems of gifted students equally well from both the perspective of giftedness and from a perspective of possible learning problems.
	K.2.D.5	Sees educational partners as important sources to explore a student's strengths and weaknesses.	V.2.D.5	Explores if the student's perspective of themselves is identical to the perspective of them held by their educational partners.
	K.2.E.5	Knows that a student's self-regulation skills may be corroded when the student's well-being decreases.	V.2.E.5	Explores the influence of the student's self-regulation skills on their self-concept.
	K.2.F.5	Knows how they can use different instruments in order to set up a personalised assessment plan.	V.2.F.5	Develops a personalised assessment plan prior to the start of an assessment.

Table B11

Domain 3, Competency 3.1

Pedagogical	Understanding Educational Needs			
	The specialist develops an understanding of how the actors in the student's ecology influence the student's strengths and weaknesses. The specialist...			
The teacher creates a safe and stimulating learning environment for every student, tuned to their individual educational needs.	K.3.A.1	Understands how the physical learning environment influences a student's academical achievements.	V.3.A.1	Creates a learning environment wherein the learning process and achievements are equally important.
	K.3.B.1	Understands the relevance of the notion of ownership to find solutions during the process of mentoring a student.	V.3.B.1	Discusses with the student how their personal behaviour can contribute to success.
	K.3.C.1	Understands how, in the change-oriented approach, the process of data interpretation takes place.	V.3.C.1	Explores the student's social functioning within the context in which it takes place.
	K.3.D.1	Knows that gifted students need instruction when new elements of the curriculum are offered.	V.3.D.1	Discusses with the student their need for instruction when new elements in the curriculum are offered.
	K.3.E.1	Knows that behaviour is the result of the interaction between the student and the ecological system.	V.3.E.1	Discusses with the student what they see as helpful learning strategies and interventions.
	K.3.F.1	Knows how a solution-focused approach contributes to the ability to see a problem with a correct perspective.	V.3.F.1	Works with the educational partners to determine when a problem is not as intense as supposed or is non-existent.

Table B12

Domain 3, Competency 3.2

Didactical	Understanding Educational Needs			
	The specialist explains how interaction between the curriculum and the student's cognitive style influences their development and instructional needs. The specialist...			
In collaboration with their colleagues, the teacher transfers curricular objectives in such a way that they become teachable in the context of school and/or district policy. They must compose the curriculum in a manner that enables their students to master the content.	K.3.A.2	Knows the curricular constructs within the different academic domains.	V.3.A.2	Determines if the pacing of the required curriculum matches the student's ability to develop knowledge and skills.
	K.3.B.2	Knows that differentiation between learning styles contributes to a student's development.	V.3.B.2	Analyses the way the student's learning style influences the student's approach to working on an assignment.
	K.3.C.2	Knows why knowledge has to have a social meaning for a student.	V.3.C.2	Determines how peers within the social environment respond to the differentiated curriculum offered to the gifted student.
	K.3.D.2	Knows how mindset influences a student's learning process.	V.3.D.2	Discusses with the student to what extent their self-efficacy related to a specific task is accurate.
	K.3.E.2	Knows how the learning process itself contributes to the development of a student's executive skills.	V.3.E.2	Evaluates whether the teacher's expectations towards a student's executive skills are realistic.
	K.3.F.2	Knows how a student's academic achievements are influenced by teacher support during the learning process.	V.3.F.2	Points out how teacher support while working on assignments influences a student's achievements.

Table B13

Domain 3, Competency 3.3

Managerial	Understanding Educational Needs			
	The specialist understands how the individual student's needs are of consequence for their classroom management. The specialist...			
The teacher provides a structured, well-organised, task-oriented classroom atmosphere that supports the students' developmental process.	K.3.A.3	Knows that gifted students need instruction on the new elements in the curriculum.	V.3.A.3	Arranges specific times to provide instruction for gifted students when they work on an individualised curriculum.
	K.3.B.3	Knows how factors in the physical educational environment influence the development of executive skills.	V.3.B.3	Creates a learning environment where the use of executive skills is an intrinsic condition during the learning process.
	K.3.C.3	Understands the relevance for gifted students to work with like-minded peers for their social development.	V.3.C.3	Creates the opportunity for gifted students to work with like-minded peers.
	K.3.D.3	Knows how interaction with the gifted student can provide understanding in order to adjust their actions to meet a student's needs.	V.3.D.3	Takes the time for teacher-student meetings to discuss the student's self-concept.
	K.3.E.3	Knows that, like other students, gifted students have to learn how to become responsible for their own actions.	V.3.E.3	Involves the student in the development of their individual intervention plan.
	K.3.F.3	Knows how the balance between the most desirable and most realistic interventions affects their own pedagogical and didactical strategies.	V.3.F.3	Clusters students with identical educational needs into small groups.

Table B14

Domain 3, Competency 3.4

Communication	Understanding Educational Needs			
	Based on the results of their exploration of the student's educational needs (academic and social-emotional), the specialist explains to the actors in the student's ecological system how their behaviour influences the student's development. The specialist...			
The teacher communicates professionally with all the educational partners who are involved with the student's education.	K.3.A.4	Knows how different levels of thinking skills may require different support strategies.	V.3.A.4	Analyses the level of thinking skills called for in an assignment.
	K.3.B.4	Knows different reflection techniques to be used by students.	V.3.B.4	Teaches the student to reflect on their own learning process.
	K.3.C.4	Knows that gifted students need like-minded peers in order for well-balanced social development.	V.3.C.4	Determines if the gifted student has ample opportunity to meet and work with like-minded peers.
	K.3.D.4	Knows how they can modify language when communicating with their students.	V.3.D.4	Communicates their exploration of the student's educational needs in a manner that is understood by, and relevant to, the gifted student.
	K.3.E.4	Has a general understanding of different theories of metacognitive skills development in children.	V.3.E.4	Provides the student with a level of responsibility that matches their developmental level.
	K.3.F.4	Knows why it is important for educational partners to be included in the assessment process of the student's educational needs.	V.3.F.4	Communicates regularly with educational partners about the student's educational development.

Table B15

Domain 3, Competency 3.5

Research and innovation	Understanding Educational Needs			
	The specialist uses theories of giftedness and RtI to interpret the results of the exploration of the student's educational needs in the context of educational development. The specialist...			
The teacher adjusts their professional knowledge, perspectives and actions if and when relevant developments in education, society and science require it.	K.3.A.5	Knows that learning behaviour is the result of the interaction between personal traits and the educational environment.	V.3.A.5	Explains to the educational partners when their interaction contributes positively to a student's development.
	K.3.B.5	Knows different theoretical models of giftedness.	V.3.B.5	Selects intervention strategies, based upon their theoretical perspective of giftedness, that are helpful in understanding the student's educational needs.
	K.3.C.5	Understands environmental influences on the development of students according to theories of giftedness.	V.3.C.5	Evaluates how the educational environment affects a student's behaviour.
	K.3.D.5	Knows which risk factors contribute to underachievement.	V.3.D.5	Identifies factors in individual students that contribute to underachievement.
	K.3.E.5	Knows how giftedness may influence a student's behaviour.	V.3.E.5	Identifies how giftedness may influence the student's behaviour.
	K.3.F.5	Understands how strengths and weaknesses can be translated into educational needs.	V.3.F.5	Describes a gifted student's educational needs as skills to be developed in the near future.

Table B16

Domain 4, Competency 4.1

Pedagogical	Responding to Educational Needs			
	The specialist encourages the student's attitude for learning in a way that stimulates an equal balance between effort, learning, and achievement. The specialist...			
The teacher creates a safe and stimulating learning environment for every student, tuned to their individual educational needs.	K.4.A.1	Knows the consequences for a student's achievement; essential stages are skipped in the learning process.	V.4.A.1	Ensures the student's active involvement during the four stages of the learning process: instruction-planning-execution-evaluation.
	K.4.B.1	Knows why it is important for a student to be able to switch between different learning styles.	V.4.B.1	Encourages the gifted student to use learning styles that are appropriate for the assignment given.
	K.4.C.1	Knows how collaboration between students can help them optimise their achievement.	V.4.C.1	Encourages gifted students to collaborate with other gifted students.
	K.4.D.1	Knows how like-minded peers can contribute to a student's development of a realistic self-concept.	V.4.D.1	Ensures that the gifted student can learn in an educational environment with other like-minded students.
	K.4.E.1	Knows how positive and negative emotions toward achievement influence a student's learning behaviour.	V.4.E.1	Coaches the student on how to deal with positive and negative emotions towards achievement and that they are a natural part of the learning process.
	K.4.F.1	Knows how a teacher's attitude contributes to the development of a student's self-concept.	V.4.F.1	Emphasises the student's efforts during the learning process and after finishing the assignment.

Table B17

Domain 4, Competency 4.2

Didactical	Responding to Educational Needs			
	The specialist adjusts the curriculum in such a way that the student has to reach for their zone of proximal development in all domains. The specialist...			
In collaboration with their colleagues, the teacher transfers curricular objectives in such a way that they become teachable in the context of school and/or district policy. They must compose the curriculum in a manner that enables their students to master the content.	K.4.A.2	Knows different techniques for curriculum compacting.	V.4.A.2	Adapts pacing in the student's curriculum by compacting.
	K.4.B.2	Knows the difference between lower and higher order thinking skills.	V.4.B.2	Presents the curriculum in such a way that higher order thinking skills are explicitly required to finish a task.
	K.4.C.2	Knows which factors during collaborative processes contribute to the student reaching their zone of proximal development.	V.4.C.2	Offers a curriculum that differentiates in learning objectives in a way that collaboration with other students becomes beneficial for the gifted student.
	K.4.D.2	Knows how a student's self-efficacy influences a student's motivation to work on an assignment.	V.4.D.2	Presents the curriculum in a way that it becomes possible for a student to realistically assess their level of self-efficacy regarding the task.
	K.4.E.2	Knows how, by adjusting their didactical strategies, they can provide educational challenges for the student.	V.4.E.2	Encourages creative and critical thinking by developing an investigative attitude in students.
	K.4.F.2	Knows curricular objectives for academic domains for each developmental stage.	V.4.F.2	Offers enrichment activities based on a student's individual educational objectives in their zone of proximal development.

Table B18

Domain 4, Competency 4.3

Managerial	Responding to Educational Needs			
	The specialist adjusts their classroom management in a way that offers the opportunity to provide instruction/support to the gifted students. The specialist...			
The teacher provides a structured, well-organised, and task-oriented classroom atmosphere that supports the students' developmental process.	K.4.A.3	Understands that enrichment activities are a natural part of a gifted learner's curriculum.	V.4.A.3	Ensures that enrichment tasks are regularly planned for the gifted student.
	K.4.B.3	Knows why there is a joint responsibility between teachers and students regarding a student's time management.	V.4.B.3	Together with the gifted student, monitors the student's weekly planning for working on classroom assignments.
	K.4.C.3	Knows how communication between students influences the pedagogical classroom climate.	V.4.C.3	Teaches students how to communicate in an open and safe manner.
	K.4.D.3	Knows how a student's success and failure shows which instruction and coaching might be beneficial for a student's learning process.	V.4.D.3	Holds regular teacher-student conferences to discuss the student's successes and challenges.
	K.4.E.3	Knows how their appreciation of a student's learning experience may support the student's academic development.	V.4.E.3	Schedules regular teacher-student conferences to monitor a student's perception of their progress on classroom assignments.
	K.4.F.3	Knows different ways to approach their classroom management to provide regular instructional support for gifted students.	V.4.F.3	Provides support for students working on enrichment tasks within the regular classroom.

Table B19

Domain 4, Competency 4.4

Communication	Responding to Educational Needs			
	The specialist uses the input of the educational partners in designing and implementing meaningful responses to educational needs for individual gifted students. The specialist...			
The teacher communicates professionally with all educational partners who are involved with the student's education.	K.4.A.4	Knows how a partnership between educational stakeholders contributes positively to the chances of a successful educational process.	V.4.A.4	Ensures clear communication between the classroom teacher and the gifted-education teacher regarding educational objectives and learning content.
	K.4.B.4	Knows how a feeling of competence, autonomy, and relationship regarding the educational objectives contributes positively to a student's motivation.	V.4.B.4	Offers students a choice in the products they deliver to show proof of their learning results.
	K.4.C.4	Sees the student as the expert of their individual learning process.	V.4.C.4	Organises group activities aimed at the inclusion of gifted students in the regular classroom.
	K.4.D.4	Knows that it is important for gifted students to take part in regular group activities.	V.4.D.4	Asks the student what kind of support they find helpful to their learning process.
	K.4.E.4	Knows how a student's emotions about achievement are influenced by their opportunity to make personal choices for the learning process and products.	V.4.E.4	Involves the student in setting educational objectives.
	K.4.F.4	Knows how communication between the regular group and a pull-out group provides for an ongoing coaching process for gifted students.	V.4.F.4	Discusses educational programming with educational partners.

Table B20

Domain 4, Competency 4.5

Research and innovation	Responding to Educational Needs			
	The specialist underpins their interventions with current theories on education and giftedness. In doing so they ensure that the interventions are workable in their daily practice. The specialist...			
The teacher adjusts their professional knowledge, perspectives and actions if and when relevant developments in education, society and science require it.	K.4.A.5	Knows how they can create a balance between the curriculum and the student's need for support.	V.4.A.5	Offers educational plans that work for both teacher and student.
	K.4.B.5	Is aware of modern didactical strategies.	V.4.B.5	Focuses on active student participation during the learning process.
	K.4.C.5	Knows how they can make use of solution-focused strategies to help a student to overcome difficulties during the learning process.	V.4.C.5	Teaches the student how to recognise difficulties during the learning process and make them the topic of conversation between teacher and student.
	K.4.D.5	Understands how different factors can positively influence a student's development.	V.4.D.5	Ensures that the role of all relevant factors in the student's ecological system is tuned in such a way that they support each other.
	K.4.E.5	Knows why it is important for a student that both learning processes and achievements are discussed.	V.4.E.5	Makes learning visible by providing feedback.
	K.4.F.5	Understands why thinking about opportunities and chances is motivating while mentoring gifted students.	V.4.F.5	Discusses the student's needs for educational support from a solution-focused perspective.

Table B21

Domain 5, Competency 5.1

Pedagogical	Assessing the Response to Intervention			
	The specialist monitors the student's intrapersonal and interpersonal functioning to determine how this development contributes to the student's ability for self-actualisation. The specialist...			
The teacher creates a safe and stimulating learning environment for every student, tuned to their individual educational needs	K.5.A.1	Knows how the exploration of the student's perspective on the curriculum can be used to develop an understanding of the student's social-emotional functioning.	V.5.A.1	Discusses with the student how the curriculum contributes to their well-being.
	K.5.B.1	Knows how they can develop an understanding of the student's cognitive style.	V.5.B.1	Uses different tools to develop an understanding of the student's personal interests, learning styles, and multiple intelligences.
	K.5.C.1	Understands how social relationships influence the learning process.	V.5.C.1	Observes how the gifted student interacts with other students in relevant educational situations.
	K.5.D.1	Understands which factors might have an influence on a gifted student's self-concept.	V.5.D.1	Analyses result of observations; student and parent meetings, and shares them with their co-workers.
	K.5.E.1	Knows how a student's self-concept influences their self-efficacy.	V.5.E.1	Discusses with the student how didactical and pedagogical changes in their education might contribute to the student's development.
	K.5.F.1	Knows how interactions with their colleagues can provide them with a broader perspective on a student's self-concept and social functioning.	V.5.F.1	Shows how the student's social-emotional functioning is documented in such a way that it is proof of ongoing development.

Table B22

Domain 5, Competency 5.2

Didactical	Assessing the Response to Intervention			
	The specialist monitors the student's academic achievements and relates these achievements to the level of support given during the learning process. The specialist...			
In collaboration with their colleagues, the teacher transfers curricular objectives in such a way that they become teachable in the context of school and/or district policy. They must compose the curriculum in a manner that enables their students to master the content.	K.5.A.2	Knows how they can assess whether educational objectives outside the regular curriculum are achieved.	V.5.A.2	Uses alternative ways to document the student's achievements and developmental process.
	K.5.B.2	Knows how support-strategies regarding the student's cognitive style affect the student's development.	V.5.B.2	Documents what support strategies are used regarding the student's cognitive style.
	K.5.C.2	Knows how they can evaluate a student's social functioning as an influencing factor on the student's learning process.	V.5.C.2	Documents the student's social functioning regarding interactions with the teacher and their fellow classmates.
	K.5.D.2	Knows how their attitude affects a student's learning process.	V.5.D.2	Maintains an intrinsic interest in their students and shows this in the way that they communicate with them.
	K.5.E.2	Knows how they can understand the relationship between a student's achievements and their self-actualisation.	V.5.E.2	Discusses with the student how their achievements relate to the set objectives.
	K.5.F.2	Knows how they can share their knowledge and skills to monitor a student's development with their colleagues.	V.5.F.2	Collaborates with colleagues, exploring alternative ways to monitor a student's achievements.

Table B23

Domain 5, Competency 5.3

Managerial	Assessing the Response to Intervention			
	The specialist plans how they can monitor and document the student's development periodically, based on a set schedule. The specialist...			
The teacher provides a structured, well-organised, and task-oriented classroom atmosphere that supports the students' developmental process.	K.5.A.3	Knows why it is relevant to discuss the student's development with the educational partners at the start of the new term.	V.5.A.3	Plans and schedules times for formal assessment.
	K.5.B.3	Knows how a student's cognitive style affects their achievements.	V.5.B.3	Selects ways to document a student's achievements that matches their cognitive style.
	K.5.C.3	Knows how social well-being affects learning and achievement.	V.5.C.3	Considers the social circumstances influencing the student's learning process and achievements.
	K.5.D.3	Knows how emotional regulation skills influence situations of formal assessment.	V.5.D.3	Explores what contributes positively to a student's self-confidence while taking a test.
	K.5.E.3	Knows how the results of assessment can contribute to a student's ability for self-actualisation.	V.5.E.3	Uses test results as an opportunity to set new educational objectives with the student.
	K.5.F.3	Knows why a joint protocol for documenting a student's development contributes positively to the quality of education.	V.5.F.3	Sets out a protocol with colleagues on how to document the student's achievements.

Table B24

Domain 5, Competency 5.4

Communication	Assessing the Response to Intervention			
	The specialist discusses the student's overall development with all the relevant actors in the student's ecological system. The specialist...			
The teacher communicates professionally with all educational partners who are involved with the student's education.	K.5.A.4	Knows how the information from parents and co-workers can help to optimise a student's curriculum.	V.5.A.4	Plans meetings with parents and teachers to discuss a student's responses to intervention programmes.
	K.5.B.4	Understands why the observations of the student's cognitive style by educational partners is helpful for planning interventions.	V.5.B.4	Offers educational partners the opportunity to share relevant observations on the student's cognitive style.
	K.5.C.4	Knows a student's social functioning can be discussed with parents and co-workers from a positive perspective.	V.5.C.4	Discusses the student's social functioning from a perspective of chances and opportunities.
	K.5.D.4	Knows how sharing knowledge of the student's achievement contributes positively to their self-concept.	V.5.D.4	Explores with the student how they can create a balance between learning, effort, and achievement.
	K.5.E.4	Sees the ability of self-actualisation as a result of a joint effort between all the actors in the student's ecological system.	V.5.E.4	Discusses how all the actors in the student's ecological system contribute to the student's development.
	K.5.F.4	Understands that the regular classroom teacher has the final responsibility regarding the student's educational development.	V.5.F.4	Works together with the regular classroom teacher to set long-term educational objectives for the student.

Table B25

Domain 5, Competency 5.5

Research and innovation	Assessing the Response to Intervention			
	Based on current theories of assessment and giftedness, the specialist designs a system of formative and summative assessment for the education of gifted students. The specialist...			
The teacher adjusts their professional knowledge, perspectives and actions if and when relevant developments in education, society and science require it.	K.5.A.5	Knows why it is important to set assessable educational objectives.	V.5.A.5	Determines how they can assess the results of the enriched curriculum.
	K.5.B.5	Knows to what extent the student's cognitive style matches the curriculum offered and how that affects the student's educational needs.	V.5.B.5	Discusses with the student how they can improve learning by switching between different cognitive styles.
	K.5.C.5	Knows how understanding the student's level of social functioning can provide insights to optimise a student's development.	V.5.C.5	In collaborative learning, finds a balance between socially challenging the student and offering the opportunity of functioning in a social comfort zone.
	K.5.D.5	Knows how, in theories about learning processes, a student's self-perception is seen as fundamental to their educational development.	V.5.D.5	Explores to what extent the student has a realistic perspective of their own abilities.
	K.5.E.5	Knows how both formative and summative assessment can contribute to a student's self-actualisation.	V.5.E.5	Is able to use assessment to explain to a student the goal valuation of a specific task.
	K.5.F.5	Knows how theories on assessment coincide with theories of learning and theories in the teacher's approach.	V.5.F.5	Works with classroom teachers to establish assessments beyond commonly used standardised tools to evaluate the student's learning process and achievements.

Appendix C

Glossary of Terms for Respondents

As experienced by students in a classroom, many of the stakeholders in gifted education, whose voices are valued, come into this study with varying levels of experience that influence their understanding of certain terminology. The following glossary of terms has been provided as an additional reference resource for participants, so that they can review definitions of terms that may be unfamiliar to those from the different stakeholder groups.

Change-oriented work	The process in which concrete efforts are made to promote the student's development and to improve the conditions under which the student develops. In this process, not only striving for change, but also weighing up whether the change will benefit the student in the longer term.
Cognitive style	The way in which the student processes and produces information. For example, preferred domains based on Gardner's intuitions and executive skills affect the student's cognitive style.
Curriculum	A plan for learning in which educational experiences are offered in a systematic way.
Ecology of the student	Factors that influence the development and functioning of the student, including: the teacher, the curriculum, the physical learning environment, peers, and the family.
Educational needs	Those things that the student needs to develop. Educational needs do not always coincide with "something the student would like". Sometimes there is a difference between what is good for the student's development and what they would like.
Educational partners	Involved in the educating and upbringing of the student. For example, parents, classroom teacher, the gifted education teacher, and administrators. An involved pedagogue/psychologist could also be an educational partner.
Educational responses	The way educational partners work to meet the educational needs of the student.
Executive skills	Higher cognitive functions needed to plan and manage actions. This can be viewed as "the control tower" for controlling personal behaviour.
Feed forward	A way of guiding student insight, through dialogue, into what is good in their work and what can be improved in relation to the intended goal. Effective feedforward gives the student a clear insight into how improvements can be achieved; it focuses on their future development rather than on feedback on their past performance.
Formative assessment	Comparing the results of a student with results that he previously achieved in order to make personal growth visible.
Homegroup/room	The class/group where the student spends most of their time during the week.
Intrapersonal functioning	Self-insight, self-image and self-efficacy and the handling of the emotions that go with it.

Interpersonal functioning	The way in which the student functions in interaction with others, such as fellow students, parents and the group teacher.
Lower and higher order thinking skills	Thinking takes place at different levels. Lower-order thinking skills are skills that often rely on automated knowledge and skills. Higher-order thinking skills require insight into the content theme and require the ability to think critically and/or creatively.
Metacognitive skills	The ability of a person to think about the way that they think and act. This is needed by people to be able to reflect on their own learning.
Performance emotions	Emotions associated with the process of learning and performing.
Preferred domain	The preferred domain of a student is the way in which the student likes to process subject material. Think, for example, of Gardner's intelligences.
Profiling educational needs	The process of systematically identifying the development needs of the student.
Pull-out groups	A group where students are offered gifted services for part of the week outside of the general classroom.
Self-actualisation	The need and the ability to make optimum use of personal possibilities; for a person to get the best out of themselves.
Self-efficacy	The ability of a person to assess for themselves if they have the right knowledge and skills to handle a task.
Self-realisation	The ability of a person to make good choices from their own behavioural possibilities and to realise, evaluate, and improve these behaviours by themself or together with others.
Social comfort zone	The peer group in which the gifted student prefers to function with relatively little effort required.
Summative assessment	Comparing a student's results with the results from other students. It is a measurement by which a teacher may take stock of the effect of their teaching.
Zone of proximal development	Areas beyond which the student can no longer progress without needing instruction or help. This is the point at which the student will experience the need for effort and perseverance; the point at which learning starts.

Appendix D

Information About Stakeholders/Respondents

Table D1

Distribution of Stakeholders by Subgroups and Per Country

Subgroups	Percentages subgroups combined N=354	Percentages subgroups Netherlands N=299	Percentages subgroups Flanders N=55
Educational professionals (EP)	37.01 (N=131)	34.45 (N=103)	50.91 (N=28)
Specialists in gifted education (SGE)	29.10 (N=103)	33.44 (N=100)	5.45 (N=3)
Parents (Pa)	23.45 (N=83)	20.74 (N=62)	38.18 (N=21)
Rest	10.45 (N=37)	11.37 (N=34)	5.45 (N=3)

Table D2

Distribution of Stakeholders Divided in Subgroups Per Country

Subgroups	Percentages subgroups Netherlands	Percentages subgroups Flanders
Educational professionals (EP) (N=131)	78.63 (N= 103)	21.37 (N=28)
Specialists in Gifted Education (SGE) (N=103)	97.09 (N=100)	2.91 (N=3)
Parents (Pa) (N=83)	74.70 (N=62)	25.30 (N=21)
Rest (N=37)	91.89 (N=34)	8.11 (N=3)

Table D3

Distribution of Subgroups by Focus Group

(N = 354)	Year 1-2 (NL) = Kindergarten (F) (N = 14)	Year 3-5 (NL) = Year 1-3 (F) (N = 40)	Year 6-8 (NL) = Year 4-6 (F) (N = 66)	all year groups (NL) = all year groups (F) (N = 212)
Educational professionals (EP) (N = 131)	4.58	13.74	22.90	64.12
Specialists in Gifted Education (SGE) (N = 103)	2.91	0.97	11.65	77.67
Parents (Pa) (N = 83)	4.82	25.30	27.71	26.51
Rest (N = 37)	2.70	0.00	2.70	70.27

Note. NL = The Netherlands, F = Flanders.

Table D4

Distribution by CPD Level – Netherlands (N=299)

Subgroups	None	Incidentally	2-5 meetings	6-10 meetings	1-2 years	NOS
Educational professionals (EP) (N=103)	6.80	21.36	8.74	9.71	33.01	0.00
Specialists in Gifted Education (SGE) (N=100)	0.00	4.00	0.00	5.00	80.00	1.00
Parents (Pa) (N=62)	58.06	14.52	6.45	4.84	6.45	1.61
Rest (N=34)	17.65	32.35	11.76	11.76	20.59	0.00

Note. CPD level = level of continuous professional development. NOS = Not otherwise specified.

Table D5

Distribution by CPD Level – Flanders (N=55)

Subgroups	None	Incidentally	2-5 meetings	6-10 meetings	1-2 years	NOS
Educational professionals (EP) (N=28)	25.00	7.14	10.71	10.71	32.14	0.00
Specialists in Gifted Education (SGE) (N=3)	0.00	33.33	0.00	0.00	66.67	0.00
Parents (Pa) (N=21)	61.90	14.29	0.00	0.00	4.76	0.00
Rest (N=3)	0.00	0.00	0.00	0.00	0.00	0.00

Note. CPD level = level of continuous professional development. NOS = Not otherwise specified.

Appendix E

Tables Domain 1, Referred to in Chapter 5

Table E1

Frequency Table of the Selection Rates of the Knowledge Indicators for Competency 1.1

Competency	Indicator	EP	SGE	Pa	R	T
K.1.1	K.1.A.1	7.89	10.03	4.82	9.01	7.91
	K.1.B.1	17.30	18.77	19.28	15.32	17.98
	K.1.C.1	12.47	8.74	15.26	9.01	11.68
	K.1.D.1	23.41	25.57	29.72	27.93	25.99
	K.1.E.1	22.14	21.04	24.10	24.32	22.50
	K.1.F.1	16.79	15.86	6.83	14.41	13.94

Note. The results are presented per indicator for each group of stakeholders. EP = educational professionals; SGE = specialists gifted education; Pa = parents; R = Rest; T = total. N = 354.

All indicators are coded. K = knowledge; the first number refers to the vertical competency domain in the competency matrix; the second letter refers to the category of indicators. The second number refers to the actual competency the indicators are matched with. See Appendix B for the match between the competencies and the indicators.

Note. The results are presented per indicator for each group of stakeholders. EP = educational professionals; SGE = specialists gifted education; Pa = parents; R = Rest; T = total. N = 354.

All indicators are coded. K = knowledge; the first number refers to the vertical competency domain in the competency matrix; the second letter refers to the category of indicators. The second number refers to the actual competency the indicators are matched with. See Appendix B for the match between the competencies and the indicators.

Table E2

Frequency Table of the Selection Rates of the Knowledge Indicators for Competency 1.2

Competency	Indicator	EP	SGE	Pa	R	T
K.1.2	K.1.A.2	17.81	17.15	19.68	23.42	18.64
	K.1.B.2	18.07	22.98	15.66	19.82	19.11
	K.1.C.2	10.43	5.83	13.65	9.01	9.70
	K.1.D.2	28.50	28.48	30.12	27.93	28.81
	K.1.E.2	17.05	20.39	14.86	15.32	17.33
	K.1.F.2	8.14	5.18	6.02	4.50	6.40

Note. The results are presented per indicator for each group of stakeholders. EP = educational professionals; SGE = specialists gifted education; Pa = parents; R = Rest; T = total. N = 354.

All indicators are coded. K = knowledge; the first number refers to the vertical competency domain in the competency matrix; the second letter refers to the category of indicators. The second number refers to the actual competency the indicators are matched with. See Appendix B for the match between the competencies and the indicators.

Table E3

Frequency Table of the Selection Rates of the Knowledge Indicators for Competency 1.3

Competency	Indicator	EP	SGE	Pa	R	T
K.1.3	K.1.A.3	17.05	24.60	12.45	17.12	18.17
	K.1.B.3	25.70	22.65	22.49	25.23	24.01
	K.1.C.3	9.92	7.12	14.06	5.41	9.60
	K.1.D.3	13.99	11.00	19.68	18.92	14.97
	K.1.E.3	19.34	20.71	16.47	19.82	19.11
	K.1.F.3	13.99	13.92	14.86	13.51	14.12

Note. The results are presented per indicator for each group of stakeholders. EP = educational professionals; SGE = specialists gifted education; Pa = parents; R = Rest; T = total. N = 354.

All indicators are coded. K = knowledge; the first number refers to the vertical competency domain in the competency matrix; the second letter refers to the category of indicators. The second number refers to the actual competency the indicators are matched with. See Appendix B for the match between the competencies and the indicators.

Table E4

Frequency Table of the Selection Rates of the Knowledge Indicators for Competency 1.4

Competency	Indicator	EP	SGE	Pa	R	T
K.1.4	K.1.A.4	25.95	27.51	21.69	27.93	25.61
	K.1.B.4	7.63	3.88	4.02	3.60	5.27
	K.1.C.4	11.45	15.53	13.25	12.61	13.18
	K.1.D.4	16.03	12.62	21.29	10.81	15.73
	K.1.E.4	18.83	14.89	20.88	24.32	18.74
	K.1.F.4	20.10	25.57	18.88	20.72	21.47

Note. The results are presented per indicator for each group of stakeholders. EP = educational professionals; SGE = specialists gifted education; Pa = parents; R = Rest; T = total. N = 354.

All indicators are coded. K = knowledge; the first number refers to the vertical competency domain in the competency matrix; the second letter refers to the category of indicators. The second number refers to the actual competency the indicators are matched with. See Appendix B for the match between the competencies and the indicators.

Table E5

Frequency Table of the Selection Rates of the Knowledge Indicators for Competency 1.5

Competency	Indicator	EP	SGE	Pa	R	T
K.1.5	K.1.A.5	19.08	15.53	18.47	17.12	17.70
	K.1.B.5	25.19	28.48	17.27	26.13	24.39
	K.1.C.5	11.96	12.30	9.24	9.01	11.11
	K.1.D.5	17.05	13.27	22.89	18.02	17.42
	K.1.E.5	13.99	17.15	16.47	16.22	15.73
	K.1.F.5	12.72	13.27	15.66	13.51	13.65

Note. The results are presented per indicator for each group of stakeholders. EP = educational professionals; SGE = specialists gifted education; Pa = parents; R = Rest; T = total. N = 354.

All indicators are coded. K = knowledge; the first number refers to the vertical competency domain in the competency matrix; the second letter refers to the category of indicators. The second number refers to the actual competency the indicators are matched with. See Appendix B for the match between the competencies and the indicators.

Table E6

Frequency Table of the Selection Rates of the Skills Indicators for Competency 1.1

Competency	Indicator	EP	SGE	Pa	R	T
V.1.1	V.1.A.1	23.41	18.12	24.10	26.13	22.32
	V.1.B.1	21.37	21.68	16.87	18.02	20.06
	V.1.C.1	7.63	4.21	8.43	7.21	6.78
	V.1.D.1	14.76	10.03	15.66	10.81	13.18
	V.1.E.1	21.63	27.51	22.09	21.62	23.45
	V.1.F.1	11.20	18.45	12.85	16.22	14.22

Note. The results are presented per indicator for each group of stakeholders. EP = educational professionals; SGE = specialists gifted education; Pa = parents; R = Rest; T = total. N = 354.

All indicators are coded. V = skills; the first number refers to the vertical competency domain in the competency matrix; the second letter refers to the category of indicators. The second number refers to the actual competency the indicators are matched with. See Appendix B for the match between the competencies and the indicators.

Table E7

Frequency Table of the Selection Rates of the Skills Indicators for Competency 1.2

Competency	Indicator	EP	SGE	Pa	R	T
V.1.2	V.1.A.2	23.16	24.92	23.29	27.03	24.11
	V.1.B.2	13.74	14.56	16.87	11.71	14.50
	V.1.C.2	14.25	7.44	11.65	9.91	11.21
	V.1.D.2	15.52	13.92	19.28	17.12	16.10
	V.1.E.2	23.16	28.16	24.10	24.32	24.95
	V.1.F.2	10.18	11.00	4.82	9.91	9.13

Note. The results are presented per indicator for each group of stakeholders. EP = educational professionals; SGE = specialists gifted education; Pa = parents; R = Rest; T = total. N = 354.

All indicators are coded. V = skills; the first number refers to the vertical competency domain in the competency matrix; the second letter refers to the category of indicators. The second number refers to the actual competency the indicators are matched with. See Appendix B for the match between the competencies and the indicators.

Table E8

Frequency Table of the Selection Rates of the Skills Indicators for Competency 1.3

Competency	Indicator	EP	SGE	Pa	R	T
V.1.3	V.1.A.3	16.79	14.89	6.43	11.71	13.28
	V.1.B.3	12.47	10.03	22.09	15.32	14.31
	V.1.C.3	17.81	17.15	17.67	14.41	17.23
	V.1.D.3	18.07	16.83	25.70	21.62	19.87
	V.1.E.3	18.32	20.06	18.88	22.52	19.40
	V.1.F.3	16.54	21.04	9.24	14.41	15.91

Note. The results are presented per indicator for each group of stakeholders. EP = educational professionals; SGE = specialists gifted education; Pa = parents; R = Rest; T = total. N = 354.

All indicators are coded. V = skills; the first number refers to the vertical competency domain in the competency matrix; the second letter refers to the category of indicators. The second number refers to the actual competency the indicators are matched with. See Appendix B for the match between the competencies and the indicators.

Table E9

Frequency Table of the Selection Rates of the Skills Indicators for Competency 1.4

Competency	Indicator	EP	SGE	Pa	R	T
V.1.4	V.1.A.4	11.20	14.56	12.05	9.91	12.24
	V.1.B.4	23.16	25.24	26.51	26.13	24.86
	V.1.C.4	26.97	28.16	19.28	21.62	24.95
	V.1.D.4	19.08	15.21	20.48	20.72	18.46
	V.1.E.4	19.34	15.86	21.69	21.62	19.11
	V.1.F.4	0.25	0.97	0.00	0.00	0.38

Note. The results are presented per indicator for each group of stakeholders. EP = educational professionals; SGE = specialists gifted education; Pa = parents; R = Rest; T = total. N = 354.

All indicators are coded. V = skills; the first number refers to the vertical competency domain in the competency matrix; the second letter refers to the category of indicators. The second number refers to the actual competency the indicators are matched with. See Appendix B for the match between the competencies and the indicators.

Table E10

Frequency Table of the Selection Rates of the Skills Indicators for Competency 1.5

Competency	Indicator	EP	SGE	Pa	R	T
V.1.5	V.1.A.5	14.25	11.65	14.46	13.51	13.47
	V.1.B.5	12.21	13.27	12.85	11.71	12.62
	V.1.C.5	20.87	13.59	18.88	18.02	17.98
	V.1.D.5	21.37	19.42	17.27	22.52	19.96
	V.1.E.5	12.47	16.18	14.06	11.71	13.84
	V.1.F.5	18.83	25.89	22.49	22.52	22.13

Note. The results are presented per indicator for each group of stakeholders. EP = educational professionals; SGE = specialists gifted education; Pa = parents; R = Rest; T = total. N = 354.

All indicators are coded. V = skills; the first number refers to the vertical competency domain in the competency matrix; the second letter refers to the category of indicators. The second number refers to the actual competency the indicators are matched with. See Appendix B for the match between the competencies and the indicators.

Table E11

Comparison of the 10 Highest Indicator Selection Rates per Competency in Domain 1 per Group of Stakeholders

Indicator	EP	Indicator	SGE	Indicator	Pa	Indicator	R
V.1.E.3	18.32	V.1.E.3	20.06	K.1.A.4	21.69	V.1.E.3	22.52
V.1.D.5	21.37	K.1.A.3	24.60	K.1.B.3	22.49	V.1.D.5/V.1.F.5	22.52
V.1.A.2/V.1.E.2	23.16	K.1.D.1	25.57	V.1.F.5	22.49	K.1.B.3	25.23
K.1.D.1	23.41	V.1.F.5	25.89	K.1.D.5	22.89	K.1.B.5	26.13
V.1.A.1	23.41	K.1.A.4	27.51	V.1.A.1	24.10	V.1.A.1	26.13
K.1.B.5	25.19	V.1.E.1	27.51	V.1.E.2	24.10	V.1.B.4	26.13
K.1.B.3	25.70	V.1.E.2	28.16	V.1.D.3	25.70	V.1.A.2	27.03
K.1.A.4	25.95	V.1.C.4	28.16	V.1.B.4	26.51	K.1.D.1	27.93
V.1.C.4	26.97	K.1.D.2	28.48	K.1.D.1	29.72	K.1.D.2	27.93
K.1.D.2	28.50	K.1.B.5	28.48	K.1.D.2	30.12	K.1.A.4	27.93

Note. EP = educational professionals; SGE = specialists gifted education; Pa = parents; R = Rest; T = total. N= 354. If two indicators matching one competency had an identical selection rate, they are ranked at the same level.

All indicators are coded. K = knowledge; V = skills; the first number refers to the vertical competency domain in the competency matrix; the second letter refers to the category of indicators. The second number refers to the actual competency the indicators are matched with. See Appendix B for the match between the competencies and the indicators.

Table E12

Comparison of the Selection Rates of the Knowledge Indicators in Domain 1 Between the Groups of Stakeholders

Indicators	Group	N	Mean	*SD*	*SE*	95% Confidence interval for mean		Min	Max
						Lower bound	Upper bound		
K.1.A.	EP	131	2.63	1.138	0.099	2.44	2.83	0	5
	SGE	103	2.84	1.091	0.108	2.63	3.06	1	5
	Pa	83	2.31	1.147	0.126	2.06	2.56	0	5
	R	37	2.84	1.014	0.167	2.50	3.18	1	5
	T	354	2.64	1.128	0.060	2.52	2.76	0	5
K.1.B.	EP	131	2.82	1.006	0.088	2.64	2.99	0	5
	SGE	103	2.90	0.965	0.095	2.71	3.09	1	5
	Pa	83	2.36	1.154	0.127	2.11	2.61	0	5
	R	37	2.70	0.812	0.133	2.43	2.97	1	5
	T	354	2.72	1.031	0.055	2.62	2.83	0	5
K.1.C.	EP	131	1.69	1.031	0.090	1.51	1.87	0	4
	SGE	103	1.49	1.074	0.106	1.28	1.70	0	4
	Pa	83	1.96	1.184	0.130	1.71	2.22	0	5
	R	37	1.35	1.033	0.170	1.01	1.70	0	5
	T	354	1.66	1.095	0.058	1.54	1.77	0	5
K.1.D.	EP	131	2.97	1.116	0.097	2.78	3.16	1	5
	SGE	103	2.73	1.214	0.120	2.49	2.97	0	5
	Pa	83	3.71	1.054	0.116	3.48	3.94	1	5
	R	37	3.11	1.149	0.189	2.72	3.49	1	5
	T	354	3.09	1.188	0.063	2.96	3.21	0	5

Note. Descriptive statistics. EP = educational professionals; SGE = specialists gifted education; Pa = parents; R = Rest; T = total. N= 354.

K.1.A. = knowledge indicator 1.A = stimulating academic development; K.1.B. = Knowledge indicator 1.B. = stimulating the effective use of different cognitive styles; K.1.C. = Knowledge indicator 1.C. = stimulating interpersonal functioning; K.1.D. = Knowledge indicator 1.D. = stimulating intrapersonal understanding.

Table E13

Comparison of the Selection Rates of the Knowledge Indicators in Domain 1 Between the Groups of Stakeholders

Indicators		Sum of Squares	*df*	Mean Square	*F*	*p*	η^2
K.1.A.	Between Groups	14.629	3	4.876	3.925	.009	.03
	Within Groups	434.809	350	1.242			
	Total	449.438	353				
K.1.B.	Between Groups	15.352	3	5.117	4.982	.002	.04
	Within Groups	359.519	350	1.027			
	Total	374.870	353				
K.1.C.	Between Groups	14.421	3	4.807	4.111	.007	.03
	Within Groups	409.220	350	1.169			
	Total	423.641	353				
K.1.D.	Between Groups	47.391	3	15.797	12.262	.000	.10
	Within Groups	450.894	350	1.288			
	Total	498.285	353				
K.1.E.	Between Groups	2.032	3	0.677	0.575	.632	.00
	Within Groups	412.126	350	1.178			
	Total	414.158	353				
K.1.F.	Between Groups	6.496	3	2.165	1.797	.147	.02
	Within Groups	421.790	350	1.205			
	Total	428.285	353				

Note. ANOVA = analysis of variance with one factor.

K.1.A. = knowledge indicator 1.A. = stimulating academic development; K.1.B. = Knowledge indicator 1.B. = stimulating the effective use of different cognitive styles; K.1.C. = Knowledge indicator 1.C. = stimulating interpersonal functioning; K.1.D. = Knowledge indicator 1.D. = stimulating intrapersonal understanding; K.1.E. = Knowledge indicator 1.E. = stimulating self-actualisation; K.1.F. = Knowledge indicator 1.F. = general knowledge and skills matching those of the (S)ENCo.

Table E14

Comparison of the Selection Rates of the Knowledge Indicators in Domain 1 Between the Groups of Stakeholders

Dependent variable	I group	J group	Mean difference (I-J)	*SE*	*p*	95% Confidence interval	
						Lower bound	Upper bound
K.1.A.	Pa	EP	-0.320	0.164	.247	-0.74	0.09
		SGE	-0.531	0.220	.008	-0.97	-0.10
		R	-0.525	0.208	.107	-1.11	0.06
K.1.B.	Pa	EP	-0.455	0.142	.009	-0.83	-0.08
		SGE	-0.541	0.149	.002	-0.94	-0.14
		R	-0.341	0.200	.536	-0.87	0.19
K.1.B.	SGE	EP	0.086	0.133	1.000	-0.27	0.44
		P	0.541	0.149	.002	0.14	0.94
		R	0.200	0.194	1.000	-0.32	0.72
K.1.C.	Pa	EP	0.277	0.152	.413	-0.13	0.68
		SGE	0.478	0.159	.017	0.06	0.90
		R	0.613	0.214	.026	0.05	1.18
K.1.D.	Pa	EP	0.741	0.159	.000	0.32	1.16
		SGE	0.983	0.167	.000	0.54	1.43
		R	0.603	0.224	.045	0.01	1.20

Note. Based on the results of the ANOVA, Bonferroni's Multiple Comparisons was used as a post hoc test for the analyses of the differences between groups of stakeholders.

EP = educational professionals; SGE = specialists gifted education; Pa = parents; R = Rest; T = total. N = 354.

K.1.A. = Knowledge indicator 1.A. = stimulating academic development; K.1.B. = Knowledge indicator 1.B. = stimulating the effective use of different cognitive styles; K.1.C. = Knowledge indicator 1.C. = stimulating interpersonal functioning; K.1.D. = Knowledge indicator 1.D. = stimulating intrapersonal understanding.

Table E15

Comparison of the Selection Rates of the Skills Indicators in Domain 1 Between the Groups of Stakeholders

						95% Confidence interval for Mean			
Indicators	Group	N	Mean	*SD*	*SE*	Lower bound	Upper bound	Min	Max
V.1.C.	EP	131	2.63	1.179	0.103	2.42	2.83	0	5
	SGE	103	2.12	1.060	0.104	1.91	2.32	0	4
	Pa	83	2.28	1.130	0.124	2.03	2.52	0	5
	R	37	2.14	1.206	0.198	1.73	2.54	0	5
	T	354	2.34	1.154	0.061	2.22	2.47	0	5
V.1.D.	EP	131	2.66	1.174	0.103	2.46	2.87	0	5
	SGE	103	2.26	1.180	0.116	2.03	2.49	0	5
	Pa	83	2.95	1.070	0.117	2.72	3.19	0	5
	R	37	2.78	1.182	0.194	2.39	3.18	1	5
	T	354	2.63	1.177	0.063	2.50	2.75	0	5
V.1.E.	EP	131	2.85	1.041	0.091	2.67	3.03	0	5
	SGE	103	3.23	0.962	0.095	3.04	3.42	1	5
	Pa	83	3.02	1.047	0.115	2.80	3.25	1	5
	R	37	3.05	1.129	0.186	2.68	3.43	0	5
	T	354	3.02	1.037	0.055	2.91	3.13	0	5
V.1.F.	EP	131	1.71	1.160	0.101	1.51	1.91	0	4
	SGE	103	2.32	0.921	0.091	2.14	2.50	0	5
	Pa	83	1.48	1.097	0.120	1.24	1.72	0	4
	R	37	1.89	1.048	0.172	1.54	2.24	0	4
	T	354	1.85	1.112	0.059	1.74	1.97	0	5

Note. Descriptive statistics. EP = educational professionals; SGE = specialists gifted education; Pa = parents; R = Rest; T = total. N= 354.

V.1.C. = Skills indicator 1.C. = stimulating interpersonal functioning; V.1.D. = Skills indicator 1.D. = stimulating intrapersonal understanding; V.1.E. = Skills indicator 1.E. = stimulating self-actualisation; V.1.F. = Skills indicator 1.F. = general knowledge and skills matching those of the (S)ENCo.

Table E16

Comparison of the Selection Rates of the Skills Indicators in Domain 1 Between the Groups of Stakeholders

Indicators		Sum of Squares	*df*	Mean Square	*F*	*p*	η^2
V.1.A.	Between Groups	3.717	3	1.239	1.097	.350	.01
	Within Groups	395.415	350	1.130			
	Total	399.133	353				
V.1.B.	Between Groups	7.813	3	2.604	2.213	.086	.02
	Within Groups	411.795	350	1.177			
	Total	419.607	353				
V.1.C.	Between Groups	17.730	3	5.910	4.574	.004	.01
	Within Groups	452.225	350	1.292			
	Total	469.955	353				
V.1.D.	Between Groups	23.558	3	7.853	5.908	.001	.03
	Within Groups	465.221	350	1.329			
	Total	488.780	353				
V.1.E.	Between Groups	8.621	3	2.874	2.710	.045	.00
	Within Groups	371.198	350	1.061			
	Total	379.819	353				
V.1.F.	Between Groups	36.667	3	12.222	10.703	.000	.06
	Within Groups	399.695	350	1.142			
	Total	436.362	353				

Note. ANOVA = analysis of variance with one factor. N = 354.

V.1.A. = Skills indicator 1.A. = stimulating academic development; V.1.B. = Skills indicator 1.B. = stimulating the effective use of different cognitive styles; V.1.C. = Skills indicator 1.C. = stimulating interpersonal functioning; V.1.D. = Skills indicator 1.D. = stimulating intrapersonal understanding; V.1.E. = Skills indicator 1.E. = stimulating self-actualisation; V.1.F. = Skills indicator 1.F. = general knowledge and skills matching those of the (S)ENCo.

Table E17

Comparison of the Selection Rates of the Skills Indicators in Domain 1 Between the Groups of Stakeholders

Dependent variable	I group	J group	Mean difference (I-J)	*SE*	*p*	95% Confidence interval Lower bound	Upper bound
V.1.C.	SGE	EP	-0.509	0.150	.004	-0.91	-0.11
		Pa	-0.161	0.168	1.000	-0.61	0.28
		R	-0.019	0.218	1.000	-0.60	0.56
V.1.D.	SGE	EP	-0.402	0.152	.051	-0.80	0.00
		Pa	-0.690	0.170	.000	-1.14	-0.24
		R	-0.522	0.221	.113	-1.11	0.06
V.1.E.	SGE	EP	0.386	0.136	.028	0.03	0.75
		Pa	0.209	0.152	1.000	-0.19	0.61
		R	0.179	0.197	1.000	-0.34	0.71
V.1.F.	SGE	EP	0.610	0.141	.000	0.24	0.98
		Pa	0.838	0.158	.000	0.42	1.26
		R	0.428	0.205	.223	-0.11	0.97

Note. Based on the results of the ANOVA, Bonferroni's Multiple Comparisons was used as a post hoc test for the analyses of the differences between groups of stakeholders.

EP = educational professionals; SGE = specialists gifted education; Pa = parents; R = Rest; T = total. N= 354.

V.1.C. = Skills indicator 1.C. = stimulating interpersonal functioning; V.1.D. = Skills indicator 1.D. = stimulating intrapersonal understanding; V.1.E. = Skills indicator 1.E. = stimulating self-actualisation; V.1.F. = Skills indicator 1.F. = general knowledge and skills matching those of the (S)ENCo.

Table E18

Knowledge Indicators for Each Competency in Domain 1 that Belonged to the Three Most Selected Indicators in All Groups of Stakeholders

1.1 The specialist in gifted education understands and envisions how their perspective on giftedness influences their ability to see, understand and respond to the pedagogical needs of gifted children.	K.1.B.1	Knows how personal traits during developmental processes influence a student's achievements.
	K.1.E.1	Knows why autonomy. competency and relationship are basic pedagogical needs for gifted students.
	K.1.D.1	Knows which risk factors might influence a student's wellbeing.
1.2 The specialist creates a coherency between the concepts of giftedness, educational objectives and educational programming and the intervention strategies to be used.	K.1.B.2	Knows how curricular adjustments can be of consequence for the way they provide instruction and support for gifted students.
	K.1.D.2	Knows different didactical strategies to challenge gifted students in their zone of proximal development.
1.3 The specialist in gifted education understands how their perspective of giftedness influences the organisation of the educational programme provided.	K.1.E.3	Knows why they have to assess the student's learning process and achievements.
	K.1.B.3	Knows why it is important to pay attention to a student's executive skills.
1.4 Based on the interaction between the educational partners, the specialist in gifted education develops a perspective of giftedness, and an approach to gifted students that matches their individual needs.	K.1.A.4	Knows why it is important to organise gifted education from the joint perspective of the educational partners.
1.5 The specialist in gifted education relates their in-depth knowledge of giftedness to the perspectives of current society in general and more specifically, to the current perspectives on education.	K.1.A.5	Knows that the interaction between aptitude and development influences a student's achievements.
	K.1.B.5	Invests in personal continuous professional development regarding theories of giftedness.

Table E19

Skills Indicators for Each Competency in Domain 1 that Belonged to the Three Most Selected Indicators in All Groups of Stakeholders

1.1 The specialist in gifted education understands and envisions how their perspective on giftedness influences their ability to see, understand and respond to the pedagogical needs of gifted children.	V.1.B.1	Discusses with the student how the student influences their personal development.
	V.1.E.1	Makes gifted students responsible for their own development, matching their individual ability to take up this responsibility.
1.2 The specialist creates a coherency between the concepts of giftedness, educational objectives and educational programming and the intervention strategies to be used.	V.1.A.2	Sets educational objectives that match the individual needs of gifted students.
	V.1.E.2	Selects interventions that are a match with the student's needs.
1.3 The specialist in gifted education understands how their perspective on giftedness influences the organisation of the educational programme provided.	V.1.E.3	Provides feed-forward daily on the gifted student's learning process and achievements, as they do for their other students as well.
1.4 Based on the interaction between the educational partners, the specialist in gifted education develops a perspective of giftedness, and an approach to gifted students that matches their individual needs.	V.1.B.4	Discusses with educational partners how they can contribute positively to the gifted student's developmental process.
1.5 The specialist in gifted education relates their in-depth knowledge of giftedness to perspectives of current society in general and more specifically, to the current perspectives on education.	V.1.D.5	Chooses interventions that can contribute to the well-being of gifted students.
	V.1.F.5	Offers feedback to their colleagues to improve the quality of gifted education.

Table E20

Mean Selection Rates of the Knowledge Indicators in Domain 1

Indicators	N	Mean	*SD*
K.1.A.	354	2.64	1.128
K.1.B.	354	2.72	1.031
K.1.C.	354	1.66	1.095
K.1.D.	354	3.09	1.188
K.1.E.	354	2.80	1.083
K.1.F.	354	2.09	1.101

Note. One-Sample Statistics. Test value 2.5.

K.1.A. = Knowledge indicator 1.A. = stimulating academic development; K.1.B. = Knowledge indicator 1.B. = stimulating the effective use of different cognitive styles; K.1.C. = Knowledge indicator 1.C. = stimulating interpersonal functioning; K.1.D. = Knowledge indicator 1.D. = stimulating intrapersonal understanding; K.1.E. = Knowledge indicator 1.E. = stimulating self-actualisation; and K.1.F. = Knowledge indicator 1.F. = general knowledge and skills matching those of the (S)ENCo.

Table E21

Selection Rates of the Knowledge Indicators per Category in Domain 1

					95% Confidence interval of the difference		
Indicators	*t*	*df*	Sig (2-tailed)	Mean difference	Lower bound	Upper bound	Cohen's *d*
K.1.A.	2.355	353	.019	0.141	0.02	0.26	.12
K.1.B.	4.074	353	.000	0.223	0.12	0.33	.21
K.1.C.	-14.458	353	.000	-0.842	-0.96	-0.72	.77
K.1.D.	9.305	353	.000	0.588	0.46	-0.71	.50
K.1.E.	5.250	353	.000	0.302	0.19	0.42	.28
K.1.F.	-7.045	353	.000	-0.412	-0.53	-0.30	.37

Note. One-Sample T test. Test value 2.5.

K.1.A. = Knowledge indicator 1.A. = stimulating academic development; K.1.B. = Knowledge indicator 1.B. = stimulating the effective use of different cognitive styles; K.1.C. = Knowledge indicator 1.C. = stimulating interpersonal functioning; K.1.D. = Knowledge indicator 1.D. = stimulating intrapersonal understanding; K.1.E. = Knowledge indicator 1.E. = stimulating self-actualisation; and K.1.F. = Knowledge indicator 1.F. = general knowledge and skills matching those of the (S)ENCo.

Table E22

Mean Selection Rates of the Skills Indicators per Category in Domain 1

Indicators	N	Mean	*SD*
V.1.A.	354	2.56	1.063
V.1.B.	354	2.59	1.090
V.1.C.	354	2.34	1.154
V.1.D.	354	2.63	1.177
V.1.E.	354	3.02	1.037
V.1.F.	354	1.85	1.112

Note. One-Sample Statistics. Test value 2.5.

V.1.A. = Skills indicator 1.A. = stimulating academic development; V.1.B. = Skills indicator 1.B. = stimulating the effective use of different cognitive styles; V.1.C. = Skills indicator 1.C. = stimulating interpersonal functioning; V.1.D. = Skills indicator 1.D. = stimulating intrapersonal understanding; V.1.E. = Skills indicator 1.E. = stimulating self-actualisation; and V.1.F. = Skills indicator 1.F. = general knowledge and skills matching those of the (S)ENCo.

Table E23

Selection Rates of the Skills Indicators per Category in Domain 1

					95% Confidence interval of the difference		
Indicators	*t*	*df*	Sig (2-tailed)	Mean difference	Lower bound	Upper bound	Cohen's *d*
V.1.A.	1.100	353	.272	0.062	-0.05	0.17	.06
V.1.B.	1.560	353	.120	0.090	-0.02	0.20	.08
V.1.C.	-2.533	353	.012	-0.155	-0.28	-0.03	.14
V.1.D.	2.033	353	.043	0.127	0.00	0.25	.11
V.1.E.	9.479	353	.000	0.523	0.41	0.63	.50
V.1.F.	-10.947	353	.000	-0.647	-0.76	-0.53	.58

Note. One-Sample T Test. Test value 2.5.

V.1.A. = Skills indicator 1.A. = stimulating academic development; V.1.B. = Skills indicator 1.B. = stimulating the effective use of different cognitive styles; V.1.C. = Skills indicator 1.C. = stimulating interpersonal functioning; V.1.D. = Skills indicator 1.D. = stimulating intrapersonal understanding; V.1.E. = Skills indicator 1.E. = stimulating self-actualisation; and V.1.F. = Skills indicator 1.F. = general knowledge and skills matching those of the (S)ENCo.

Table E24

Indicator Category with the Highest Mean Selection Rate per Group of Stakeholders in Domain 1

Indicators	Educational professionals	Specialists in gifted education	Parents	Rest
Knowledge	D	B	D	D
Skills	E	E	E	E

Note. B = stimulating the effective use of different cognitive styles; D = stimulating intrapersonal understanding; E = stimulating self-actualisation.

Table E25

The Comparison of Selection Rates Between Two Clusters of Knowledge Indicators in Domain 1: ADE-Indicators vs BCF-Indicators

Clusters of Indicators	Mean	N	*SD*	*SE*
Dom1_K_ADE	2.8437	354	0.44888	0.02386
Dom1_K_BCF	2.1563	354	0.44888	0.02386

Note. Paired Samples Statistics. Dom1_K_ADE = Domain 1, cluster of three categories of knowledge indicators A, D, E compared with Dom1_K_BCF = Domain 1, cluster of three categories of knowledge indicators B, C, F.

Table E26

The Comparison of Selection Rates Between Two Clusters of Knowledge Indicators in Domain 1: ADE-Indicators vs BCF-Indicators

				95% Confidence interval of the difference					
Clusters of Indicators	Mean	*SD*	*SE*	Lower bound	Upper bound	*t*	*df*	Sig (2-tailed)	Cohen's *d*
Dom1_K_ADE – Dom1_K_BCF	0.68738	0.89777	0.04772	0.59354	0.78123	14.406	353	.000	0.77

Note. Paired Samples T test. Dom1_K_ADE = Domain 1, cluster of three categories of knowledge indicators A, D, E compared with Dom_1_K_BCF = Domain 1, cluster of three categories of knowledge indicators B, C, F.

Table E27

The Comparison of Selection Rates Between Two Clusters of Skills Indicators in Domain 1: ADE-Indicators vs BCF- Indicators

Clusters of Indicators	Mean	N	*SD*	*SE*
Dom1_V_ADE	2.7373	354	0.46234	0.02457
Dom1_V_BCF	2.2627	354	0.46234	0.02457

Note. Paired Samples Statistics. Dom1_V_ADE = Domain 1, cluster of three categories of skills indicators A, D, E compared with Dom1_V_BCF = Domain 1, cluster of three categories of skills indicators B, C, F.

Table E28

The Comparison of Selection Rates Between Two Clusters of Skills Indicators in Domain 1: ADE-Indicators vs BCF- Indicators

				95% Confidence interval of the difference					
Clusters of Indicators	Mean	*SD*	*SE*	Lower bound	Upper bound	*t*	*df*	Sig (2-tailed)	Cohen's *d*
Dom1_V_ADE – Dom1_V_BCF	0.47458	0.92468	0.04915	0.37792	0.57123	9.656	353	.000	.51

Note. Paired Samples T test. Dom1_V_ADE = Domain 1, cluster of three categories of skills indicators A, D, E compared with Dom1_V_BCF = Domain 1, cluster of three categories of skills indicators B, C, F.

Table E29

Knowledge Indicators in Domain 1 with the Highest Selection Rates Amongst All Groups of Stakeholders

Competency	Code	Indicator
1.1 The specialist in gifted education understands and envisions how their perspective on giftedness influences their ability to see, understand and respond to the pedagogical needs of gifted children.	K.1.D.1	Knows which risk factors might influence a student's wellbeing.
1.2 The specialist creates a coherency between the concepts of giftedness, educational objectives and educational programming and the intervention strategies to be used.	K.1.D.2	Knows different didactical strategies to challenge gifted students in their zone of proximal development.
1.4 Based on the interaction between the educational partners, the specialist in gifted education develops a perspective of giftedness, and an approach to gifted students that matches their individual needs.	K.1.A.4	Knows why it is important to organise gifted education from a joint perspective with educational partners.

Table E30

Knowledge Indicators in Domain 1 with the Lowest Selection Rates Amongst All Groups of Stakeholders

Competency	Code	Indicator
1.2 The specialist creates a coherency between the concepts of giftedness, educational objectives and educational programming and the intervention strategies to be used.	K.1.F.2	Knows that a different perspective on giftedness also means a different focus on the ultimate educational objectives.
1.4 Based on the interaction between the educational partners, the specialist in gifted education develops a perspective of giftedness, and an approach to gifted students that matches their individual needs.	K.1.B.4	Knows what theoretical perspectives there are on the learning process of gifted students.
1.5 The specialist in gifted education relates their in-depth knowledge of giftedness to the perspectives of current society in general and more specifically, to the current perspectives on education.	K.1.C.5	Is aware of different theoretical perspectives on the social functioning of gifted students.

Table E31

Comparison of Mean Selection Rates in Domain 1 per Category of Knowledge Indicators Between Dutch and Flemish Stakeholders

Indicators	Country	N	Mean	*SD*	*SE*
K.1.A.	NL	299	2.69	1.109	0.064
	F	55	2.40	1.211	0.163
K.1.B.	NL	299	2.80	0.993	0.057
	F	55	2.31	1.136	0.153
K.1.C.	NL	299	1.60	1.102	0.064
	F	55	1.96	1.018	0.137
K.1.D.	NL	299	3.04	1.182	0.068
	F	55	3.33	1.203	0.162
K.1.E.	NL	299	2.82	1.055	0.061
	F	55	2.69	1.230	0.166
K.1.F.	NL	299	2.05	1.104	0.064
	F	55	2.31	1.069	0.144

Note. Descriptive Statistics. NL = The Netherlands. F = Flanders. N=354.

K.1.A. = Knowledge indicator 1.A. = stimulating academic development; K.1.B. = Knowledge indicator 1.B. = stimulating the effective use of different cognitive styles; K.1.C. = Knowledge indicator 1.C. = stimulating interpersonal functioning; K.1.D. = Knowledge indicator 1.D. = stimulating intrapersonal understanding; K.1.E. = Knowledge indicator 1.E. = stimulating self-actualisation; and K.1.F. = Knowledge indicator 1.F. = general knowledge and skills matching those of the (S)ENCo.

Table E32

Comparison of Mean Selection Rates in Domain 1 per Category of Knowledge Indicators Between Dutch and Flemish Stakeholders

		Levene's test for Equality of variances		t-test for Equality of means							
									95% Confidence Interval of the difference		
Indicators		*F*	*p*	*t*	*df*	Sig. (2-tailed)	Mean difference	*SE* difference	Lower bound	Upper bound	Cohen's *d*
K.1.A.	Equal variances assumed	0.810	.369	1.730	352	.084	0.286	0.165	-0.039	0.610	.25
	Equal variances not assumed			1.628	71.632	.108	0.286	0.175	-0.064	0.635	
K.1.B.	Equal variances assumed	2.586	.109	3.287	352	.001	0.490	0.149	0.197	0.784	.48
	Equal variances not assumed			2.996	69.988	.004	0.490	0.164	0.164	0.817	
K.1.C.	Equal variances assumed	3.644	.057	-2.263	352	.024	-0.362	0.160	-0.676	-0.047	.33
	Equal variances not assumed			-2.390	79.120	.019	-0.362	0.151	-0.663	-0.060	
K.1.D.	Equal variances assumed	0.534	.465	-1.632	.52	.104	-0.284	0.174	-0.626	0.058	.24
	Equal variances not assumed			-1.613	74.472	.111	-0.284	0.176	-0.634	0.067	
K.1.E.	Equal variances assumed	2.910	.089	0.829	352	.408	0.132	0.159	-0.181	0.445	.12
	Equal variances not assumed			0.746	69.361	.458	0.132	0.177	-0.221	0.484	
K.1.F.	Equal variances assumed	0.178	.673	-1.627	352	.105	-0.262	0.161	-0.579	0.055	.24
	Equal variances not assumed			-1.663	76.727	.100	0.262	0.158	-0.576	0.052	

Note. Independent Samples Test. N=354.

K.1.A. = Knowledge indicator 1.A. = stimulating academic development; K.1.B. = Knowledge indicator 1.B. = stimulating the effective use of different cognitive styles; K.1.C. = Knowledge indicator 1.C. = stimulating interpersonal functioning; K.1.D. = Knowledge indicator 1.D. = stimulating intrapersonal understanding; K.1.E. = Knowledge indicator 1.E. = stimulating self-actualisation; and K.1.F. = Knowledge indicator 1.F. = general knowledge and skills matching those of the (S)ENCo.

Table E33

Comparison of Mean Selection Rates in Domain 1 per Category of Skills Indicators Between Dutch and Flemish Stakeholders

Indicators	Country	N	Mean	*SD*	*SE*
V.1.A.	NL	299	2.58	1.063	0.061
	F	55	2.47	1.069	0.144
V.1.B.	NL	299	2.61	1.070	0.062
	F	55	2.49	1.200	0.162
V.1.C.	NL	299	2.29	1.144	0.066
	F	55	2.64	1.176	0.159
V.1.D.	NL	299	2.59	1.185	0.069
	F	55	2.82	1.124	0.152
V.1.E.	NL	299	3.02	1.038	0.060
	F	55	3.02	1.045	0.141
V.1.F.	NL	299	1.91	1.083	0.063
	F	55	1.56	1.229	0.166

Note. Descriptive Statistics. NL = The Netherlands. F = Flanders. N = 354.

V.1.A. = Skills indicator 1.A. = stimulating academic development; V.1.B. = Skills indicator 1.B. = stimulating the effective use of different cognitive styles; V.1.C. = Skills indicator 1.C. = stimulating interpersonal functioning; V.1.D. = Skills indicator 1.D. = stimulating intrapersonal understanding; V.1.E. = Skills indicator 1.E. = stimulating self-actualisation; and V.1.F. = Skills indicator 1.F. = general knowledge and skills matching those of the (S)ENCo.

Table E34

Comparison of Mean Selection Rates in Domain 1 per Category of Skills Indicators Between Dutch and Flemish Stakeholders

	Levene's test for Equality of variances			t-test for Equality of means							
									95% Confidence interval of the difference		
Indicators		*F*	*p*	*t*	*df*	Sig. (2-tailed)	Mean difference	*SE* difference	Lower bound	Upper bound	Cohen's *d*
V.1.A.	Equal variances assumed	0.002	.962	0.678	352	.498	0.106	0.156	-0.201	0.413	. 01
	Equal variances not assumed			0.676	74.992	.501	0.106	0.157	-0.206	0.418	
V.1.B.	Equal variances assumed	1.821	.178	0.736	352	.462	0.118	0.160	-0.197	0.433	.11
	Equal variances not assumed			0.680	70.681	.499	0.118	0.173	-0.228	0.463	
V.1.C.	Equal variances assumed	0.019	.890	-2.050	352	.041	-0.345	0.169	-0.677	-0.014	.30
	Equal variances not assumed			-2.010	73.999	.048	-0.345	0.172	-0.688	-0.003	
V.1.D.	Equal variances assumed	0.318	.573	-1.312	352	.191	-0.226	0.172	-0.565	0.113	.19
	Equal variances not assumed			-1.360	77.750	.178	-0.226	0.166	-0.557	0.105	
V.1.E.	Equal variances assumed	0.303	.582	0.034	352	.973	0.005	0.152	-0.295	0.305	.01
	Equal variances not assumed			0.034	74.911	.973	0.005	0.153	-0.300	0.310	
V.1.F.	Equal variances assumed	3.637	.057	2.111	352	.035	0.343	0.162	0.023	0.662	.31
	Equal variances not assumed			1.935	70.269	.057	0.343	0.177	-0.011	0.696	

Note. Independent Samples Test. N =354.

V.1.A. = Skills indicator 1.A. = stimulating academic development; V.1.B. = Skills indicator 1.B. = stimulating the effective use of different cognitive styles; V.1.C. = Skills indicator 1.C. = stimulating interpersonal functioning; V.1.D. = Skills indicator 1.D. = stimulating intrapersonal understanding; V.1.E. = Skills indicator 1.E. = stimulating self-actualisation; and V.1.F. = Skills indicator 1.F. = general knowledge and skills matching those of the (S)ENCo.

Table E35

Domain 1 Knowledge Indicators. Mean per Category per Competency per Group of Stakeholders

Indicators	EP	Indicators	SGE	Indicators	Pa	Indicators	R
K.1.C.	1.69	K.1.C.	1.49	K.1.F.	1.87	K.1.C.	1.35
K.1.F.	2.15	K.1.F.	2.15	K.1.C.	1.96	K.1.F.	2.00
K.1.A.	2.63	K.1.D.	2.73	K.1.A.	2.31	K.1.B.	2.72
K.1.E.	2.74	K.1.E.	2.74	K.1.B.	2.36	K.1.A.	2.84
K.1.B.	2.82	K.1.A.	2.84	K.1.E.	2.78	K.1.E.	3.00
K.1.D.	2.97	K.1.B.	2.90	K.1.D.	3.71	K.1.D.	3.11

Note. Test value 2.5. EP = educational professionals; SGE = specialists gifted education; Pa = parents; R = rest. N = 354.

K.1.A. = Knowledge indicator 1.A. = stimulating academic development; K.1.B. = Knowledge indicator 1.B. = stimulating the effective use of different cognitive styles; K.1.C. = Knowledge indicator 1.C. = stimulating interpersonal functioning; K.1.D. = Knowledge indicator 1.D. = stimulating intrapersonal understanding; K.1.E. = Knowledge indicator 1.E. = stimulating self-actualisation; and K.1.F. = Knowledge indicator 1.F. = general knowledge and skills matching those of the (S)ENCo.

Table E36

Domain 1 Skills Indicators. Mean per Category per Competency per Group of Stakeholders

Indicators	EP	Indicators	SGE	Indicators	Pa	Indicators	R
V.1.F.	1.17	V.1.C.	2.12	V.1.F.	1.48	V.1.F	1.89
V.1.B.	2.49	V.1.D.	2.26	V.1.C.	2.28	V.1.C.	2.14
V.1.C.	2.63	V.1.F.	2.32	V.1.A.	2.41	V.1.B.	2.49
V.1.A.	2.66	V.1.A.	2.52	V.1.B.	2.86	V.1.A.	2.65
V.1.D.	2.66	V.1.B.	2.54	V.1.D.	2.95	V.1.D.	2.78
V.1.E.	2.85	V.1.E.	3.23	V.1.E.	3.02	V.1.E.	3.05

Note. Test value 2.5. EP = educational professionals; SGE = specialists gifted education; Pa = parents; R = rest. N = 354.

V.1.A. = Skills indicator 1.A. = stimulating academic development; V.1.B = Skills indicator 1.B. = stimulating the effective use of different cognitive styles; V.1.C. = Skills indicator 1.C. = stimulating interpersonal functioning; V.1.D. = Skills indicator 1.D. = stimulating intrapersonal understanding; V.1.E. = Skills indicator 1.E. = stimulating self-actualisation; and V.1.F. = Skills indicator 1.F. = general knowledge and skills matching those of the (S)ENCo.

Table E37

Ranking of the Knowledge and Skills Indicators of Domain 1

Indicators	N	%	Indicators	N	%
V.1.F.4	4	0.04	K.1.E.2	184	1.73
K.1.B.4	56	0.53	K.1.D.5	185	1.74
K.1.F.2	68	0.64	K.1.A.5	188	1.77
V.1.C.1	72	0.68	K.1.B.1	191	1.80
K.1.A.1	84	0.79	V.1.C.5	191	1.80
V.1.F.2	97	0.91	K.1.A.3	193	1.82
K.1.C.3	102	0.96	V.1.D.4	196	1.85
K.1.C.2	103	0.97	K.1.A.2	198	1.86
K.1.C.5	118	1.11	K.1.E.4	199	1.87
V.1.C.2	119	1.12	K.1.B.2	203	1.91
K.1.C.1	124	1.17	K.1.E.3	203	1.91
V.1.A.4	130	1.22	V.1.E.4	203	1.91
V.1.B.5	134	1.26	V.1.E.3	206	1.94
V.1.D.1	140	1.32	V.1.D.3	211	1.99
K.1.C.4	140	1.32	V.1.D.5	212	2.00
V.1.A.3	141	1.33	V.1.B.1	213	2.01
V.1.A.5	143	1.35	K.1.F.4	228	2.15
K.1.F.5	145	1.37	V.1.F.5	235	2.21
V.1.E.5	147	1.38	V.1.A.1	237	2.23
K.1.F.1	148	1.39	K.1.E.1	239	2.25
K.1.F.3	150	1.41	V.1.E.1	249	2.34
V.1.F.1	151	1.42	K.1.B.3	255	2.40
V.1.B.3	152	1.43	V.1.A.2	256	2.41
V.1.B.2	154	1.45	K.1.B.5	259	2.44
K.1.D.3	159	1.50	V.1.B.4	264	2.49
K.1.D.4	167	1.57	V.1.E.2	265	2.50
K.1.E.5	167	1.57	V.1.C.4	265	2.50
V.1.F.3	169	1.59	K.1.A.4	272	2.56
V.1.D.2	171	1.61	K.1.D.1	276	2.60
V.1.C.3	183	1.72	K.1.D.2	306	2.88

Note. This ranking is based on selection rates for the entire group of stakeholders, N = 354.

All indicators are coded. K = knowledge; V = skills; the first number refers to the vertical competency domain in the competency matrix; the second letter refers to the category of indicators. The second number refers to the actual competency the indicators are matched with. See Appendix B for the match between the competencies and the indicators.

Appendix F

Tables Domain 2, Referred to in Chapter 6

Table F1

Frequency Table of the Selection Rates of Knowledge Indicators for Competency 2.1

Competency	Indicator	EP	SGE	Pa	R	Total
2.1	K.2.A.1	18.58	21.36	18.88	15.32	19.11
	K.2.B.1	13.49	11.00	16.87	14.41	13.65
	K.2.C.1	15.27	15.53	16.06	18.02	15.82
	K.2.D.1	27.23	26.21	26.51	27.93	26.84
	K.2.E.1	12.72	11.33	10.44	9.91	11.49
	K.2.F.1	12.72	14.56	11.24	14.41	13.09

Note. The results are presented per indicator for each group of stakeholders. EP = educational professionals; SGE = specialist gifted education; Pa = parents; R = Rest; T = total. N = 354.

All indicators are coded. K = knowledge; the first number refers to the vertical competency domain in the competency matrix; the second letter refers to the category of indicators. The second number refers to the actual competency the indicators are matched with. See Appendix B for the match between the competencies and the indicators.

Table F2

Frequency Table of the Selection Rates of Knowledge Indicators for Competency 2.2

Competency	Indicator	EP	SGE	Pa	R	Total
2.2	K.2.A.2	12.47	12.62	17.27	9.91	13.37
	K.2.B.2	11.45	9.06	8.03	11.71	9.98
	K.2.C.2	12.21	21.04	11.65	11.71	14.60
	K.2.D.2	18.83	12.62	22.49	18.92	17.89
	K.2.E.2	24.43	22.98	20.08	24.32	22.98
	K.2.F.2	20.61	21.68	20.48	23.42	21.19

Note. The results are presented per indicator for each group of stakeholders. EP = educational professionals; SGE = specialist gifted education; Pa = parents; R = Rest; T = total. N = 354.

All indicators are coded. K = knowledge; the first number refers to the vertical competency domain in the competency matrix; the second letter refers to the category of indicators. The second number refers to the actual competency the indicators are matched with. See Appendix B for the match between the competencies and the indicators.

Table F3

Frequency Table of the Selection Rates of Knowledge Indicators for Competency 2.3

Competency	Indicator	EP	SGE	Pa	R	Total
2.3	K.2.A.3	14.76	15.53	15.66	12.61	14.97
	K.2.B.3	20.87	16.83	16.47	19.82	18.55
	K.2.C.3	14.50	17.80	16.06	16.22	16.01
	K.2.D.3	13.99	7.77	16.87	10.81	12.52
	K.2.E.3	19.08	19.09	22.49	23.42	20.34
	K.2.F.3	16.79	22.98	12.45	17.12	17.61

Note. The results are presented per indicator for each group of stakeholders. EP = educational professionals; SGE = specialist gifted education; Pa = parents; R = Rest; T = total. N = 354.

All indicators are coded. K = knowledge; the first number refers to the vertical competency domain in the competency matrix; the second letter refers to the category of indicators. The second number refers to the actual competency the indicators are matched with. See Appendix B for the match between the competencies and the indicators.

Table F4

Frequency Table of the Selection Rates of Knowledge Indicators for Competency 2.4

Competency	Indicator	EP	SGE	Pa	R	Total
2.4	K.2.A.4	22.39	23.30	21.69	21.62	22.41
	K.2.B.4	21.12	18.12	22.09	14.41	19.77
	K.2.C.4	19.34	21.36	19.68	21.62	20.24
	K.2.D.4	18.83	18.45	17.67	18.92	18.46
	K.2.E.4	10.94	9.06	14.46	13.51	11.49
	K.2.F.4	7.38	9.71	4.42	9.91	7.63

Note. The results are presented per indicator for each group of stakeholders. EP = educational professionals; SGE = specialist gifted education; Pa = parents; R = Rest; T = total. N = 354.

All indicators are coded. K = knowledge; the first number refers to the vertical competency domain in the competency matrix; the second letter refers to the category of indicators. The second number refers to the actual competency the indicators are matched with. See Appendix B for the match between the competencies and the indicators.

Table F5

Frequency Table of the Selection Rates of Knowledge Indicators for Competency 2.5

Competency	Indicator	EP	SGE	Pa	R	Total
2.5	K.2.A.5	18.32	13.92	21.29	15.32	17.42
	K.2.B.5	18.83	14.56	12.05	14.41	15.54
	K.2.C.5	19.08	20.71	24.10	22.52	21.09
	K.2.D.5	18.83	21.68	12.05	18.92	18.08
	K.2.E.5	12.47	13.59	17.67	16.22	14.41
	K.2.F.5	12.47	15.53	12.85	12.61	13.47

Note. The results are presented per indicator for each group of stakeholders. EP = educational professionals; SGE = specialist gifted education; Pa = parents; R = Rest; T = total. N = 354.

All indicators are coded. K = knowledge; the first number refers to the vertical competency domain in the competency matrix; the second letter refers to the category of indicators. The second number refers to the actual competency the indicators are matched with. See Appendix B for the match between the competencies and the indicators.

Table F6

Frequency Table of the Selection Rates of Skills Indicators for Competency 2.1

Competency	Indicator	EP	SGE	Pa	R	Total
2.1	V.2.A.1	22.90	22.01	23.69	18.92	22.41
	V.2.B.1	22.14	17.15	16.87	17.12	18.93
	V.2.C.1	17.05	22.33	20.88	20.72	19.87
	V.2.D.1	10.18	11.65	10.84	11.71	10.92
	V.2.E.1	16.54	11.00	16.06	15.32	14.69
	V.2.F.1	11.20	15.86	11.65	16.22	13.18

Note. The results are presented per indicator for each group of stakeholders. EP = educational professionals; SGE = specialist gifted education; Pa = parents; R = Rest; T = total. N = 354.

All indicators are coded. V = skills; the first number refers to the vertical competency domain in the competency matrix; the second letter refers to the category of indicators. The second number refers to the actual competency the indicators are matched with. See Appendix B for the match between the competencies and the indicators.

Table F7

Frequency Table of the Selection Rates of Skills Indicators for Competency 2.2

Competency	Indicator	EP	SGE	Pa	R	Total
2.2	V.2.A.2	16.03	17.80	14.86	14.41	16.10
	V.2.B.2	14.76	16.83	20.08	16.22	16.76
	V.2.C.2	4.33	2.27	5.62	4.50	4.05
	V.2.D.2	13.99	13.59	15.26	14.41	14.22
	V.2.E.2	23.16	21.36	19.68	22.52	21.75
	V.2.F.2	27.74	28.16	24.50	27.93	27.12

Note. The results are presented per indicator for each group of stakeholders. EP = educational professionals; SGE = specialist gifted education; Pa = parents; R = Rest; T = total. N = 354.

All indicators are coded. V = skills; the first number refers to the vertical competency domain in the competency matrix; the second letter refers to the category of indicators. The second number refers to the actual competency the indicators are matched with. See Appendix B for the match between the competencies and the indicators.

Table F8

Frequency Table of the Selection Rates of Skills Indicators for Competency 2.3

Competency	Indicator	EP	SGE	Pa	R	Total
2.3	V.2.A.3	16.79	15.21	12.45	14.41	15.07
	V.2.B.3	18.58	14.24	18.88	15.32	17.04
	V.2.C.3	17.30	18.77	19.28	18.02	18.27
	V.2.D.3	18.32	15.86	16.87	16.22	17.04
	V.2.E.3	20.10	22.65	18.88	25.23	21.09
	V.2.F.3	8.91	13.27	13.65	10.81	11.49

Note. The results are presented per indicator for each group of stakeholders. EP = educational professionals; SGE = specialist gifted education; Pa = parents; R = Rest; T = total. N = 354.

All indicators are coded. V = skills; the first number refers to the vertical competency domain in the competency matrix; the second letter refers to the category of indicators. The second number refers to the actual competency the indicators are matched with. See Appendix B for the match between the competencies and the indicators.

Table F9

Frequency Table of the Selection Rates of Skills Indicators for Competency 2.4

Competency	Indicator	EP	SGE	Pa	R	Total
2.4	V.2.A.4	8.40	12.62	16.06	15.32	12.15
	V.2.B.4	16.03	14.24	14.86	14.41	15.07
	V.2.C.4	12.98	11.33	10.44	10.81	11.68
	V.2.D.4	26.72	25.89	26.10	27.03	26.37
	V.2.E.4	23.66	22.65	20.08	22.52	22.41
	V.2.F.4	12.21	13.27	12.45	9.91	12.34

Note. The results are presented per indicator for each group of stakeholders. EP = educational professionals; SGE = specialist gifted education; Pa = parents; R = Rest; T = total. N = 354.

All indicators are coded. V = skills; the first number refers to the vertical competency domain in the competency matrix; the second letter refers to the category of indicators. The second number refers to the actual competency the indicators are matched with. See Appendix B for the match between the competencies and the indicators.

Table F10

Frequency Table of the Selection Rates of Skills Indicators for Competency 2.5

Competency	Indicator	EP	SGE	Pa	R	Total
2.5	V.2.A.5	14.76	13.27	18.47	18.02	15.54
	V.2.B.5	23.41	18.45	21.69	23.42	21.56
	V.2.C.5	21.88	24.27	16.47	21.62	21.28
	V.2.D.5	18.07	19.42	18.07	14.41	18.08
	V.2.E.5	14.25	11.97	17.67	15.32	14.50
	V.2.F.5	7.63	12.62	7.63	7.21	9.04

Note. The results are presented per indicator for each group of stakeholders. EP = educational professionals; SGE = specialist gifted education; Pa = parents; R = Rest; T = total. N = 354.

All indicators are coded. V = skills; the first number refers to the vertical competency domain in the competency matrix; the second letter refers to the category of indicators. The second number refers to the actual competency the indicators are matched with. See Appendix B for the match between the competencies and the indicators.

Table F11

Comparison of the 10 Highest Indicator Selection Rates per Competency in Domain 2 per Group of Stakeholders

Indicator	EP	Indicator	SGE	Indicator	Pa	Indicator	R
K.2.C.5	19.08	K.2.D.5	21.68	V.2.C.3	19.28	V.2.C.1	20.72
V.2.E.3	20.10	V.2.C.1	22.33	V.2.B.5	21.69	K.2.C.4 / K.2.A.4	21.62
K.2.B.3	20.87	V.2.E.3	22.65	K.2.B.4	22.09	K.2.C.5	22.52
K.2.A.4	22.39	K.2.E.2	22.98	K.2.D.2	22.49	K.2.E.3	23.42
V.2.A.1	22.90	K.2.F.3	22.98	K.2.E.3	22.49	V.2.B.5	23.42
V.2.B.5	23.41	K.2.A.4	23.30	V.2.A.1	23.69	K.2.E.2	24.32
K.2.E.2	24.43	V.2.C.5	24.27	K.2.C.5	24.10	V.2.E.3	25.23
V.2.D.4	26.72	V.2.D.4	25.89	V.2.F.2	24.50	V.2.D.4	27.03
K.2.D.1	27.23	K.2.D.1	26.21	V.2.D.4	26.10	K.2.D.1	27.93
V.2.F.2	27.74	V.2.F.2	28.16	K.2.D.1	26.51	V.2.F.2	27.93

Note. EP = educational professionals; SGE = specialist gifted education; Pa = parents; R = Rest; T = total. N = 354. If two indicators matching one competency had an identical selection rate, they are ranked at the same level.

All indicators are coded. K = knowledge; V = skills; the first number refers to the vertical competency domain in the competency matrix; the second letter refers to the category of indicators. The second number refers to the actual competency the indicators are matched with. See Appendix B for the match between the competencies and the indicators.

Table F12

Comparison of the Selection Rates of the Knowledge Indicators in Domain 2 Between the Groups of Stakeholders

Indicators	Group	N	Mean	*SD*	*SE*	95% Confidence interval for mean Lower bound	Upper bound	Min	Max
K.2.B.	EP	131	2.94	1.013	0.089	2.76	3.11	0	5
	SGE	103	2.99	0.089	0.088	2.82	3.16	1	5
	Pa	83	2.29	1.110	0.122	2.05	2.53	0	5
	R	37	2.78	0.821	0.135	2.51	3.06	1	4
	T	354	2.79	1.021	0.054	2.68	2.89	0	5
K.2.C.	EP	131	1.54	1.017	0.089	1.37	1.72	0	4
	SGE	103	1.35	1.007	0.099	1.15	1.55	0	4
	Pa	83	1.76	1.175	0.129	1.50	2.02	0	5
	R	37	1.30	0.968	0.159	0.97	1.62	0	4
	T	354	1.51	1.057	0.056	1.40	1.62	0	5
K.2.D.	EP	131	2.71	1.167	0.102	2.51	2.91	0	5
	SGE	103	2.26	1.212	0.119	2.03	2.50	0	5
	Pa	83	3.29	1.185	0.130	3.03	3.55	0	5
	R	37	2.59	1.189	0.196	2.20	2.99	0	5
	T	354	2.70	1.239	0.066	2.57	2.83	1	5

Note. Descriptive Statistics. EP = educational professionals; SGE = specialist gifted education; Pa = parents; R = Rest; T = total. N = 354.

K.2.B. = Knowledge indicator 2.B. = stimulating the effective use of different cognitive styles; K.2.C. = Knowledge indicator 2.C. = stimulating interpersonal functioning; K.2.D. = Knowledge indicator 2.D. = stimulating intrapersonal understanding.

Table F13

Comparison of the Selection Rates of the Knowledge Indicators in Domain 2 Between the Groups of Stakeholders

Indicators		Sum of Squares	*df*	Mean Square	*F*	*p*	η^2
K.2.A.	Between Groups	5.710	3	1.903	1.707	.165	.01
	Within Groups	390.380	350	1.115			
	Total	396.090	353				
K.2.B.	Between Groups	27.851	3	9.284	9.562	.000	.08
	Within Groups	339.832	350	0.971			
	Total	367.684	353				
K.2.C.	Between Groups	9.608	3	3.203	2.913	.034	.02
	Within Groups	384.847	350	1.100			
	Total	394.455	353				
K.2.D.	Between Groups	48.977	3	16.326	11.593	.000	.09
	Within Groups	492.879	350	1.408			
	Total	541.856	353				
K.2.E.	Between Groups	6.379	3	2.126	1.882	.132	.02
	Within Groups	395.451	350	1.130			
	Total	401.831	353				
K.2.F.	Between Groups	5.804	3	1.935	1.624	.183	.01
	Within Groups	416.931	350	1.191			
	Total	422.734	353				

Note. ANOVA = analyses of variance with one factor.

K.2.A. = knowledge indicator 2.A. = stimulating academic development; K.2.B. = Knowledge indicator 2.B. = stimulating the effective use of different cognitive styles; K.2.C. = Knowledge indicator 2.C. = stimulating interpersonal functioning; K.2.D. = Knowledge indicator 2.D. = stimulating intrapersonal understanding; K.2.E. = Knowledge indicator 2.E. = stimulating self-actualisation; K.2.F. = Knowledge indicator 2.F. = general knowledge and skills matching those of the (S)ENCO.

Table F14

Comparison of the Selection Rates of the Knowledge Indicators in Domain 2 Between the Groups of Stakeholders

Dependent variable	I group	J group	Mean difference (I-J)	*SE*	*p*	95% Confidence interval	
						Lower bound	Upper bound
K.2.B.	Pa	EP	-0.650	0.138	.000	-1.02	-0.28
		SGE	-0.701	0.145	.000	-1.09	-0.32
		R	-0.495	0.195	.069	-1.01	0.02
K.2.C.	Pa	EP	0.217	0.138	.846	-0.17	0.61
		SGE	0.410	0.155	.051	0.00	0.82
		R	0.462	0.201	.159	-0.09	1.01
K.2.D.	Pa	EP	0.579	0.166	.003	0.14	1.02
		SGE	1.027	0.175	.000	0.56	1.49
		R	0.695	0.235	.020	0.57	1.32
K.2.D.	EP	SGE	0.448	0.156	.027	0.03	0.86
		Pa	-0.579	0.166	.003	-1.02	-0.14
		R	0.115	0.221	1.000	-0.47	0.70

Note. Based on the results of the ANOVA, Bonferroni's Multiple Comparisons was used as a post hoc test for the analyses of the differences between groups of stakeholders.

EP = educational professionals; SGE = specialist gifted education; Pa = parents; R = Rest; T = total. N = 354.

K.2.B. = Knowledge indicator 2.B. = stimulating the effective use of different cognitive styles; K.2.C. = Knowledge indicator 2.C. = stimulating interpersonal functioning; K.2.D. = Knowledge indicator 2.D. = stimulating intrapersonal understanding.

Table F15

Comparison of the Selection Rates of the Skills Indicators in Domain 2 Between the Groups of Stakeholders

Indicators	Group	N	Mean	*SD*	*SE*	95% Confidence interval for Mean		Min	Max
						Lower bound	Upper bound		
V.2.B.	EP	131	2.25	1.139	0.100	2.06	2.45	0	5
	SGE	103	2.22	0.969	0.096	2.03	2.41	0	5
	Pa	83	2.86	1.149	0.126	2.60	3.11	0	5
	R	37	2.38	1.037	0.170	2.03	2.72	1	4
	T	354	2.40	1.110	0.059	2.28	2.51	0	5
V.2.C.	EP	131	2.85	1.151	0.101	2.66	3.05	0	5
	SGE	103	2.46	1.118	0.110	2.24	2.67	0	5
	Pa	83	2.51	1.152	0.126	2.25	2.76	0	5
	R	37	2.46	1.016	0.167	2.12	2.80	0	4
	T	354	2.62	1.139	0.061	2.50	2.73	0	5
V.2.D.	EP	131	3.04	1.133	0.099	2.84	3.23	0	5
	SGE	103	2.75	1.169	0.115	2.52	2.98	0	5
	Pa	83	3.28	1.063	0.117	3.05	3.51	1	5
	R	37	3.30	1.222	0.201	2.89	3.70	1	5
	T	354	3.04	1.152	0.061	2.92	3.16	0	5
V.2.F.	EP	131	1.76	1.151	0.101	1.56	1.95	0	4
	SGE	103	2.20	0.964	0.095	2.02	2.39	0	4
	Pa	83	1.43	1.038	0.114	1.21	1.66	0	4
	R	37	1.84	0.986	0.162	1.51	2.17	0	4
	T	354	1.82	1.089	0.058	1.71	1.93	0	4

Note. Descriptive Statistics. EP = educational professionals; SGE = specialist gifted education; Pa = parents; R = Rest; T = total. N = 354.

V.2.B. = Skills indicator 2.B. = stimulating the effective use of different cognitive styles; V.2.C. = Skills indicator 2.C. = stimulating interpersonal functioning; V.2.D. = Skills indicator 2.D. = stimulating intrapersonal understanding; V.2.F. = Skills indicator 2.F. = general knowledge and skills matching those of the (S)ENCo.

Table F16

Comparison of the Selection Rates of the Skills Indicators in Domain 2 Between the Groups of Stakeholders

Indicators		Sum of Squares	*df*	Mean Square	*F*	*p*	η^2
V.2.A.	Between Groups	7.363	3	2.454	2.032	.109	.02
	Within Groups	422.730	350	1.208			
	Total	430.093	353				
V.2.B.	Between Groups	23.320	3	7.773	6.611	.000	.05
	Within Groups	411.519	350	1.176			
	Total	434.839	353				
V.2.C.	Between Groups	12.018	3	4.006	3.145	.025	.03
	Within Groups	445.734	350	1.274			
	Total	457.751	353				
V.2.D.	Between Groups	15.920	3	5.307	4.104	.007	.03
	Within Groups	452.602	350	1.293			
	Total	468.523	353				
V.2.E.	Between Groups	1.705	3	.568	0.560	.642	.00
	Within Groups	355.292	350	1.015			
	Total	356.997	353				
V.2.F.	Between Groups	28.115	3	9.372	8.404	.000	.07
	Within Groups	390.314	350	1.115			
	Total	418.429	353				

Note. ANOVA = analyses of variance with one factor. N = 354.

V.2.A. = Skills indicator 2.A. = stimulating academic development; V.2.B. = Skills indicator 2.B. = stimulating the effective use of different cognitive styles; V.2.C. = Skills indicator 2.C. = stimulating interpersonal functioning; V.2.D. = Skills indicator 2.D. = stimulating intrapersonal understanding; V.2.E. = Skills indicator 2.E. = stimulating self-actualisation; V.2.F. = Skills indicator 2.F. = general knowledge and skills matching those of the (S)ENCo.

Table F17

Comparison of the Selection Rates of the Skills Indicators in Domain 2 Between the Groups of Stakeholders

Dependent variable	I group	J group	Mean difference (I-J)	*SE*	*p*	95% Confidence interval Lower bound	Upper bound
V.2.B.	Pa	EP	0.604	0.152	.001	0.20	1.01
		SGE	0.632	0.160	.001	0.21	1.06
		R	0.477	0.214	.160	-0.09	1.05
V.2.C.	EP	SGE	0.399	0.149	.046	0.00	0.79
		Pa	0.349	0.158	.169	-0.07	0.77
		R	0.396	0.210	.364	-0.16	0.95
V.2.D.	Pa	EP	0.239	0.218	.811	-0.18	0.66
		SGE	0.530	0.160	.010	0.08	0.97
		R	-0.020	0.168	1.000	-0.62	0.58
V.2.D.	SGE	EP	0.448	0.139	.008	-0.69	0.11
		Pa	0.770	0.156	.000	-0.97	-0.08
		R	0.366	0.202	.428	-1.13	0.03

Note. Based on the results of the ANOVA, Bonferroni's Multiple Comparisons was used as a post hoc test for the analyses of the differences between groups of stakeholders.

EP = educational professionals; SGE = specialist gifted education; Pa = parents; R = Rest; T = total. N = 354.

V.2.B. = Skills indicator 2.B. = stimulating the effective use of different cognitive styles; V.2.C. = Skills indicator 2.C. = stimulating interpersonal functioning; V.2.D. = Skills indicator 2.D. = stimulating intrapersonal understanding.

Table F18

Knowledge Indicators of Each Competency in Domain 2 that Belonged to the Three Most Selected Indicators in all Groups of Stakeholders

Competency	Code	Indicator
2.1 The specialist explores the strengths and weaknesses of the gifted student within the intrapersonal and interpersonal developmental domain.	K.2.A.1	Knows how the curriculum offered affects a student's well-being.
	K.2.D.1	Knows what types of behaviour by the gifted student may be a sign for the development of negative self-concept.
2.2 The specialist assesses the student's developmental levels in all academic domains offered, considering the student's cognitive style.	K.2.E.2	Knows how executive skills affect the developmental process of gifted students.
	K.2.F.2	Knows the limitations of school achievements as a reliable indicator for giftedness.
2.3 The specialist takes on a systematic approach to the process of profiling educational needs and manages their time and actions in a way that contributes to the optimal results of this process.	K.2.E.3	Knows how they can use observation techniques as a way to explore the student's ability for self-regulation.
2.4 The specialist approaches all actors in the student's ecological system as equal partners and values their information about the student's educational needs as such.	K.2.A.4	Knows which information the educational partners need in order to provide a matching curriculum for the student.
	K.2.C.4	Knows how information about the student's social functioning can contribute to a student's well-being.
2.5 The specialist is able to develop a personal toolkit for profiling the educational needs of gifted students flexibly and appropriately, using different instruments in different contexts.	K.2.C.5	Knows why it is important to take on different perspectives when exploring a gifted student's problematic social-emotional behaviour.

Table F19

Skills Indicators of Each Competency in Domain 2 that Belonged to the Three Most Selected Indicators by All Groups of Stakeholders

Competency	Code	Indicator
2.1 The specialist explores the strengths and weaknesses of the gifted student within the intrapersonal and interpersonal developmental domain.	V.2.A.1	Has teacher-student conversations about how the gifted student experiences and values the curriculum offered.
	V.2.B.1	Stimulates students to work in groups where they complement each other and support each other's strengths.
	V.2.C.1	Observes how gifted students use their strengths when interacting with classmates.
2.2 The specialist assesses the student's developmental levels in all academic domains offered, considering the student's cognitive style.	V.2.E.2	Profiles a gifted student's strengths and weaknesses regarding their executive skills.
	V.2.F.2	Takes other indicators besides school achievements into consideration when estimating if the student may be gifted.
2.3 The specialist takes on a systematic approach to the process of profiling educational needs and manages their time and actions in a way that contributes to the optimal results of this process.	V.2.E.3	Observes the student's self-regulating skills in a situation where the student is required to use these skills to the best of their ability.
2.4 The specialists approaches all actors in the student's ecological system as equal partners and values their information about the student's educational needs as such.	V.2.D.4	Is focused on solutions and learnable skills instead of being focused on problems when talking with a student.
	V.2.E.4	Consults the student when prioritising intervention strategies.
2.5 The specialist is able to develop a personal toolkit for profiling the educational needs of gifted students flexibly and appropriately, using different instruments in different contexts.	V.2.B.5	Profiles strengths and weaknesses in the student's use of thinking skills.

Table F20

Mean Selection Rates of the Knowledge Indicators per Category in Domain 2

Indicators	N	Mean	*SD*
K.2.A.	354	3.07	1.059
K.2.B.	354	2.79	1.021
K.2.C.	354	1.51	1.057
K.2.D.	354	2.70	1.239
K.2.E.	354	2.83	1.067
K.2.F.	354	2.10	1.094

Note. One-Sample Statistics. Test value 2.5.

K.2.A. = Knowledge indicator 2.A. = stimulating academic development; K.2.B. = Knowledge indicator 2.B. = stimulating the effective use of different cognitive styles; K.2.C. = Knowledge indicator 2.C. = stimulating interpersonal functioning; K.2.D = Knowledge indicator 2.D. = stimulating intrapersonal understanding; K.2.E. = Knowledge indicator 2.E. = stimulating self-actualisation; and K.2.F. = Knowledge indicator 2.F. = general knowledge and skills matching those of the (S)ENCo.

Table F21

Selection Rates of the Knowledge Indicators per Category in Domain 2

					95% Confidence interval of the difference		
Indicators	*t*	*df*	Sig (2-tailed)	Mean difference	Lower bound	Upper bound	Cohen's *d*
K.2.A.	10.186	353	.000	0.573	0.46	0.68	.54
K.2.B.	5.260	353	.000	0.285	0.18	0.39	.28
K.2.C.	-17.598	353	.000	-0.989	-1.10	-0.88	.94
K.2.D.	3.089	353	.002	0.203	0.07	0.33	.16
K.2.E.	5.828	353	.000	0.331	0.22	0.44	.31
K.2.F.	-6.945	353	.000	-0.404	-0.52	-0.29	.37

Note. One-Sample T test. Test value 2.5. N = 354.

K.2.A. = Knowledge indicator 2.A. = stimulating academic development; K.2.B. = Knowledge indicator 2.B. = stimulating the effective use of different cognitive styles; K.2.C. = Knowledge indicator 2.C. = stimulating interpersonal functioning; K.2.D = Knowledge indicator 2.D. = stimulating intrapersonal understanding; K.2.E. = Knowledge indicator 2.E. = stimulating self-actualisation; and K.2.F. = Knowledge indicator 2.F. = general knowledge and skills matching those of the (S)ENCo.

Table F22

Mean Selection Rates of the Skills Indicators per Category in Domain 2

Indicators	N	Mean	*SD*
V.2.A.	354	2.47	1.104
V.2.B.	354	2.40	1.110
V.2.C.	354	2.62	1.139
V.2.D.	354	3.04	1.152
V.2.E.	354	2.66	1.006
V.2.F.	354	1.82	1.089

Note. One-Sample Statistics. Test value 2.5.

V.2.A. = Skills indicator 2.A. = stimulating academic development; V.2.B. = Skills indicator 2.B. = stimulating the effective use of different cognitive styles; V.2.C. = Skills indicator 2.C. = stimulating interpersonal functioning; V.2.D. = Skills indicator 2.D. = stimulating intrapersonal understanding; V.2.E. = Skills indicator 2.E. = stimulating self-actualisation; and V.2.F. = Skills indicator 2.F. = general knowledge and skills matching those of the (S)ENCo.

Table F23

Selection Rates of the Skills indicators per Category in Domain 2

					95% Confidence interval of the difference		
Indicators	*t*	*df*	Sig (2-tailed)	Mean difference	Lower bound	Upper bound	Cohen's *d*
V.2.A.	-0.578	353	.564	-0.034	-0.15	0.08	.03
V.2.B.	-1.724	353	.086	-0.102	-0.22	0.01	.09
V.2.C.	1.914	353	.056	0.116	0.00	0.23	.11
V.2.D.	8.765	353	.000	0.537	0.42	0.66	.47
V.2.E.	3.065	353	.002	0.164	0.06	0.27	.16
V.2.F.	-11.765	353	.000	-0.681	-0.79	-0.57	.62

Note. One-Sample Test. Test value 2.5. N = 354.

V.2.A. = Skills indicator 2.A. = stimulating academic development; V.2.B. = Skills indicator 2.B. = stimulating the effective use of different cognitive styles; V.2.C. = Skills indicator 2.C. = stimulating interpersonal functioning; V.2.D. = Skills indicator 2.D. = stimulating intrapersonal understanding; V.2.E. = Skills indicator 2.E. = stimulating self-actualisation; and V.2.F. = Skills indicator 2.F. = general knowledge and skills matching those of the (S)ENCO.

Table F24

Indicator Category with Highest Mean Selection Rate in Domain 2 per Group of Stakeholders

Indicators	Educational professionals	Specialists in gifted education	Parents	Rest
Knowledge	A	A	D	A
Skills	D	D/E	D	D

Note. A = stimulating academic development; D = stimulating intrapersonal understanding; E = stimulating self-actualisation;

Table F25

The Comparison of Selection Rates Between Two Clusters of Knowledge Indicators in Domain 2: ADE-indicators vs BCF-indicators

Clusters of Indicators	Mean	N	*SD*	*SE*
Dom2_K_ADE	2.8691	354	0.45593	0.02423
Dom2_K_BCF	2.1309	354	0.45593	0.02423

Note: Paired Samples Statistics.

Dom2_K_ADE = Domain 2, cluster of three categories of knowledge indicators A, D, E; compared with Dom2_K_BCF = Domain 2, cluster of three categories of knowledge indicators B, C, F.

Table F26

The Comparison of Selection Rates Between Two Clusters of Knowledge Indicators in Domain 2: ADE-Indicators Vs BCF- Indicators

				95% Confidence interval of the difference					
Clusters of Indicators	Mean	*SD*	*SE*	Lower bound	Upper bound	*t*	*df*	Sig (2-tailed)	Cohen's *d*
Dom2_K_ADE – Dom2_K_BCF	0.73823	0.91187	0.04847	0.64291	0.83355	15.232	353	.000	.81

Note. Paired Samples T test. N = 354.

Dom2_K_ADE = Domain 2, cluster of three categories of knowledge indicators A, D, E; compared to Dom2_K_BCF = Domain 2, cluster of three categories of knowledge indicators B, C, F.

Table F27

The Comparison of Selection Rates Between Two Clusters of Skills Indicators in Domain 2: ADE-indicators vs BCF-indicators

Clusters of Indicators	Mean	N	*SD*	*SE*
Dom2_V_ADE	2.7222	354	0.45894	0.02439
Dom2_V_BCF	2.2778	354	0.45894	0.02439

Note: Paired Samples Statistics.

Dom2_V_ADE = Domain 2, cluster of three categories of skills indicators A, D, E; compared to Dom2_V_BCF = Domain 2, cluster of three categories of skills indicators B, C, F.

Table F28

The Comparison of Selection Rates Between Two Clusters of Skills Indicators in Domain 2: ADE-indicators vs BCF-indicators

				95% Confidence interval of the difference					
Clusters of Indicators	Mean	*SD*	*SE*	Lower bound	Upper bound	*t*	*df*	Sig (2-tailed)	Cohen's *d*
Dom2_V_ADE – Dom2_V_BCF	0.4444	0.91789	0.04879	0.34850	0.54039	9.110	353	.000	.48

Note. Paired Samples T test. N = 354.

Dom2_V_ADE = Domain 2, cluster of three categories of skills indicators A, D, E; compared to Dom2_V_BCF = Domain 2, cluster of three categories of skills indicators B, C, F.

Table F29

Knowledge Indicators in Domain 2 With the Highest Selection Rates Amongst all Groups of Stakeholders

Competency	Code	Indicator
2.1 The specialist explores the strengths and weaknesses of the gifted student within the intrapersonal and interpersonal developmental domain.	K.2.D.1	Knows what types of behaviour by the gifted student may be a sign for the development of a negative self-concept.

Table F30

Skills Indicators in Domain 2 With the Highest Selection Rates Amongst all Groups of Stakeholders

Competency	Code	Indicator
2.2 The specialist assesses the student's developmental levels in all academic domains offered, considering the student's cognitive style.	V.2.F.2	Takes other indicators besides school achievements in consideration when estimating if the student may be gifted.
2.4 The specialist approaches all actors in the student's ecological system as equal partners and values their information about the student's educational needs as such.	V.2.D.4	Is focused on solutions and learnable skills instead of being focused on problems when talking with a student.

Table F31

Knowledge Indicators in Domain 2 With the Lowest Selection Rates Amongst all Groups of Stakeholders

Competency	Code	Indicator
2.4 The specialist approaches all actors in the student's ecological system as equal partners and values their information about the student's educational needs as such.	K.2.F.4	Understands why it is important to explain the objectives of the assessment of educational needs to the educational partners.

Table F32

Comparison of Mean Selection Rates in Domain 2 per Category of Knowledge Indicators Between Dutch and Flemish Stakeholders

Indicators	Country	N	Mean	*SD*	*SE*
K.2.A.	NL	299	3.13	1.061	0.061
	F	55	2.78	1.013	0.137
K.2.B.	NL	299	2.90	0.965	0.056
	F	55	2.18	1.107	0.149
K.2.C.	NL	299	1.44	1.045	0.060
	F	55	1.91	1.041	0.140
K.2.D.	NL	299	2.64	1.225	0.071
	F	55	3.07	1.260	0.170
K.2.E.	NL	299	2.85	1.027	0.059
	F	55	2.73	1.269	0.171
K.2.F.	NL	299	2.05	1.076	0.062
	F	55	2.33	1.171	0.158

Note. Descriptive Statistics. NL = The Netherlands. F = Flanders. N =354.

K.2.A. = Knowledge indicator 2.A. = stimulating academic development; K.2.B. = Knowledge indicator 2.B. = stimulating the effective use of different cognitive styles; K.2.C. = Knowledge indicator 2.C. = stimulating interpersonal functioning; K.2.D. = Knowledge indicator 2.D. = stimulating intrapersonal understanding; K.2.E. = Knowledge indicator 2.E. = stimulating self-actualisation; and K.2.F. = Knowledge indicator 2.F. = general knowledge and skills matching those of the (S)ENCo.

Table F33

Comparison of Mean Selection Rates in Domain 2 per Category of Knowledge Indicators Between Dutch and Flemish Stakeholders

		Levene's test for Equality of variances		t-test for Equality of means							
									95% Confidence interval of the difference		
Indicators		*F*	*p*	*t*	*df*	Sig. (2-tailed)	Mean difference	*SE difference*	Lower bound	Upper bound	Cohen's *d*
K.2.A.	Equal variances assumed	0.036	.850	2.234	352	.026	0.345	0.155	0.041	0.649	.33
	Equal variances not assumed			2.306	77.413	.024	0.345	0.150	0.047	0.643	
K.2.B.	Equal variances assumed	1.368	.243	4.926	352	.000	0.715	0.145	0.429	1.000	.72
	Equal variances not assumed			4.483	69.919	.000	0.715	0.159	0.397	1.032	
K.2.C.	Equal variances assumed	0.212	.646	-3.073	352	.002	-0.471	0.153	-0.772	-0.170	.45
	Equal variances not assumed			-3.081	75.407	.003	-0.471	0.153	-0.775	-0.166	
K.2.D.	Equal variances assumed	0.038	.845	-2.422	352	.016	-0.437	0.181	-0.792	-0.082	.36
	Equal variances not assumed			-2.376	74.014	.020	-0.437	0.184	0.804	-0.071	
K.2.E.	Equal variances assumed	5.907	.016	0.780	352	.436	0.122	0.157	-0.186	0.430	.11
	Equal variances not assumed			0.675	67.622	.502	0.122	0.181	-0.239	0.484	
K.2.F.	Equal variances assumed	4.175	.042	-1.710	352	.088	-0.274	0.160	-0.589	0.041	.25
	Equal variances not assumed			-1.612	71.754	.111	-0.274	0.170	-0.612	0.065	

Note. Independent Samples Test. N = 354.

K.2.A. = Knowledge indicator 2.A. = stimulating academic development; K.2.B. = Knowledge indicator 2.B. = stimulating the effective use of different cognitive styles; K.2.C. = Knowledge indicator 2.C. = stimulating interpersonal functioning; K.2.D. = Knowledge indicator 2.D. = stimulating intrapersonal understanding; K.2.E. = Knowledge indicator 2.E. = stimulating self-actualisation; and K.2.F. = Knowledge indicator 2.F. = general knowledge and skills matching those of the (S)ENCo.

Table F34

Comparison of Mean Selection Rates in Domain 2 per Category of Skills Indicators Between Dutch and Flemish Stakeholders

Indicators	Country	N	Mean	*SD*	*SE*
V.2.A.	NL	299	2.51	1.082	0.063
	F	55	2.22	1.197	0.161
V.2.B.	NL	299	2.38	1.091	0.063
	F	55	2.49	1.215	0.164
V.2.C.	NL	299	2.57	1.152	0.067
	F	55	2.85	1.044	0.141
V.2.D.	NL	299	3.01	1.159	0.067
	F	55	3.20	1.112	0.150
V.2.E.	NL	299	2.64	0.995	0.058
	F	55	2.82	1.056	0.142
V.2.F.	NL	299	1.89	1.082	0.063
	F	55	1.42	1.049	0.141

Note. Descriptive Statistics. NL = The Netherlands. F = Flanders. N = 354.

V.2.A. = Skills indicator 2.A. = stimulating academic development; V.2.B. = Skills indicator 2.B. = stimulating the effective use of different cognitive styles; V.2.C. = Skills indicator 2.C. = stimulating interpersonal functioning; V.2.D. = Skills indicator 2.D. = stimulating intrapersonal understanding; V.2.E. = Skills indicator 2.E. = stimulating self-actualisation; and V.2.F. = Skills indicator 2.F. = general knowledge and skills matching those of the (S)ENCo.

Table F35

Comparison of Mean Selection Rates in Domain 2 per Category of Skills Indicators Between Dutch and Flemish Stakeholders

		Levene's test for Equality of variances		t-test for Equality of means							
									95% Confidence interval of the difference		
Indicators		*F*	*p*	*t*	*df*	Sig. (2-tailed)	Mean difference	*SE difference*	Lower bound	Upper bound	Cohen's *d*
V.2.A.	Equal variances assumed	0.911	.340	1.818	352	.070	0.294	0.161	-0.024	0.611	.27
	Equal variances not assumed			1.696	71.150	.094	0.294	0.173	-0.052	0.639	
V.2.B.	Equal variances assumed	1.175	.279	-0.673	352	.502	-0.110	0.163	-0.430	0.211	.10
	Equal variances not assumed			-0.624	70.908	.534	-0.110	0.176	-0.460	0.240	
V.2.C.	Equal variances assumed	3.972	.047	-1.696	352	.091	-0.283	0.167	-0.610	0.045	.25
	Equal variances not assumed			-1.815	80.159	.073	-0.283	0.156	-0.593	0.027	
V.2.D.	Equal variances assumed	0.002	.962	-1.144	352	.253	-0.193	0.169	-0.526	0.139	.17
	Equal variances not assumed			-1.177	77.150	.243	-0.193	0.164	-0.520	0.134	
V.2.E.	Equal variances assumed	0.000	.984	-1.239	352	.216	-0.183	0.147	-0.473	0.107	.18
	Equal variances not assumed			-1.190	72.754	.238	-0.183	0.154	-0.489	0.123	
V.2.F.	Equal variances assumed	0.038	.846	3.006	352	.003	0.475	0.158	0.164	0.785	.44
	Equal variances not assumed			3.071	76.667	.003	0.475	0.155	0.167	0.783	

Note. Independent Samples Test. N =354.

V.2.A. = Skills indicator 2.A. = stimulating academic development; V.2.B. = Skills indicator 2.B. = stimulating the effective use of different cognitive styles; V.2.C. = Skills indicator 2.C. = stimulating interpersonal functioning; V.2.D. = Skills indicator 2.D. = stimulating intrapersonal understanding; V.2.E. = Skills indicator 2.E. = stimulating self-actualisation; and V.2.F. = Skills indicator 2.F. = general knowledge and skills matching those of the (S)ENCo.

Table F36

Domain 2 Knowledge Indicators. Mean per Category per Competency per Group of Stakeholders

Indicators	EP	Indicators	SGE	Indicators	Pa	Indicators	R
K.2.C.	1.54	K.2.C.	1.35	K.2.C.	1.76	K.2.C.	1.30
K.2.F.	1.98	K.2.D.	2.26	K.2.F.	2.05	K.2.F.	2.05
K.2.D.	2.71	K.2.F.	2.29	K.2.B.	2.29	K.2.D.	2.59
K.2.E.	2.73	K.2.B.	2.99	K.2.E.	2.72	K.2.B.	2.76
K.2.B.	2.94	K.2.E.	3.02	K.2.A.	2.89	K.2.E.	2.92
K.2.A.	3.10	K.2.A.	3.09	K.2.D.	3.29	K.2.A.	3.35

Note. Test value 2.5. EP = educational professionals; SGE = specialists gifted education; Pa = parents; R = rest. N = 354.

K.2.A. = Knowledge indicator 2.A. = stimulating academic development; K.2.B. = Knowledge indicator 2.B. = stimulating the effective use of different cognitive styles; K.2.C. = Knowledge indicator 2.C. = stimulating interpersonal functioning; K.2.D. = Knowledge indicator 2.D. = stimulating intrapersonal understanding; K.2.E = Knowledge indicator 2.E. = stimulating self-actualisation; and K.2.F. = Knowledge indicator 2.F. = general knowledge and skills matching those of the (S)ENCo.

Table F37

Domain 2 Skills Indicators. Mean per Category per Competency per Group of Stakeholders

Indicators	EP	Indicators	SGE	Indicators	Pa	Indicators	R
V.2.F.	1.76	V.2.F.	2.20	V.2.F.	1.43	V.2.F.	1.84
V.2.B.	2.25	V.2.B.	2.22	V.2.A.	2.25	V.2.A.	2.32
V.2.A.	2.52	V.2.C.	2.46	V.2.C.	2.51	V.2.B.	2.38
V.2.E.	2.58	V.2.A.	2.62	V.2.E.	2.67	V.2.C.	2.46
V.2.C.	2.85	V.2.D.	2.75	V.2.B.	2.86	V.2.E.	2.70
V.2.D.	3.04	V.2.E.	2.75	V.2.D.	3.28	V.2.D.	3.30

Note. Test value 2.5. EP = educational professionals; SGE = specialists gifted education; Pa = parents; R = rest. N = 354.

V.2.A. = Skills indicator 2.A. = stimulating academic development; V.2.B. = Skills indicator 2.B. = stimulating the effective use of different cognitive styles; V.2.C. = Skills indicator 2.C. = stimulating interpersonal functioning; V.2.D. = Skills indicator 2.D. = stimulating intrapersonal understanding; V.2.E. = Skills indicator 2.E. = stimulating self-actualisation; and V.2.F. = Skills indicator 2.F. = general knowledge and skills matching those of the (S)ENCo.

Table F38

Ranking of the Knowledge and Skills Indicators of Domain 2

Indicators	N	%	Indicators	N	%
V.2.C.2	43	0.40	V.2.B.2	178	1.68
K.2.F.4	81	0.76	V.2.B.3	181	1.70
V.2.F.5	96	0.90	V.2.D.3	181	1.70
K.2.B.2	106	1.00	K.2.A.5	185	1.74
V.2.D.1	116	1.09	K.2.F.3	187	1.76
K.2.E.1	122	1.15	K.2.D.2	190	1.79
K.2.E.4	122	1.15	K.2.D.5	192	1.81
V.2.F.3	122	1.15	V.2.D.5	192	1.81
V.2.C.4	124	1.17	V.2.C.3	194	1.83
V.2.A.4	129	1.21	K.2.D.4	196	1.85
V.2.F.4	131	1.23	K.2.B.3	197	1.85
K.2.D.3	133	1.25	V.2.B.1	201	1.89
K.2.F.1	139	1.31	K.2.A.1	203	1.91
V.2.F.1	140	1.32	K.2.B.4	210	1.98
K.2.A.2	142	1.34	V.2.C.1	211	1.99
K.2.F.5	143	1.35	K.2.C.4	215	2.02
K.2.B.1	145	1.37	K.2.E.3	216	2.03
V.2.D.2	151	1.42	K.2.C.5	224	2.11
K.2.E.5	153	1.44	V.2.E.3	224	2.11
V.2.E.5	154	1.45	K.2.F.2	225	2.12
K.2.C.2	155	1.46	V.2.C.5	226	2.13
V.2.E.1	156	1.47	V.2.B.5	229	2.16
K.2.A.3	159	1.50	V.2.E.2	231	2.18
V.2.A.3	160	1.51	K.2.A.4	238	2.24
V.2.B.4	160	1.51	V.2.A.1	238	2.24
V.2.A.5	165	1.55	V.2.E.4	238	2.24
K.2.B.5	165	1.55	K.2.E.2	244	2.30
K.2.C.1	168	1.58	V.2.D.4	280	2.64
K.2.C.3	170	1.60	K.2.D.1	285	2.68
V.2.A.2	171	1.61	V.2.F.2	288	2.71

Note. This ranking is based on selection rates of the entire group of stakeholders. N =354.

All indicators are coded. K = knowledge; V = skills; the first number refers to the vertical competency domain in the competency matrix; the second letter refers to the category of indicators. The second number refers to the actual competency the indicators are matched with. See Appendix B for the match between the competencies and the indicators.

Appendix G

Tables Domain 3, Referred to in Chapter 7

Table G1

Frequency Table of the Selection Rates of Knowledge Indicators for Competency 3.1

Competency	Indicator	EP	SGE	Pa	R	Total
3.1	K.3.A.1	13.23	13.27	13.25	13.51	13.28
	K.3.B.1	22.39	22.01	19.68	23.42	21.75
	K.3.C.1	7.63	7.12	10.84	14.41	8.95
	K.3.D.1	20.61	20.06	21.29	16.22	20.15
	K.3.E.1	20.61	19.74	18.07	16.22	19.30
	K.3.F.1	15.52	17.80	16.87	16.22	16.57

Note. The results are presented per indicator for each group of stakeholders. EP = educational professionals; SGE = specialist gifted education; Pa = parents; R = Rest; T = total. N = 354.

All indicators are coded. K = knowledge; the first number refers to the vertical competency domain in the competency matrix; the second letter refers to the category of indicators. The second number refers to the actual competency the indicators are matched with. See Appendix B for the match between the competencies and the indicators.

Table G2

Frequency Table of the Selection Rates of Knowledge Indicators for Competency 3.2

Competency	Indicator	EP	SGE	Pa	R	T
3.2	K.3.A.2	11.96	9.39	8.03	14.41	10.55
	K.3.B.2	16.03	16.18	17.27	15.32	16.29
	K.3.C.2	6.62	6.15	9.24	9.01	7.34
	K.3.D.2	29.26	26.21	26.10	25.23	27.21
	K.3.E.2	18.07	20.06	16.06	15.32	17.89
	K.3.F.2	18.07	22.01	23.29	20.72	20.72

Note. The results are presented per indicator for each group of stakeholders. EP = educational professionals; SGE = specialist gifted education; Pa = parents; R = Rest; T = total. N = 354.

All indicators are coded. K = knowledge; the first number refers to the vertical competency domain in the competency matrix; the second letter refers to the category of indicators. The second number refers to the actual competency the indicators are matched with. See Appendix B for the match between the competencies and the indicators.

Table G3

Frequency Table of the Selection Rates of Knowledge Indicators for Competency 3.3

Competency	Indicator	EP	SGE	Pa	R	Total
3.3	K.3.A.3	20.36	22.01	19.68	20.72	20.72
	K.3.B.3	8.91	8.41	9.64	8.11	8.85
	K.3.C.3	19.85	19.74	19.68	21.62	19.96
	K.3.D.3	17.05	13.59	16.47	16.22	15.82
	K.3.E.3	22.39	19.09	23.69	18.02	21.28
	K.3.F.3	11.45	17.15	10.84	15.32	13.37

Note. The results are presented per indicator for each group of stakeholders. EP = educational professionals; SGE = specialist gifted education; Pa = parents; R = Rest; T = total. N = 354.

All indicators are coded. K = knowledge; the first number refers to the vertical competency domain in the competency matrix; the second letter refers to the category of indicators. The second number refers to the actual competency the indicators are matched with. See Appendix B for the match between the competencies and the indicators.

Table G4

Frequency Table of the Selection Rates of Knowledge Indicators for Competency 3.4

Competency	Indicator	EP	SGE	Pa	R	Total
3.4	K.3.A.4	23.66	15.86	21.29	19.82	20.43
	K.3.B.4	15.01	14.56	14.06	20.72	15.25
	K.3.C.4	16.28	17.48	20.08	11.71	17.04
	K.3.D.4	10.18	10.36	15.26	10.81	11.49
	K.3.E.4	21.63	21.04	15.26	24.32	20.24
	K.3.F.4	13.23	20.71	14.06	12.61	15.54

Note. The results are presented per indicator for each group of stakeholders. EP = educational professionals; SGE = specialist gifted education; Pa = parents; R = Rest; T = total. N = 354.

All indicators are coded. K = knowledge; the first number refers to the vertical competency domain in the competency matrix; the second letter refers to the category of indicators. The second number refers to the actual competency the indicators are matched with. See Appendix B for the match between the competencies and the indicators.

Table G5

Frequency Table of the Selection Rates of Knowledge Indicators for Competency 3.5

Competency	Indicator	EP	SGE	Pa	R	Total
3.5	K.3.A.5	11.20	10.68	7.63	11.71	10.26
	K.3.B.5	13.74	12.62	8.03	9.91	11.68
	K.3.C.5	10.69	12.62	8.84	17.12	11.49
	K.3.D.5	22.65	19.74	22.09	18.92	21.28
	K.3.E.5	19.08	19.09	27.31	19.82	21.09
	K.3.F.5	22.65	25.24	26.10	22.52	24.20

Note. The results are presented per indicator for each group of stakeholders. EP = educational professionals; SGE = specialist gifted education; Pa = parents; R = Rest; T = total. N = 354.

All indicators are coded. K = knowledge; the first number refers to the vertical competency domain in the competency matrix; the second letter refers to the category of indicators. The second number refers to the actual competency the indicators are matched with. See Appendix B for the match between the competencies and the indicators.

Table G6

Frequency Table of the Selection Rates of Skills Indicators for Competency 3.1

Competency	Indicator	EP	SGE	Pa	R	Total
3.1	V.3.A.1	22.65	17.80	20.08	16.22	19.96
	V.3.B.1	15.52	12.62	19.68	20.72	16.20
	V.3.C.1	17.30	15.86	15.66	15.32	16.29
	V.3.D.1	9.92	10.36	14.86	9.91	11.21
	V.3.E.1	24.68	25.24	22.49	27.03	24.58
	V.3.F.1	9.92	18.12	7.23	10.81	11.77

Note. The results are presented per indicator for each group of stakeholders. EP = educational professionals; SGE = specialist gifted education; Pa = parents; R = Rest; T = total. N = 354.

All indicators are coded. V = skills; the first number refers to the vertical competency domain in the competency matrix; the second letter refers to the category of indicators. The second number refers to the actual competency the indicators are matched with. See Appendix B for the match between the competencies and the indicators.

Table G7

Frequency Table of the Selection Rates of Skills Indicators for Competency 3.2

Competency	Indicator	EP	SGE	Pa	R	T
3.2	V.3.A.2	19.59	21.68	19.28	22.52	20.43
	V.3.B.2	17.56	17.48	16.06	12.61	16.67
	V.3.C.2	5.34	2.91	6.83	3.60	4.80
	V.3.D.2	19.34	18.45	18.88	18.92	18.93
	V.3.E.2	21.12	23.30	23.29	23.42	22.50
	V.3.F.2	17.05	16.18	15.66	18.92	16.67

Note. The results are presented per indicator for each group of stakeholders. EP = educational professionals; SGE = specialist gifted education; Pa = parents; R = Rest; T = total. N = 354.

All indicators are coded. V = skills; the first number refers to the vertical competency domain in the competency matrix; the second letter refers to the category of indicators. The second number refers to the actual competency the indicators are matched with. See Appendix B for the match between the competencies and the indicators.

Table G8

Frequency Table of the Selection Rates of Skills Indicators for Competency 3.3

Competency	Indicator	EP	SGE	Pa	R	Total
3.3	V.3.A.3	19.34	20.06	18.47	16.22	19.02
	V.3.B.3	22.90	22.98	25.30	25.23	23.73
	V.3.C.3	13.49	15.86	13.25	9.01	13.65
	V.3.D.3	21.12	15.21	17.27	21.62	18.55
	V.3.E.3	14.76	14.24	17.27	18.02	15.54
	V.3.F.3	8.40	11.65	8.43	9.91	9.51

Note. The results are presented per indicator for each group of stakeholders. EP = educational professionals; SGE = specialist gifted education; Pa = parents; R = Rest; T = total. N = 354.

All indicators are coded. V = skills; the first number refers to the vertical competency domain in the competency matrix; the second letter refers to the category of indicators. The second number refers to the actual competency the indicators are matched with. See Appendix B for the match between the competencies and the indicators.

Table G9

Frequency Table of the Selection Rates of Skills Indicators for Competency 3.4

Competency	Indicator	EP	SGE	Pa	R	Total
3.4	V.3.A.4	10.69	9.71	11.65	11.71	10.73
	V.3.B.4	24.17	22.01	18.07	21.62	21.85
	V.3.C.4	7.63	6.80	7.63	9.91	7.63
	V.3.D.4	11.70	11.00	10.44	12.61	11.30
	V.3.E.4	22.90	24.92	26.91	21.62	24.29
	V.3.F.4	22.90	25.57	25.30	22.52	24.20

Note. The results are presented per indicator for each group of stakeholders. EP = educational professionals; SGE = specialist gifted education; Pa = parents; R = Rest; T = total. N = 354.

All indicators are coded. V = skills; the first number refers to the vertical competency domain in the competency matrix; the second letter refers to the category of indicators. The second number refers to the actual competency the indicators are matched with. See Appendix B for the match between the competencies and the indicators.

Table G10

Frequency Table of the Selection Rates of Skills Indicators for Competency 3.5

Competency	Indicator	EP	SGE	Pa	R	Total
3.5	V.3.A.5	18.07	21.36	22.89	19.82	20.34
	V.3.B.5	8.40	7.77	5.62	9.01	7.63
	V.3.C.5	11.45	14.89	12.05	13.51	12.81
	V.3.D.5	15.01	5.83	14.06	11.71	11.77
	V.3.E.5	21.37	22.98	23.29	20.72	22.22
	V.3.F.5	25.70	27.18	22.09	25.23	25.24

Note. The results are presented per indicator for each group of stakeholders. EP = educational professionals; SGE = specialist gifted education; Pa = parents; R = Rest; T = total. N = 354.

All indicators are coded. V = skills; the first number refers to the vertical competency domain in the competency matrix; the second letter refers to the category of indicators. The second number refers to the actual competency the indicators are matched with. See Appendix B for the match between the competencies and the indicators.

Table G11

Comparison of the 10 Highest Indicator Selection Rates per Competency in Domain 3 per Group of Stakeholders

Indicator	EP	Indicator	SGE	Indicator	Pa	Indicator	R
V.3.E.2	21.12	K.3.E.4	21.04	K.3.A.4	21.29	K.3.C.3	21.62
K.3.B.1	22.39	K.3.B.1	22.01	K.3.D.1	21.29	K.3.F.5	22.52
K.3.E.3	22.39	K.3.A.3	22.01	V.3.E.1	22.49	V.3.F.4	22.52
K.3.D.5 / K.3.F.5	22.65	V.3.B.3	22.98	V.3.E.2	23.29	K.3.B.1	23.42
V.3.B.3	22.90	V.3.E.2	23.30	V.3.E.5	23.29	V.3.E.2	23.42
K.3.A.4	23.66	K.3.F.5	25.24	K.3.E.3	23.69	K.3.E.4	24.32
V.3.B.4	24.17	V.3.E.1	25.24	V.3.B.3	25.30	K.3.D.2	25.23
V.3.E.1	24.68	V.3.F.4	25.57	K.3.D.2	26.10	V.3.B.3	25.23
V.3.F.5	25.70	K.3.D.2	26.21	V.3.E.4	26.91	V.3.F.5	25.23
K.3.D.2	29.26	V.3.F.5	27.18	K.3.E.5	27.31	V.3.E.1	27.03

Note. EP = educational professionals; SGE = specialist gifted education; Pa = parents; R = Rest; T = total. N = 354. If two indicators matching one competency had an identical selection rate, they are ranked at the same level.

All indicators are coded. K = knowledge; V = skills; the first number refers to the vertical competency domain in the competency matrix; the second letter refers to the category of indicators. The second number refers to the actual competency the indicators are matched with. See Appendix B for the match between the competencies and the indicators.

Table G12

Comparison of the Selection Rates of the Knowledge Indicators in Domain 3 Between the Groups of Stakeholders

Indicators	Group	N	Mean	*SD*	*SE*	95% Confidence interval for mean		Min	Max
						Lower bound	Upper bound		
K.3.B.	EP	131	2.81	1.068	0.093	2.62	2.99	0	5
	SGE	103	2.74	0.928	0.091	2.56	2.92	1	5
	Pa	83	2.33	1.094	0.120	2.09	2.56	0	5
	R	37	2.54	0.869	0.143	2.25	2.83	1	4
	T	354	2.65	1.301	0.055	2.54	2.75	0	5
K.3.D.	EP	131	2.32	1.308	0.114	2.09	2.55	0	5
	SGE	103	1.83	1.271	0.125	1.58	2.07	0	5
	Pa	83	2.96	1.204	0.132	2.70	3.23	0	5
	R	37	2.27	1.140	0.231	1.80	2.74	0	5
	T	354	2.32	1.344	0.071	2.18	2.46	0	5
K.3.F.	EP	131	2.05	1.143	0.100	1.85	2.24	0	5
	SGE	103	2.47	1.274	0.126	2.22	2.72	0	5
	Pa	83	2.01	1.215	0.133	1.75	2.28	0	5
	R	37	2.22	1.109	0.182	1.85	2.59	0	4
	T	354	2.18	1.206	0.064	2.05	2.30	0	5

Note. Descriptive Statistics. EP = educational professionals; SGE = specialist gifted education; Pa = parents; R = Rest; T = total. N = 354.

K.3.B. = Knowledge indicator 3.B. = stimulating the effective use of different cognitive styles; K.3.D. = Knowledge indicator 3.D. = stimulating intrapersonal understanding; K.3.F. = Knowledge indicator 3.F. = general knowledge and skills matching those of the (S)ENCo.

Table G13

Comparison of the Selection Rates of the Knowledge Indicators in Domain 3 Between the Groups of Stakeholders

Indicators		Sum of Squares	*df*	Mean Square	*F*	*p*	η^2
K.3.A.	Between Groups	7.276	3	2.425	2.314	.076	.019
	Within Groups	366.792	350	1.048			
	Total	374.068	353				
K.3.B.	Between Groups	13.304	3	4.435	4.293	.005	.037
	Within Groups	361.557	350	1.033			
	Total	374.862	353				
K.3.C.	Between Groups	7.570	3	2.523	2.104	.099	.018
	Within Groups	419.800	350	1.199			
	Total	427.370	353				
K.3.D.	Between Groups	59.711	3	19.904	12.061	.000	.094
	Within Groups	577.578	350	1.650			
	Total	637.288	353				
K.3.E.	Between Groups	7.741	3	2.580	2.509	.059	.021
	Within Groups	360.013	350	1.029			
	Total	367.754	353				
K.3.F.	Between Groups	13.174	3	4.391	3.070	.028	.026
	Within Groups	500.614	350	1.430			
	Total	513.788	353				

Note. ANOVA = analyses of variance with one factor.

K.3.A. = knowledge indicator 3.A. = stimulating academic development; K.3.B. = Knowledge indicator 3.B. = stimulating the effective use of different cognitive styles; K.3.C. = Knowledge indicator 3.C. = stimulating interpersonal functioning; K.3.D. = Knowledge indicator 3.D. = stimulating intrapersonal understanding; K.3.E. = Knowledge indicator 3.E. = stimulating self-actualisation; K.3.F. = Knowledge indicator 3.F. = general knowledge and skills matching those of the (S)ENCo.

Table G14

Comparison of the Selection Rates of the Knowledge Indicators in Domain 3 Between the Groups of Stakeholders

Dependent variable	I group	J group	Mean difference (I-J)	*SE*	*p*	95% Confidence interval	
						Lower bound	Upper bound
K.3.B.	Pa	EP	-0.484	0.143	.005	-0.86	-0.11
		SGE	-0.413	0.150	.037	-0.81	-0.01
		R	-0.215	0.201	1.000	-0.75	0.32
K.3.D.	Pa	EP	0.643	0.180	.002	0.17	1.12
		SGE	1.139	0.189	.000	0.64	1.64
		R	0.694	0.254	.040	0.02	1.37
K.3.D.	SGE	EP	-0.495	0.169	.022	-0.94	-0.05
		Pa	-1.139	0.189	.000	-1.64	-0.64
		R	-0.445	0.246	.429	-1.10	0.21
K.3.F.	SGE	EP	0.420	0.157	.048	0.00	0.84
		Pa	0.454	0.176	.063	-0.01	0.92
		R	0.250	0.229	1.000	-0.36	0.86

Note. Based on the results of the ANOVA, Bonferroni's Multiple Comparisons was used as a post hoc test for the analyses of the differences between groups of stakeholders.

EP = educational professionals; SGE = specialist gifted education; Pa = parents; R = Rest; T = total. N = 354.

K.3.B. = Knowledge indicator 3.B. = stimulating the effective use of different cognitive styles; K.3.D. = Knowledge indicator 3.D. = stimulating intrapersonal understanding; K.3.F. = Knowledge indicator 3.F. = general knowledge and skills matching those of the (S)ENCo.

Table G15

Comparison of the Selection Rates of the Skills Indicators in Domain 3 Between the Groups of Stakeholders

Indicators	Group	N	Mean	*SD*	*SE*	95% Confidence interval for mean		Min	Max
						Lower bound	Upper bound		
V.3.B.	EP	131	2.18	1.058	0.092	2.00	2.37	0	5
	SGE	103	2.06	1.065	0.105	1.81	2.27	0	5
	Pa	83	2.59	1.220	0.134	2.32	2.86	0	5
	R	37	2.38	0.924	0.152	2.07	2.69	1	4
	T	354	2.26	1.102	0.059	2.15	2.38	0	5
V.3.D.	EP	131	3.14	1.135	0.099	2.94	3.33	0	5
	SGE	103	2.71	1.160	0.114	2.48	2.94	0	5
	Pa	83	3.37	1.079	0.118	3.14	3.61	1	5
	R	37	3.35	1.184	0.195	2.96	3.75	1	5
	T	354	3.09	1.160	0.062	2.97	3.21	0	5
V.3.F.	EP	131	2.07	1.054	0.092	1.89	2.25	0	4
	SGE	103	2.52	0.989	0.097	2.33	2.72	0	4
	Pa	83	1.90	1.031	0.113	1.68	2.13	0	4
	R	37	2.24	1.038	0.171	1.90	2.59	0	4
	T	354	2.18	1.052	0.056	2.07	2.29	0	4

Note. Descriptive Statistics. EP = educational professionals; SGE = specialist gifted education; Pa = parents; R = Rest; T = total. N = 354.

V.3.B. = Knowledge indicator 3.B. = stimulating the effective use of different cognitive styles; V.3.D. = Knowledge indicator 3.D. = stimulating intrapersonal understanding; V.3.F. = Knowledge indicator 3.F. = general knowledge and skills matching those of the (S)ENCo.

Table G16

Comparison of the Selection Rates of the Skills Indicators in Domain 3 Between the Groups of Stakeholders

Indicators		Sum of Squares	*df*	Mean Square	*F*	*p*	η^2
V.3.A.	Between Groups	6.134	3	2.045	1.518	.210	.01
	Within Groups	471.519	350	1.347			
	Total	477.653	353				
V.3.B.	Between Groups	14.539	3	4.846	4.097	.007	.03
	Within Groups	414.029	350	1.183			
	Total	428.568	353				
V.3.C.	Between Groups	8.229	3	2.743	2.356	.072	.02
	Within Groups	407.522	350	1.164			
	Total	415.751	353				
V.3.D.	Between Groups	24.464	3	8.155	6.334	.000	.05
	Within Groups	450.643	350	1.288			
	Total	475.107	353				
V.3.E.	Between Groups	.607	3	0.202	.170	.916	.00
	Within Groups	416.026	350	1.189			
	Total	416.633	353				
V.3.F.	Between Groups	20.319	3	6.773	6.405	.000	.05
	Within Groups	370.111	350	1.057			
	Total	390.429	353				

Note. ANOVA = analyses of variance with one factor. N = 354.

V.3.A. = knowledge indicator 3.A. = stimulating academic development; V.3.B. = Knowledge indicator 3.B. = stimulating the effective use of different cognitive styles; V.3.C. = Knowledge indicator 3.C. = stimulating interpersonal functioning; V.3.D. = Knowledge indicator 3.D. = stimulating intrapersonal understanding; V.3.E. = Knowledge indicator 3.E. = stimulating self-actualisation; V.3.F. = Knowledge indicator 3.F. = general knowledge and skills matching those of the (S)ENCo.

Table G17

Comparison of the Selection Rates of the Skills Indicators in Domain 3 Between the Groups of Stakeholders

Dependent variable	I group	J group	Mean Difference (I-J)	*SE*	*p*	95% Confidence interval Lower bound	Upper bound
V.3.B.	P	EP	0.407	0.153	.048	0.00	0.81
		SGE	0.532	0.160	.006	0.11	0.96
		R	0.212	0.215	1.000	-0.36	0.78
V.3.D.	SGE	EP	-0.429	0.149	.026	-0.83	-0.03
		Pa	-0.665	0.167	.001	-1.11	-0.22
		R	-0.643	0.217	.020	-1.22	-0.07
V.3.F.	SGE	EP	0.456	0.191	.005	0.10	0.81
		Pa	0.621	0.152	.000	0.22	1.02
		R	0.281	0.197	.929	-0.24	0.80

Note. Based on the results of the ANOVA, Bonferroni's Multiple Comparisons was used as a post hoc test for the analyses of the differences between groups of stakeholders.

EP = educational professionals; SGE = specialist gifted education; Pa = parents; R = Rest; T = total. N = 354. V.3.B. = Knowledge indicator 3.B. = stimulating the effective use of different cognitive styles; V.3.D. = Knowledge indicator 3.D. = stimulating intrapersonal understanding; V.3.F. = Knowledge indicator 3.F. = general knowledge and skills matching those of the (S)ENCo.

Table G18

Knowledge Indicators of Each Competency in Domain 3 that Belonged to the Three Most Selected Indicators by all Groups of Stakeholders

Competency	Code	Indicator
3.1 The specialist develops an understanding how the actors in the student's ecology influence the student's strengths and weaknesses.	K.3.B.1	Understands the relevance of the notion of ownership to find solutions during the process of mentoring a student.
	K.3.D.1	Knows that gifted students need instruction when new elements of the curriculum are offered.
	K.3.E.1	Knows that behaviour is the result of the interaction between the student and the ecological system.
3.2 The specialist explains how the interplay between the curriculum and the student's cognitive style influences their development and instructional needs.	K.3.D.2	Knows how mindset influences a student's learning process.
	K.3.F.2	Knows how a student's academic achievements are influenced by teacher support during the learning process.
3.3 The specialist understands how the individual student's needs are of consequence for their classroom management.	K.3.A.3	Knows that gifted student's need instruction on the new elements in the curriculum.
	K.3.C.3	Understands the relevance for gifted students to work with like-minded peers for their social development.
	K.3.E.3	Knows that, like other students, gifted students have to learn how to become responsible for their own actions.
3.4 Based on the results of their exploration of the student's educational needs, the specialist explains to the actors in the student's ecological system how their behaviour influences the student's development.	K.3.E.4	Has a general understanding of different theories of metacognitive skill development in children.
3.5 The specialist uses theory on giftedness and RtI to interpret the results of the exploration of the student's educational needs in the context of educational development.	K.3.D.5	Knows which risk factors contribute to underachievement.
	K.3.E.5	Knows how giftedness may influence a student's behaviour.
	K.3.F.5	Understands how strengths and weaknesses can be translated into educational needs.

Table G19

Skills Indicators of Each Competency in Domain 3 that Belonged to the Three Most Selected Indicators by All Groups of Stakeholders

Competency	Code	Indicator
3.1 The specialist develops an understanding of how the actors in the student's ecology influence the student's strengths and weaknesses.	V.3.A.1	Creates a learning environment wherein the learning process and achievements are equally important.
	V.3.E.1	Discusses with the student what they see as helpful learning strategies and interventions.
3.2 The specialist explains how the interplay between the curriculum and the student's cognitive style influences their development and instructional needs.	V.3.A.2	Determines if the pacing of the required curriculum matches the student's ability to develop knowledge and skills.
	V.3.D.2	Discusses with the student to what extent their self-efficacy related to a specific task is accurate.
	V.3.E.2	Evaluates whether the teacher's expectations towards a student's executive skills are realistic.
3.3 The specialist understands how the individual student's needs are of consequence for their classroom management.	V.3.B.3	Creates a learning environment where the use of executive skills is an intrinsic condition during the learning process.
	V.3.D.3	Takes the time for teacher-student meetings to discuss the student's self-concept.
3.4 Based on the results of their exploration of the student's educational needs, the specialist explains to the actors in the student's ecological system how their behaviour influences the student's development.	V.3.B.4	Teaches the student to reflect on their own learning process.
	V.3.E.4	Provides the student with a level of responsibility that matches their developmental level.
	V.3.F.4	Communicates regularly with educational partners about the student's educational development.
3.5 The specialist uses theory on giftedness and RtI to interpret the results of the exploration of the student's educational needs in the context of educational development.	V.3.A.5	Explains to the educational partners when their interaction contributes positively to a student's development.
	V.3.E.5	Identifies how giftedness may influence the student's behaviour.
	V.3.F.5	Describes a gifted student's educational needs as skills to be developed in the near future.

Table G20

Mean Selection Rates of the Knowledge Indicators per Category in Domain 3

Indicators	N	Mean	*SD*
K.3.A.	354	3.24	1.029
K.3.B.	354	2.65	1.031
K.3.C.	354	1.56	1.100
K.3.D.	354	2.32	1.344
K.3.E.	354	3.06	1.021
K.3.F.	354	2.18	1.206

Note. One-Sample Statistics. Test value 2.5.

K.3.A. = knowledge indicator 3.A. = stimulating academic development; K.3.B. = Knowledge indicator 3.B. = stimulating the effective use of different cognitive styles; K.3.C. = Knowledge indicator 3.C. = stimulating interpersonal functioning; K.3.D. = Knowledge indicator 3.D. = stimulating intrapersonal understanding; K.3.E. = Knowledge indicator 3.E. = stimulating self-actualisation; K.3.F. = Knowledge indicator 3.F. = general knowledge and skills matching those of the (S)ENCo.

Table G21

Selection Rates of the Knowledge Indicators per Category in Domain 3

					95% Confidence interval of the difference		
Indicators	*t*	*df*	Sig (2-tailed)	Mean difference	Lower bound	Upper bound	Cohen's *d*
K.3.A.	13.476	353	.000	0.737	0.63	0.84	.72
K.3.B.	2.682	353	.008	0.147	0.04	0.25	.15
K.3.C.	-16.134	353	.008	-0.944	-1.06	-0.83	.85
K.3.D.	-2.492	353	.013	-0.178	-0.32	-0.04	.13
K.3.E.	10.310	353	.000	0.559	0.45	0.67	.55
K.3.F.	-5.022	353	.000	-0.322	-0.45	-0.20	.27

Note. One-Sample T test. Test value 2.5. N = 354.

K.3.A. = knowledge indicator 3.A. = stimulating academic development; K.3.B. = Knowledge indicator 3.B. = stimulating the effective use of different cognitive styles; K.3.C. = Knowledge indicator 3.C. = stimulating interpersonal functioning; K.3.D. = Knowledge indicator 3.D. = stimulating intrapersonal understanding; K.3.E. = Knowledge indicator 3.E. = stimulating self-actualisation; K.3.F. = Knowledge indicator 3.F. = general knowledge and skills matching those of the (S)ENCo.

Table G22

Mean Selection Rates of the Skills Indicators per Category in Domain 3

Indicators	N	Mean	*SD*
V.3.A.	354	2.14	1.163
V.3.B.	354	2.26	1.102
V.3.C.	354	2.72	1.085
V.3.D.	354	3.09	1.160
V.3.E.	354	2.60	1.086
V.3.F.	354	2.18	1.052

Note. One-Sample Statistics. Test value 2.5.

V.3.A. = Skills indicator 3.A. = stimulating academic development; V.3.B. = Skills indicator 3.B. = stimulating the effective use of different cognitive styles; V.3.C. = Skills indicator 3.C. = stimulating interpersonal functioning; V.3.D. = Skills indicator 3.D. = stimulating intrapersonal understanding; V.3.E. = Skills indicator 3.E. = stimulating self-actualisation; and V.3.F. = Skills indicator 3.F. = general knowledge and skills matching those of the (S)ENCo.

Table G23

Selection Rates of the Skills Indicators per Category in Domain 3

					95% Confidence interval of the difference		
Indicators	*t*	*df*	Sig (2-tailed)	Mean difference	Lower bound	Upper bound	Cohen's *d*
V.3.A.	-5.757	353	.000	-0.356	-0.48	-0.23	.31
V.3.B.	-4.052	353	.000	-0.237	-0.35	-0.12	.22
V.3.C.	3.771	353	.000	0.218	0.10	0.33	.20
V.3.D.	9.575	353	.000	0.590	0.47	0.71	.51
V.3.E.	1.810	353	.071	0.105	-0.01	0.22	.09
V.3.F.	-5.711	353	.000	-0.319	-0.43	-0.21	.30

Note. One-Sample Test. Test Value 2.5. N = 354.

V.3.A. = Skills indicator 3.A. = stimulating academic development; V.3.B. = Skills indicator 3.B. = stimulating the effective use of different cognitive styles; V.3.C. = Skills indicator 3.C. = stimulating interpersonal functioning; V.3.D. = Skills indicator 3.D. = stimulating intrapersonal understanding; V.3.E. = Skills indicator 3.E. = stimulating self-actualisation; and V.3.F. = Skills indicator 3.F. = general knowledge and skills matching those of the (S)ENCo.

Table G24

Indicator Category with Highest Mean Selection Rate in Domain 3 per Group of Stakeholders

Indicators	Educational professionals	Specialists in gifted education	Parents	Rest
Knowledge	A	A	A/E	A
Skills	D	C	D	D

Note. A = stimulating academic development; C = stimulating interpersonal functioning; D = stimulating intrapersonal understanding; E = stimulating self-actualisation.

Table G25

The Comparison of Selection Rates Between Two Clusters of Knowledge Indicators in Domain 3: ADE-Indicators vs BCF-Indicators

Clusters of Indicators	Mean	N	*SD*	*SE*
Dom3_K_ADE	2.8729	354	0.50040	0.02660
Dom3_K_BCF	2.1271	354	0.50400	0.02660

Note: Paired Samples Statistics.

Dom3_K_ADE = Domain 3, cluster of three categories of knowledge indicators A, D, E; compared with Dom3_K_BCF = Domain 3, cluster of three categories of knowledge indicators B, C, F.

Table G26

The Comparison of Selection Rates Between Two Clusters of Knowledge Indicators in Domain 3: ADE-Indicators vs BCF-Indicators

				95% Confidence interval of the difference					
Clusters of Indicators	Mean	*SD*	*SE*	Lower bound	Upper bound	*t*	*df*	Sig (2-tailed)	Cohen's *d*
Dom3_K_ADE – Dom3_K_BCF	0.74576	1.00080	0.05319	0.64115	0.85038	14.020	353	.000	.75

Note. Paired Samples T test. N = 354.

Dom3_K_ADE = Domain 3, cluster of three categories of knowledge indicators A, D, E; compared to Dom3_K_BCF = Domain 3, cluster of three categories of knowledge indicators B, C, F.

Table G27

The Comparison of Selection Rates Between Two Clusters of Skills Indicators in Domain 3: ADE-Indicators vs BCF- Indicators

Clusters of Indicators	Mean	N	*SD*	*SE*
Dom3_V_ADE	2.6130	354	0.46122	0.02451
Dom3_V_BCF	2.3870	354	0.46122	0.02451

Note: Paired Samples Statistics.

Dom3_V_ADE = Domain 3, cluster of three categories of skills indicators A, D, E; compared to Dom3_V_BCF = Domain 3, cluster of three categories of skills indicators B, C, F.

Table G28

The Comparison of Selection Rates Between Two Clusters of Skills Indicators in Domain 3: ADE-Indicators vs BCF- Indicators

				95% Confidence interval of the difference					
Clusters of Indicators	Mean	*SD*	*SE*	Lower bound	Upper bound	*t*	*df*	Sig (2-tailed)	Cohen's *d*
Dom3_V_ADE – Dom3_V_BCF	0.22599	0.2244	0.04903	0.12957	0.32241	4.609	353	.000	.25

Note. Paired Samples T test. N = 354.

Dom3_V_ADE = Domain 3, cluster of three categories of skills indicators A, D, E; compared to Dom3_V_BCF = Domain 3, cluster of three categories of skills indicators B, C, F.

Table G29

Knowledge Indicators in Domain 3 With the Highest Selection Rates Amongst all Groups of Stakeholders

Competency	Code	Indicator
3.2 The specialist explains how the interplay between the curriculum and the student's cognitive style influences their development and instructional needs.	K.3.D.2	Knows how mindset influences a student's learning process.

Table G30

Skills Indicators in Domain 3 With the Highest Selection Rates Amongst all Groups of Stakeholders

Competency	Code	Indicator
3.1 The specialist develops an understanding of how the actors in the student's ecology influence the student's strengths and weaknesses.	V.3.E.1	Discusses with the student what they see as helpful learning strategies and interventions.
3.2 The specialist explains how the interplay between the curriculum and the student's cognitive style influences their development and instructional needs.	V.3.E.2	Evaluates whether the teacher's expectations towards a student's executive skills are realistic.
3.3 The specialist understands how the individual student's needs are of consequence for their classroom management.	V.3.B.3	Creates a learning environment where the use of executive skills is an intrinsic condition during the learning process.

Table G31

Knowledge Indicators in Domain 3 With the Lowest Selection Rates Amongst all Groups of Stakeholders

Competency	Code	Indicator
3.3 The specialist understands how the individual student's needs are of consequence for their classroom management.	K.3.B.3	Knows how factors in the physical educational environment influence the development of executive skills.

Table G32

Skills Indicators in Domain 3 With the Lowest Selection Rates Amongst all Groups of Stakeholders

Competency	Code	Indicator
3.2 The specialist explains how the interplay between the curriculum and the student's cognitive style influences their development and instructional needs.	V.3.C.2	Determines how peers within the social environment respond to the differentiated curriculum offered to the gifted student.
3.4 Based on the results of their exploration of the student's educational needs, the specialist explains to the actors in the student's ecological system how their behaviour influences the student's development.	V.3.C.4	Determines if the gifted student has ample opportunity to meet and work with like-minded peers.
3.5 The specialist uses theory on giftedness and RtI to interpret the results of the exploration of the student's educational needs in the context of educational development.	V.3.B.5	Selects intervention strategies, based upon their theoretical perspective of giftedness, that are helpful in understanding the student's educational needs.

Table G33

Comparison of Mean Selection Rates in Domain 3 per Category of Knowledge Indicators Between Dutch and Flemish Stakeholders

Indicators	Country	N	Mean	*SD*	*SE*
K.3.A.	NL	299	3.29	1.027	0.059
	F	55	2.93	0.997	0.134
K.3.B.	NL	299	2.76	0.974	0.056
	F	55	2.04	1.122	0.151
K.3.C.	NL	299	1.48	1.069	0.062
	F	55	1.96	1.186	0.160
K.3.D.	NL	299	2.25	1.324	0.077
	F	55	2.69	1.399	0.189
K.3.E.	NL	299	3.06	1.010	0.058
	F	55	3.04	1.088	0.147
K.3.F.	NL	299	2.15	1.201	0.069
	F	55	2.35	1.236	0.167

Note. Descriptive Statistics. NL = The Netherlands. F = Flanders. N = 354.

K.3.A. = Knowledge indicator 3.A. = stimulating academic development; K.3.B. = Knowledge indicator 3.B. = stimulating the effective use of different cognitive styles; K.3.C. = Knowledge indicator 3.C. = stimulating interpersonal functioning; K.3.D. = Knowledge indicator 3.D. = stimulating intrapersonal understanding; K.3.E. = Knowledge indicator 3.E. = stimulating self-actualisation; and K.3.F. = Knowledge indicator 3.F. = general knowledge and skills matching those of the (S)ENCo.

Table G34

Comparison of Mean Selection Rates in Domain 3 per Category of Knowledge Indicators Between Dutch and Flemish Stakeholders

		Levene's test for Equality of variances		t-test for Equality of means							
									95% Confidence interval of the difference		
Indicators		*F*	*p*	*t*	*df*	Sig. (2-tailed)	Mean difference	*SE* difference	Lower bound	Upper bound	Cohen's *d*
K.3.A.	Equal variances assumed	2.343	.127	2.447	352	.015	0.367	0.150	0.072	0.662	.36
	Equal variances not assumed			2.497	76.578	.015	0.367	0.147	0.074	0.660	
K.3.B.	Equal variances assumed	0.167	.683	4.937	352	.000	0.723	0.146	0.435	1.011	.72
	Equal variances not assumed			4.479	69.776	.000	0.723	0.161	0.401	1.045	
K.3.C.	Equal variances assumed	0.433	.511	-3.020	352	.003	-0.482	0.160	-0.796	-0.168	.44
	Equal variances not assumed			-2.812	71.071	.006	-0.482	0.171	-0.824	-0.140	
K.3.D.	Equal variances assumed	0.343	.558	-2.228	352	.027	-0.437	0.196	-0.822	-0.051	.33
	Equal variances not assumed			-2.145	72.901	.035	-0.437	0.204	-0.843	-0.031	
K.3.E	Equal variances assumed	0.306	.580	0.181	352	.856	0.027	0.150	-0.268	0.322	.03
	Equal variances not assumed			0.172	72.134	.864	0.027	0.158	-0.288	0.342	
K.3.F.	Equal variances assumed	0.397	.529	-1.121	352	.263	-0.198	0.177	-0.546	0.150	.16
	Equal variances not assumed			1.099	73.981	.275	-0.198	0.180	-0.558	0.161	

Note. Independent Samples Test. N = 354.

K.3.A. = Knowledge indicator 3.A. = stimulating academic development; K.3.B. = Knowledge indicator 3.B. = stimulating the effective use of different cognitive styles; K.3.C. = Knowledge indicator 3.C. = stimulating interpersonal functioning; K.3.D. = Knowledge indicator 3.D. = stimulating intrapersonal understanding; K.3.E. = Knowledge indicator 3.E. = stimulating self-actualisation; and K.3.F. = Knowledge indicator 3.F. = general knowledge and skills matching those of the (S)ENCo.

Table G35

Comparison of Mean Selection Rates in Domain 3 per Category of Skills Indicators Between Dutch and Flemish Stakeholders

Indicators	Country	N	Mean	*SD*	*SE*
V.3.A.	NL	299	2.16	1.136	0.066
	F	55	2.05	1.311	0.177
V.3.B.	NL	299	2.24	1.104	0.064
	F	55	2.36	1.095	0.148
V.3.C.	NL	299	2.72	1.097	0.063
	F	55	2.73	1.027	0.138
V.3.D.	NL	299	3.03	1.185	0.069
	F	55	3.42	0.956	0.129
V.3.E.	NL	299	2.62	1.050	0.061
	F	55	2.55	1.274	0.172
V.3.F.	NL	299	2.23	1.064	0.062
	F	55	1.89	0.936	0.126

Note. Descriptive Statistics. NL = The Netherlands. F = Flanders. N = 354.

V.3.A. = Skills indicator 3.A. = stimulating academic development; V.3.B. = Skills indicator 3.B. = stimulating the effective use of different cognitive styles; V.3.C. = Skills indicator 3.C. = stimulating interpersonal functioning; V.3.D. = Skills indicator 3.D. = stimulating intrapersonal understanding; V.3.E. = Skills indicator 3.E. = stimulating self-actualisation; and V.3.F. = Skills indicator 3.F. = general knowledge and skills matching those of the (S)ENCo.

Table G36

Comparison of Mean Selection Rates in Domain 3 per Category of Skills Indicators Between Dutch and Flemish Stakeholders

		Levene's test for Equality of variances		t-test for Equality of means							
									95% Confidence interval of the difference		
Indicators		*F*	*p*	*t*	*df*	Sig. (2-tailed)	Mean difference	*SE* difference	Lower bound	Upper bound	Cohen's *d*
V.3.A.	Equal variances assumed	0.761	.384	0.620	352	.535	0.106	0.171	-0.230	0.442	.09
	Equal variances not assumed			0.562	69.689	.576	0.106	0.189	-0.270	0.482	
V.3.B.	Equal variances assumed	0.143	.705	-0.739	352	.461	-0.119	0.162	-0.438	0.199	.11
	Equal variances not assumed			-0.743	75.608	.460	-0.119	0.161	-0.440	0.201	
V.3.C.	Equal variances assumed	1.305	.254	-0.072	352	.942	-0.012	0.159	-0.325	0.302	.01
	Equal variances not assumed			-0.076	78.456	.940	-0.012	0.152	-0.315	0.292	
V.3.D.	Equal variances assumed	1.692	.194	-2.294	352	.022	-0.388	0.169	-0.721	-0.055	.34
	Equal variances not assumed			-2.657	87.571	.009	-0.388	0.146	-0.678	-0.098	
V.3.E.	Equal variances assumed	6.511	.011	0.438	352	.661	0.070	0.160	-0.244	0.384	.06
	Equal variances not assumed			0.384	68.156	702	0.070	0.182	-0.294	0.433	
V.3.F.	Equal variances assumed	2.398	.122	2.237	352	.026	0.343	0.153	0.041	0.645	.33
	Equal variances not assumed			2.443	81.883	.017	0.343	0.140	0.064	0.623	

Note. Independent Samples Test. N = 354.

V.3.A. = Skills indicator 3.A. = stimulating academic development; V.3.B. = Skills indicator 3.B. = stimulating the effective use of different cognitive styles; V.3.C. = Skills indicator 3.C. = stimulating interpersonal functioning; V.3.D. = Skills indicator 3.D. = stimulating intrapersonal understanding; V.3.E. = Skills indicator 3.E. = stimulating self-actualisation; and V.3.F. = Skills indicator 3.F. = general knowledge and skills matching those of the (S)ENCo.

Table G37

Domain 3 Knowledge Indicators. Mean per Category per Competency per Group of Stakeholders

Indicators	EP	Indicators	SGE	Indicators	Pa	Indicators	R
K.3.C.	1.66	K.3.C.	1.40	K.3.C.	1.70	K.3.C.	1.32
K.3.F.	2.05	K.3.D.	1.83	K.3.F.	2.01	K.3.F.	2.22
K.3.D.	2.32	K.3.F.	2.47	K.3.B.	2.33	K.3.D.	2.27
K.3.B.	2.81	K.3.B.	2.74	K.3.D.	2.96	K.3.B.	2.54
K.3.E.	2.91	K.3.E.	3.25	K.3.A.	3.00	K.3.E.	3.19
K.3.A.	3.26	K.3.A.	3.32	K.3.E.	3.00	K.3.A.	3.46

Note. Test value 2.5. EP = educational professionals; SGE = specialist gifted education; Pa = parents; R = rest. N = 354.

K.3.A. = Knowledge indicator 3.A. = stimulating academic development; K.3.B. = Knowledge indicator 3.B. = stimulating the effective use of different cognitive styles; K.3.C. = Knowledge indicator 3.C. = stimulating interpersonal functioning; K.3.D. = Knowledge indicator 3.D. = stimulating intrapersonal understanding; K.3.E. = Knowledge indicator 3.E. = stimulating self-actualisation; and K.3.F. = Knowledge indicator 3.F. = general knowledge and skills matching those of the (S)ENCo.

Table G38

Domain 3 Skills Indicators. Mean per Category per Competency per Group of Stakeholders

Indicators	EP	Indicators	SGE	Indicators	Pa	Indicators	R
V.3.F.	2.07	V.3.B.	2.06	V.3.F.	1.90	V.3.A.	1.81
V.3.B.	2.18	V.3.A.	2.25	V.3.A.	2.07	V.3.F.	2.24
V.3.A.	2.20	V.3.F.	2.52	V.3.C.	2.51	V.3.B.	2.38
V.3.E.	2.62	V.3.E.	2.59	V.3.E.	2.55	V.3.C.	2.51
V.3.C.	2.79	V.3.D.	2.71	V.3.B.	2.59	V.3.E.	2.70
V.3.D.	3.14	V.3.C.	2.86	V.3.D.	3.37	V.3.D.	3.35

Note. Test value 2.5. EP = educational professionals; SGE = specialist gifted education; Pa = parents; R = rest. N = 354.

V.3.A. = Skills indicator 3.A. = stimulating academic development; V.3.B. = Skills indicator 3.B. = stimulating the effective use of different cognitive styles; V.3.C. = Skills indicator 3.C. = stimulating interpersonal functioning; V.3.D. = Skills indicator 3.D. = stimulating intrapersonal understanding; V.3.E. = Skills indicator 3.E. = stimulating self-actualisation; and V.3.F. = Skills indicator 3.F. = general knowledge and skills matching those of the (S)ENCo.

Table G39

Ranking of the Knowledge and Skills Indicators of Domain 3

Indicators	N	%	Indicators	N	%
V.3.C.2	51	0.48	V.3.F.2	177	1.67
K.3.C.2	78	0.73	K.3.C.4	181	1.70
V.3.B.5	81	0.76	K.3.E.2	190	1.79
V.3.C.4	81	0.76	V.3.D.3	197	1.85
K.3.B.3	94	0.89	V.3.D.2	201	1.89
K.3.C.1	95	0.89	V.3.A.3	202	1.90
V.3.F.3	101	0.95	K.3.E.1	205	1.93
K.3.A.5	109	1.03	V.3.A.1	212	2.00
K.3.A.2	112	1.05	K.3.C.3	212	2.00
V.3.A.4	114	1.07	K.3.D.1	214	2.02
V.3.D.1	119	1.12	K.3.E.4	215	2.02
V.3.D.4	120	1.13	V.3.A.5	216	2.03
K.3.C.5	122	1.15	K.3.A.4	217	2.04
K.3.D.4	122	1.15	V.3.A.2	217	2.04
K.3.B.5	124	1.17	K.3.A.3	220	2.07
V.3.D.5	125	1.18	K.3.F.2	220	2.07
V.3.F.1	125	1.18	K.3.E.5	224	2.11
V.3.C.5	136	1.28	K.3.D.5	226	2.13
K.3.A.1	141	1.33	K.3.E.3	226	2.13
K.3.F.3	142	1.34	K.3.B.1	231	2.18
V.3.C.3	145	1.37	V.3.B.4	232	2.18
K.3.B.4	162	1.53	V.3.E.5	236	2.22
V.3.E.3	165	1.55	V.3.E.2	239	2.25
K.3.F.4	165	1.55	V.3.B.3	252	2.37
K.3.D.3	168	1.58	K.3.F.5	257	2.42
V.3.B.1	172	1.62	V.3.F.4	257	2.42
K.3.B.2	173	1.63	V.3.E.4	258	2.43
V.3.C.1	173	1.63	V.3.E.1	261	2.46
K.3.F.1	176	1.66	V.3.F.5	268	2.52
V.3.B.2	177	1.67	K.3.D.2	289	2.72

Note. This ranking is based on selection rates of the entire group of stakeholders, N =354.

All indicators are coded. K = knowledge; V = skills; the first number refers to the vertical competency domain in the competency matrix; the second letter refers to the category of indicators. The second number refers to the actual competency the indicators are matched with. See Appendix B for the match between the competencies and the indicators.

Appendix H

Tables Domain 4. Referred to in Chapter 8

Table H1

Frequency Table of the Selection Rates of Knowledge Indicators for Competency 4.1

Competency	Indicator	EP	SGE	Pa	NOS	Total
4.1	K.4.A.1	16.28	17.15	14.46	18.92	16.38
	K.4.B.1	10.69	9.71	11.24	8.11	10.26
	K.4.C.1	17.81	10.03	10.04	12.61	13.18
	K.4.D.1	14.25	21.36	21.69	18.92	18.55
	K.4.E.1	16.54	18.45	20.88	21.62	18.64
	K.4.F.1	24.43	23.30	21.69	19.82	22.98

Note. The results are presented per indicator for each group of stakeholders. EP = educational professionals; SGE = specialist gifted education; Pa = parents; R = Rest; T = total. N = 354.

All indicators are coded. K = knowledge; the first number refers to the vertical competency domain in the competency matrix; the second letter refers to the category of indicators. The second number refers to the actual competency the indicators are matched with. See Appendix B for the match between the competencies and the indicators.

Table H2

Frequency Table of the Selection Rates of Knowledge Indicators for Competency 4.2

Competency	Indicator	EP	SGE	Pa	R	Total
4.2	K.4.A.2	16.03	15.86	17.67	18.02	16.57
	K.4.B.2	20.61	21.04	14.46	18.92	19.11
	K.4.C.2	9.67	8.74	9.24	9.91	9.32
	K.4.D.2	14.76	18.45	23.29	21.62	18.55
	K.4.E.2	23.16	22.33	24.90	18.02	22.79
	K.4.F.2	15.78	13.59	10.44	13.51	13.65

Note. The results are presented per indicator for each group of stakeholders. EP = educational professionals; SGE = specialist gifted education; Pa = parents; R = Rest; T = total. N = 354.

All indicators are coded. K = knowledge; the first number refers to the vertical competency domain in the competency matrix; the second letter refers to the category of indicators. The second number refers to the actual competency the indicators are matched with. See Appendix B for the match between the competencies and the indicators.

Table H3

Frequency Table of the Selection Rates of Knowledge Indicators for Competency 4.3

Competency	Indicator	EP	SGE	Pa	R	Total
4.3	K.4.A.3	22.14	24.27	24.10	20.72	23.07
	K.4.B.3	12.21	13.92	20.08	15.32	14.88
	K.4.C.3	16.28	9.71	10.04	13.51	12.62
	K.4.D.3	16.54	15.21	17.27	15.32	16.20
	K.4.E.3	15.52	16.18	13.65	18.02	15.54
	K.4.F.3	17.30	20.71	14.86	17.12	17.70

Note. The results are presented per indicator for each group of stakeholders. EP = educational professionals; SGE = specialist gifted education; Pa = parents; R = Rest; T = total. N= 354.

All indicators are coded. K = knowledge; the first number refers to the vertical competency domain in the competency matrix; the second letter refers to the category of indicators. The second number refers to the actual competency the indicators are matched with. See Appendix B for the match between the competencies and the indicators.

Table H4

Frequency Table of the Selection Rates of Knowledge Indicators for Competency 4.4

Competency	Indicator	EP	SGE	Pa	R	Total
4.4	K.4.A.4	13.99	15.86	10.04	11.71	13.37
	K.4.B.4	11.20	9.39	12.45	14.41	11.30
	K.4.C.4	21.12	18.12	25.30	16.22	20.72
	K.4.D.4	12.98	8.09	9.24	6.31	9.98
	K.4.E.4	24.68	26.86	28.11	28.83	26.55
	K.4.F.4	16.03	21.68	14.86	22.52	18.08

Note. The results are presented per indicator for each group of stakeholders. EP = educational professionals; SGE = specialist gifted education; Pa = parents; R = Rest; T = total. N= 354.

All indicators are coded. K = knowledge; the first number refers to the vertical competency domain in the competency matrix; the second letter refers to the category of indicators. The second number refers to the actual competency the indicators are matched with. See Appendix B for the match between the competencies and the indicators.

Table H5

Frequency Table of the Selection Rates of Knowledge Indicators for Competency 4.5

Competency	Indicator	EP	SGE	Pa	R	Total
4.5	K.4.A.5	20.36	17.80	20.48	21.62	19.77
	K.4.B.5	8.65	6.80	8.03	9.01	8.00
	K.4.C.5	15.78	15.86	21.69	18.02	17.42
	K.4.D.5	11.45	14.56	11.65	14.41	12.71
	K.4.E.5	22.14	21.04	19.68	18.92	20.90
	K.4.F.5	21.63	23.95	18.47	18.02	21.19

Note. The results are presented per indicator for each group of stakeholders. EP = educational professionals; SGE = specialist gifted education; Pa = parents; R = Rest; T = total. N= 354.

All indicators are coded. K = knowledge; the first number refers to the vertical competency domain in the competency matrix; the second letter refers to the category of indicators. The second number refers to the actual competency the indicators are matched with. See Appendix B for the match between the competencies and the indicators.

Table H6

Frequency Table of the Selection Rates of Skills Indicators for Competency 4.1

Competency	Indicator	EP	SGE	Pa	R	Total
4.1	V.4.A.1	20.87	21.04	22.09	25.23	21.66
	V.4.B.1	20.61	19.74	24.50	16.22	20.81
	V.4.C.1	6.87	5.83	4.42	6.31	5.93
	V.4.D.1	10.18	7.77	11.24	10.81	9.79
	V.4.E.1	22.39	22.65	21.29	22.52	22.22
	V.4.F.1	19.08	22.98	16.47	18.92	19.59

Note. The results are presented per indicator for each group of stakeholders. EP = educational professionals; SGE = specialist gifted education; Pa = parents; R = Rest; T = total. N= 354.

All indicators are coded. V = skills; the first number refers to the vertical competency domain in the competency matrix; the second letter refers to the category of indicators. The second number refers to the actual competency the indicators are matched with. See Appendix B for the match between the competencies and the indicators.

Table H7

Frequency Table of the Selection Rates of Skills Indicators for Competency 4.2

Competency	Indicator	EP	SGE	Pa	R	Total
4.2	V.4.A.2	11.96	16.50	11.24	9.91	12.90
	V.4.B.2	19.85	18.12	23.29	19.82	20.15
	V.4.C.2	15.52	15.86	15.66	21.62	16.29
	V.4.D.2	9.67	9.39	7.63	11.71	9.32
	V.4.E.2	26.21	22.33	26.10	26.13	25.05
	V.4.F.2	16.79	17.80	16.06	10.81	16.29

Note. The results are presented per indicator for each group of stakeholders. EP = educational professionals; SGE = specialist gifted education; Pa = parents; R = Rest; T = total. N= 354.

All indicators are coded. V = skills; the first number refers to the vertical competency domain in the competency matrix; the second letter refers to the category of indicators. The second number refers to the actual competency the indicators are matched with. See Appendix B for the match between the competencies and the indicators.

Table H8

Frequency Table of the Selection Rates of Skills Indicators for Competency 4.3

Competency	Indicator	EP	SGE	Pa	R	Total
4.3	V.4.A.3	19.34	26.21	22.09	27.03	22.79
	V.4.B.3	10.18	10.03	14.06	14.41	11.49
	V.4.C.3	14.76	8.41	10.84	12.61	11.77
	V.4.D.3	20.10	15.21	22.09	16.22	18.74
	V.4.E.3	21.63	19.42	13.25	11.71	17.98
	V.4.F.3	13.99	20.71	17.67	18.02	17.23

Note. The results are presented per indicator for each group of stakeholders. EP = educational professionals; SGE = specialist gifted education; Pa = parents; R = Rest; T = total. N= 354.

All indicators are coded. V = skills; the first number refers to the vertical competency domain in the competency matrix; the second letter refers to the category of indicators. The second number refers to the actual competency the indicators are matched with. See Appendix B for the match between the competencies and the indicators.

Table H9

Frequency Table of the Selection Rates of Skills Indicators for Competency 4.4

Competency	Indicator	EP	SGE	Pa	R	Total
4.4	V.4.A.4	15.52	16.50	14.86	19.82	16.10
	V.4.B.4	13.74	12.62	13.65	14.41	13.47
	V.4.C.4	16.54	17.80	17.67	15.32	17.04
	V.4.D.4	18.58	14.56	20.08	20.72	17.98
	V.4.E.4	21.63	23.30	16.87	16.22	20.43
	V.4.F.4	13.99	15.21	16.87	13.51	14.97

Note. The results are presented per indicator for each group of stakeholders. EP = educational professionals; SGE = specialist gifted education; Pa = parents; R = Rest; T = total. N= 354.

All indicators are coded. V = skills; the first number refers to the vertical competency domain in the competency matrix; the second letter refers to the category of indicators. The second number refers to the actual competency the indicators are matched with. See Appendix B for the match between the competencies and the indicators.

Table H10

Frequency Table of the Selection Rates of Skills Indicators for Competency 4.5

Competency	Indicator	EP	SGE	Pa	R	Total
4.5	V.4.A.5	17.30	20.06	20.48	18.02	18.93
	V.4.B.5	20.36	11.33	13.25	14.41	15.44
	V.4.C.5	17.05	16.50	22.89	13.51	17.89
	V.4.D.5	12.21	16.50	13.25	13.51	13.84
	V.4.E.5	15.27	15.53	12.45	19.82	15.16
	V.4.F.5	17.81	20.06	17.67	20.72	18.74

Note. The results are presented per indicator for each group of stakeholders. EP = educational professionals; SGE = specialist gifted education; Pa = parents; R = Rest; T = total. N= 354.

All indicators are coded. V = skills; the first number refers to the vertical competency domain in the competency matrix; the second letter refers to the category of indicators. The second number refers to the actual competency the indicators are matched with. See Appendix B for the match between the competencies and the indicators.

Table H11

Comparison of the 10 Highest Indicator Selection Rates per Competency in Domain 4 per Group of Stakeholders

Indicator	EP	Indicator	SGE	Indicator	Pa	Indicator	R
V.4.B.5	20.36	V.4.A.5	20.06	V.4.D.4	20.08	K.4.A.3	20.72
V.4.E.3	21.63	V.4.F.5	20.06	K.4.D.1/K.4.F.1	21.69	V.4.D.4	20.72
V.4.E.4	21.63	K.4.E.2	22.33	K.4.C.5	21.69	V.4.F.5	20.72
K.4.A.3	22.14	V.4.E.2	22.33	V.4.A.3/V.4.D.3	22.09	K.4.E.1	21.62
K.4.E.5	22.14	V.4.F.1	22.98	V.4.C.5	22.89	K.4.D.2	21.62
V.4.E.1	22.39	K.4.F.1/V.4.E.4	23.30	K.4.A.3	24.10	K.4.A.5	21.62
K.4.E.2	23.16	K.4.F.5	23.95	V.4.B.1	24.50	V.4.A.1	25.23
K.4.F.1	24.43	K.4.A.3	24.27	K.4.E.2	24.90	V.4.E.2	26.13
K.4.E.4	24.68	V.4.A.3	26.21	V.4.E.2	26.10	V.4.A.3	27.03
V.4.E.2	26.21	K.4.E.4	26.86	K.4.E.4	28.11	K.4.E.4	28.83

Note. EP = educational professionals; SGE = specialist gifted education; Pa = parents; R = Rest; T = total. N= 354. If two indicators matching one competency had an identical selection rate, they are ranked at the same level.

All indicators are coded. K = knowledge / V = skills; the first number refers to the vertical competency domain in the competency matrix; the second letter refers to the category of indicators. The second number refers to the actual competency the indicators are matched with. See Appendix B for the match between the competencies and the indicators.

Table H12

Comparison of the Selection Rates of the Knowledge Indicators in Domain 4 Between the Groups of Stakeholders

Indicators	Group	N	Mean	*SD*	*SE*	95% Confidence interval for mean		Min	Max
						Lower bound	Upper bound		
K.4.A.	EP	131	3.25	0.979	0.086	3.08	3.42	1	5
	SGE	103	3.03	1.107	0.109	2.81	3.25	0	5
	Pa	83	2.82	0.939	0.103	2.61	3.02	0	5
	R	37	3.30	0.878	0.144	3.00	3.59	1	5
	T	354	3.09	1.011	0.054	2.98	3.20	0	5
K.4.D.	EP	131	2.44	1.223	0.107	2.23	2.65	0	5
	SGE	103	2.00	1.343	0.132	1.74	2.26	0	5
	Pa	83	3.14	1.211	0.133	2.88	3.41	0	5
	R	37	2.35	1.358	0.223	1.90	2.80	1	5
	T	354	2.47	1.332	0.071	2.33	2.61	0	5
K.4.E.	EP	131	2.88	1.157	0.101	2.68	3.08	0	5
	SGE	103	3.23	0.931	0.092	3.05	3.41	1	5
	Pa	83	3.07	1.033	0.113	2.85	3.30	1	5
	R	37	3.27	0.990	0.163	2.94	3.60	1	5
	T	354	3.07	1.057	0.056	2.96	3.18	0	5
K.4.F.	EP	131	2.12	1.137	0.099	1.93	2.32	0	5
	SGE	103	2.68	1.139	0.112	2.46	2.90	0	5
	Pa	83	1.84	1.215	0.133	1.58	2.11	0	5
	R	37	2.24	1.065	0.175	1.89	2.60	0	4
	T	354	2.23	1.187	0.063	2.11	2.36	0	5

Note. Descriptive Statistics. EP = educational professionals; SGE = specialist gifted education; Pa = parents; R = Rest; T = total. N= 354.

K.4.A. = Knowledge indicator 4.A. = stimulating academic development; K.4.D. = Knowledge indicator 4.D. = stimulating intrapersonal understanding; K.4.E. = Knowledge indicator 4.E. = stimulating self-actualisation; K.4.F. = Knowledge indicator 4.F. = general knowledge and skills matching those of the (S)ENCo.

Table H13

Comparison of the Selection Rates of the Knowledge Indicators in Domain 4 Between the Groups of Stakeholders

Indicators		Sum of Squares	*df*	Mean Square	*F*	*p*	η^2
K.4.A.	Between Groups	11.489	3	3.830	3.834	.010	.03
	Within Groups	349.619	350	0.999			
	Total	361.107	353				
K.4.B.	Between Groups	1.037	3	0.346	0.338	.798	.00
	Within Groups	358.115	350	1.023			
	Total	359.153	353				
K.4.C.	Between Groups	3.675	3	1.225	0.997	.394	.01
	Within Groups	430.008	350	1.229			
	Total	433.684	353				
K.4.D.	Between Groups	61.140	3	20.380	12.624	.000	.10
	Within Groups	565.018	350	1.614			
	Total	626.158	353				
K.4.E.	Between Groups	9.056	3	3.019	2.742	.043	.02
	Within Groups	385.317	350	1.101			
	Total	394.373	353				
K.4.F.	Between Groups	34.758	3	11.586	8.773	.000	.07
	Within Groups	462.248	350	1.321			
	Total	497.006	353				

Note. ANOVA = analyses of variance with one factor.

K.4.A. = Knowledge indicator 4.A. = stimulating academic development; K.4.B. = Knowledge indicator 4.B. = stimulating the effective use of different cognitive styles; K.4.C. = Knowledge indicator 4.C. = stimulating interpersonal functioning; K.4.D. = Knowledge indicator 4.D. = stimulating intrapersonal understanding; K.4.E. = Knowledge indicator 4.E. = stimulating self-actualisation; K.4.F. = Knowledge indicator 4.F. = general knowledge and skills matching those of the (S)ENCo.

Table H14

Comparison of the Selection Rates of the Knowledge Indicators in Domain 4 Between the Groups of Stakeholders

Dependent variable	I group	J group	Mean difference (I-J)	*SE*	*p*	95% Confidence interval Lower bound	95% Confidence interval Upper bound
	Pa	EP	-0.433	0.140	.013	-0.80	-0.06
K.4.A.		SGE	-0.210	0.147	.933	-0.60	0.18
		R	-0.478	0.198	.096	-1.00	0.05
	Pa	EP	0.702	0.178	.001	0.23	1.17
K.4.D.		SGE	1.145	0.187	.000	0.65	1.64
		R	0.793	0.251	.010	0.13	1.46
	SGE	EP	0.557	0.151	.002	0.16	1.02
K.4.F.		Pa	0.836	0.170	.000	0.39	0.15
		R	0.436	0.220	.290	-0.15	-0.39

Note. Based on the results of the ANOVA, Bonferroni's Multiple Comparisons was used as a post hoc test for the analyses of the differences between groups of stakeholders.

EP = educational professionals; SGE = specialist gifted education; Pa = parents; R = Rest; T = total. N= 354.

K.4.A. = Knowledge indicator 4.A. = stimulating academic development; K.4.D. = Knowledge indicator 4.D. = stimulating intrapersonal understanding; K.4.F. = Knowledge indicator 4.F. = general knowledge and skills matching those of the (S)ENCo.

Table H15

Comparison of the Selection Rates of the Skills Indicators in Domain 4 Between the Groups of Stakeholders

Indicators	Group	N	Mean	*SD*	*SE*	95% Confidence interval for mean Lower bound	95% Confidence interval for mean Upper bound	Min	Max
V.4.D.	EP	131	3.02	1.240	0.108	2.80	3.23	0	5
	SGE	103	2.44	1.143	0.113	2.21	2.66	0	5
	Pa	83	3.11	1.137	0.125	2.86	3.36	0	5
	R	37	3.03	1.166	0.192	2.64	3.42	0	5
	T	354	2.87	1.209	0.064	2.74	3.00	0	5
V.4.F.	EP	131	2.08	1.012	0.057	1.90	2.25	0	4
	SGE	103	2.58	0.985	0.088	2.39	2.78	0	4
	Pa	83	2.00	1.012	0.097	1.78	2.22	0	4
	R	37	2.32	0.884	0.111	2.03	2.62	1	4
	T	354	2.23	1.017	0.145	2.13	2.34	0	4

Note. Descriptive Statistics. EP = educational professionals; SGE = specialist gifted education; Pa = parents; R = Rest; T = total. N= 354.

V.4.D. = Skills indicator 4.D. = stimulating intrapersonal understanding; V.4.F. = Skills indicator 4.F. = general skills and skills matching those of the (S)ENCo.

Table H16

Comparison of the Selection Rates of the Skills Indicators in Domain 4 Between the Groups of Stakeholders

Indicators		Sum of Squares	*df*	Mean Square	*F*	*p*	η^2
V.4.A.	Between Groups	7.741	3	2.580	1.765	.154	.01
	Within Groups	511.810	350	1.462			
	Total	519.551	353				
V.4.B.	Between Groups	2.598	3	0.866	0.729	.535	.01
	Within Groups	415.605	350	1.187			
	Total	418.203	353				
V.4.C.	Between Groups	8.858	3	2.953	2.472	.062	.02
	Within Groups	418.072	350	1.194			
	Total	426.929	353				
V.4.D.	Between Groups	27.716	3	9.239	6.622	.000	.05
	Within Groups	488.306	350	1.395			
	Total	516.023	353				
V.4.E.	Between Groups	0.931	3	0.310	0.271	.847	.00
	Within Groups	401.329	350	1.147			
	Total	402.260	353				
V.4.F.	Between Groups	20.612	3	6.871	6.983	.000	.06
	Within Groups	344.393	350	0.984			
	Total	365.006	353				

Note. ANOVA = analysis of variance with one factor. N = 354.

V.4.A. = Skills indicator 4.A. = stimulating academic development; V.4.B. = Skills indicator 4.B. = stimulating the effective use of different cognitive styles; V.4.C. = Skills indicator 4.C. = stimulating interpersonal functioning; V.4.D. = Skills indicator 4.D. = stimulating intrapersonal understanding; V.4.E. = Skills indicator 4.E. = stimulating self-actualisation; V.4.F. = Skills indicator 4.F. = general skills and skills matching those of the (S)ENCo.

Table H17

Comparison of the Selection Rates of the Skills Indicators in Domain 4 Between the Groups of Stakeholders

						95% Confidence interval	
Dependent variable	I group	J group	Mean difference (I-J)	*SE*	p	Lower bound	Upper bound
V.4.D.	SGE	EP	-0.578	0.156	.001	-0.99	-0.17
		Pa	-0.672	0.174	.001	-1.13	-0.21
		R	-0.590	0.226	.057	-1.19	-0.01
V.4.F.	SGE	EP	0.506	0.131	.001	0.16	0.85
		Pa	0.583	0.146	.001	0.19	0.97
		R	0.258	0.190	1.000	-0.25	0.76

Note. Based on the results of the ANOVA, Bonferroni's Multiple Comparisons was used as a post hoc test for the analyses of the differences between groups of stakeholders.

EP = educational professionals; SGE = specialist gifted education; Pa = parents; R = Rest; T = total. N = 354.

V.4.D. = Skills indicator 4.D. = stimulating intrapersonal understanding; V.4.F. = Skills indicator 4.F. = general skills and skills matching those of the (S)ENCo.

Table H18

Knowledge Indicators of Each Competency in Domain 4 that Belonged to the Three Most Selected Indicators by all Groups of Stakeholders

Competency	Code	Indicator
4.1 The specialist encourages the student's attitude for learning in a way that stimulates an equal balance between effort, learning and achievement.	K.4.F.1	Knows how a teacher's attitude contributes to the development of a student's self-concept.
4.2 The specialist adjusts the curriculum in such a way that the student has to reach for their zone of proximal development in all domains.	K.4.D.2	Knows how a student's self-efficacy influences a student's motivation to work on an assignment.
	K.4.E.2	Knows how. by adjusting their didactical strategies, they can provide educational challenges for the student.
4.3 The specialist adjusts their classroom management in a way that offers the opportunity to provide instruction/support to the gifted students.	K.4.A.3	Understands that enrichment activities are a natural part of a gifted learner's curriculum.
4.4 The specialist uses the input of the educational partners in designing and implementing meaningful responses to educational needs for individual gifted students.	K.4.C.4	Knows that it is important for gifted students to take part in regular group activities.
	K.4.E.4	Knows how a feeling of competence, autonomy, and relationship regarding the educational objectives contributes positively to a student's motivation.
	K.4.F.4	Knows how a partnership between educational stakeholders contributes positively to the chances of a successful educational process.
4.5 The specialist underpins their interventions with current theories on education and giftedness. In doing so, they ensure that the interventions are workable in their daily practice.	K.4.A.5	Knows how they can create a balance between the curriculum and the student's need for support.
	K.4.E.5	Knows why it is important for a student that both learning processes and achievements are discussed.

Table H19

Skills Indicators of Each Competency in Domain 4 That Belonged to the Three Most Selected Indicators by All Groups of Stakeholders

Competency	Code	Indicator
4.1 The specialist encourages the student's attitude for learning in a way that stimulates an equal balance between effort, learning and achievement.	V.4.A.1	Ensures the student's active involvement during the four stages of the learning process: instruction-planning-execution-evaluation.
	V.4.E.1	Coaches the student on how to deal with positive and negative emotions toward achievement and that they are a natural part of the learning process.
4.2 The specialist adjusts the curriculum in such a way that the student has to reach for their zone of proximal development in all domains.	V.4.B.2	Presents the curriculum in a way that higher order thinking skills are explicitly required to finish a task.
	V.4.E.2	Encourages creative and critical thinking by developing an investigative attitude in students.
4.3 The specialist adjusts their classroom management in a way that offers the opportunity to provide instruction/support to the gifted students.	V.4.A.3	Ensures that enrichment tasks are regularly planned for the gifted student.
4.4 The specialist uses the input of the educational partners in designing and implementing meaningful responses to educational needs for individual gifted students.	V.4.E.4	Involves the student in setting educational objectives.
4.5 The specialist underpins their interventions with current theories on education and giftedness. In doing so, they ensure that the interventions are workable in their daily practice.	V.4.A.5	Offers educational plans that work for both teacher and student.
	V.4.F.5	Discusses the student's needs for educational support from a solution-focused perspective.

Table H20

Mean Selection Rates of the Knowledge Indicators per Category in Domain 4

Indicators	N	Mean	*SD*
K.4.A.	354	3.09	1.011
K.4.B.	354	2.36	1.009
K.4.C.	354	1.79	1.108
K.4.D.	354	2.47	1.332
K.4.E.	354	3.07	1.057
K.4.F.	354	2.23	1.187

Note. One-Sample Statistics. Test value 2.5.

K.4.A. = Knowledge indicator 4.A. = stimulating academic development; K.4.B. = Knowledge indicator 4.B. = stimulating the effective use of different cognitive styles; K.4.C. = Knowledge indicator 4.C. = stimulating interpersonal functioning; K.4.D. = Knowledge indicator 4.D. = stimulating intrapersonal understanding; K.4.E. = Knowledge indicator 4.E. = stimulating self-actualisation; K.4.F. = Knowledge indicator 4.F. = general knowledge and skills matching those of the (S)ENCo.

Table H21

Selection Rates of the Knowledge Indicators per Category in Domain 4

					95% Confidence interval of the difference		
Indicators	*t*	*df*	Sig (2-tailed)	Mean difference	Lower bound	Upper bound	Cohen's *d*
K.4.A.	10.983	353	.000	0.590	0.48	0.70	.58
K.4.B.	-2.687	353	.008	-0.144	-0.25	-0.04	.14
K.4.C.	-12.132	353	.000	-0.715	-0.83	-0.60	.64
K.4.D.	-0.439	353	.661	-0.031	-0.17	0.11	.02
K.4.E.	10.107	353	.000	0.568	0.46	0.68	.54
K.4.F.	-4.255	353	.000	-0.268	-0.39	-0.14	.23

Note. One-Sample T test. Test Value 2.5. N = 354.

K.4.A. = Knowledge indicator 4.A. = stimulating academic development; K.4.B. = Knowledge indicator 4.B. = stimulating the effective use of different cognitive styles; K.4.C. = Knowledge indicator 4.C. = stimulating interpersonal functioning; K.4.D. = Knowledge indicator 4.D. = stimulating intrapersonal understanding; K.4.E. = Knowledge indicator 4.E. = stimulating self-actualisation; K.4.F. = Knowledge indicator 4.F. = general knowledge and skills matching those of the (S)ENCo.

Table H22

Mean Selection Rates of the Skills Indicators per Category in Domain 4

Indicators	N	Mean	*SD*
V.4.A.	354	2.19	1.213
V.4.B.	354	2.39	1.088
V.4.C.	354	2.68	1.100
V.4.D.	354	2.87	1.209
V.4.E.	354	2.63	1.067
V.4.F.	354	2.23	1.017

Note. One-Sample Statistics. Test value 2.5.

V.4.A. = Skills indicator 4.A. = stimulating academic development; V.4.B. = Skills indicator 4.B. = stimulating the effective use of different cognitive styles; V.4.C. = Skills indicator 4.C. = stimulating interpersonal functioning; V.4.D. = Skills indicator 4.D. = stimulating intrapersonal understanding; V.4.E. = Skills indicator 4.E. = stimulating self-actualisation; and V.4.F. = Skills indicator 4.F. = general knowledge and skills matching those of the (S)ENCo.

Table H23

Selection Rates of the Skills Indicators per Category in Domain 4

					95% Confidence interval of the difference		
Indicators	*t*	*df*	Sig (2-tailed)	Mean difference	Lower bound	Upper bound	Cohen's *d*
V.4.A.	-4.731	353	.000	-0.305	-0.43	-0.18	.26
V.4.B.	-1.904	353	.058	-0.110	-0.22	0.00	.10
V.4.C.	3.093	353	.002	0.181	0.07	0.30	.16
V.4.D.	5.579	353	.000	0.370	0.24	0.50	.31
V.4.E.	2.340	353	.020	0.133	0.02	0.24	.12
V.4.F.	-4.965	353	.000	-0.268	-0.37	-0.16	.27

Note. One-Sample Test. Test value 2.5. N = 354.

V.4.A. = Skills indicator 4.A. = stimulating academic development; V.4.B. = Skills indicator 4.B. = stimulating the effective use of different cognitive styles; V.4.C. = Skills indicator 4.C. = stimulating interpersonal functioning; V.4.D. = Skills indicator 4.D. = stimulating intrapersonal understanding; V.4.E. = Skills indicator 4.E. = stimulating self-actualisation; and V.4.F. = Skills indicator 4.F. = general knowledge and skills matching those of the (S)ENCo.

Table H24

Indicator Category with Highest Mean Selection Rate in Domain 4 per Group of Stakeholders

Indicator	Educational professionals	Specialists in gifted education	Parents	Rest
Knowledge	A	E	D	E
Skills	D	C	D	D

Note. A = stimulating academic development; C = stimulating interpersonal functioning; D = stimulating intrapersonal understanding; E = stimulating self-actualisation.

Table H25

The Comparison of Selection Rates Between Two Clusters of Knowledge Indicators in Domain 4: ADE-Indicators vs BCF-Indicators

Clusters of Indicators	Mean	N	*SD*	*SE*
Dom4_K_ADE	2.8757	354	0.50080	0.02662
Dom4_K_BCF	2.1243	354	0.50080	0.02662

Note: Paired Samples Statistics.

Dom4_K_ADE = Domain 4, cluster of three categories of knowledge indicators A, D, E; compared to Dom4_K_BCF = Domain 4, cluster of three categories of knowledge indicators B, C, F.

Table H26

The Comparison of Selection Rates Between Two Clusters of Knowledge Indicators in Domain 4: ADE-Indicators vs BCF-Indicators

				95% Confidence interval of the difference					
Clusters of Indicators	Mean	*SD*	*SE*	Lower bound	Upper bound	*t*	*df*	Sig (2-tailed	Cohen's *d*
Dom4_K_ADE – Dom4_K_BCF	0.75141	1.00159	0.05323	0.64672	0.85611	14.115	353	.000	.75

Note. Paired Samples T test. N = 354.

Dom4_K_ADE = Domain 4, cluster of three categories of knowledge indicators A, D, E; compared to Dom4_K_BCF = Domain 4, cluster of three categories of knowledge indicators B, C, F.

Table H27

The Comparison of Selection Rates Between Two Clusters of Skills Indicators in Domain 4: ADE-Indicators vs BCF-Indicators

Clusters of Indicators	Mean	N	*SD*	*SE*
Dom4_V_ADE	2.5659	354	0.49124	0.02611
Dom4_V_BCF	2.4341	354	0.49124	0.02611

Note: Paired Samples Statistics.

Dom4_V_ADE = Domain 4, cluster of three categories of skills indicators A, D, E; compared to Dom4_V_BCF = Domain 4, cluster of three categories of skills indicators B, C, F.

Table H28

The Comparison of Selection Rates Between Two Clusters of Skills Indicators in Domain 4: ADE-Indicators vs BCF-Indicators

				95% Confidence interval of the difference					
Clusters of Indicators	Mean	*SD*	*SE*	Lower bound	Upper bound	*t*	*df*	Sig (2-tailed	Cohen's *d*
Dom4_K_ADE – Dom4_K_BCF	0.13183	0.98248	0.05222	0.02913	0.23452	2.525	353	0.12	.13

Note. Paired Samples T test. N = 354.

Dom4_V_ADE = Domain 4, cluster of three categories of skills indicators A, D, E; compared to Dom4_V_BCF = Domain 4, cluster of three categories of skills indicators B, C, F.

Table H29

Knowledge Indicators in Domain 4 With the Highest Selection Rates Amongst all Groups of Stakeholders

Competency	Code	Indicator
4.3 The specialist adjusts their classroom management in a way that offers the opportunity to provide instruction/support to the gifted students.	K.4.A.3	Understands that enrichment activities are a natural part of a gifted learner's curriculum.
4.4 The specialist uses the input of the educational partners in designing and implementing meaningful responses to educational needs for individual gifted students.	K.4.E.4	Knows how a feeling of competence, autonomy, and relationship regarding the educational objectives contributes positively to a student's motivation.

Table H30

Skills Indicators in Domain 4 With the Highest Selection Rates Amongst all Groups of Stakeholders

Competency	Code	Indicator
4.2 The specialist adjusts the curriculum in such a way that the student has to reach for their zone of proximal development in all domains.	V.4.E.2	Encourages creative and critical thinking by developing an investigative attitude in students.

Table H31

Knowledge Indicators in Domain 4 With the Lowest Selection Rates Amongst all Groups of Stakeholders

Competency	Code	Indicator
4.2 The specialist adjusts the curriculum in such a way that the student has to reach for their zone of proximal development in all domains.	K.4.C.2	Knows which factors during collaborative processes contribute to the student reaching their zone of proximal development.
4.5 The specialist underpins their interventions with current theories on education and giftedness. In doing so, they ensure that the interventions are workable in their daily practice	K.4.B.5	Is aware of modern didactical strategies.

Table H32

Skills Indicators in Domain 4 With the Lowest Selection Rates Amongst all Groups of Stakeholders

Competency	Code	Indicator
4.1 The specialist encourages the student's attitude for learning in a way that stimulates an equal balance between effort, learning, and achievement.	V.4.C.1	Encourages gifted students to collaborate with other gifted students.

Table H33

Comparison of Mean Selection Rates in Domain 4 per Category of Knowledge Indicators Between Dutch and Flemish Stakeholders

Indicators	Country	N	Mean	*SD*	*SE*
K.4.A.	NL	299	3.09	1.029	0.059
	F	55	3.07	0.920	0.124
K.4.B.	NL	299	2.45	0.969	0.056
	F	55	1.85	1.079	0.145
K.4.C.	NL	299	1.72	1.082	0.063
	F	55	2.16	1.183	0.159
K.4.D.	NL	299	2.43	1.335	0.077
	F	55	2.69	1.303	0.176
K.4.E.	NL	299	3.07	1.062	0.061
	F	55	3.04	1.036	0.140
K.4.F.	NL	299	2.24	1.185	0.069
	F	55	2.18	1.203	0.162

Note. Descriptive Statistics. NL = The Netherlands. F = Flanders. N=354.

K.4.A. = Knowledge indicator 4.A. = stimulating academic development; K.4.B. = Knowledge indicator 4.B. = stimulating the effective use of different cognitive styles; K.4.C. = Knowledge indicator 4.C. = stimulating interpersonal functioning; K.4.D. = Knowledge indicator 4.D. = stimulating intrapersonal understanding; K.4.E. = Knowledge indicator 4.E. = stimulating self-actualisation; and K.4.F. = Knowledge indicator 4.F. = general knowledge and skills matching those of the (S)ENCo.

Table H34

Comparison of Mean Selection Rates in Domain 4 per Category of Knowledge Indicators Between Dutch and Flemish Stakeholders

		Levene's test for Equality of variances		t-test for Equality of means							
									95% Confidence interval of the difference		
Indicators		*F*	*p*	*t*	*df*	Sig. (2-tailed)	Mean difference	*SE* difference	Lower bound	Upper bound	Cohen's *d*
K.4.A.	Equal variances assumed	3.175	.076	0.141	352	.888	0.021	0.149	-0.271	0.313	.02
	Equal variances not assumed			0.152	80.919	.880	0.021	0.139	-0.253	0.295	
K.4.B.	Equal variances assumed	0.150	.699	4.100	352	.000	0.594	0.145	0.309	0.878	.60
	Equal variances not assumed			3.808	70.946	.000	0.594	0.156	0.283	0.904	
K.4.C.	Equal variances assumed	0.085	.771	-2.780	352	.006	-0.448	0.161	-0.765	-0.131	.41
	Equal variances not assumed			-2.615	71.599	.011	-0.448	0.171	-0.789	-0.106	
K.4.D.	Equal variances assumed	0.140	.708	-1.347	352	.179	-0.263	0.195	0.647	0.121	.20
	Equal variances not assumed			-1.369	76340	.175	-0.263	0.192	0.645	0.119	
K.4.E.	Equal variances assumed	0.613	.434	0.240	352	.811	0.037	0.155	0.268	0.343	.04
	Equal variances not assumed			0.244	76.409	.808	0.037	0.153	0.267	0.341	
K.4.F.	Equal variances assumed	0.006	.938	0.338	352	.735	0.059	0.174	0.284	0.402	.05
	Equal variances not assumed			0.335	74.567	.739	0.059	0.176	0.292	0.410	

Note. Independent Samples Test. N = 354.

K.4.A. = Knowledge indicator 4.A. = stimulating academic development; K.4.B. = Knowledge indicator 4.B. = stimulating the effective use of different cognitive styles; K.4.C. = Knowledge indicator 4.C. = stimulating interpersonal functioning; K.4.D. = Knowledge indicator 4.D. = stimulating intrapersonal understanding; K.4.E. = Knowledge indicator 4.E. = stimulating self-actualisation; and K.4.F. = Knowledge indicator 4.F. = general knowledge and skills matching those of the (S)ENCo.

Table H35

Comparison of Mean Selection Rates in Domain 4 per Category of Skills Indicators Between Dutch and Flemish Stakeholders

Indicators	Country	N	Mean	*SD*	*SE*
V.3.A.	NL	299	2.22	1.195	0.069
	F	55	2.05	1.311	0.177
V.3.B.	NL	299	2.37	1.081	0.062
	F	55	2.47	1.136	0.153
V.3.C.	NL	299	2.68	1.103	0.064
	F	55	2.67	1.090	0.147
V.3.D.	NL	299	2.78	1.188	0.069
	F	55	3.35	1.220	0.165
V.3.E.	NL	299	2.63	1.043	0.060
	F	55	2.67	1.203	0.162
V.3.F.	NL	299	2.31	1.017	0.059
	F	55	1.78	0.896	0.121

Note. Descriptive Statistics. NL = The Netherlands. F = Flanders. N=354.

V.4.A. = Skills indicator 4.A. = stimulating academic development; V.4.B. = Skills indicator 4.B. = stimulating the effective use of different cognitive styles; V.4.C. = Skills indicator 4.C. = stimulating interpersonal functioning; V.4.D. = Skills indicator 4.D. = stimulating intrapersonal understanding; V.4.E. = Skills indicator 4.E. = stimulating self-actualisation; and V.4.F. = Skills indicator 4.F. = general knowledge and skills matching those of the (S)ENCo.

Table H36

Comparison of Mean Selection Rates in Domain 4 per Category of Skills Indicators Between Dutch and Flemish Stakeholders

		Levene's test for Equality of variances		t-test for Equality of means							
									95% Confidence interval of the difference		
Indicators		*F*	*p*	*t*	*df*	Sig. (2-tailed)	Mean difference	*SE* difference	Lower bound	Upper bound	Cohen's *d*
V.4.A.	Equal variances assumed	0.783	.377	0.934	352	.351	0.166	0.178	-0.184	0.516	.14
	Equal variances not assumed			0.876	71.455	.384	0.166	0.190	-0.212	0.545	
V.4.B.	Equal variances assumed	0.433	.511	-0.614	352	.540	-0.098	0.160	-0.412	0.216	.09
	Equal variances not assumed			-0.593	73.103	.555	-0.098	0.165	-0.428	0.232	
V.4.C.	Equal variances assumed	0.012	.914	0.059	352	.953	0.010	0.162	-0.308	0.327	.01
	Equal variances not assumed			0.060	75.806	.953	0.010	0.160	-0.309	0.329	
V.4.D.	Equal variances assumed	0.105	.746	-3.215	352	.001	-0.563	0.175	-0.907	-0.218	.47
	Equal variances not assumed			-3.156	74.070	.002	-0.563	0.178	-0.918	-0.207	
V.4.E.	Equal variances assumed	2.587	.109	-0.302	352	.763	-0.047	0.157	-0.356	0.261	.04
	Equal variances not assumed			0.273	69.726	.785	-0.047	0.173	-0.392	0.298	
V.4.F.	Equal variances assumed	2.964	.086	3.631	352	.000	0.533	0.147	0.244	0.821	.53
	Equal variances not assumed			3.962	81.796	.000	0.533	0.134	0.265	0.800	

Note. Independent Samples Test. N=354

V.4.A. = Skills indicator 4.A. = stimulating academic development; V.4.B. = Skills indicator 4.B. = stimulating the effective use of different cognitive styles; V.4.C. = Skills indicator 4.C. = stimulating interpersonal functioning; V.4.D. = Skills indicator 4.D. = stimulating intrapersonal understanding; V.4.E. = Skills indicator 4.E. = stimulating self-actualisation; and V.4.F. = Skills indicator 4.F. = general knowledge and skills matching those of the (S)ENCo.

Table H37

Domain 4 Knowledge Indicators. Mean per Category per Competency per Group of Stakeholders

Indicators	EP	Indicators	SGE	Indicators	Pa	Indicators	R
K.4.C.	1.89	K4C	1.70	K4C	1.81	K4C	1.59
K.4.F.	2.12	K4D	2.00	K4F	1.84	K4B	2.24
K.4.B.	2.41	K4B	2.46	K4B	2.31	K4F	2.24
K.4.D.	2.44	K4F	2.68	K4A	2.82	K4D	2.35
K.4.E.	2.88	K4A	3.03	K4E	3.07	K4E	3.27
K.4.A.	3.25	K4E	3.23	K4D	3.14	K4A	3.30

Note. Test value 2.5. EP = educational professionals; SGE = specialist gifted education; Pa = parents; R = rest. N = 354.

K.4.A. = Knowledge indicator 4.A. = stimulating academic development; K.4.B. = Knowledge indicator 4.B. = stimulating the effective use of different cognitive styles; K.4.C. = Knowledge indicator 4.C. = stimulating interpersonal functioning; K.4.D. = Knowledge indicator 4.D. = stimulating intrapersonal understanding; K.4.E. = Knowledge indicator 4.E. = stimulating self-actualisation; and K.4.F. = Knowledge indicator 4.F. = general knowledge and skills matching those of the (S)ENCo.

Table H38

Domain 4 Skills Indicators. Mean per Category per Competency per Group of Stakeholders

Indicators	EP	Indicators	SGE	Indicators	Pa	Indicators	R
V.4.F.	2.08	V4B	2.26	V4F	2.32	V4A	1.84
V.4.A.	2.14	V4A	2.27	V4A	2.35	V4F	2.32
V.4.B.	2.44	V4D	2.44	V4B	2.42	V4B	2.51
V.4.E.	2.64	V4E	2.56	V4C	2.46	V4C	2.57
V.4.C.	2.69	V4F	2.58	V4E	2.73	V4E	2.73
V.4.D.	3.02	V4C	2.88	V4D	3.11	V4D	3.03

Note. Test value 2.5. EP = educational professionals; SGE = specialist gifted education; Pa = parents; R = rest. N = 354.

V.4.A. = Skills indicator 4.A. = stimulating academic development; V.4.B. = Skills indicator 4.B. = stimulating the effective use of different cognitive styles; V.4.C. = Skills indicator 4.C. = stimulating interpersonal functioning; V.4.D. = Skills indicator 4.D. = stimulating intrapersonal understanding; V.4.E. = Skills indicator 4.E. = stimulating self-actualisation; and V.4.F. = Skills indicator 4.F. = general knowledge and skills matching those of the (S)ENCo.

Table H39

Ranking of the Knowledge and Skills Indicators of Domain 4

Indicators	N	%	Indicators	N	%
V.4.C.1	63	0.59	V.4.F.3	183	1.72
K.4.B.5	85	0.80	K.4.C.5	185	1.74
K.4.C.2	99	0.93	K.4.F.3	188	1.77
V.4.D.2	99	0.93	V.4.C.5	190	1.79
V.4.D.1	104	0.98	V.4.D.4	191	1.80
K.4.D.4	106	1.00	V.4.E.3	191	1.80
K.4.B.1	109	1.03	K.4.F.4	192	1.81
K.4.B.4	120	1.13	K.4.D.1	197	1.85
V.4.B.3	122	1.15	K.4.D.2	197	1.85
V.4.C.3	125	1.18	K.4.E.1	198	1.86
K.4.C.3	134	1.26	V.4.D.3	199	1.87
K.4.D.5	135	1.27	V.4.F.5	199	1.87
V.4.A.2	137	1.29	V.4.A.5	201	1.89
K.4.C.1	140	1.32	K.4.B.2	203	1.91
K.4.A.4	142	1.34	V.4.F.1	208	1.96
V.4.B.4	143	1.35	K.4.A.5	210	1.98
K.4.F.2	145	1.37	V.4.B.2	214	2.02
V.4.D.5	147	1.38	V.4.E.4	217	2.04
K.4.B.3	158	1.49	K.4.C.4	220	2.07
V.4.F.4	159	1.50	V.4.B.1	221	2.08
V.4.E.5	161	1.52	K.4.E.5	222	2.09
V.4.B.5	164	1.54	K.4.F.5	225	2.12
K.4.E.3	165	1.55	V.4.A.1	230	2.17
V.4.A.4	171	1.61	V.4.E.1	236	2.22
K.4.D.3	172	1.62	V.4.A.3	242	2.28
V.4.C.2	173	1.63	K.4.E.2	242	2.28
V.4.F.2	173	1.63	K.4.F.1	244	2.30
K.4.A.1	174	1.64	K.4.A.3	245	2.31
K.4.A.2	176	1.66	V.4.E.2	266	2.50
V.4.C.4	181	1.70	K.4.E.4	282	2.66

Note. This ranking is based on selection rates of the entire group of stakeholders, N =354.

All indicators are coded. K = knowledge; V = skills; the first number refers to the vertical competency domain in the competency matrix; the second letter refers to the category of indicators. The second number refers to the actual competency the indicators are matched with. See Appendix B for the match between the competencies and the indicators.

Appendix I

Tables Domain 5. Referred to in Chapter 9

Table I1

Frequency Table of the Selection Rates of Knowledge Indicators for Competency 5.1

Competency	Indicator	EP	SGE	Pa	R	T
5.1	K.5.A.1	14.25	17.80	18.88	18.92	16.85
	K.5.B.1	10.18	5.18	10.84	7.21	8.57
	K.5.C.1	15.01	10.36	17.27	9.91	13.65
	K.5.D.1	22.65	23.95	16.47	17.12	21.00
	K.5.E.1	20.10	23.30	22.49	25.23	22.13
	K.5.F.1	17.81	19.42	14.06	21.62	17.80

Note. The results are presented per indicator for each group of stakeholders. EP = educational professionals; SGE = specialist gifted education; Pa = parents; R = Rest; T = total. N = 354.

All indicators are coded. K = knowledge; the first number refers to the vertical competency domain in the competency matrix; the second letter refers to the category of indicators. The second number refers to the actual competency the indicators are matched with. See Appendix B for the match between the competencies and the indicators.

Table I2

Frequency Table of the Selection Rates of Knowledge Indicators for Competency 5.2

Competency	Indicator	EP	SGE	Pa	R	T
5.2	K.5.A.2	19.08	18.45	20.48	17.12	19.02
	K.5.B.2	12.98	13.92	14.06	12.61	13.47
	K.5.C.2	11.45	6.15	11.24	6.31	9.32
	K.5.D.2	21.12	20.06	18.47	18.92	19.96
	K.5.E.2	16.54	18.45	15.66	22.52	17.51
	K.5.F.2	18.83	22.98	20.08	22.52	20.72

Note. The results are presented per indicator for each group of stakeholders. EP = educational professionals; SGE = specialist gifted education; Pa = parents; R = Rest; T = total. N = 354.

All indicators are coded. K = knowledge; the first number refers to the vertical competency domain in the competency matrix; the second letter refers to the category of indicators. The second number refers to the actual competency the indicators are matched with. See Appendix B for the match between the competencies and the indicators.

Table I3

Frequency Table of the Selection Rates of Knowledge Indicators for Competency 5.3

Competency	Indicator	EP	SGE	Pa	R	T
5.3	K.5.A.3	19.85	23.62	22.49	22.52	21.85
	K.5.B.3	12.21	11.00	14.06	9.01	11.96
	K.5.C.3	18.32	19.42	20.48	16.22	18.93
	K.5.D.3	13.23	13.92	16.87	15.32	14.50
	K.5.E.3	15.01	8.09	9.24	15.32	11.68
	K.5.F.3	21.37	23.95	16.87	21.62	21.09

Note. The results are presented per indicator for each group of stakeholders. EP = educational professionals; SGE = specialist gifted education; Pa = parents; R = Rest; T = total. N = 354.

All indicators are coded. K = knowledge; the first number refers to the vertical competency domain in the competency matrix; the second letter refers to the category of indicators. The second number refers to the actual competency the indicators are matched with. See Appendix B for the match between the competencies and the indicators.

Table I4

Frequency Table of the Selection Rates of Knowledge Indicators for Competency 5.4

Competency	Indicator	EP	SGE	Pa	R	T
5.4	K.5.A.4	26.97	25.89	28.11	24.32	26.65
	K.5.B.4	9.67	10.36	10.44	8.11	9.89
	K.5.C.4	16.03	16.50	20.88	18.92	17.61
	K.5.D.4	21.63	16.18	18.47	19.82	19.11
	K.5.E.4	16.03	20.39	16.06	16.22	17.33
	K.5.F.4	9.67	10.68	6.02	12.61	9.42

Note. The results are presented per indicator for each group of stakeholders. EP = educational professionals; SGE = specialist gifted education; Pa = parents; R = Rest; T = total. N = 354.

All indicators are coded. K = knowledge; the first number refers to the vertical competency domain in the competency matrix; the second letter refers to the category of indicators. The second number refers to the actual competency the indicators are matched with. See Appendix B for the match between the competencies and the indicators.

Table I5

Frequency Table of the Selection Rates of Knowledge Indicators for Competency 5.5

Competency	Indicator	EP	SGE	Pa	R	T
5.5	K.5.A.5	18.83	23.95	18.88	17.12	20.15
	K.5.B.5	14.25	12.30	21.29	9.91	14.88
	K.5.C.5	17.81	14.89	18.07	14.41	16.67
	K.5.D.5	18.32	18.77	15.66	20.72	18.08
	K.5.E.5	19.08	19.09	14.86	18.92	18.08
	K.5.F.5	11.70	11.00	11.24	18.92	12.15

Note. The results are presented per indicator for each group of stakeholders. EP = educational professionals; SGE = specialist gifted education; Pa = parents; R = Rest; T = total. N = 354.

All indicators are coded. K = knowledge; the first number refers to the vertical competency domain in the competency matrix; the second letter refers to the category of indicators. The second number refers to the actual competency the indicators are matched with. See Appendix B for the match between the competencies and the indicators.

Table I6

Frequency Table of the Selection Rates of Skills Indicators for Competency 5.1

Competency	Indicator	EP	SGE	Pa	R	T
5.1	V.5.A.1	16.03	16.50	21.29	14.41	17.23
	V.5.B.1	14.25	15.53	16.06	14.41	15.07
	V.5.C.1	11.70	8.41	9.24	8.11	9.79
	V.5.D.1	21.12	24.27	19.28	24.32	21.94
	V.5.E.1	23.92	24.60	24.10	24.32	24.20
	V.5.F.1	12.98	10.68	10.04	14.41	11.77

Note. The results are presented per indicator for each group of stakeholders. EP = educational professionals; SGE = specialist gifted education; Pa = parents; R = Rest; T = total. N = 354.

All indicators are coded. V = skills; the first number refers to the vertical competency domain in the competency matrix; the second letter refers to the category of indicators. The second number refers to the actual competency the indicators are matched with. See Appendix B for the match between the competencies and the indicators.

Table I7

Frequency Table of the Selection Rates of Skills Indicators for Competency 5.2

Competency	Indicator	EP	SGE	Pa	R	T
5.2	V.5.A.2	13.49	14.89	14.06	16.22	14.31
	V.5.B.2	14.25	10.36	13.65	7.21	12.24
	V.5.C.2	15.01	9.71	12.85	13.51	12.81
	V.5.D.2	24.43	22.33	29.72	27.03	25.33
	V.5.E.2	21.37	27.83	18.07	25.23	22.88
	V.5.F.2	11.45	14.89	11.65	10.81	12.43

Note. The results are presented per indicator for each group of stakeholders. EP = educational professionals; SGE = specialist gifted education; Pa = parents; R = Rest; T = total. N = 354.

All indicators are coded. V = skills; the first number refers to the vertical competency domain in the competency matrix; the second letter refers to the category of indicators. The second number refers to the actual competency the indicators are matched with. See Appendix B for the match between the competencies and the indicators.

Table I8

Frequency Table of the Selection Rates of Skills Indicators for Competency 5.3

Competency	Indicator	EP	SGE	Pa	R	T
5.3	V.5.A.3	14.76	13.59	10.44	9.01	12.81
	V.5.B.3	12.72	15.21	12.05	12.61	13.28
	V.5.C.3	14.50	13.27	15.66	11.71	14.12
	V.5.D.3	18.58	15.86	23.69	23.42	19.49
	V.5.E.3	20.87	20.06	20.08	21.62	20.53
	V.5.F.3	18.58	22.01	18.07	21.62	19.77

Note. The results are presented per indicator for each group of stakeholders. EP = educational professionals; SGE = specialist gifted education; Pa = parents; R = Rest; T = total. N = 354.

All indicators are coded. V = skills; the first number refers to the vertical competency domain in the competency matrix; the second letter refers to the category of indicators. The second number refers to the actual competency the indicators are matched with. See Appendix B for the match between the competencies and the indicators.

Table I9

Frequency Table of the Selection Rates of Skills Indicators for Competency 5.4

Competency	Indicator	EP	SGE	Pa	R	T
5.4	V.5.A.4	23.16	22.98	22.49	16.22	22.22
	V.5.B.4	9.67	12.94	12.45	13.51	11.68
	V.5.C.4	15.27	14.89	15.26	15.32	15.16
	V.5.D.4	19.85	14.89	22.89	23.42	19.49
	V.5.E.4	18.07	20.06	13.65	21.62	17.98
	V.5.F.4	13.99	14.24	13.25	9.91	13.47

Note. The results are presented per indicator for each group of stakeholders. EP = educational professionals; SGE = specialist gifted education; Pa = parents; R = Rest; T = total. N = 354.

All indicators are coded. V = skills; the first number refers to the vertical competency domain in the competency matrix; the second letter refers to the category of indicators. The second number refers to the actual competency the indicators are matched with. See Appendix B for the match between the competencies and the indicators.

Table I10

Frequency Table of the Selection Rates of Skills Indicators for Competency 5.5

Competency	Indicator	EP	SGE	Pa	R	T
5.5	V.5.A.5	13.74	18.12	10.84	9.91	13.94
	V.5.B.5	17.30	13.27	22.49	14.41	17.04
	V.5.C.5	15.27	16.50	18.47	18.02	16.67
	V.5.D.5	18.32	16.83	19.68	21.62	18.55
	V.5.E.5	16.28	13.27	10.04	13.51	13.65
	V.5.F.5	19.08	22.01	18.47	22.52	20.15

Note. The results are presented per indicator for each group of stakeholders. EP = educational professionals; SGE = specialist gifted education; Pa = parents; R = Rest; T = total. N = 354.

All indicators are coded. V = skills; the first number refers to the vertical competency domain in the competency matrix; the second letter refers to the category of indicators. The second number refers to the actual competency the indicators are matched with. See Appendix B for the match between the competencies and the indicators.

Table I11

Comparison of the 10 Highest Indicator Selection Rates per Competency in Domain 5 per Group of Stakeholders

Indicator	EP	Indicator	SGE	Indicator	Pa	Indicator	R
K.5.E.5	19.08	V.5.F.3	22.01	K.5.A.2	20.48	K.5.D.5	20.72
V.5.F.5	19.08	V.5.F.5	22.01	K.5.B.5	21.29	K.5.E.2/ K.5.F.2	22.52
V.5.E.3	20.87	K.5.F.2	22.98	K.5.E.1	22.49	K.5.A.3	22.52
K.5.D.2	21.12	V.5.A.4	22.98	K.5.A.3	22.49	V.5.F.5	22.52
K.5.F.3	21.37	K.5.D.1	23.95	V.5.B.5	22.49	V.5.D.3	23.42
K.5.D.1	22.65	K.5.F.3	23.95	V.5.D.4	22.89	V.5.D.4	23.42
V.5.A.4	23.16	K.5.A.5	23.95	V.5.D.3	23.69	K.5.A.4	24.32
V.5.E.1	23.92	V.5.E.1	24.60	V.5.E.1	24.10	V.5.D.1/ V.5.E.1	24.32
V.5.D.2	24.43	K.5.A.4	25.89	K.5.A.4	28.11	K.5.E.1	25.23
K.5.A.4	26.97	V.5.E.2	27.83	V.5.D.2	29.72	V.5.D.2	27.03

Note. EP = educational professionals; SGE = specialist gifted education; Pa = parents; R = Rest; T = total. N = 354. If two indicators matching one competency had an identical selection rate, they are ranked at the same level.

All indicators are coded. K = knowledge / V = skills ; the first number refers to the vertical competency domain in the competency matrix; the second letter refers to the category of indicators. The second number refers to the actual competency the indicators are matched with. See Appendix B for the match between the competencies and the indicators.

Table I12

Comparison of the Selection Rates of the Knowledge Indicators in Domain 5 Between the Groups of Stakeholders

Indicators	Group	N	Mean	*SD*	*SE*	95% Confidence Interval for Mean		Min	Max
						Lower bound	Upper bound		
K.5.C.	EP	131	2.36	1.060	0.093	2.18	3.05	0	5
	SGE	103	2.08	1.100	0.108	1.86	2.54	0	5
	Pa	83	1.99	1.076	0.118	1.75	2.29	0	5
	R	37	1.86	1.110	0.182	1.49	2.22	0	4
	T	354	2.14	1.091	0.058	2.02	2.23	0	5
K.5.D.	EP	131	2.53	1.198	0.105	2.33	2.25	0	5
	SGE	103	2.08	1.273	0.125	1.83	2.74	0	5
	Pa	83	3.12	1.224	0.134	2.85	2.33	0	5
	R	37	2.65	1.207	0.198	2.25	3.39	1	5
	T	354	2.55	1.279	0.068	2.42	3.05	0	5
K.5.E.	EP	131	2.89	1.076	0.094	2.71	2.68	1	5
	SGE	103	3.26	0.907	0.089	3.08	3.08	1	5
	Pa	83	3.10	1.007	0.111	2.88	3.44	0	5
	R	37	3.19	1.126	0.185	2.81	3.32	0	5
	T	354	3.08	1.026	0.055	2.97	3.56	0	5
K.5.F.	EP	131	1.53	1.025	0.090	1.35	3.19	0	4
	SGE	103	1.94	1.065	0.105	1.73	1.70	0	5
	Pa	83	1.28	1.051	0.115	1.05	2.15	0	4
	R	37	1.62	0.924	0.152	1.31	1.51	0	3
	T	354	1.60	1.058	0.056	1.49	1.93	0	5

Note. Descriptive Statistics. EP = educational professionals; SGE = specialist gifted education; Pa = parents; R = Rest; T = total. N = 354.

K.5.C. = Knowledge indicator 5.C. = stimulating interpersonal functioning; K.5.D. = Knowledge indicator 5.D. = stimulating intrapersonal understanding; K.5.E. = Knowledge indicator 5.E. = stimulating self-actualisation; K.5.F. = Knowledge indicator 5.F. = general knowledge and skills matching those of the (S)ENCo.

Table I13

Comparison of the Selection Rates of the Knowledge Indicators in Domain 5 Between the Groups of Stakeholders

Indicators		Sum of Squares	*df*	Mean Square	*F*	*p*	η^2
K.5.A.	Between Groups	4.396	3	1.465	1.396	.244	.01
	Within Groups	367.423	350	1.050			
	Total	371.819	353				
K.5.B.	Between Groups	1.080	3	0.360	0.337	.799	.00
	Within Groups	373.791	350	1.068			
	Total	374.870	353				
K.5.C.	Between Groups	11.389	3	3.796	3.250	.022	.03
	Within Groups	408.828	350	1.168			
	Total	420.218	353				
K.5.D.	Between Groups	50.383	3	16.794	11.149	.000	.09
	Within Groups	527.202	350	1.506			
	Total	577.585	353				
K.5.E.	Between Groups	8.455	3	2.818	2.715	.045	.02
	Within Groups	363.331	350	1.038			
	Total	371.785	353				
K.5.F.	Between Groups	21.403	3	7.134	6.683	.000	.05
	Within Groups	373.636	350	1.068			
	Total	395.040	353				

Note. ANOVA = analyses of variance with one factor.

K.5.A. = Knowledge indicator 5.A. = stimulating academic development; K.5.B. = Knowledge indicator 5.B. = stimulating the effective use of different cognitive styles; K.5.C. = Knowledge indicator 5.C. = stimulating interpersonal functioning; K.5.D. = Knowledge indicator 5.D. = stimulating intrapersonal understanding; K.5.E. = Knowledge indicator 5.E. = stimulating self-actualisation; K.5.F. = Knowledge indicator 5F = general knowledge and skills matching those of the (S)ENCo.

Table I14

Comparison of the Selection Rates of the Knowledge Indicators in Domain 5 Between the Groups of Stakeholders

Dependent variable	I group	J group	Mean difference (I-J)	*SE*	*p*	95% Confidence interval Lower bound	Upper bound
K.5.D.	Pa	EP	0.586	0.172	.004	0.13	1.04
		SGE	1.043	0.181	.000	0.56	1.52
		R	0.472	0.243	.316	-0.17	1.12
K.5.D.	SGE	EP	-0.457	0.162	.030	-0.89	-0.03
		Paa	-1.043	0.181	.000	-1.52	-0.56
		R	-0.571	0.235	.094	-1.20	0.05
K.5.E.	SGE	EP	0.369	0.134	.038	0.01	0.56
		Pa	0.166	0.150	1.000	-0.23	0.59
		R	0.073	0.195	1.000	-0.45	0.58
K.5.F.	SGE	EP	0.415	0.136	.015	0.05	0.78
		Pa	0.665	0.152	.000	0.26	1.07
		R	0.320	0.198	.641	-0.21	0.85

Note. Based on the results of the ANOVA, Bonferroni's Multiple Comparisons was used as a post hoc test for the analyses of the differences between groups of stakeholders.

EP = educational professionals; SGE = specialist gifted education; Pa = parents; R = Rest; T = total. N = 354.

K.5.D. = Knowledge indicator 5.D. = stimulating intrapersonal understanding; K.5.E. = Knowledge indicator 5.E. = stimulating self-actualisation; K.5.F. = Knowledge indicator 5.F. = general knowledge and skills matching those of the (S)ENCo.

Table I15

Comparison of the Selection Rates of the Skills Indicators in Domain 5 Between the Groups of Stakeholders

Indicators	Group	N	Mean	*SD*	*SE*	95% Confidence interval for mean Lower bound	Upper bound	Min	Max
V.5.D.	EP	131	3.01	1.199	0.105	2.80	3.21	0	5
	SGE	103	2.53	1.145	0.113	2.31	2.76	0	5
	Pa	83	3.02	1.147	0.126	2.77	3.27	0	5
	R	37	2.97	1.258	0.207	2.55	3.39	0	5
	T	354	2.87	1.193	0.063	2.75	2.99	0	5
V.5.F.	EP	131	2.29	1.113	0.097	2.10	2.48	0	5
	SGE	103	2.84	1.091	0.108	2.63	3.06	0	5
	Pa	83	2.13	1.102	0.121	1.89	2.37	0	4
	R	37	2.62	1.010	0.166	2.28	2.96	1	5
	T	354	2.45	1.126	0.060	2.33	2.57	0	5

Note. Descriptive Statistics. EP = educational professionals; SGE = specialist gifted education; Pa = parents; R = Rest; T = total. N = 354.

V.5.D. = Skills indicator 5.D. = stimulating intrapersonal understanding; V.5.F. = Skills indicator 5.F. = general knowledge and skills matching those of the (S)ENCo.

Table I16

Comparison of the Selection Rates of the Skills Indicators in Domain 5 Between the Groups of Stakeholders

Indicators		Sum of Squares	*df*	Mean Square	*F*	*p*	η^2
V.5.A.	Between Groups	5.380	3	1.793	1.294	.276	.01
	Within Groups	485	350	1.386			
	Total	490.500	353				
V.5.B.	Between Groups	6.551	3	2.184	1.839	.140	.02
	Within Groups	415.516	350	1.187			
	Total	422.068	353				
V.5.C.	Between Groups	3.618	3	1.056	0.831	.478	.01
	Within Groups	444.778	350	1.271			
	Total	447.946	353				
V.5.D.	Between Groups	16.474	3	5.491	3.958	.009	.03
	Within Groups	485.548	350	1.387			
	Total	502.023	353				
V.5.E.	Between Groups	0.631	3	0.210	0.189	.904	.00
	Within Groups	390.603	350	1.116			
	Total	391.234	353				
V.5.F.	Between Groups	28.848	3	9.616	8.038	.000	.06
	Within Groups	418.737	350	1.196			
	Total	447.585	353				

Note. ANOVA = analysis of variance with one factor. N = 354.

V.5.A. = Skills indicator 5.A. = stimulating academic development; V.5.B. = Skills indicator 5.B. = stimulating the effective use of different cognitive styles; V.5.C. = Skills indicator 5.C. = stimulating interpersonal functioning; V.5.D. = Skills indicator 5.D. = stimulating intrapersonal understanding; V.5.E. = Skills indicator 5.E. = stimulating self-actualisation; V.5.F. = Skills indicator 5.F. = general knowledge and skills matching those of the (S)ENCo.

Table I17

Comparison of the Selection Rates of the Skills Indicators in Domain 5 Between the Groups of Stakeholders

Dependent variable	I group	J group	Mean difference (I-J)	*SE*	*p*	95% Confidence interval	
						Lower bound	Upper bound
V.5.D.	SGE	EP	-0.474	0.155	.015	-0.89	-0.06
		Pa	-0.490	0.174	.030	-0.95	-0.03
		R	-0.439	0.226	.316	-1.04	0.16
V.5.F.	SGE	EP	0.555	0.144	.001	0.17	0.94
		Pa	0.715	0.161	.000	0.28	1.14
		R	0.223	0.210	1.000	-0.33	0.78

Note. Based on the results of the ANOVA, Bonferroni's Multiple Comparisons was used as a post hoc test for the analyses of the differences between groups of stakeholders.

EP = educational professionals; SGE = specialist gifted education; Pa = parents; R = Rest; T = total. N = 354.

V.5.D. = Skills indicator 5.D. = stimulating intrapersonal understanding; V.5.F. = Skills indicator 5F = general knowledge and skills matching those of the (S)ENCo.

Table I18

Knowledge Indicators of Each Competency in Domain 5 that Belonged to the Three Most Selected Indicators by all Groups of Stakeholders

Competency	Code	Indicator
5.1 The specialist monitors the student's intra- and interpersonal functioning to determine how this development contributes to the student's ability for self-actualisation.	K.5.E.1	Knows how a student's self-concept influences their self-efficacy.
5.2 The specialist monitors the student's academic achievements and relates these achievements to the level of support given during the learning process.	K.5.A.2	Knows how they can assess whether educational objectives outside the regular curriculum are achieved.
	K.5.D.2	Knows how their attitude affects a student's learning process.
	K.5.F.2	Knows how they can share their knowledge and skills to monitor a student's development with their colleagues.
5.3 The specialist plans how they can monitor and document the student's development periodically, based on a set schedule.	K.5.A.3	Knows why it is relevant to discuss the student's development with the educational partners at the start of the new term.
	K.5.C.3	Knows how social wellbeing affects learning and achievement.
	K.5.F.3	Knows how the results of assessment can contribute to a student's ability for self-actualisation.
5.4 The specialist discusses the student's overall development with all the relevant actors in the student's ecological system.	K.5.A.4	Knows how the information from parents and co-workers can help to optimise a student's curriculum.
	K.5.C.4	Knows a student's social functioning can be discussed with parents and co-workers from a positive perspective.
5.5 Based on current theories of assessment and giftedness, the specialist designs a system of formative and summative assessment for the education of gifted students.	K.5.A.5	Knows why it is important to set assessable educational objectives.

Table I19

Skills Indicators of Each Competency in Domain 5 that Belonged to the Three Most Selected Indicators by all Groups of Stakeholders

Competency	Code	Indicators
5.2 The specialist monitors the student's academic achievements and relates these achievements to the level of support given during the learning process.	V.5.D.2	Maintains an intrinsic interest in their students and shows this in the way that they communicate with them.
	V.5.E.2	Discusses with the student how their achievements relate to the set objectives.
5.4 The specialist discusses the student's overall development with all the relevant actors in the student's ecological system.	V.5.A.4	Plans meetings with parents and teachers to discuss a student's responses to intervention programs.
	V.5.D.4	Explores with the student how they can create a balance between learning, effort, and achievement.
5.5 Based on current theories of assessment and giftedness, the specialist designs a system of formative and summative assessment for the education of gifted students.	V.5.D.5	Explores to what extent the student has a realistic perspective of their own abilities.
	V.5.F.5	Works with classroom teachers to establish assessments beyond commonly used standardised tools, to evaluate the student's learning process and achievements.

Table I20

Mean Selection Rates of the Knowledge Indicators per Category in Domain 5

Indicators	N	Mean	*SD*
K.5.A.	354	2.69	1.026
K.5.B.	354	2.94	1.031
K.5.C.	354	2.14	1.091
K.5.D.	354	2.55	1.279
K.5.E.	354	3.08	1.026
K.5.F.	354	1.60	1.058

Note. One-Sample Statistics. Test value 2.5.

K.5.A. = Knowledge indicator 5.A. = stimulating academic development; K.5.B. = Knowledge indicator 5.B. = stimulating the effective use of different cognitive styles; K.5.C. = Knowledge indicator 5.C. = stimulating interpersonal functioning; K.5.D. = Knowledge indicator 5.D. = stimulating intrapersonal understanding; K.5.E. = Knowledge indicator 5.E = stimulating self-actualisation; K.5.F. = Knowledge indicator 5.F. = general knowledge and skills matching those of the (S)ENCo.

Table I21

Selection Rates of the Knowledge Indicators per Category in Domain 5

Indicators	*t*	*df*	Sig (2-tailed)	Mean difference	95% Confidence interval of the difference Lower bound	Upper bound	Cohen's *d*
K.5.A.	3.470	353	.001	0.189	0.08	0.30	.19
K.5.B.	8.097	353	.000	0.444	0.34	0.55	.43
K.5.C.	-6.235	353	.000	-0.362	-0.48	-0.25	.33
K.5.D.	0.748	353	.455	0.051	-0.08	0.18	.04
K.5.E.	10.617	353	.000	0.579	0.47	0.69	.57
K.5.F.	-16.027	353	.000	-0.901	-1.01	-0.79	.85

Note. One-Sample T test. Test Value 2.5. N = 354.

K.5.A. = Knowledge indicator 5.A. = stimulating academic development; K.5.B. = Knowledge indicator 5.B. = stimulating the effective use of different cognitive styles; K.5.C. = Knowledge indicator 5.C. = stimulating interpersonal functioning; K.5.D. = Knowledge indicator 5.D. = stimulating intrapersonal understanding; K.5.E. = Knowledge indicator 5.E. = stimulating self-actualisation; K.5.F. = Knowledge indicator 5.F. = general knowledge and skills matching those of the (S)ENCo.

Table I22

Mean Selection Rates of the Skills Indicators per Category in Domain 5

Indicators	N	Mean	*SD*
V.5.A.	354	2.50	1.179
V.5.B.	354	2.24	1.093
V.5.C.	354	2.54	1.126
V.5.D.	354	2.87	1.193
V.5.E.	354	2.40	1.053
V.5.F.	354	2.45	1.126

Note. One-Sample Statistics. Test value 2.5.

V.5.A. = Skills indicator 5.A. = stimulating academic development; V.5.B. = Skills indicator 5.B. = stimulating the effective use of different cognitive styles; V.5.C. = Skills indicator 5.C. = stimulating interpersonal functioning; V.5.D. = Skills indicator 5.D. = stimulating intrapersonal understanding; V.5.E. = Skills indicator 5.E. = stimulating self-actualisation; and V.5.F. = Skills indicator 5.F. = general knowledge and skills matching those of the (S)ENCo.

Table I23

Selection Rates of the Skills Indicators per Category in Domain 5

Indicators	*t*	*df*	Sig (2-tailed)	Mean difference	95% Confidence interval of the difference		Cohen's *d*
					Lower	Upper	
V.5.A.	0.000	353	1.000	0.000	-0.12	0.12	.00
V.5.B.	-4.520	353	.000	-0.263	-0.38	-0.15	.24
V.5.C.	0.661	353	.509	0.040	-0.08	0.16	.04
V.5.D.	5.838	353	.000	0.370	0.25	0.49	.31
V.5.E.	-1.717	353	.087	-0.096	-0.21	0.01	.09
V.5.F.	-0.885	353	.396	-0.051	-0.17	0.07	.04

Note. One-Sample Test. Test Value 2.5. N = 354.

V.5.A. = Skills indicator 5.A. = stimulating academic development; V.5.B. = Skills indicator 5.B. = stimulating the effective use of different cognitive styles; V.5.C. = Skills indicator 5.C. = stimulating interpersonal functioning; V.5.D. = Skills indicator 5.D. = stimulating intrapersonal understanding; V.5.E. = Skills indicator 5.E. = stimulating self-actualisation; and V.5.F. = Skills indicator 5.F. = general knowledge and skills matching those of the (S)ENCo.

Table I24

Indicator Category with Highest Mean Selection Rate in Domain 5 per Group of Stakeholders

Indicator	Educational Professionals	Specialists in Gifted Education	Parents	Rest
Knowledge	E	E	D	E
Skills	D	F	D	D

Note. D = stimulating intrapersonal understanding; E = stimulating self-actualisation; F = general knowledge and skills matching those of the (S)ENCo.

Table I25

The Comparison of Selection Rates Between Two Clusters of Knowledge Indicators in Domain 5: ADE-Indicators vs BCF-Indicators

Clusters of Indicators	Mean	N	*SD*	*SE*
Dom5_K_ADE	2.7731	354	0.46904	0.02493
Dom5_K_BCF	2.2269	354	0.46904	0.02493

Note: Paired Samples Statistics.

Dom5_K_ADE = Domain 5, cluster of three categories of knowledge indicators A, D, E; compared to Dom5_K_BCF = Domain 5, cluster of three categories of knowledge indicators B, C, F.

Table I26

The Comparison of Selection Rates Between Two Clusters of Knowledge Indicators in Domain 5: ADE-Indicators vs BCF-Indicators

Clusters of Indicators	Mean	*SD*	*SE*	95% Confidence interval of the difference		*t*	*df*	Sig (2-tailed)	Cohen's *d*
				Lower bound	Upper bound				
Dom5_K_ADE – Dom5_K_BCF	0.54614	0.93808	0.04986	0.44808	0.64420	10.954	353	.000	.58

Note: Paired Samples T test. N = 354.

Dom5_K_ADE = Domain 5, cluster of three categories of knowledge indicators A, D, E; compared to Dom5_K_BCF = Domain 5, cluster of three categories of knowledge indicators B, C, F.

Table I27

The Comparison of Selection Rates Between Two Clusters of Skills Indicators in Domain 5: ADE-Indicators vs BCF-Indicators

Clusters of Indicators	Mean	N	*SD*	*SE*
Dom5_V_ADE	2.5913	354	0.49927	0.02654
Dom5_V_BCF	2.4087	354	0.49927	0.02654

Note: Paired Samples Statistics.

Dom5_V_ADE = Domain 5, cluster of three categories of skills indicators A, D, E; compared to Dom5_V_BCF = Domain 5, cluster of three categories of skills indicators B, C, F.

Table I28

The Comparison of Selection Rates Between Two Clusters of Skills Indicators in Domain 5: ADE-Indicators vs BCF-Indicators

Clusters of Indicators	Mean	*SD*	*SE*	95% Confidence interval of the difference		*t*	*df*	Sig (2-tailed)	Cohen's *d*
				Lower bound	Upper bound				
Dom5_V_ADE – Dom5_V_BCF	0.18267	0.99853	0.05307	0.07830	0.28705	3.442	353	.001	.18

Note: Paired Samples T test. N = 354.

Dom5_V_ADE = Domain 5, cluster of three categories of skills indicators A, D, E; compared to Dom5_V_BCF = Domain 5, cluster of three categories of skills indicators B, C, F.

Table I29

Knowledge Indicators in Domain 5 With the Highest Selection Rates Amongst all Groups of Stakeholders

Competency	Code	Indicator
5.4 The specialist discusses the student's overall development with all the relevant actors in the student's ecological system.	K.5.A.4	Knows how the information from parents and co-workers can help to optimise a student's curriculum.

Table I30

Skills Indicators in Domain 5 With the Highest Selection Rates Amongst all Groups of Stakeholders

Competency	Code	Indicator
5.1 The specialist monitors the student's intrapersonal and interpersonal functioning to determine how this development contributes to the student's ability for self-actualisation.	V.5.E.1	Discusses with the student how didactical and pedagogical changes in their education might contribute to the student's development.

Table I31

Knowledge Indicators in Domain 5 With the Lowest Selection Rates Amongst all Groups of Stakeholders

Competency	Code	Indicator
5.1 The specialist monitors the student's intrapersonal and interpersonal functioning to determine how this development contributes to the student's self-actualisation.	K.5.B.1	Knows how they can develop an understanding of the student's cognitive style.
5.2 The specialist monitors the student's academic achievements and relates these achievements to the level of support given during the learning process.	K.5.C.2	Knows how they can evaluate a student's social functioning as an influencing factor on the student's learning process.

Table I32

Comparison of Mean Selection Rates in Domain 5 per Category of Knowledge Indicators Between Dutch and Flemish Stakeholders

Indicators	Country	N	Mean	*SD*	*SE*
K.5.A.	NL	299	2.71	1.042	0.060
	F	55	2.58	0.937	0.126
K.5.B.	NL	299	3.02	0.974	0.056
	F	55	2.51	1.215	0.164
K.5.C.	NL	299	2.07	1.087	0.063
	F	55	2.49	1.052	0.142
K.5.D.	NL	299	2.49	1.270	0.073
	F	55	2.87	1.292	0.174
K.5.E.	NL	299	3.10	1.030	0.060
	F	55	2.98	1.009	0.136
K.5.F.	NL	299	1.61	1.058	0.061
	F	55	1.56	1.067	0.144

Note. Descriptive Statistics. NL = The Netherlands. F = Flanders. N = 354.

K.5.A. = Knowledge indicator 5.A. = stimulating academic development; K.5.B. = Knowledge indicator 5B = stimulating the effective use of different cognitive styles; K.5.C. = Knowledge indicator 5.C. = stimulating interpersonal functioning; K.5.D. = Knowledge indicator 5.D. = stimulating intrapersonal understanding; K.5.E. = Knowledge indicator 5.E. = stimulating self-actualisation; and K.5.F. = Knowledge indicator 5.F. = general knowledge and skills matching those of the (S)ENCo.

Table I33

Comparison of Mean Selection Rates in Domain 5 per Category of Knowledge Indicators Between Dutch and Flemish Stakeholders

		Levene's test for Equality of variances		t-test for Equality of means							
									95% Confidence interval of the difference		
Indicators		*F*	*p*	*t*	*df*	Sig. (2-tailed)	Mean difference	*SE* difference	Lower bound	Upper bound	Cohen's *d*
K.5.A.	Equal variances assumed	0.720	0.397	0.844	352	.399	0.127	0.151	-0.169	0.423	.12
	Equal variances not assumed			0.909	80.635	.366	0.127	0.140	-0.151	0.406	
K.5.B.	Equal variances assumed	8.156	0.005	3.454	352	.001	0.514	0.149	0.221	0.807	.51
	Equal variances not assumed			2.968	67.352	.004	0.514	0.173	0.168	0.860	
K.5.C.	Equal variances assumed	0.369	0.544	-2.629	352	.009	-0.417	0.159	-0.730	-0.105	.39
	Equal variances not assumed			-2.690	76.782	.009	-0.417	0.155	-0.726	-0.108	
K.5.D.	Equal variances assumed	0.099	0.754	-2.040	352	.042	-0.381	0.187	-0.749	-0.014	.30
	Equal variances not assumed			-2.016	74.486	.047	-0.381	0.189	-0.758	-0.004	
K.5.E.	Equal variances assumed	0.170	0.681	0.764	352	.445	0.115	0.151	-0.181	0.411	.11
	Equal variances not assumed			0.775	76.179	.440	0.115	0.149	-0.181	0.411	
K.5.F.	Equal variances assumed	0.011	0.915	0.268	352	.789	0.042	0.155	-0.264	0.347	.04
	Equal variances not assumed			0.267	74.828	.790	0.042	0.158	-0.270	0.353	

Note. Independent Samples Test. N = 354.

K.5.A. = Knowledge indicator 5.A. = stimulating academic development; K.5.B. = Knowledge indicator 5.B. = stimulating the effective use of different cognitive styles; K.5.C. = Knowledge indicator 5.C. = stimulating interpersonal functioning; K.5.D. = Knowledge indicator 5.D. = stimulating intrapersonal understanding; K.5.E. = Knowledge indicator 5.E. = stimulating self-actualisation; and K.5.F. = Knowledge indicator 5.F. = general knowledge and skills matching those of the (S)ENCo.

Table I34

Comparison of Mean Selection Rates in Domain 5 per Category of Skills Indicators Between Dutch and Flemish Stakeholders

Indicators	Country	N	Mean	*SD*	*SE*
V.5.A.	NL	299	2.51	1.154	0.067
	F	55	2.45	1.317	0.178
V.5.B.	NL	299	2.22	1.093	0.063
	F	55	2.31	1.103	0.149
V.5.C.	NL	299	2.54	1.139	0.066
	F	55	2.55	1.068	0.144
V.5.D.	NL	299	2.81	1.184	0.068
	F	55	3.18	1.203	0.162
V.5.E.	NL	299	2.38	1.024	0.059
	F	55	2.51	1.200	0.162
V.5.F.	NL	299	2.53	1.130	0.065
	F	55	2.00	1.000	0.135

Note. Descriptive Statistics. NL = The Netherlands. F = Flanders. N = 354.

V.5.A. = Skills indicator 5.A. = stimulating academic development; V.5.B. = Skills indicator 5.B. = stimulating the effective use of different cognitive styles; V.5.C. = Skills indicator 5.C. = stimulating interpersonal functioning; V.5.D. = Skills indicator 5.D. = stimulating intrapersonal understanding; V.5.E. = Skills indicator 5.E. = stimulating self-actualisation; and V.5.F. = Skills indicator 5.F. = general knowledge and skills matching those of the (S)ENCo.

Table I35

Comparison of Mean Selection Rates in Domain 5 per Category of Skills Indicators Between Dutch and Flemish Stakeholders

		Levene's test for Equality of variances		t-test for Equality of means							
									95% Confidence interval of the difference		
Indicators		*F*	*p*	*t*	*df*	Sig. (2-tailed)	Mean difference	*SE* difference	Lower bound	Upper bound	Cohen's *d*
V.5.A.	Equal variances assumed	2.620	.106	0.311	352	.756	0.054	0.173	-0.287	0.394	.05
	Equal variances not assumed			0.284	70.076	.777	0.054	0.190	-0.324	0.432	
V.5.B.	Equal variances assumed	0.122	.728	-0.529	352	.597	-0.085	0.161	-0.401	0.231	.08
	Equal variances not assumed			-0.526	74.810	.601	-0.085	0.162	-0.407	0.237	
V.5.C.	Equal variances assumed	0.782	.377	-0.042	352	.966	-0.007	0.166	-0.333	0.319	.01
	Equal variances not assumed			-0.044	78.300	.965	-0.007	0.158	-0.322	0.308	
V.5.D.	Equal variances assumed	0.051	.822	-2.120	352	.035	-0.369	0.174	-0.712	-0.027	.31
	Equal variances not assumed			-2.096	74.508	.039	-0.369	0.176	-0.720	-0.018	
V.5.E.	Equal variances assumed	4.078	.044	-0.805	352	.421	-0.124	0.155	-0.428	0.179	.12
	Equal variances not assumed			-0.722	69.228	.472	-0.124	0.172	-0.468	0.219	
V.5.F.	Equal variances assumed	6.030	.015	3.262	352	.001	0.532	0.163	0.211	0.852	.48
	Equal variances not assumed			3.549	81.526	.001	0.532	0.150	0.234	0.830	

Note. Independent Samples Test. N = 354.

V.5.A. = Skills indicator 5.A. = stimulating academic development; V.5.B. = Skills indicator 5.B. = stimulating the effective use of different cognitive styles; V.5.C. = Skills indicator 5.C. = stimulating interpersonal functioning; V.5.D. = Skills indicator 5.D. = stimulating intrapersonal understanding; V.5.E. = Skills indicator 5.E. = stimulating self-actualisation; and V.5.F. = Skills indicator 5.F. = general knowledge and skills matching those of the (S)ENCo.

Table I36

Domain 5 Knowledge Indicators. Mean per Category per Competency per Group of Stakeholders

Indicators	EP	Indicators	SGE	Indicators	P	Indicators	R
K.5.F.	1.53	K.5.F.	1.94	K.5.F.	1.28	K.5.F.	1.62
K.5.C.	2.36	K.5.C.	2.08	K.5.C.	1.99	K.5.C.	1.86
K.5.D.	2.53	K.5.D.	2.08	K.5.A.	2.53	K.5.D.	2.65
K.5.A.	2.81	K.5.A.	2.64	K.5.B.	2.99	K.5.A.	2.76
K.5.B.	2.88	K.5.B.	3.00	K.5.E.	3.10	K.5.B.	2.92
K.5.E.	2.89	K.5.E.	3.26	K.5.D.	3.12	K.5.E.	3.19

Note. Test value 2.5. EP = educational professionals; SGE = specialist gifted education; Pa = parents; R = rest. N = 354.

K.5.A. = Knowledge indicator 5.A. = stimulating academic development; K.5.B. = Knowledge indicator 5.B. = stimulating the effective use of different cognitive styles; K.5.C. = Knowledge indicator 5.C. = stimulating interpersonal functioning; K.5.D. = Knowledge indicator 5.D. = stimulating intrapersonal understanding; K.5.E. = Knowledge indicator 5.E. = stimulating self-actualisation; and K.5.F. = Knowledge indicator 5.F. = general knowledge and skills matching those of the (S)ENCo.

Table I37

Domain 5 Skills Indicators. Mean per Category per Competency per Group of Stakeholders

Indicators	EP	Indicators	SGE	Indicators	P	Indicators	R
V.5.F.	2.29	V5B	2.05	V5F	2.13	V5B	2.16
V.5.B.	2.37	V5E	2.36	V5B	2.29	V5A	2.19
V.5.E.	2.39	V5A	2.53	V5E	2.45	V5E	2.49
V.5.A.	2.47	V5D	2.53	V5C	2.47	V5C	2.57
V.5.C.	2.47	V5C	2.68	V5A	2.64	V5F	2.62
V.5.D.	3.01	V5F	2.84	V5D	3.02	V5D	2.97

Note. Test value 2.5. EP = educational professionals; SGE = specialist gifted education; Pa = parents; R = rest. N = 354.

V.5.A. = Skills indicator 5.A. = stimulating academic development; V.5.B. = Skills indicator 5.B. = stimulating the effective use of different cognitive styles; V.5.C. = Skills indicator 5.C. = stimulating interpersonal functioning; V.5.D. = Skills indicator 5.D. = stimulating intrapersonal understanding; V.5.E. = Skills indicator 5.E. = stimulating self-actualisation; and V.5.F. = Skills indicator 5.F. = general knowledge and skills matching those of the (S)ENCo.

Table I38

Ranking of the Knowledge and Skills Indicators of Domain 5

Indicators	N	%	Indicators	N	%
K.5.B.1	91	0.86	V.5.A.1	183	1.72
K.5.C.2	99	0.93	K.5.E.4	184	1.73
K.5.F.4	100	0.94	K.5.E.2	186	1.75
V.5.C.1	104	0.98	K.5.C.4	187	1.76
K.5.B.4	105	0.99	K.5.F.1	189	1.78
V.5.B.4	124	1.17	V.5.E.4	191	1.80
K.5.E.3	124	1.17	K.5.D.5	192	1.81
V.5.F.1	125	1.18	K.5.E.5	192	1.81
K.5.B.3	127	1.20	V.5.D.5	197	1.85
K.5.F.5	129	1.21	K.5.C.3	201	1.89
V.5.B.2	130	1.22	K.5.A.2	202	1.90
V.5.F.2	132	1.24	K.5.D.4	203	1.91
V.5.A.3	136	1.28	V.5.D.3	207	1.95
V.5.C.2	136	1.28	V.5.D.4	207	1.95
V.5.B.3	141	1.33	V.5.F.3	210	1.98
K.5.B.2	143	1.35	K.5.D.2	212	2.00
V.5.F.4	143	1.35	K.5.A.5	214	2.02
K.5.C.1	145	1.37	V.5.F.5	214	2.02
V.5.E.5	145	1.37	V.5.E.3	218	2.05
V.5.A.5	148	1.39	K.5.F.2	220	2.07
V.5.C.3	150	1.41	K.5.D.1	223	2.10
V.5.A.2	152	1.43	K.5.F.3	224	2.11
K.5.D.3	154	1.45	K.5.A.3	232	2.18
K.5.B.5	158	1.49	V.5.D.1	233	2.19
V.5.B.1	160	1.51	K.5.E.1	235	2.21
V.5.C.4	161	1.52	V.5.A.4	236	2.22
K.5.C.5	177	1.67	V.5.E.2	243	2.29
V.5.C.5	177	1.67	V.5.E.1	257	2.42
K.5.A.1	179	1.69	V.5.D.2	269	2.53
V.5.B.5	181	1.70	K.5.A.4	283	2.66

Note. This ranking is based on selection rates of the entire group of stakeholders. N =354.

All indicators are coded. K = knowledge; V = skills; the first number refers to the vertical competency domain in the competency matrix; the second letter refers to the category of indicators. The second number refers to the actual competency the indicators are matched with. See Appendix B for the match between the competencies and the indicators.

Appendix J

Tables Data Analyses at Level 3. Referred to in Chapter 10

Table J1

Percentage of Respondents who Selected a Specific Indicator

Indicator	R	% R	Indicator	R	% R	Indicator	R	% R	Indicator	R	% R	Indicator	R	% R	Indicator	R	%R
V.1.F.4	4	1.13	K.5.E.3	124	35.03	V.5.A.2	152	42.94	V.5.B.5	181	51.13	K.5.A.2	202	57.06	K.1.F.4	228	64.41
V.2.C.2	43	12.15	V.4.C.3	125	35.31	V.1.B.3	152	42.94	K.3.C.4	181	51.13	K.2.A.1	203	57.34	V.2.B.5	229	64.69
V.3.C.2	51	14.41	V.3.D.5	125	35.31	K.2.E.5	153	43.22	V.4.C.4	181	51.13	K.1.B.2	203	57.34	V.4.A.1	230	64.97
K.1.B.4	56	15.82	V.3.F.1	125	35.31	V.1.B.2	154	43.50	V.2.D.3	181	51.13	K.4.B.2	203	57.34	K.3.B.1	231	65.25
V.4.C.1	63	17.80	V.5.F.1	125	35.31	K.5.D.3	154	43.50	V.5.A.1	183	51.69	K.5.D.4	203	57.34	V.2.E.2	231	65.25
K.1.F.2	68	19.21	K.5.B.3	127	35.88	V.2.E.5	154	43.50	V.1.C.3	183	51.69	K.1.E.3	203	57.34	K.5.A.3	232	65.54
V.1.C.1	72	20.34	V.2.A.4	129	36.44	K.2.C.2	155	43.79	V.4.F.3	183	51.69	V.1.E.4	203	57.34	V.3.B.4	232	65.54
K.3.C.2	78	22.03	K.5.F.5	129	36.44	V.2.E.1	156	44.07	K.1.E.2	184	51.98	K.3.E.1	205	57.91	V.5.D.1	233	65.82
V.3.B.5	81	22.88	V.1.A.4	130	36.72	K.4.B.3	158	44.63	K.5.E.4	184	51.98	V.1.E.3	206	58.19	K.5.E.1	235	66.38
V.3.C.4	81	22.88	V.5.B.2	130	36.72	K.5.B.5	158	44.63	K.2.A.5	185	52.26	V.5.D.3	207	58.47	V.1.F.5	235	66.38
K.2.F.4	81	22.88	V.2.F.4	131	37.01	K.2.A.3	159	44.92	K.4.C.5	185	52.26	V.5.D.4	207	58.47	V.5.A.4	236	66.67
K.1.A.1	84	23.73	V.5.F.2	132	37.29	K.1.D.3	159	44.92	K.1.D.5	185	52.26	V.4.F.1	208	58.76	V.3.E.5	236	66.67
K.4.B.5	85	24.01	K.2.D.3	133	37.57	V.4.F.4	159	44.92	K.5.E.2	186	52.54	K.4.A.5	210	59.32	V.4.E.1	236	66.67
K.5.B.1	91	25.71	V.1.B.5	134	37.85	V.2.A.3	160	45.20	K.5.C.4	187	52.82	K.2.B.4	210	59.32	V.1.A.1	237	66.95
K.3.B.3	94	26.55	K.4.C.3	134	37.85	V.2.B.4	160	45.20	K.2.F.3	187	52.82	V.5.F.3	210	59.32	K.2.A.4	238	67.23
K.3.C.1	95	26.84	K.4.D.5	135	38.14	V.5.B.1	160	45.20	K.1.A.5	188	53.11	V.2.C.1	211	59.60	V.2.A.1	238	67.23
V.2.F.5	96	27.12	V.5.A.3	136	38.42	V.5.C.4	161	45.48	K.4.F.3	188	53.11	V.1.D.3	211	59.60	V.2.E.4	238	67.23
V.1.F.2	97	27.40	V.3.C.5	136	38.42	V.4.E.5	161	45.48	K.5.F.1	189	53.39	V.3.A.1	212	59.89	K.1.E.1	239	67.51
K.4.C.2	99	27.97	V.5.C.2	136	38.42	K.3.B.4	162	45.76	V.4.C.5	190	53.67	K.3.C.3	212	59.89	V.3.E.2	239	67.51
K.5.C.2	99	27.97	V.4.A.2	137	38.70	V.4.B.5	164	46.33	K.2.D.2	190	53.67	V.1.D.5	212	59.89	V.4.A.3	242	68.36
V.4.D.2	99	27.97	K.2.F.1	139	39.27	V.2.A.5	165	46.61	K.3.E.2	190	53.67	K.5.D.2	212	59.89	K.4.E.2	242	68.36
K.5.F.4	100	28.25	K.1.C.4	140	39.55	K.2.B.5	165	46.61	K.1.B.1	191	53.95	V.1.B.1	213	60.17	V.5.E.2	243	68.64
V.3.F.3	101	28.53	K.4.C.1	140	39.55	V.3.E.3	165	46.61	V.1.C.5	191	53.95	K.5.A.5	214	60.45	K.2.E.2	244	68.93
K.1.C.3	102	28.81	V.1.D.1	140	39.55	K.4.E.3	165	46.61	V.4.D.4	191	53.95	V.4.B.2	214	60.45	K.4.F.1	244	68.93

Indicator	R	% R	Indicator	R	% R	Indicator	R	% R	Indicator	R	% R	Indicator	R	% R	Indicator	R	%R
K.1.C.2	103	29.10	V.2.F.1	140	39.55	K.3.F.4	165	46.61	V.4.E.3	191	53.95	K.3.D.1	214	60.45	K.4.A.3	245	69.21
V.5.C.1	104	29.38	V.1.A.3	141	39.83	K.1.D.4	167	47.18	V.5.E.4	191	53.95	V.5.F.5	214	60.45	V.1.E.1	249	70.34
V.4.D.1	104	29.38	K.3.A.1	141	39.83	K.1.E.5	167	47.18	K.2.D.5	192	54.24	K.2.C.4	215	60.73	V.3.B.3	252	71.19
K.5.B.4	105	29.66	V.5.B.3	141	39.83	K.2.C.1	168	47.46	V.2.D.5	192	54.24	K.3.E.4	215	60.73	K.1.B.3	255	72.03
K.2.B.2	106	29.94	K.2.A.2	142	40.11	K.3.D.3	168	47.46	K.5.D.5	192	54.24	V.3.A.5	216	61.02	V.1.A.2	256	72.32
K.4.D.4	106	29.94	K.4.A.4	142	40.11	V.1.F.3	169	47.74	K.5.E.5	192	54.24	K.2.E.3	216	61.02	V.5.E.1	257	72.60
K.3.A.5	109	30.79	K.3.F.3	142	40.11	K.2.C.3	170	48.02	K.4.F.4	192	54.24	K.3.A.4	217	61.30	K.3.F.5	257	72.60
K.4.B.1	109	30.79	V.1.A.5	143	40.40	V.2.A.2	171	48.31	K.1.A.3	193	54.52	V.3.A.2	217	61.30	V.3.F.4	257	72.60
K.3.A.2	112	31.64	V.4.B.4	143	40.40	V.4.A.4	171	48.31	V.2.C.3	194	54.80	V.4.E.4	217	61.30	V.3.E.4	258	72.88
V.3.A.4	114	32.20	K.5.B.2	143	40.40	V.1.D.2	171	48.31	V.1.D.4	196	55.37	V.5.E.3	218	61.58	K.1.B.5	259	73.16
V.2.D.1	116	32.77	K.2.F.5	143	40.40	V.3.B.1	172	48.59	K.2.D.4	196	55.37	K.3.A.3	220	62.15	V.3.E.1	261	73.73
K.1.C.5	118	33.33	V.5.F.4	143	40.40	K.4.D.3	172	48.59	K.2.B.3	197	55.65	K.4.C.4	220	62.15	V.1.B.4	264	74.58
V.1.C.2	119	33.62	K.2.B.1	145	40.96	K.3.B.2	173	48.87	V.3.D.3	197	55.65	K.3.F.2	220	62.15	V.1.C.4	265	74.86
V.3.D.1	119	33.62	V.3.C.3	145	40.96	V.3.C.1	173	48.87	K.4.D.1	197	55.65	K.5.F.2	220	62.15	V.1.E.2	265	74.86
K.4.B.4	120	33.90	K.5.C.1	145	40.96	V.4.C.2	173	48.87	K.4.D.2	197	55.65	V.4.B.1	221	62.43	V.4.E.2	266	75.14
V.3.D.4	120	33.90	V.5.E.5	145	40.96	V.4.F.2	173	48.87	V.5.D.5	197	55.65	K.4.E.5	222	62.71	V.3.F.5	268	75.71
V.4.B.3	122	34.46	K.1.F.5	145	40.96	K.4.A.1	174	49.15	K.1.A.2	198	55.93	K.5.D.1	223	62.99	V.5.D.2	269	75.99
K.3.C.5	122	34.46	K.4.F.2	145	40.96	K.4.A.2	176	49.72	K.4.E.1	198	55.93	K.2.C.5	224	63.28	K.1.A.4	272	76.84
K.3.D.4	122	34.46	V.4.D.5	147	41.53	K.3.F.1	176	49.72	V.4.D.3	199	56.21	V.2.E.3	224	63.28	K.1.D.1	276	77.97
K.2.E.1	122	34.46	V.1.E.5	147	41.53	V.3.B.2	177	50.00	K.1.E.4	199	56.21	K.3.E.5	224	63.28	V.2.D.4	280	79.10
K.2.E.4	122	34.46	V.5.A.5	148	41.81	K.5.C.5	177	50.00	V.4.F.5	199	56.21	K.5.F.3	224	63.28	K.4.E.4	282	79.66
V.2.F.3	122	34.46	K.1.F.1	148	41.81	V.5.C.5	177	50.00	V.4.A.5	201	56.78	K.2.F.2	225	63.56	K.5.A.4	283	79.94
K.3.B.5	124	35.03	V.5.C.3	150	42.37	V.3.F.2	177	50.00	V.2.B.1	201	56.78	K.4.F.5	225	63.56	K.2.D.1	285	80.51
V.5.B.4	124	35.03	K.1.F.3	150	42.37	V.2.B.2	178	50.28	K.5.C.3	201	56.78	V.2.C.5	226	63.84	V.2.F.2	288	81.36
K.1.C.1	124	35.03	V.2.D.2	151	42.66	K.5.A.1	179	50.56	V.3.D.2	201	56.78	K.3.D.5	226	63.84	K.3.D.2	289	81.64
V.2.C.4	124	35.03	V.1.F.1	151	42.66	V.2.B.3	181	51.13	V.3.A.3	202	57.06	K.3.E.3	226	63.84	K.1.D.2	306	86.44

Note. N 354. R = the number of respondents that selected an indicator; %R = percentage of respondents that selected an indicator.

Table J2

Indicators with Overall Lowest Selection Rates

N	%	Indicator	Description
4	1.13	V.1.F.4	Creates a joint perspective on giftedness between the educational partners.
43	12.15	V.2.C.2	Explores the influences of classmates on the gifted student's achievements.
51	14.41	V.3.C.2	Determines how peers within the social environment respond to the differentiated curriculum offered to the gifted student.
56	15.82	K.1.B.4	Knows what the theoretical perspectives are on the learning process of gifted students.
63	17.80	V.4.C.1	Encourages gifted students to collaborate with other gifted students.
68	19.21	K.1.F.2	Knows that a different perspective on giftedness also means a different focus on the ultimate educational objectives.
72	20.34	V.1.C.1	Stimulates gifted students to interact with gifted peers
78	22.03	K.3.C.2	Knows why knowledge has to have a social meaning for a student.
81	22.88	V.3.B.5	Selects intervention strategies, based upon their theoretical perspective of giftedness, that are helpful in understanding the student's educational needs.
81	22.88	V.3.C.4	Determines if the gifted student has ample opportunity to meet and work with like-minded peers.
81	22.88	K.2.F.4	Understands why it is important to explain the objectives of the assessment of educational needs to the educational partners.
84	23.73	K.1.A.1	Knows that the interaction between aptitude and development influences a student's achievements.
85	24.01	K.4.B.5	Is aware of modern didactical strategies.
91	25.71	K.5.B.1	Knows how they can develop an understanding of the student's cognitive style.
94	26.55	K.3.B.3	Knows how factors in the physical educational environment influence the development of executive skills.
95	26.84	K.3.C.1	Understands how, in the change-oriented approach, the process of data interpretation takes place.
96	27.12	V.2.F.5	Develops a personalised assessment plan prior to the start of an assessment.
97	27.40	V.1.F.2	Sets educational objectives that are fit for the school's perspective of giftedness.

Table J3

Indicators with Overall Highest Selection Rates

N	%	Indicator	Description
256	72.32	V.1.A.2	Sets educational objectives that match the individual needs of gifted students.
257	72.60	V.5.E.1	Discusses with the student how didactical and pedagogical changes in their education might contribute to the student's development.
257	72.60	K.3.F.5	Understands how strengths and weaknesses can be translated into educational needs.
257	72.60	V.3.F.4	Communicates regularly with educational partners about the student's educational development.
258	72.88	V.3.E.4	Provides the student with a level of responsibility that matches their developmental level.
259	73.16	K.1.B.5	Invests in personal continuous professional development regarding theories of giftedness
261	73.73	V.3.E.1	Discusses with the student what they see as helpful learning strategies and interventions.
264	74.58	V.1.B.4	Discusses with the educational partners how they can contribute positively to the gifted student's developmental process.
265	74.86	V.1.C.4	Encourages open communication between all educational partners involved.
265	74.86	V.1.E.2	Selects interventions that are a match with the student's needs.
266	75.14	V.4.E.2	Encourages creative and critical thinking by developing an investigative attitude in students.
268	75.71	V.3.F.5	Describes a gifted student's educational needs as skills to be developed in the near future.
269	75.99	V.5.D.2	Maintains an intrinsic interest in their students and shows this in the way that they communicate with them.
272	76.84	K.1.A.4	Knows why it is important to organise gifted education from the joint perspective of the educational partners.
276	77.97	K.1.D.1	Knows which risk factors might influence a student's wellbeing.
280	79.10	V.2.D.4	Is focused on solutions and learnable skills instead of being focused on problems when talking with a student.
282	79.66	K.4.E.4	Knows how a feeling of competence, autonomy, and relationship regarding the educational objectives contributes positively to a student's motivation.
283	79.94	K.5.A.4	Knows how the information from parents and co-workers can help to optimise a student's curriculum.
285	80.51	K.2.D.1	Knows what types of behaviour by the gifted student may be a sign for the development of a negative self-concept.
288	81.36	V.2.F.2	Takes other indicators besides school achievements into consideration when estimating if the student may be gifted.
289	81.64	K.3.D.2	Knows how mindset influences a student's learning process.
306	86.44	K.1.D.2	Knows different didactical strategies to challenge gifted students in their zone of proximal development.

Table J4

Mean Selection Rates of Knowledge Indicators at General Level

Knowledge indicator	N	Mean	*SD*	*SE*
A	354	14.73	4.360	0.232
B	354	13.45	4.204	0.223
C	354	8.65	4.717	0.251
D	354	13.13	5.820	0.309
E	354	14.84	4.406	0.234
F	354	10.19	4.906	0.261

Note. One-Sample Statistics. Test value = 12.5.

A = stimulating academic development; B = stimulating the effective use of different cognitive styles; C = stimulating interpersonal functioning; D = stimulating intrapersonal understanding; E = stimulating self-actualisation; F = general knowledge and skills matching those of the (S)ENCo.

Table J5

Mean Selection Rates of Knowledge Indicators at General Level

					95% Confidence interval of the difference		
Knowledge indicator	*t*	*df*	Sig (2-tailed)	Mean difference	Lower bound	Upper bound	Cohen's *d*
A	9.631	353	.000	2.232	1.78	2.69	.51
B	4.274	353	.000	0.955	0.52	1.39	.23
C	-15.357	353	.000	-3.850	-4.34	-3.36	.82
D	2.045	353	.042	0.633	0.02	1.24	.10
E	9.989	353	.000	2.339	1.88	2.80	.53
F	-8.852	353	.000	2.308	-2.82	-1.80	.47

Note. One-Sample T test. Test value 12.5. N = 354.

A = stimulating academic development; B = stimulating the effective use of different cognitive styles; C = stimulating interpersonal functioning; D = stimulating intrapersonal understanding; E = stimulating self-actualisation; F = general knowledge and skills matching those of the (S)ENCo.

Table J6

Mean Selection Rates of Skills Indicators at General Level

Skills indicator	N	Mean	*SD*	*SE*
A	354	11.87	4.902	0.261
B	354	11.88	4.462	0.237
C	354	12.90	4.669	0.248
D	354	14.49	5.183	0.275
E	354	13.33	4.381	0.233
F	354	10.53	4.559	0.242

Note. One-Sample Statistics. Test value = 12.5.

A = stimulating academic development; B = stimulating the effective use of different cognitive styles; C = stimulating interpersonal functioning; D = stimulating intrapersonal understanding; E = stimulating self-actualisation; F = general knowledge and skills matching those of the (S)ENCo.

Table J7

Mean Selection Rates of Skills Indicators at General Level

					95% Confidence interval of the difference		
Skills indicator	*t*	*df*	Sig (2-tailed)	Mean difference	Lower	Upper	Cohen's *d*
A	-2.429	353	.016	-0.633	-1.15	-0.12	.12
B	-2.621	353	.009	-0.621	-1.09	-0.16	.14
C	1.605	353	.109	0.398	-0.09	0.89	.08
D	7.239	353	.000	1.994	1.45	2.54	.38
E	3.555	353	.000	0.828	0.37	1.29	.19
F	-8.115	353	.000	-1.966	-2.44	-1.49	.43

Note. One-Sample T test. Test Value 12.5. N = 354.

A = stimulating academic development; B = stimulating the effective use of different cognitive styles; C = stimulating interpersonal functioning; D = stimulating intrapersonal understanding; E = stimulating self-actualisation; F = general knowledge and skills matching those of the (S)ENCo.

Table J8

Comparison Selection Rates of Ownership & Voicing, Educational Partnership and Peer Group Education with REST Indicators

Groups		Mean	N	*SD*	*SE*
Pair 1	PERC_Ownership	54.6429	354	12.34298	0.65602
	PERC_REST	49.6873	354	2.04567	0.10873
Pair 2	PERC_Educational partnership	50.2119	354	14.10746	0.74980
	PERC_REST	49.6873	354	2.04567	0.10873
Pair 3	PERC_Peer group education	37.8531	354	21.86544	1.16213
	PERC_REST	49.6873	354	2.04567	0.10873

Note. Paired Samples Statistics.

PERC_Ownership: percentage of stakeholders selecting indicators referring to ownership and voicing; PERC_REST: percentage of stakeholders selecting indicators *not* referring to ownership and voicing, educational partnership or peer group education; PERC_Educational Partnership: percentage of stakeholders selecting items referring to educational partnership; PERC_Peer group education: percentage of stakeholders selecting items referring to Peer group education.

Table J9

Comparison of Selection Rates for Knowledge Indicators Between the Netherlands & Flanders

Knowledge indicator	Country	N	Mean	*SD*	*SE*
A	NL	299	14.91	4.376	0.253
	F	55	13.76	4.176	0.563
B	NL	299	13.93	3.950	0.228
	F	55	10.89	4.637	0.625
C	NL	299	8.31	4.636	0.625
	F	55	10.49	4.772	0.643
D	NL	299	12.85	5.756	0.333
	F	55	14.65	5.988	0.807
E	NL	299	14.91	4.351	0.252
	F	55	14.47	4.717	0.636
F	NL	299	10.09	4.902	0.284
	F	55	10.73	4.935	0.665

Note. Descriptive Statistics. NL = The Netherlands. F = Flanders. N = 354.

A = stimulating academic development; B = stimulating the effective use of different cognitive styles; C = stimulating interpersonal functioning; D = stimulating intrapersonal understanding; E = stimulating self-actualisation; and F = general knowledge and skills matching those of the (S)ENCo.

Table J10

Comparison of Selection Rates for Knowledge Indicators Between the Netherlands & Flanders

		Levene's test for Equality of variances		t-test for Equality of means							
									95% Confidence interval of the difference		
Knowledge Indicators		*F*	*p*	*t*	*df*	Sig (2-tailed)	Mean difference	*SE* difference	Lower bound	Upper bound	Cohen's *d*
A	Equal variances assumed	0.603	.438	1.797	352	.073	1.146	0.683	-0.108	2.400	.26
	Equal variances not assumed			1.856	77.440	.067	1.146	0.617	-0.083	2.375	
B	Equal variances assumed	2.981	.085	5.093	352	.000	3.036	0.596	1.863	4.208	.75
	Equal variances not assumed			4.560	69.147	.000	3.036	0.666	1.707	4.364	
C	Equal variances assumed	0.179	.673	-3.190	352	.002	-2.180	0.683	-3.524	-0.836	.47
	Equal variances not assumed			-3.127	73.970	.003	-2.180	0.697	-3.569	-0.791	
D	Equal variances assumed	0.288	.592	-2.120	352	.035	-1.802	0.850	-3.473	-0.130	.31
	Equal variances not assumed			-2.063	73.526	.043	-1.802	0.873	-3.542	-0.130	
E	Equal variances assumed	0.190	.663	0.670	352	.503	0.434	0.647	-0.839	1.706	.10
	Equal variances not assumed			0.634	71.903	.528	0.434	0.684	-0.930	1.797	
F	Equal variances assumed	0.038	.845	-0.880	352	.379	-0.634	0.720	-2.050	0.782	.13
	Equal variances not assumed			-0.876	74.938	.384	-0.634	0.723	-2.074	0.807	

Note. Independent Samples Test. N = 354.

A = stimulating academic development; B = stimulating the effective use of different cognitive styles; C = stimulating interpersonal functioning; D = stimulating intrapersonal understanding; E = stimulating self-actualisation; F = general knowledge and skills matching those of the (S)ENCo.

Table J11

Comparison of Selection Rates for Skills Indicators Between the Netherlands & Flanders

Skills Indicators	Country	N	Mean	*SD*	*SE*
A	NL	299	11.98	4.778	0.276
	F	55	11.25	5.535	0.746
B	NL	299	11.83	4.376	0.253
	F	55	12.13	4.941	0.666
C	NL	299	12.80	4.700	0.272
	F	55	13.44	4.504	0.607
D	NL	299	14.22	5.197	0.301
	F	55	15.96	4.895	0.660
E	NL	299	13.28	4.242	0.245
	F	55	13.56	5.105	0.688
F	NL	299	10.88	4.536	0.262
	F	55	8.65	4.248	0.573

Note. Descriptive Statistics. NL = The Netherlands. F = Flanders. N = 354.

A = stimulating academic development; B = stimulating the effective use of different cognitive styles; C = stimulating interpersonal functioning; D = stimulating intrapersonal understanding; E = stimulating self-actualisation; F = general knowledge and skills matching those of the (S)ENCo.

Table J12

Comparison of Selection Rates for Skills Indicators Between the Netherlands & Flanders

		Levene's test for Equality of variances		t-test for Equality of means							
									95% Confidence interval of the difference		
Skills Indicators		*F*	*p*	*t*	*df*	Sig (2-tailed)	Mean difference	*SE* difference	Lower bound	Upper bound	Cohen's *d*
A	Equal variances assumed	3.178	.075	1.009	352	.314	0.725	0.719	-0.696	2.140	.15
	Equal variances not assumed			0.911	69.582	.365	0.725	0.796	-0.862	2.313	
B	Equal variances assumed	1.531	.217	-0.449	352	.353	-0.294	0.655	-1.583	0.994	.07
	Equal variances not assumed			-0.413	70.441	.681	-0.294	0.713	-1.716	1.127	
C	Equal variances assumed	0.599	.440	-0.930	352	.353	-0.637	0.685	-1.985	0.711	.14
	Equal variances not assumed			-0.957	77.233	.341	-0.637	0.665	-1.962	0.688	
D	Equal variances assumed	0.490	.484	-2.301	352	.022	-1.740	0.756	-3.226	-0.253	.34
	Equal variances not assumed			-2.399	78.110	.019	-1.740	0.725	-3.183	-0.296	
E	Equal variances assumed	4.172	.042	-0.434	352	.664	-0.279	0.643	-1.545	0.986	.06
	Equal variances not assumed			-0.382	68.387	.703	-0.279	0.731	-1.738	1.179	
F	Equal variances assumed	0.433	.511	3.375	352	.001	2.225	0.659	0.929	3.522	.50
	Equal variances not assumed			3.532	78.403	.001	2.225	0.630	0.971	3.479	

Note. Independent Samples Test. N = 354.

A = stimulating academic development; B = stimulating the effective use of different cognitive styles; C = stimulating interpersonal functioning; D = stimulating intrapersonal understanding; E = stimulating self-actualisation; F = general knowledge and skills matching those of the (S)ENCo.

Table J13

Comparison of Selection Rates for Categories Ownership & Voicing, Educational Partnership, and Peer Group Education with REST Indicators

Pairs	Mean	*SD*	*SE*	95% Confidence interval of the difference		*t*	*df*	Sig (2-tailed)	Cohen's *d*
				Lower bound	Upper bound				
Pair 1 Perc_Ownership – Perc_REST	4.95561	13.19478	0.70129	3.57637	6.33486	7.066	353	.000	.37
Pair 2 Perc_Educational partnership – Perc_REST	0.52457	15.12875	0.80408	-1.05683	2.10597	0.652	353	.515	.03
Pair 3 Perc_Peer group education – Perc_REST	-11.83419	22.38939	1.18998	-14.17453	-9.49384	-9.945	353	.000	.53

Note. Paired Samples T test. N = 354.

PERC_Ownership: percentage of stakeholders selecting indicators referring to ownership and voicing; PERC_REST: percentage of stakeholders selecting indicators *not* referring to ownership and voicing, educational partnership or peer group education; PERC_Educational partnership: percentage of stakeholders selecting items referring to educational partnership; PERC_Peer group education: percentage of stakeholders selecting items referring to peer group education.

Table J14

Comparison of Selection Rates for Categories Ownership & Voicing, Educational Partnership, Peer Group Education, and REST Indicators

		Sum of squares	*df*	Mean square	*F*	*p*	η^2
PERC_ Ownership	Between Groups	1543.693	3	514.564	3.448	.017	.00
	Within Groups	52235.526	350	149.244			
	Total	533779.219	353				
PERC_Educational partnership	Between Groups	4408.225	3	1469.408	7.811	.000	.06
	Within Groups	65845.947	350	188.131			
	Total	70254.172	353				
PERC_Peer group education	Between Groups	1600.985	3	533.662	1.117	.342	.00
	Within Groups	167167.377	350	477.621			
	Total	168768.362	353				
PERC_REST	Between Groups	37.060	3	12.353	3.002	.031	.02
	Within Groups	1440.169	350	4.115			
	Total	1477.228	353				

Note. ANOVA = analyses of variance with one factor. N= 354.

PERC_Ownership: percentage of stakeholders selecting indicators referring to ownership and voicing; PERC_REST: percentage of stakeholders selecting indicators *not* referring to ownership and voicing, educational partnership or peer group education; PERC_Educational partnership: percentage of stakeholders selecting items referring to educational partnership; PERC_Peer group education: percentage of stakeholders selecting items referring to peer group education.

Table J15

Comparison of Selection Rates for Categories Ownership & Voicing, Educational Partnership, and REST Indicators

Dependent variable	I group	J group	Mean difference (I-J)	*SE*	*p*	95% Confidence interval	
						Lower bound	Upper bound
PERC_ Ownership	SGE	EP	-2.81894	1.60880	.484	-7.0876	1.4498
		P	-5.56910	1.80197	.013	-10.3503	-0.7879
		NOS	-4.44193	2.34150	.352	-10.6547	1.7709
PERC_ Educational partnership	SGE	EP	-7.51238	1.80628	.000	2.8197	12.4050
		P	7.61238	2.02316	.000	3.0656	13.8019
		NOS	8.43373	2.62891	.278	-1.7201	12.2306
PERC_REST	EP	SGE	0.75467	0.26713	.030	0.0459	1.4635
		P	0.55995	0.28458	.299	-0.1951	1.3150
		NOS	0.56904	0.37765	.797	-0.4330	1.5711

Note. Based on the results of the ANOVA, Bonferroni's Multiple Comparisons was used as a post hoc test for the analyses of the differences between groups of stakeholders. N = 354.

PERC_Ownership: percentage of stakeholders selecting indicators referring to ownership and voicing; PERC_REST: percentage of stakeholders selecting indicators *not* referring to ownership and voicing, educational partnership or peer group education; PERC_Educational partnership: percentage of stakeholders selecting items referring to educational partnership.

Table J16

Comparison of Selection Rates for Categories Ownership & Voicing, Educational Partnership, Peer Group Education, and REST Indicators Between Dutch and Flemish Stakeholders

Pairs	Country	N	Mean	*SD*	*SE*	Cohen's *d*
PERC_Ownership	NL	299	54.8323	12.20405	0.70578	.10
	F	55	53.6131	13.14196	1.77206	
PERC_Educational partnership	NL	299	50.8082	14.18010	0.82006	.27
	F	55	46.9697	13.36751	1.80247	
PERC_Peer group education	NL	299	36.5552	21.21730	1.22703	.39
	F	55	44.9091	24.10415	3.25020	
PERC_REST	NL	299	49.6134	1.95748	0.11320	.23
	F	55	50.0888	2.45280	0.33074	

Note. Descriptive Statistics. NL = The Netherlands. F = Flanders. N=354.

PERC_Ownership: percentage of stakeholders selecting indicators referring to ownership and voicing; PERC_REST: percentage of stakeholders selecting indicators *not* referring to ownership and voicing, educational partnership or peer group education; PERC_Educational partnership: percentage of stakeholders selecting items referring to educational partnership; PERC_Peer group education: percentage of stakeholders selecting items referring to peer group education.

Table J17

Comparison of Selection Rates Indicators Regarding Peer Group Education (PGE) Between Dutch and Flemish Stakeholders

		Levene's test for Equality of variances		t-test for Equality of means						
									95% Confidence interval of the difference	
		F	Sig	*t*	*df*	Sig. (2-tailed)	Mean difference	*SE* difference	Lower bound	Upper bound
PERC_PGE	Equal variances assumed	2.837	.093	-2.626	352	.009	-8.35391	3.18161	-14.61126	-2.09655
	Equal variances not assumed			-2.405	70.231	0.19	-8.35391	3.47411	-15.28239	-1.42542

Note. N = 354.

Table J18

Comparison of the Selection Rates of Knowledge Indicators at a General Level Between the Groups of Stakeholders

Knowledge Indicators	Group	N	Mean	*SD*	*SE*	95% Confidence interval for mean Lower bound	Upper bound	Min	Max
A	EP	131	15.05	4.419	0.386	14.29	15.82	4	25
	SGE	103	14.92	4.633	0.457	14.02	15.83	3	25
	Pa	83	13.55	4.079	0.448	12.66	14.44	2	22
	R	37	15.70	3.534	0.581	14.52	16.88	7	22
	T	354	14.73	4.360	0.232	14.28	15.19	2	25
B	EP	131	13.85	4.279	0.374	13.12	14.59	3	24
	SGE	103	13.99	3.682	0.363	13.27	14.71	5	24
	Pa	83	12.28	4.781	0.525	11.23	13.32	1	22
	R	37	13.19	3.463	0.569	12.03	14.34	6	20
	T	354	13.45	4.204	0.223	13.02	13.89	1	24
C	EP	131	9.14	4.463	0.390	8.37	9.91	0	20
	SGE	103	8.01	4.706	0.464	7.09	8.93	1	20
	Pa	83	9.22	5.177	0.568	8.09	10.35	1	25
	R	37	7.43	4.259	0.700	6.01	8.85	0	19
	T	354	8.65	4.717	0.251	8.16	9.14	0	25
D	EP	131	12.98	5.414	0.473	12.04	13.91	3	25
	SGE	103	10.89	5.773	0.569	9.76	12.02	0	23
	Pa	83	16.23	5.204	0.571	15.09	17.37	5	25
	R	37	12.97	5.800	0.953	11.04	14.91	5	25
	T	354	13.13	5.820	0.309	12.52	13.74	0	25
E	EP	131	14.15	4.544	0.397	13.36	14.93	0	25
	SGE	103	15.59	3.777	0.372	14.85	16.33	6	24
	Pa	83	14.67	4.981	0.547	13.59	15.76	3	25
	R	37	15.57	3.855	0.634	14.28	16.85	8	24
	T	354	14.84	4.406	0.234	14.38	15.30	0	25
F	EP	131	9.83	4.630	0.040	9.03	10.63	0	24
	SGE	103	11.59	4.991	0.492	10.62	12.57	0	22
	Pa	83	9.05	5.104	0.560	7.93	10.16	0	22
	R	37	10.14	4.423	0.727	8.66	11.61	2	19
	T	354	10.19	4.906	0.261	9.68	10.70	0	24

Note. Descriptive Statistics. EP = educational professionals; SGE = specialist gifted education; Pa = parents; R = Rest; T = total. N = 354.

A = stimulating academic development; B = stimulating the effective use of different cognitive styles; C = stimulating interpersonal functioning; D = stimulating intrapersonal understanding; E = stimulating self-actualisation; F = general knowledge and skills matching those of the (S)ENCo.

Table J19

Comparison of the Selection Rates of Knowledge Indicators at a General Level Between the Groups of Stakeholders

Knowledge Indicators		Sum of squares	*df*	Mean square	*F*	*p*	η^2
A	Between Groups	167.265	3	55.755	2.983	.031	.02
	Within Groups	6542.240	350	18.692			
	Total	6709.506	353				
B	Between Groups	168.240	3	56.080	3.234	.022	.02.
	Within Groups	6069.537	350	17.342			
	Total	6237.777	353				
C	Between Groups	154.870	3	51.623	2.347	.073	.02
	Within Groups	7699.694	350	21.999			
	Total	7854.565	353				
D	Between Groups	1316.380	3	438.793	14.431	.000	.01
	Within Groups	10642.380	350	30.407			
	Total	11958.760	353				
E	Between Groups	143.406	3	47.802	2.494	.060	.02
	Within Groups	6708.416	350	19.167			
	Total	6851.822	353				
F	Between Groups	327.627	3	109.209	4.680	.003	.04
	Within Groups	8167.311	350	23.335			
	Total	8494.938	353				

Note. ANOVA = analysis of variance with one factor. N = 354.

A = stimulating academic development; B = stimulating the effective use of different cognitive styles; C = stimulating interpersonal functioning; D = stimulating intrapersonal understanding; E = stimulating self-actualisation; F = general knowledge and skills matching those of the (S)ENCo.

Table J20

Comparison of the Selection Rates of Knowledge Indicators at a General Level Between the Groups of Stakeholders

Dependent variable	I group	J group	Mean difference (I-J)	*SE*	*p*	95% Confidence interval	
						Lower bound	Upper bound
B	EP	SGE	-0.135	0.548	1.000	-1.59	1.32
		Pa	1.578	0.584	.044	0.03	3.13
		R	0.666	0.775	1.000	-1.39	2.72
	SGE	EP	0.135	0.548	1.000	-1.32	1.59
		Pa	1.713	0.614	.033	0.08	3.34
		R	0.801	0.798	1.000	-1.32	2.92
	Pa	EP	-1.578	0.584	.044	-3.13	-0.03
		SGE	-1.713	0.614	.033	-3.34	-0.08
		R	-0.912	0.823	1.000	-3.10	1.27
	R	EP	-0.666	0.775	1.000	-2.72	1.39
		SGE	-0.801	0.798	1.000	-2.92	1.32
		Pa	0.912	0.823	1.000	-1.27	3.10
D	EP	SGE	2.084	0.726	.026	0.16	4.01
		Pa	-3.252	0.774	.000	-5.30	-1.20
		R	0.004	1.027	1.000	-2.72	2.73
	SGE	EP	-2.084	0.726	.026	-4.01	-0.16
		P	-5.336	0.813	.000	-7.49	-3.18
		R	-2.080	1.057	.299	-4.88	0.72
	Pa	EP	3.252	0.774	.000	1.20	5.30
		SGE	5.336	0.813	.000	3.18	7.49
		R	3.256	1.090	.018	0.36	6.15
	R	EP	-0.004	1.027	1.000	-2.73	2.72
		SGE	2.080	1.057	.299	-0.72	4.88
		Pa	-3.256	1.090	.018	-6.15	-0.36
F	EP	SGE	-1.760	0.636	.036	-3.45	-0.07
		P	0.784	0.678	1.000	-1.01	2.58
		R	-0.303	0.899	1.000	-2.69	2.08
	SGE	EP	1.760	0.636	.036	0.07	3.45
		Pa	2.544	0.713	.002	0.65	4.43
		R	1.457	0.926	.699	-1.00	3.91
	Pa	EP	-0.784	0.678	1.000	-2.58	1.01
		SGE	-2.544	0.713	.002	-4.43	-0.65
		R	-1.087	0.955	1.000	-3.62	1.45
	R	EP	0.303	0.899	1.000	-2.08	2.69
		SGE	-1.457	0.926	.699	-3.91	1.00
		Pa	1.087	0.955	1.000	-1.45	3.62

Note. Based on the results of the ANOVA, Bonferroni's Multiple Comparisons was used as a post hoc test for the analyses of the differences between groups of stakeholders.

EP = educational professionals; SGE = specialist gifted education; Pa = parents; R = Rest; T = total. N = 354.

B = stimulating the effective use of different cognitive styles; D = stimulating intrapersonal understanding; F = general knowledge and skills matching those of the (S)ENCo.

Table J21

Comparison of the Selection Rates of Skills Indicators at General Level Between the Groups of Stakeholders

Skills indicators	Group	N	Mean	*SD*	*SE*	95% Confidence interval for mean Lower bound	Upper bound	Min	Max
A	EP	131	11.99	4.763	0.416	1.17	12.82	1	22
	SGE	103	12.20	4.788	0.472	11.27	13.14	1	24
	Pa	83	11.72	5.271	0.579	10.57	12.87	1	22
	R	37	10.81	4.887	0.803	9.18	12.44	3	24
	T	354	11.87	4.902	0.261	11.35	12.38	1	24
B	EP	131	11.73	4.518	0.395	10.95	12.51	1	24
	SGE	103	11.14	4.099	0.404	10.33	11.94	3	23
	Pa	83	13.01	4.910	0.539	11.94	14.08	0	23
	R	37	11.92	3.789	0.623	10.66	13.18	3	19
	T	354	11.88	4.462	0.237	11.41	12.34	0	24
C	EP	131	13.44	4.679	0.409	12.63	14.24	3	25
	SGE	103	13.00	4.357	0.429	12.15	13.85	2	24
	Pa	83	12.22	4.929	0.541	11.14	13.29	0	23
	R	37	12.24	4.798	0.789	10.64	13.84	2	21
	T	354	12.90	4.669	0.248	12.41	13.39	0	25
D	EP	131	14.86	5.159	0.451	13.97	15.75	0	25
	SGE	103	12.69	5.062	0.499	11.70	13.68	2	22
	Pa	83	15.73	4.722	0.518	14.70	16.77	4	25
	R	37	15.43	5.449	0.896	13.62	17.25	3	24
	T	354	14.49	5.183	0.275	13.95	15.04	0	25
E	EP	131	13.08	4.413	0.386	12.31	13.84	5	24
	SGE	103	13.50	4.268	0.421	12.66	14.33	3	24
	Pa	83	13.36	4.568	0.501	12.36	14.36	5	22
	R	37	13.68	4.276	0.703	12.25	15.10	6	25
	T	354	13.33	4.381	0.233	12.87	13.79	3	25
F	EP	131	9.90	4.594	0.401	9.11	10.69	0	21
	SGE	103	12.48	4.031	0.397	11.69	13.26	2	21
	Pa	83	8.95	4.542	0.499	7.96	9.94	0	19
	R	37	10.92	4.044	0.665	9.57	12.27	4	19
	T	354	10.53	4.559	0.242	10.06	1.01	0	21

Note. Descriptive Statistics. EP = educational professionals; SGE = specialist gifted education; Pa = parents; R = Rest; T = total. N = 354.

A = stimulating academic development; B = stimulating the effective use of different cognitive styles; C = stimulating interpersonal functioning; D = stimulating intrapersonal understanding; E = stimulating self-actualisation; F = general knowledge and skills matching those of the (S)ENCo.

Table J22

Comparison of the Selection Rates of Skills Indicators at a General Level Between the Groups of Stakeholders

Skills Indicator		Sum of squares	df	Mean square	*F*	*p*	η^2
A	Between Groups	56.747	3	18.916	.786	.503	.01
	Within Groups	8426.013	350	24.074			
	Total	8482.760	353				
B	Between Groups	166.286	3	55.429	2.827	.039	.02
	Within Groups	6861.491	350	19.604			
	Total	7027.777	353				
C	Between Groups	93.233	3	31.078	1.431	.234	.01
	Within Groups	7603.106	350	21.723			
	Total	7696.339	353				
D	Between Groups	513.654	3	171.218	6.680	.000	.05
	Within Groups	8970.835	350	25.631			
	Total	9484.489	353				
E	Between Groups	15.740	3	5.247	.272	.846	.00
	Within Groups	6758.249	350	19.309			
	Total	6773.989	353				
F	Between Groups	654.130	3	218.043	11.421	.000	.09
	Within Groups	6681.963	350	19.091			
	Total	7336.093	353				

Note. ANOVA = analysis of variance with one factor. N = 354.

A = stimulating academic development; B = stimulating the effective use of different cognitive styles; C = stimulating interpersonal functioning; D = stimulating intrapersonal understanding; E = stimulating self-actualisation; F = general knowledge and skills matching those of the (S)ENCo.

Table J23

Comparison of the Selection Rates of Skills Indicators at a General Level Between the Groups of Stakeholders

Dependent variable	I group	J group	Mean difference (I-J)	*SE*	*p*	95% Confidence interval	
						Lower bound	Upper bound
B	EP	SGE	0.597	0.583	1.000	-0.95	2.14
		Pa	-1.279	0.621	.241	-2.93	0.37
		R	-0.186	0.824	1.000	-2.37	2.00
	SGE	EP	-0.597	0.583	1.000	-2.14	0.95
		Pa	-1.876	0.653	.026	-3.61	-0.14
		R	-0.783	0.849	1.000	-3.03	1.47
	Pa	EP	1.279	0.621	.241	-0.37	2.93
		SGE	1.876	0.653	.026	0.14	3.61
		R	1.093	0.875	1.000	-1.23	3.42
	R	EP	0.186	0.824	1.000	2.00	2.37
		SGE	0.783	0.849	1.000	-1.47	3.03
		Pa	-1.093	0.875	1.000	-3.42	1.23
D	EP	SGE	2.173	0.667	.007	0.40	3.94
		Pa	-0.872	0.710	1.000	-2.76	1.01
		R	-0.570	0.943	1.000	-3.07	1.93
	SGE	EP	-2.173	0.667	.007	-3.94	-0.40
		Pa	-3.046	0.747	.000	-5.03	-1.06
		R	-2.743	0.970	.030	-5.32	-0.17
	Pa	EP	0.872	0.710	1.000	-1.01	2.76
		SGE	3.046	0.747	.000	1.06	5.03
		R	0.303	1.001	1.000	-2.35	2.96
	R	EP	0.570	0.943	1.000	-1.93	3.07
		SGE	2.743	0.970	.030	0.17	5.32
		Paa	-0.303	1.001	1.000	-2.96	2.35
F	EP	SGE	-2.575	0.575	.000	-4.10	-1.05
		P	0.949	0.613	.735	-0.68	2.58
		R	-1.018	0.813	1.000	-3.18	1.14
	SGE	EP	2.575	0.575	.000	1.05	4.10
		Pa	3.524	0.644	.000	1.81	5.23
		R	1.557	0.837	.383	-0.67	3.78
	Pa	EP	-0.949	0.613	.735	-2.58	0.68
		SGE	-3.524	0.644	.000	-5.23	-1.81
		R	-1.967	0.864	.140	-4.26	0.32
	R	EP	1.018	0.813	1.000	-1.14	3.18
		SGE	-1.557	0.837	.383	3.78	0.67
		Pa	1.967	0.864	.140	-0.32	4.26

Note. Based on the results of the ANOVA, Bonferroni's Multiple Comparisons was used as a post hoc test for the analyses of the differences between groups of stakeholders.

EP = educational professionals; SGE = specialist gifted education; Pa = parents; R = Rest; T = total. N = 354.

B = stimulating the effective use of different cognitive styles; D = stimulating intrapersonal understanding; F = general knowledge and skills matching those of the (S)ENCo.

Table J24

Selection Rate of Indicators for the Entire Competency Matrix

Indicator (K+S)	N	Mean	*SD*	*SE* Mean
A	354	26.60	7.658	0.407
B	354	25.33	6.855	0.364
C	354	21.55	7.618	0.405
D	354	27.63	9.678	0.514
E	354	28.17	6.993	0.372
F	354	20.73	7.807	0.415

Note. Descriptive Statistics. (K+S) = Knowledge and Skills indicators. Test value 25.

A = stimulating academic development; B = stimulating the effective use of different cognitive styles; C = stimulating interpersonal functioning; D = stimulating intrapersonal understanding; E = stimulating self-actualisation; F = general knowledge and skills matching those of the (S)ENCo.

Table J25

Selection Rate of Indicators for the Entire Competency Matrix

Indicator (K+S)	*t*	*df*	Sig (2-tailed)	Mean difference	95% Confidence interval of the difference		Cohen's *d*
					Lower	Upper	
A	3.928	353	.000	1.599	0.80	2.40	.21
B	0.915	353	.361	0.333	-0.38	1.05	.05
C	-8.525	353	.000	-3.452	-4.25	-2.66	.45
D	5.108	353	.000	2.627	1.62	3.64	.27
E	8.520	353	.000	3.167	2.44	3.90	.45
F	-10.301	353	.000	-4.274	-5.09	-3.46	.55

Note. One-Sample T test. (K+S) = Knowledge and Skills indicators. Test value 25.

A = stimulating academic development; B = stimulating the effective use of different cognitive styles; C = stimulating interpersonal functioning; D = stimulating intrapersonal understanding; E = stimulating self-actualisation; F = general knowledge and skills matching those of the (S)ENCo.

Table J26

The Comparison of Selection Rates Between Two Clusters of Indicators for the Entire Competency Matrix: ADE-Indicators vs BCF- Indicators

Clusters of indicators	Mean	N	*SD*	*SE*
ADE	27.46	354	3.269	0.174
BCF	22.54	354	3.269	0.174

Note: Paired Samples Statistics. ADE = cluster of three categories of knowledge indicators A, D, E; compared to BCF = cluster of three categories of knowledge indicators B, C, F.

Table J27

The Comparison of Selection Rates Between Two Clusters of Indicators for the Entire Competency Matrix: ADE-Indicators vs BCF- Indicators

Clusters of indicators	Mean	*SD*	*SE* Mean	95% Confidence interval of the difference		*t*	*df*	Sig (2-tailed)	Cohen's *d*
				Lower bound	Upper bound				
ADE –BCF	4.928	6.538	0.347	4.245	5.612	14.184	353	.000	.75

Table J28

Comparison of Mean Selection Rates of indicators for the Entire Competency Matrix Between Dutch and Flemish Stakeholders

	Country	N	Mean	SD	*SE* Mean
Indicator A (K+S)	NL	299	26.89	7.624	0.441
	F	55	25.02	7.716	1.040
Indicator B (K+S)	NL	299	25.76	6.554	0.379
	F	55	23.02	7.978	1.076
Indicator C (K+S)	NL	299	21.11	7.582	0.438
	F	55	23.93	7.438	1.003
Indicator D (K+S)	NL	299	27.08	9.603	0.555
	F	55	30.62	9.618	1.297
Indicator E (K+S)	NL	299	28.19	6.950	0.402
	F	55	28.04	7.283	0.982
Indicator F (K+S)	NL	299	20.97	7.780	0.450
	F	55	19.38	7.887	1.063

Note. Descriptive Statistics. NL = The Netherlands. F = Flanders. N=354.

(K+S) = Knowledge and Skills indicators: indicator A = stimulating academic development; indicator B = stimulating the effective use of different cognitive styles; indicator C = stimulating interpersonal functioning; indicator D = stimulating intrapersonal understanding; indicator E = stimulating self-actualisation; and indicator F = general knowledge and skills matching those of the (S)ENCo.

Table J29

Comparison of Mean Selection Rates of indicators for the Entire Competency Matrix Between Dutch and Flemish Stakeholders

		Levene's test for Equality of variances					t-test for Equality of means		95% Confidence interval of the difference		
		F	*p*	*t*	*df*	Sig (2-tailed)	Mean difference	*SE* difference	Lower bound	Upper bound	Cohen's *d*
Indicator A (K+S)	Equal variances assumed	0.065	0.799	1.670	352	.096	1.871	1.121	-0.333	4.076	.24
	Equal variances not assumed			1.656	74.703	.102	1.871	1.130	-0.380	4.123	
Indicator B (K+S)	Equal variances assumed	5.373	0.021	2.751	352	.006	2.741	0.996	0.781	4.701	.40
	Equal variances not assumed			2.403	68.050	.019	2.741	1.141	0.465	5.017	
Indicator C (K+S)	Equal variances assumed	0.289	0.591	-2.540	352	.012	-2.817	1.109	-4.998	-0.635	.37
	Equal variances not assumed			-2.573	76.111	.012	-2.817	1.095	-4.997	-0.637	
Indicator D (K+S)	Equal variances assumed	0.001	0.979	-2.513	352	.012	-3.541	1.409	-6.313	-0.770	.37
	Equal variances not assumed			-2.510	75.166	.014	-3.541	1.411	-6.352	-0.731	
Indicator E (K+S)	Equal variances assumed	0.002	0.966	0.150	352	.881	0.154	1.027	-1.866	2.175	.00
	Equal variances not assumed			0.145	73.239	.885	0.154	1.061	-1.960	2.269	
Indicator F (K+S)	Equal variances assumed	0.196	0.658	1.391	352	.165	1.591	1.144	0.658	3.841	.20
	Equal variances not assumed			1.378	74.626	.172	1.591	1.155	0.709	3.892	

Note. Independent Samples Test. N = 354.

(K+S) = Knowledge and Skills indicators: indicator A = stimulating academic development; indicator B = stimulating the effective use of different cognitive styles; indicator C = stimulating interpersonal functioning; indicator D = stimulating intrapersonal understanding; indicator E = stimulating self-actualisation; and indicator F = general knowledge and skills matching those of the (S)ENCo.

Table J30

Comparison of the Selection Rates of the Indicators for the Entire Competency Matrix Between the Groups of Stakeholders

Indicators (K+S)	Group	N	Mean	SD	SE	95% Confidence interval for mean Lower bound	Upper bound	Min	Max
A (K+S)	EP	131	27.05	7.419	0.648	25.76	28.33	10	42
	SGE	103	27.13	8.357	0.823	25.49	28.76	8	49
	Pa	83	25.28	7.451	0.818	23.65	26.90	7	41
	R	37	26.51	6.809	1.119	24.24	28.78	16	43
	T	354	26.60	7.658	0.407	25.80	27.40	7	49
B (K+S)	EP	131	25.59	7.150	0.625	24.35	26.82	6	41
	SGE	103	25.13	5.881	0.579	23.98	26.28	12	43
	Pa	83	25.29	8.036	0.882	23.53	27.04	4	45
	R	37	25.11	5.537	0.910	23.26	26.95	12	35
	T	354	25.33	6.855	0.364	24.62	26.05	4	45
C (K+S)	EP	131	22.57	7.501	0.655	21.28	23.87	3	41
	SGE	103	21.01	7.560	0.745	19.53	22.49	6	39
	Pa	83	21.43	8.103	0.889	19.66	23.20	6	40
	R	37	19.68	6.795	1.117	17.41	21.94	9	36
	T	354	21.55	7.618	0.405	20.75	22.34	3	41
D (K+S)	EP	131	27.84	9.266	0.810	26.24	29.44	7	50
	SGE	103	23.58	9.322	0.918	21.76	25.40	4	42
	Pa	83	31.96	8.565	0.940	30.09	33.83	10	47
	R	37	28.41	10.202	1.677	25.00	31.81	8	47
	T	354	27.63	9.678	0.514	26.62	28.64	4	50
E (K+S)	EP	131	27.22	6.929	0.605	26.02	28.42	5	45
	SGE	103	29.09	6.314	0.622	27.85	30.32	13	43
	Pa	83	28.04	7.903	0.867	26.31	29.76	11	44
	R	37	29.24	6.635	1.091	27.03	31.46	14	43
	T	354	28.17	6.993	0.372	27.44	28.90	5	45
F (K+S)	EP	131	19.73	7.795	0.681	18.39	21.08	0	42
	SGE	103	24.07	7.237	0.713	22.65	25.48	5	39
	Pa	83	18.00	7.859	0.863	16.28	19.72	2	35
	R	37	21.05	6.023	0.990	19.05	23.06	8	33
	T	354	20.73	7.807	0.415	19.91	21.54	0	42

Note. Descriptive Statistics. EP = educational professionals; SGE = specialist gifted education; Pa = parents; R = Rest; T = total. N = 354.

(K+S) = Knowledge and Skills indicators: A = stimulating academic development; B = stimulating the effective use of different cognitive styles; C = stimulating interpersonal functioning; D = stimulating intrapersonal understanding; E = stimulating self-actualisation; F = general knowledge and skills matching those of the (S)ENCo.

Table J31

Comparison of the Selection Rates of the Indicators for the Entire Competency Matrix Between the Groups of Stakeholders

Indicators		Sum of squares	*df*	Mean square	*F*	*p*	η^2
Indicator A (K+S)	Between Groups	200.085	3	66.695	1.139	.333	.01
	Within Groups	20500.954	350	58.574			
	Total	20701.040	353				
Indicator B (K+S)	Between Groups	14.939	3	4.980	0.105	.957	.00
	Within Groups	16571.727	350	47.348			
	Total	16586.667	353				
Indicator C (K+S)	Between Groups	298.139	3	99.380	1.723	.162	.01
	Within Groups	20189.545	350	57.684			
	Total	20487.684	353				
Indicator D (K+S)	Between Groups	3274.287	3	1091.429	12.825	.000	.10
	Within Groups	29786.493	350	85.104			
	Total	33060.780	353				
Indicator E (K+S)	Between Groups	248.671	3	82.890	1.705	.166	.01
	Within Groups	17012.496	350	48.607			
	Total	17261.167	353				
Indicator F (K+S)	Between Groups	1900.356	3	633.452	11.305	.000	.09
	Within Groups	19612.065	350	56.034			
	Total	21512.421	353				

Note. ANOVA = analysis of variance with one factor. N = 354.

(K+S) = Knowledge and Skills indicators: A = stimulating academic development; B = stimulating the effective use of different cognitive styles; C = stimulating interpersonal functioning; D = stimulating intrapersonal understanding; E = stimulating self-actualisation; F = general knowledge and skills matching those of the (S)ENCo.

Table J32

Comparison of the Selection Rates of the Indicators for the Entire Competency Matrix Between the Groups of Stakeholders

Dependent variable	I group	J group	Mean difference (I-J)	*SE*	*p*	95% Confidence interval	
						Lower bound	Upper bound
B	EP	SGE	0.462	0.906	1.000	-1.94	2.87
		Pa	0.299	0.965	1.000	-2.26	2.86
		R	0.480	1.281	1.000	-2.92	3.88
	SGE	EP	-0.462	0.906	1.000	-2.87	1.94
		Pa	-0.163	1.015	1.000	-2.86	2.53
		R	0.018	1.139	1.000	-3.48	3.52
	Pa	EP	-0.299	0.965	1.000	-2.86	2.26
		SGE	0.163	1.015	1.000	-2.53	2.86
		R	0.181	1.360	1.000	-3.43	3.79
	R	EP	-0.480	1.281	1.000	-3.88	2.92
		SGE	-0.018	1.319	1.000	-3.52	3.48
		Pa	-0.161	1.360	1.000	-3.79	3.43
D	EP	SGE	4.257	1.215	.003	1.03	7.46
		Pa	-4.124	1.294	.009	-7.56	-0.89
		R	-0.566	1.717	1.000	-5.12	3.99
	SGE	EP	-4.257	1.215	.003	-7.48	-1.03
		Paa	-8.381	1.361	.000	-11.99	-4.77
		R	-4.823	1.768	.040	-9.51	-0.13
	Pa	EP	4.124	1.294	.009	0.69	7.56
		SGE	8.381	1.361	.000	4.77	11.99
		R	3.558	1.824	.311	-1.28	8.40
	R	EP	0.566	1.717	1.000	-3.99	5.12
		SGE	4.823	1.768	.040	0.13	9.51
		Pa	-3.558	1.824	.311	-8.40	1.28
F	EP	SGE	-4.335	0.986	.000	-6.95	-1.72
		Pa	1.733	1.050	.599	-1.05	4.52
		R	-1.321	1.394	1.000	-5.02	2.38
	SGE	EP	4.335	0.986	.000	1.72	6.95
		Pa	6.068	1.104	.000	3.14	9.00
		R	3.014	1.435	.218	-0.79	6.82
	Pa	EP	-1.733	1.050	.599	-4.52	1.05
		SGE	-6.068	1.104	.000	-9.00	-3.14
		R	-3.054	1.480	.239	-6.98	0.87
	R	EP	1.321	1.394	1.000	-2.38	5.02
		SGE	-3.014	1.435	.218	-6.82	0.79
		Pa	3.054	1.480	.239	-0.87	6.98

Note. Based on the results of the ANOVA, Bonferroni's Multiple Comparisons was used as a post hoc test for the analyses of the differences between groups of stakeholders.

EP = educational professionals; SGE = specialist gifted education; Pa = parents; R = Rest; T = total. N = 354. Indicator B = stimulating the effective use of different cognitive styles; indicator D = stimulating intrapersonal understanding; indicator F = general knowledge and skills matching those of the (S)ENCo.

Dankwoord

In 2015 zag ik op het congres van de Worldcouncil for Gifted and Talented Children in Odense (Denemarken) voor het eerst sinds lange tijd twee collega's terug, Tessa Kieboom en Kathleen Venderickx. Op een van de middagen raakten we tijdens de lunch in gesprek over onze wederzijdse werkzaamheden. Het werd een lange lunch en bij de theepauze van de overige deelnemers drong tot ons door hoelang het gesprek had geduurd. We spraken af om elkaar na het congres nog weer eens te ontmoeten en verder te praten. Het duurde tot oktober 2016 voor dat dit ervan kwam. In dat tweede gesprek daagden zij mij uit om mijn eerder gemaakte werk over de competenties van de specialist begaafdheid nog eens onder de loep te nemen en op basis daarvan tot een promotievoorstel te komen. Hoewel me dat in eerste instantie een goed plan leek, had ik op weg naar huis onmiddellijk koudwatervrees. Ik legde de uitdaging naast me neer met het idee dat ik promoveren niet nodig had om mijn rol als lerarenopleider te kunnen vervullen. Maar bloed kruipt waar het niet gaan kan en zo begon ik in het voorjaar van 2017 verder na te denken of het toch niet een goed plan zou zijn om de uitdaging op te pakken. In de zomer van 2018 schreef ik het promotievoorstel en uiteindelijk ging mijn aanstelling voor het verrichten van het promotieonderzoek in op 1 november 2018. Een strikte planning en regelmatige gesprekken over het verloop van het onderzoek maakte het mogelijk om op 6 juli 2020 een conceptversie van het manuscript in te sturen. En hier sta ik nu, elf hoofdstukken verder en met een goedgekeurde dissertatie! Zonder die eerste uitdaging was het nooit zover gekomen en ik kan alleen maar dankbaar zijn voor het bieden van de uitdaging en het gegeven vertrouwen van Tessa Kieboom en Kathleen Venderickx. Zelf zullen zij zich beslist wel eens afgevraagd hebben waaraan ze begonnen waren. Als ik op "aan" stond, ontvingen ze veel vragen en lange hoofdstukken en hadden ze daardoor veel huiswerk te doen. Ik ben al ruim 25 jaar werkzaam als lerarenopleider in het domein begaafdheidsonderwijs en heb door mijn werkzaamheden een groot gevoel voor autonomie ontwikkeld. Ik ben gewend om zelfsturend op te treden en aan een ruime professionele vrijheid. Dat beide hoogleraren het maakproces van deze elf hoofdstukken overleefd hebben is daarom een buiging waard.

Omdat ik graag ook iemand met een onderwijskundige blik over mijn schouder wilde laten meekijken naar het onderzoek, is Dr. Margaret Sutherland benaderd. Zij heeft als derde lid van de doctoraatscommissie een waardevolle bijdrage aan de totstandkoming van mijn dissertatie geleverd. Elk hoofdstuk voorzag zij nauwgezet van feedforward. Ik vroeg haar "te schieten, zo hard als ze kon" en geloof me, dat heeft ze gedaan. Haar kritische blik heeft me op gezonde wijze menigmaal aan het twijfelen gebracht, me breder laten kijken, nog meer laten lezen en laten nadenken over te maken en gemaakte keuzes. Margaret, thank you. I needed your support and I could not have done this without you.

Iedereen die onderzoek doet, iedereen die schrijft, iedereen die theorie en praktijk wil verbinden, heeft behoefte aan critical friends. Voor mij waren dat Annemieke Weterings en Alja de Bruin. Zij begrepen waar mijn behoefte vandaan kwam om die "s" van mijn titel af te schaven. Menigmaal zijn tijdens onze ontmoetingsmomenten mijn plannen voor het onderzoek besproken, gewogen, over de heg gegooid en weer uit de prullenmand gehaald om bij te schaven. Zij keken steeds kritisch mee naar wat ik bedacht en wilde. Zij stonden altijd klaar om mij met een gezond-kritische blik naar mezelf en mijn werk te laten kijken. Annemieke heeft veel, heel veel gelezen. Haar grootste taak was elke tekst te fileren en me te wijzen op mijn omissies. Van die taak heeft zij zich met verve gekweten. Zij zorgde voor de bijna wekelijkse portie tegengas die ik nodig had om te kunnen blijven schrijven. Alja de Bruin heeft veel gebeld, heel veel gebeld. Steeds vroeg ze naar de vooruitgang en maakte ze me zichtbaar dat kleine stappen vooruit ook progressie was. Dat hielp me doorzetten als ik eigenlijk zin had om er gewoon maar mee te stoppen.

Dit onderzoek had nooit kunnen plaatsvinden zonder de hulp van Lex Bouts. Lex en ik gaan heel lang terug, naar het midden van de jaren tachtig toen ik in Nijmegen studeerde en het begin van de jaren negentig toen ik in Nijmegen werkte. Eerst als mijn docent en later als mijn collega leerde Lex me de betekenis van getallen. Wie mij een beetje kent, weet dat getallen mij alleen iets zeggen als ze mijn bankrekening betreffen; Lex heeft daar verandering in gebracht. Ons contact werd na mijn vertrek aan de universiteit een contact op afstand en de intensiteit van het contact nam af tot incidenteel niveau. Maar toen ik hem in 2018 via e-mail vroeg of hij me op weg wilde helpen met het statistische deel van mijn onderzoek kreeg ik binnen een kwartier een oprecht "ja" terug. Sinds die tijd hebben we menige lunch samen genoten; Lex heeft elke stap van het opzetten van het empirisch onderzoek en de daaruit voortvloeiende dataverwerking begeleid. Met engelengeduld legde hij uit wat ik moest doen, waarom ik dat moest doen en hoe ik dat moest doen. Als het echt te moeilijk werd dan sprong hij in. Ik mocht altijd bellen, appen et cetera. We belegden tweemansconferenties waarbij we altijd eerst elkaars persoonlijk wel en wee bespraken en dan samen aan het werk gingen. Lex, die tweemansconferenties hebben we verheven tot een traditie, laten we die in ere houden.

Engels is niet mijn eerste taal en daarom ben ik gaan zoeken naar een professional die in staat zou zijn om de enorme klus aan te gaan al mijn teksten te corrigeren en daar waar nodig was op te poetsen naar Engels Engels. Geraldine Glover was die persoon. Ik schat in dat ik haar menigmaal tot absolute waanzin heb gedreven onder de tijdsdruk waarin we de correcties van het manuscript hebben uitgevoerd. APA 7 heeft ons beiden nachtmerries bezorgd maar uiteindelijk hebben we dat overleefd. Zonder haar was het allemaal niet gelukt. Dear Geraldine, I know that my English is currently better than your Dutch, so if you ever want a crash course Dutch... I'll be there for you!

Omdat boeken maken nu eenmaal een vak is en een manuscript nog geen boek is, hebben Sabine Kokee, Marita Weener en Sabrina Wakker mij geholpen om van het manuscript een boek te maken. Dat ze daarbij hun gebruikelijke creativiteit achter de deur moesten zetten omdat APA 7 de kaders bepaalde, namen ze voor lief. Dat ze onder tijdsdruk hun werk moesten doen, namen ze ook voor lief. Dames, heel erg dank je wel!

Mijn kinderen volgden mijn onderzoek van een afstand. Dat was een verstandige keuze want ik weet dat ik in deze jaren niet altijd gemakkelijk mentaal "bereikbaar" was. Omdat ik niet elk verjaardagsfeestje en elke gezinsontmoeting wilde beïnvloeden door het over de voortgang van mijn onderzoek te hebben, moesten ze soms informatie uit me trekken. Maar geloof me, er zijn genoeg momenten geweest dat ik ze dodelijk heb verveeld met mijn verhalen. Ze doorstonden ze dapper en op een aantal momenten hebben ze me ook over drempels heen geholpen door heel wijs tegen me te appen: "Het is normaal, mam, dat je er zo genoeg van hebt, maar doorzetten moet je leren en hoort erbij. Niet opgeven hoor!"

Lieve Eugenio, zonder jou was ik er nooit aan begonnen. Jij bent de vleugels voor mijn vlucht. Jij moedigde me aan om mijn eigen weg te gaan, je luisterde, je lachte, je moedigde me aan, je droogde tranen, je gaf tegengas en hebt zonder mopperen heel veel avonden heel erg laat gegeten omdat ik eerst nog wat moest afmaken. Je bood me de ruimte om in Londen in alle rust te kunnen schrijven en mijn taken op Ham House te verrichten. Die tijd was kostbaar, heel kostbaar, dat is iets wat ik weet nu we hier staan. Deze ervaring was een wijze les en maakt dat ik weet dat de jaren die nu voor ons liggen me nog veel kostbaarder en dierbaarder zullen zijn. Het succes van het afronden van dit proefschrift rust geheel op jouw schouders. Zonder jou was ik niet wie ik nu ben.